Major Appliances

Major Appliances

Operation, Maintenance, Troubleshooting, and Repair

Billy C. Langley

Regents/Prentice Hall
Englewood Cliffs, New Jersey 07632

Library of Congress Cataloging-in-Publication Data

Langley, Billy C.
   Major appliances : operation, maintenance, troubleshooting, and
repair / by Billy C. Langley.
     p.   cm.
   Includes index.
   ISBN 0-13-544834-4
   1. Household appliances, Electric—Maintenance and repair. 2. Gas
appliances—Maintenance and repair. I. Title.
TK7018.L33 1992
683'.83—dc20   91-25974
                                      CIP

Acquisitions: Robert Koehler
Editorial/production supervision and interior design: Tally Morgan,
   WordCrafters Editorial Services, Inc.
Cover design: Marianne Frasco
Prepress buyer: Ilene Levy
Manufacturing buyer: Ed O'Dougherty

©1993 by Regents/Prentice Hall
A Division of Simon & Schuster
Englewood Cliffs, New Jersey 07632

Printed in the United States of America
10 9 8 7 6 5 4 3 2 1

ISBN 0-13-544834-4
#24174712

Prentice-Hall International (UK) Limited, *London*
Prentice-Hall of Australia Pty. Limited, *Sydney*
Prentice-Hall Canada Inc., *Toronto*
Prentice-Hall Hispanoamericana, S.A., *Mexico*
Prentice-Hall of India Private Limited, *New Delhi*
Prentice-Hall of Japan, Inc., *Tokyo*
Simon & Schuster Asia Pte. Ltd., *Singapore*
Editora Prentice-Hall do Brazil, Ltda., *Rio de Janeiro*

Contents

Preface

TO THE INSTRUCTOR

This textbook has been designed to provide a thorough and practical study of the major appliances used in most residences. This book will provide a basic knowledge of electricity, electronics, refrigeration, and major appliance operation. This book is designed to be a complete text for anyone wanting to learn this exciting and rewarding field, especially those in a post-secondary training program.

The text emphasizes the more practical aspects of the subject matter. Wherever possible, sections are included that cover the operation, maintenance, troubleshooting, and repair of major appliances.

APPROACH AND PHILOSOPHY

1. The fundamentals necessary for successful servicing of major appliances are covered first, along with chapters on safety, the service occupation, customer relations, tools and instruments, service techniques, installation, basic electricity and electronics, proper lubrication procedures, refrigeration principles, air control, water chemistry, and combustion.

2. The operation chapters contain in-depth information on how each appliance is designed to work. The theory of operation is presented in clear, familiar terms so that the beginning learner is not discouraged or confused.

3. This textbook is written using common terminology and language pertaining to the field.

4. Each chapter is concluded with review exercises (where applicable) that the reader should be able to answer after completing the chapter.

CHAPTER ORGANIZATION

Chapter 1 provides the reader with the common safety procedures used in the major appliance industry. This chapter covers the safety of the worker, safety of the equipment, and safety of the contents, in addition to welding safety, safety in handling refrigerants, safety in working with electricity and electronics, and safety in working with solvents.

Chapter 2 covers the occupation of major appliance servicing. Also presented are the nature of the work, the responsibilities and different classifications of appliance technicians, problem solving techniques, customer relations, advancement, effective use of time, and the need to continue learning the business and covers the basics of customer relations involved in the major appliance service business. Also covered are appearance, work habits, and the ability to communicate effectively.

Chapter 3 provides an overview of the tools and instruments used in the trade. Each of the tools and instruments common to the field of appliance service is presented separately, so the reader will have knowledge of its use when it becomes necessary to use it.

Chapter 4 introduces the reader to basic electricity and electronics. The electron theory, sources of electricity, resistance, electrical and electronic circuits, diagrams, charts, P and N materials, current carriers, diodes, transistors, IC circuits, circuit boards, and modules are covered, along with control problems and unusual service problems.

Chapter 5 covers the reasons for proper lubrication, how lubrication works, bearing types, bearing failure, and bearing replacement procedures.

Chapter 6 introduces the reader to the principles of refrigeration. Also presented are heat flow, heat laws, pressure, temperature, pressure-temperature relationships, the refrigeration cycle, flow control devices, trouble diagnosis, and service procedures.

Chapter 7 covers the basic control of air in buildings. Air movement, pollutants, filtering, germicidal and ozone lamps, the effects of temperature and humidity, and air system maintenance are also presented.

Chapter 8 presents the chemistry of water, water testing, detergents, and the different qualities of water to the reader in familiar terms so that the material can be easily understood.

Chapter 9 covers the fundamentals of combustion. The different types of fuels used with appliances, the combustion process, the different types of flames, and the upper and lower limits of flammability are also presented.

Chapter 10 covers the operation of room air conditioning units. Sizing room units, air systems, electrical diagrams, the refrigeration cycle, and troubleshooting charts are also presented.

Chapter 11 covers domestic refrigeration systems, their operation, location, automatic ice makers, electrical problems, electrical diagrams, and troubleshooting charts.

Chapter 12 presents domestic freezers, their operation, location, refrigerant cycle, electrical components, electrical diagrams, and troubleshooting charts.

Chapter 13 covers the use and operation of dehumidifiers, along with the different types and their capacities. Chapter 13 also discusses the need for humidification equipment in buildings. The effects of different types of water, air flow, water flow, and maintenance are also presented to the reader.

Chapter 14 covers dishwashers. The installation, operation, electrical power supply, water supply, use, operation, and service procedures are also presented. Troubleshooting charts and electrical diagrams are presented to aid the reader in understanding dishwashers.

Chapter 15 presents electric and gas dryer installation, operation, component replacement, electrical diagrams, component location, burner adjustments, and troubleshooting charts to the reader.

Chapter 16 covers gas laundry centers. Washer and dryer construction and operation are covered. Also included are maintenance and component replacement procedures to help the reader understand these operations.

Chapter 17 presents the theory of microwave ovens. Use, safety procedures, installation, service procedures, electrical diagrams, test instruments, and troubleshooting charts are also covered.

Chapter 18 covers electric wall oven operation and maintenance. Self-cleaning ovens, electrical diagrams, and troubleshooting charts are also provided to the reader.

Chapter 19 discusses the operation and maintenance of gas ranges. Also presented are control adjustments, burner adjustments, and gas ignition.

ACKNOWLEDGMENTS

The following people contributed greatly to the production of this textbook. Their help, assistance, and moral support made the completion of this project much easier:

Tally Morgan and the staff at WordCrafters Editorial Services, Inc.

Tony Evans of WCI Major Appliance Group
Don Koehler of WCI Major Appliance Group
Al Wilderman of WCI Major Appliance Group
Michelle Johnson of Prentice Hall, Inc.

TO THE STUDENTS

It is difficult to express the amount of self-satisfaction realized when a dedicated appliance service technician successfully completes a service procedure. This satisfaction comes from knowing that a correct and economical service has been provided to a fellow human being.

To achieve this type of success requires dedication to learning and use of this knowledge in performing appliance service work. The knowledge is gained through study and understanding information such as that presented in this textbook. The more a person knows about a given machine, the easier the troubleshooting, maintenance, and repair of that machine becomes.

Each brand and model of appliance will be a little different from the rest. However, when the basics are known, they can be applied to aid in determining the cause of the problem and the steps required in repairing it. This textbook provides that basic information.

There are several steps that you can take to make certain that you are getting the most from this textbook and course. The first step is to become aware that learning requires active participation by you, both in the classroom and when you are studying the material in this textbook. You must take an active part in the theory as well as in the laboratory exercises.

There are several habits that must be developed. First, you must regularly attend class. While in class, do not sit back and let someone else do the projects. To learn the most be active in class participation.

You should take time to learn the terminology used in the appliance industry. You need to know these terms so that you can apply them correctly at the proper time. Take time to commit the meaning of new terms to memory.

Follow the procedures and examples given in the textbook. This will help you develop the analytical skills that are so important in appliance service work. Remember that practice, in addition to study, will help you tremendously in developing service skills.

You should realize by now that there is a lot involved in being an active learner. However, the knowledge gained will be well worth the extra effort. Your skills as an appliance service technician will be indicated by your increased ability to service appliances. This increased ability will also be rewarded by better pay and recognition. My best wishes go with you always.

Billy C. Langley

Major Appliances

1
Safety

There is a general feeling that safety is only a state of mind. Some think that it is an attitude that individuals have while they are at work, play, or at home. When safety is involved, we are generally our own worst enemy. Sometimes injuries occur due to a lack of knowledge; however, injuries are caused more often because we do not practice what we know about safety.

We should use safety aids to protect ourselves from injuries. Gloves, safety shoes, and other proper clothing have prevented many serious injuries. Using the proper tools, the posting of safety rules in prominent areas of the workplace, and having the ability to recognize hazards have also prevented serious injuries. However, the most important thing is to have the proper mental attitude about safety while we are going about our daily activities. The proper mental attitude causes us to be more alert to effective ways to accomplish out tasks in a safe manner.

SAFE SERVICING PRACTICES

To avoid the possibility of personal injury and/or property damage, it is important that safe servicing practices be observed. The following are examples, but without limitation, of such practices:

1. Do not attempt a product repair if you have any doubts as to your ability to complete it in a safe and satisfactory manner.
2. Before servicing or moving an appliance:
   a. remove power cord from electric outlet, trip circuit breaker to OFF, or remove fuse.
   b. turn off gas supply.
   c. turn off water supply.
3. Never interfere with the proper operation of any safety device.

4. USE ONLY REPLACEMENT PARTS CATALOGED FOR THIS APPLIANCE. SUBSTITUTIONS MAY DEFEAT COMPLIANCE WITH SAFETY STANDARDS SET FOR HOME APPLIANCES.

5. GROUNDING: The standard color coding for safety ground wires is GREEN or GREEN with YELLOW STRIPES. Ground leads are not to be used as current carrying conductors. IT IS EXTREMELY IMPORTANT THAT THE SERVICE TECHNICIAN RE-ESTABLISH ALL SAFETY GROUNDS PRIOR TO COMPLETION OF SERVICE. FAILURE TO DO SO WOULD CREATE A POTENTIAL HAZARD.

6. Prior to returning the product to service ensure that:
   a. all electric, gas, and water connections are correctly and securely connected.
   b. all gas and water connections are tested for leaks. DO NOT TEST FOR GAS LEAKS WITH A FLAME.
   c. all electrical leads are properly dressed and secured away from sharp edges, high-temperature components and moving parts.
   d. all uninsulated electrical terminals, connectors, heaters, etc. have adequate spacing from all metal parts and panels.
   e. all safety grounds (both internal and external to the product) are correctly and securely connected.
   f. all panels are properly and securely reassembled.

SAFETY PROCEDURES

The saying "safety doesn't cost, it pays" is as true today as it was when it was first used. Following the proper safety procedures cannot be overemphasized, especially by workers in the appliance field. These workers generally work alone and help may be difficult to obtain if a serious injury were to happen.

Safety may be divided into three general categories: safety of the worker, safety of the equipment, and safety of the contents.

Safety of the Worker

This is without a doubt the most important of all three. Machinery and materials are replaceable, but a human life is only a one-time thing. The proper use of tools and machinery represents very little danger to the worker. When lifting heavy objects, always use the leg muscles. Never use the back muscles when lifting. When an object weighing more than 30 pounds is to be lifted, ask for help. The floor should always be kept clear of water and oil to prevent slipping. To slip when carrying heavy objects will almost always result in a serious injury.

Major appliances use electricity as a source of power for operation. Make certain that all electrical circuits are turned off before working on them. The disconnect switch or the circuit breaker will usually break the circuit. Never work a "hot" circuit.

Always wear safety goggles when working with refrigerants. Should liquid refrigerant get into the eye, a frozen eye will result. Liquid refrigerant should be washed from the eye with a gentle stream of tap water. When the eye has been flushed for about 15 minutes, have a physician properly medicate the eye. Also, the skin should be protected from frostbite when working with refrigerants.

Safety of the Equipment

Safety of the equipment is always of great concern to the technician. Many of the components on major appliances are easily broken. Some of these components are very expensive and all of them take time to replace, usually at a lot of expense. When tightening bolts and nuts, use the proper torque to prevent breaking them. Use the proper sequence with tightening bolts and nuts. Be sure that all moving parts are free to rotate and all belts are clear of objects before starting the equipment.

Safety of the Contents

This is a basic requirement and will depend to a great extent on the care given to the equipment during the installation and servicing processes. It is the responsibility of the service technician to make certain that the inside of the appliance is kept at the desired temperature. It is the technician's responsibility to know what the desired conditions are and to make sure that the equipment can produce these conditions. Most manufacturers provide tables and charts showing the desired operating conditions and the required control settings.

WELDING SAFETY (OXY-ACETYLENE)

The following section is presented with the permission of Uniweld Products, Inc., Fort Lauderdale, Florida.

Caution: Always read the instruction sheets provided with the equipment before operating the equipment.

Safety Precautions

The safe and effective use of flame tool equipment depends on the user fully understanding and carefully following practical time-tested safety and operating instructions to prevent and avoid unnecessary painful injuries and costly property damage.

For adequate personal safety, the user must be fully aware at all times that the torch flame can reach almost 6000°F and the work piece can reach a high heat of almost 3000°F, which may produce flying sparks, slag, fumes, and intense light rays, all of which can be harmful to the user or nearby persons without proper precautions and protection.

Practical proper "head-to-toe" protection includes hair and head coverage, welding-and-cutting-type tempered lens eye goggles (shade 5 is standard). Body coverage includes proper gloves and proper shoes. Avoid wearing anything flammable or clothing that has been exposed to flammables (oil, grease, solvents, etc.). Sparks and molten materials have a way of finding unprotected areas, so be prepared before starting work by being properly protected.

Adequate ventilation must be provided, especially in confined areas, to remove harmful fumes and provide an adequate air supply for the user.

Important: Oxygen rapidly increases the burning rate of almost any ignited material, especially oil and grease; therefore, it must never be allowed to saturate a confined work area. Oxy/fuel or air/fuel concentrations in confined areas can also be hazardous and explosive. Never use a torch on any kind of container or pipe until it has been properly cleaned, purged, or vented.

Fire protection must be provided for the work area. The user must be fully aware of the impact that the torch flame, sparks, and molten materials can have on both close and surrounding areas; sparks can fly over 35 feet. Keep hoses clear of sparks and slag. Remove all flammables and carefully cover or shield, with fireproof materials, anything that could possibly catch fire, explode, or both.

Note: Concrete can chip explosively when overheated by slag and molten metal. Carefully check out the work area after the work has been completed for places where sparks or molten metal could light and smolder. Douse any doubtful areas with water or a proper fire extinguisher. Have a bucket of water and a proper fire extinguisher available in the work area at all times. A water bucket is also handy for leak-testing hose and torches and for cooling the work.

Important: To avoid a dangerous pressure imbalance and cylinder contamination due to the reverse flow of gases, do not allow cylinders (especially oxygen) to become empty in use. Check for adequate gas supplies before starting work. Also, install reverse-flow check valves in the welding hoses.

Checking out the Equipment

New equipment in the original packaging has been inspected and tested and is ready for connecting. It is standard safety practice, however, to check out and leak-test all connection points from the cylinder valves to the tips to assure safe, leak-tight service before using the equipment. This checkup and leak-testing is especially important after the equipment has been used and has been exposed to dust, dirt, oil, grease, and possible damage. The following guidelines for checking out, cleaning, and testing the equipment before using it are designed to cover normal working conditions. The proper procedures should be followed each time the equipment is set up and used.

Cylinders

The following procedures should be followed when working with cylinders:

1. The cylinders must be secured to keep them from falling over.
2. Fuel gas cylinders must be used in the upright position.
3. Do not store cylinders and equipment in unventilated, confined spaces, closed vehicles, or near any source of heat or ignition.
4. The cylinder valves must be securely closed when not in use and also when empty.
5. To avoid a dangerous pressure imbalance and cylinder contamination, do not allow cylinders, especially oxygen, to become completely empty when in use. Check for adequate gas supplies before starting work. For added protection install external-type reverse-flow check valves on regulators and or torches. They are easy to install, test, or replace.
6. Valve protection caps should be replaced when the cylinders are empty, stored, or when the regulator is removed.
7. Do not expose cylinders to torch flames or electrical arcs because they can seriously damage the cylinders.
8. Do not use a cylinder, full or empty, as a roller or support. The cylinder wall may be damaged and result in a rupture or an explosion.
9. Do not use a cylinder gas without a suitable pressure-reducing regulator.
10. Do not allow oil or grease to come in contact with cylinder valves, especially oxygen. Do not use the cylinder if oil is on the cylinder valve; return the cylinder to the supplier.
11. Do not use a cylinder with a leaking valve. Move it away from any source of ignition, preferably outdoors, and notify the supplier.
12. Handle all cylinders with care and protect the valves from damage.

Regulators

The following procedures should be used when working with regulators:

1. Use the regulators for the gas or gases for which they were intended. Oxygen regulators must be used only for oxygen service.
2. Regulators and cylinder connections must be made free of dirt, dust, grease, and oil. Remember that oxygen reacts with oil or grease. Inspect the regulator and cylinder valve connections for damaged threads, dirt, dust, oil, or grease. Remove any dust and dirt with a clean cloth. Do not use the regulator if it is damaged or has oil or grease on it. Have your authorized repair station or Uniweld clean the regulator or repair any damage. Do not use cylinders with damaged or dirty valves. Return such cylinders to the supplier.
3. Do not crack fuel gas or oxygen cylinders near an open flame or any source of ignition. Make sure that you are in a well-ventilated area, and stand clear of the valve outlet when opening it.
4. Carefully crack (open briefly) the cylinder valves, one at a time, to clean the valves of dust, dirt, or foreign matter, then retighten.
5. To connect the regulator, attach the inlet connection nut to the cylinder valve. Make sure that the threads engage properly. Tighten the nut with a wrench, but do not use excessive force. Excessive force could damage the nut and valve threads.
6. Turn the pressure-adjusting screw counterclockwise until the tension is fully released, to shut off the regulator. *Note:* The regulator should always be shut off when not in use to help in avoiding gas loss if the cylinder or torch valves leak.
7. Do not stand directly in front or in back of the regulators when opening them. Open the cylinder valve slowly. Open fully only after the contents gauge stops moving. *Note:* Open the acetylene valves $\frac{3}{4}$ to 1 turn only, all others are fully opened to seal the valve packing.

Instructions for Cutting, Welding, and Brazing

Use the following instructions when performing any of these procedures.

Cutting steel Position the preheat cone just above the work (about $\frac{1}{8}$ inch) and hold it steady until the metal turns bright orange. Then slowly press the cutting lever and move the torch along steadily to avoid overheating and excess melting. To restart the cut, release the lever and preheat the metal before pressing the lever again.

Piercing For piercing holes or cutting into confined spaces, the tip should be tilted over to keep the metal spatter from plugging the tip. Return the tip upright after the cut is started.

Cutting thin steel Tilt the tip over so that the preheat flames hit the surface at an angle which can be increased or decreased as needed to vary the heat. Move the tip fast enough to avoid overheating and excess melting of the metal. Use the correctly sized tip for the job.

Welding steel *Note:* Oxy-acetylene must be used for welding steel. Clean and assemble the parts with good fit-up. Position the flame cone just above the surface until a puddle forms and penetrates well into the joint. Then dip the rod into and out of the puddle to melt it as needed to fill the joint and move the puddle along the joint. When welding metal under ⅛ inch, melt the rod ahead of the flame. Tilt the tip to decrease or increase the heat to the joint and slightly withdraw the flame as needed to avoid overheating the puddle. For metal over ⅛ inch, bevel the edge of the joints approximately 30° for good weld penetration.

Brass welding Brazing uses metals that melt and flow below the melting point of the base metal (brass at about 1600°F). When brazing, the proper filler rod and flux must be used. The joints must be clean and assembled with good fit-up. Be sure to provide proper ventilation to remove any harmful fumes. Heat the tip of the brazing rod and dip it into the flux (not necessary for prefluxed rods). Preheat the metal to a dull red color and apply the fluxed rod to the base metal and melt off a small amount of the rod. If it flows on the base metal, the heat is right, continue to heat and flow the rod on the joints. *Note:* Phos-copper brazing rod does not need fluxing when used on copper. Flux must be used on dissimilar metals.

Silver brazing Silver alloys melt at lower temperatures (about 1200°F) than brass alloys. The joints must be clean and have a good fit-up. The joint must be coated with flux and the joint heated evenly. When the flux bubbles and melts clear, start applying the rod to the joint. If it flows, the heat is right, and the solder will flow quickly in and around the joint. *Do not* overheat the joint. Remove the flame when the solder has flowed into the joint.

Soft soldering Lead solders melt at around 500°F. The joints must be clean, have a good fit-up, and be well covered with soft solder flux. Heat the joint until the flux boils, then apply the solder. If it flows, the heat is right, and the solder will flow quickly in and around the joint. *Do not* overheat the joint. Remove the flame when the solder has flowed into the joint. Always heat the joint and not the rod.

Flame Heat Control

The oxy-acetylene torch flame has an inner cone with a widening brush outer flame. The highest heat transfer zone is just past the inner cone (about ⅛ inch) which is used to melt metal for welding or to preheat steel to a bright orange color, so the cutting oxygen can start burning and melting (about ⅔ burning of the iron to form iron-oxide slag and ⅓ melting caused by the heat of the burning iron, more melting if the travel is too slow). The outer brush flame

Table 1-1 Flame temperatures (neutral flames)

Oxygen-acetylene	5600°F (3100°C)
Air-acetylene	4200°F (2315°C)
Oxygen-mapp	5100°F (2600°C)
Air-mapp	3500°F (2095°C)
Oxygen-propane	4500°F (2450°C)
Air-propane	3500°F (1927°C)

The flame temperatures of natural gas and propylene are in about the same range as propane. Only the neutral oxy-acetylene flame can be used for welding steel—other gas flames become too oxidizing at higher temperatures. (Courtesy of Uniweld Products, Inc., Fred L. Stettner, Senior Technical Advisor.)

spreads the heat out as it moves away from the inner cone. The flame heat can be controlled by moving the tip in or out from the work and also by moving it around so that it heats the joint evenly. Soft solder joints require less heat than brazed joints and brazed joints require less heat than welded joints. Larger tips and rosebud heating tips heat faster and can be used if the flame heat is controlled properly. Use larger-size tips for large tubing and for fuel gas brazing.

Oxy-acetylene provides the most versatile all-purpose flame tool for maximum heat and speed in getting the job done when the flame heat transferred to the work is controlled properly.

Air/acetylene, especially the high-velocity acetylene thruster-type tip, is the next best, followed by the LP thruster-type tip with air/mapp or air/propane.

The flame temperature table, Table 1-1, shows the differences between the flame temperatures.

Safety and Setup Checklist

Use the following steps to ensure a safe setup:

1. Review the previous information before using the equipment and use the outlined precautions and procedures.

2. *Safety precautions* The safety precautions include personnel protection, adequate ventilation, and fire protection and precautions.

3. *Cylinders* Cylinders should be secured and upright with an adequate gas supply to avoid dangerous empty cylinder conditions due to reversed gas flow, for safe storage, the valves should be closed when not in use or empty; use protective caps on stored or empty cylinders.

4. *Regulators* Regulators should have leak-tight connections; they should be turned off before opening the cylinder valve slowly and closed after the work is completed to avoid leaks from the cylinder or torch valves.

5. *Torches and tips* Before using torches and tips, make all connections wrench-tight and leak-test them; use proper operating pressures and clean, efficient tips. Purge the oxygen before each torch use. *Purge oxygen before each torch lighting and use.* Use the proper size and tip to fit the torch, job, and the fuel gas. *Do not* use damaged or plugged tips—repair or replace such tips.

6. *Operations* The torch connections should be checked for leaks before the torch is lit. Follow the proper tip lighting procedure with "no-smoke" fuel gas flame for adequate gas flow; maintain the proper neutral flame adjustment; use the proper tip angle and distance to decrease or increase the heat to the work for cutting, welding, or brazing; use the proper speed to avoid excess heating and melting on thin material when cutting or welding. Purge the oxygen before each torch lighting and use the "sniff test." Purge the torch and the hoses after each cylinder change to vent out possible mixed gases.

7. *Shutdown procedure* Shut the oxygen off first, fuel gas last and avoid backfire. The flame cannot burn in the torch without oxygen. Close the cylinder valves, drain off each gas separately, close the regulators and avoid leaks from the cylinder or the torch valves, close all torch valves, store the cylinders and equipment safely away from heat and ignition sources in a well-ventilated area.

8. Use the proper size tip cleaners to keep the tips effective and efficient for every job.

Caution: The withdrawal rate of an individual acetylene cylinder should not exceed one-seventh (15 percent) of the cylinder contents per hour. Use an adequate size cylinder or an adequate acetylene manifold system to supply the necessary volume for large usage and to avoid dangerous reverse flow of gases due to unbalanced pressures. *Do not allow the gas cylinder (especially oxygen) to become completely empty while in use.*

SAFETY IN HANDLING REFRIGERANTS

In operating refrigeration equipment, it is necessary that only the type of refrigerant for which the machine was designed be used in it. When other types of gases are used in the system, it will not operate as it was designed. In some cases the pressures may become excessive.

Personal Protection

Most of the refrigerants used today are nontoxic and present no personal hazard unless these is sufficient concentration to displace the oxygen in the air inside the space. Without oxygen the air will not support life. The refrigerant vapors are three to five times heavier than air and will, therefore, tend to collect in a strong concentration near the floor or in low places and displace the air. The stronger the concentration of refrigerant, the less air there is to breathe and a person can actually suffocate.

The deliberate inhalation of refrigerant vapors can cause a cardiac sensitization and death. This is an extremely dangerous act because death can occur without warning. When entering tanks or other places where there is the possibility of a strong concentration of refrigerant vapors, wear an air mask, use the buddy system, and attach a lifeline.

At times it is difficult to smell refrigerant vapors except in strong concentration. They have a faintly sweet odor which is difficult to detect. When you are working in an enclosed space where there has been a refrigerant leak or refrigerant has been purged into the air and you begin to feel a little dizzy, immediately get out of the space and into fresh air before it is too late.

Should refrigerant be drawn into the mouth, do not induce vomiting. When vomiting is induced, there is a possibility that the refrigerant can be drawn into the lungs. Should refrigerant be drawn into the lungs, seek medical attention immediately. This condition could cause chemical pneumonitis, pulmonary edema, and hemorrhage.

Most of the refrigerants used in modern equipment are in the vapor form at room temperature. These vapors have very little if any effect on the skin or eyes. However, those refrigerants that remain liquid at room temperature, such as CFC-11 and CFC-113, have a tendency to dissolve the protective fat from the skin which will cause a dryness and some irritation after prolonged contact or repeated contact with the liquid form. Any time there is the possibility of coming in contact with liquid refrigerant, protective clothing should be worn. Always wear protective eye goggles when working with refrigerants to prevent their entrance into the eye. Should refrigerant get into the eyes, flush the eye with water, and seek medical attention immediately. Liquid refrigerant will freeze the moisture in the eye and possibly cause blindness.

Should CFC-12, CFC-22, or any of the medium-temperature refrigerants come into contact with the skin, frostbite is probable. If treatment is started within 20 to 30 minutes after contact with the liquid refrigerant, soak the exposed area in lukewarm water. Never use hot or cold water. If the situation prevents starting of the treatment after 30 minutes have passed, apply a light coat of Vaseline petroleum jelly to the area. If the frostbite occurred in a place where it would be difficult to apply the petroleum jelly, then apply a light bandage to the affected area. Seek medical attention as soon as possible to reduce the chance of gangrene.

Handling Refrigerants Under Pressure

Some of the hazards of handling refrigerants under pressure are as follows:

1. Do not overfill a refrigerant container. Should an overfilled container be subjected to temperatures greater than 125°F, the container may become "liquid full" causing an immediately dangerous hydrostatic pressure to build up inside the container. The container may then rupture.

2. Do not overheat or store a refrigerant container in the direct sunlight. There is the possibility of a buildup of hydrostatic pressure inside a correctly filled container when heated above 125°F.

3. Never connect a refrigerant container to the discharge side of a refrigeration system. The pressure inside the container may rise above the capacity of the relief device and possibly cause the container to rupture. This is very possible with the 14-ounce CFC-12 cans used to charge automobile air conditioners. It is possible that they will explode when connected to the discharge side of the system.

Cylinder abuse is the most common cause of pressure-related problems with refrigerants. Four recommendations are as follows:

1. Never heat cylinders above 125°F.
2. Never tamper with the valves or the safety devices.
3. Never refill disposable cylinders with refrigerant or convert them into compressed air tanks. The empty cylinders should be vented and discarded. Reusing disposable cylinders is a dangerous practice because of the possibility of corrosion and weakening of the cylinder walls. The corrosion is greater when it is caused by water vapor in the refrigerant or in the compressed air. The greatest danger is that the damage may not be visible until the cylinder explodes. It is then too late.
4. Disposable containers should always be stored in a dry place to prevent rusting and corrosion of the cylinder.

Some Do's and Don'ts for Handling Refrigerant Cylinders

The following are some of the more important procedures for handling refrigerant cylinders:

1. Make certain that the cylinder is properly connected to the system.
2. Open the valves very slowly to avoid rupturing a hose or ruining a gauge.
3. Store the cylinder in a dry place to protect it from rusting.
4. Do not tamper with the pressure relief devices. To do so may cause the cylinder to rupture when subjected to excessive pressures.
5. Do not drop or otherwise abuse the cylinders.
6. Do not use disposable cylinders for air tanks. The moisture in the air will cause the cylinder to corrode from the inside out. It may then rupture when least expected and cause serious injury.
7. Do not refill disposable cylinders. They are designed for one-time use only and any moisture in the refrigerant will cause corrosion on the inside of the cylinder.
8. Do not force connections. This weakens the connection which may blow out when subjected to pressure and cause serious injury.
9. Do not heat the cylinder above 125°F. To do so may cause the cylinder to become liquid full and a hydrostatic pressure will build up rapidly and possibly rupture the cylinder.
10. Make certain that the refrigerant, the cylinder color, and the label match.
11. The caps which are provided for valve protection should always be on the valve except when the cylinder is in use.
12. Never mix refrigerant gas in a cylinder.
13. When a cylinder is emptied, immediately close the cylinder valve to prevent the entrance of moisture, air, and dirt into the cylinder.

14. Never use refrigerant cylinders for rollers, supports, or for any purpose other than to carry refrigerant.
15. Never attempt to repair or alter cylinders or valves.

Brazing or Welding on Refrigeration System Piping

The following are some of the precautions used when attempting these operations:

1. Remove all of the liquid refrigerant from the part of the system that is to be heated.
2. Be sure to purge all refrigerant out of doors.
3. Provide plenty of ventilation to the space. Use an auxiliary vent fan if necessary.
4. Purge with nitrogen the part of the system that is to be heated to force as much of the refrigerant vapor out of the system as possible.
5. Leave the system open during the welding or brazing process to prevent a buildup of pressure inside the system and decomposition of the refrigerant. Decomposed refrigerant may cause toxic and irritating compounds, such as hydrogen chloride and hydrogen fluoride.
6. Do not breathe the vapors produced when welding or brazing on refrigerant lines. These vapors are acidic and will irritate your nose and throat. Anyone exposed to these vapors should be taken to fresh air immediately and given medical attention.
7. Some silver-brazing alloys contain cadmium. These alloys are very dangerous to use. Most silver-brazing alloys contain large quantities of cadmium and when these alloys are melted they emit cadmium-oxide fumes into the air. These fumes are very toxic and must be avoided. An increase in the temperature above the molten state will cause an increase in the amount of fumes produced. Because oxyacetylene gas burns at a temperature of about 6000°F, higher concentrations of the cadmium-oxide fumes are produced than with other types of welding and heating fuels.

You cannot smell cadmium-oxide fumes, thus making it even more dangerous. A lethal dose does not necessarily need to be irritating enough to cause discomfort until after enough has been absorbed by the worker to place his or her life in immediate danger. Some of the more common symptoms of cadmium poisoning are headache, fever, chills, irritation of the throat, vomiting, weakness, and diarrhea. These symptoms may not appear for many hours after the exposure to the fumes. The primary injury is to the respiratory system.

Those who breathe the air within 5 feet of where the silver brazing is done are in the most dangerous area. When these materials are used in production work, precautions should be taken to protect workers from the fumes. However, when appliance technicians are working in the field, these precautions are not available. It is, therefore, up to the technicians to protect themselves. Lunches should not be stored or eaten in the area where silver brazing with these materials is done. The workers should be sure to wash both their hands and faces before eating, smoking, or leaving work.

The nice thing is that today no one must use these types of silver-brazing materials. There are materials available that do not contain cadmium. However, technicians must make certain that the type they are using is safe. There are several companies that manufacture cadmium-free silver-brazing materials.

SAFETY IN WORKING WITH HEATING EQUIPMENT

Heating equipment is probably the most dangerous of all the equipment used in the industry, especially natural and liquefied petroleum (LP) gases. The danger is in the combustion process that provides heat for the operation of the appliance. Electricity is probably the least dangerous of all the heat sources used for appliance operation. However, all of these heat sources must be treated with respect. Therefore, appliance service and installation technicians must know the combustion process and they must practice safety procedures to prevent loss of life, property, and equipment.

The following are some of the more common safety procedures to follow when working on heating equipment:

1. Do not tamper with the safety controls. If a problem arises with one, replace it. Tampering may cause delayed ignition which could cause an explosion or an overheated condition which could cause a fire.

2. Do not light the pilot until the firebox has been cleared of all unburned fuel.

3. Follow the manufacturer's directions carefully. The manufacturer has spent much time and money developing these directions and they should be heeded.

4. Always follow the local building safety codes.

5. Be sure the unit is grounded electrically.

6. Do not make adjustments without knowing what the results will be.

7. Do not alter the venting system in any way.

8. Never reduce the size of the vent pipe.

9. Increase the vent pipe size only when it is needed to provide proper ventilation of the products of combustion.

10. Be sure that the proper ventilation air is supplied to the appliance.

11. Never allow the unit to be operated when a bad heat exchanger is suspected. The products of combustion will enter the building and possibly cause death or illness of the occupants. A bad heat exchanger should be suspected when the occupants of the building complain of having headaches which are relieved when they leave the building for a short time.

12. Never leave a malfunctioning appliance in operation. To do so is inviting an explosion, fire, or other damage to the building and/or its contents.

13. Always check the venting system and remove any obstructions before lighting the burners. If an appliance is fired with a restricted venting system, the products of combustion will be forced into the building.

14. All safety devices must be in excellent working condition and be properly adjusted. Never leave a safety device bypassed.

15. It is never a good idea to work on "hot" electrical systems. However, at times it is necessary to find a problem with the equipment. Take precautions to avoid becoming grounded and receiving an electrical shock.

16. The equipment and its components must be installed according to local, state, and national electrical codes. The electrical circuits should be installed by a licensed electrician. If there is a local licensing code for appliance technicians, it must be adhered to. The equipment should have Underwriters' Laboratory (UL) approval.

17. Leak-testing should be done with soap bubbles or a liquid leak detector. Never check for leaks with an open flame. If a leak is present there will probably be a flame out of control.

18. Always allow the appliance to stand open for at least 5 minutes before lighting the pilot or burner. This is to allow any accumulation of fuel to escape to the atmosphere without an explosion.

19. Always exercise care when working around heating appliances to prevent burns because almost all of the surfaces are hot when the appliance is operating.

20. When in doubt about a heating appliance, check the vent products with one of the meters which are available for this purpose. Be safe not sorry.

21. When servicing heating appliances, before leaving the job, check all safety devices to make certain that they are operating properly.

SAFETY IN WORKING WITH ELECTRICITY AND ELECTRONICS

Appliance technicians are involved in the installation, service, and maintenance of equipment in which dangerously high voltages are present. This work is often done in confined spaces. Among the hazards of this work are electrical shock, electrical fires, harmful gases which are sometimes generated by faulty electrical and electronic devices, and injuries which may be caused by the improper use of tools.

Because of these dangers, appliance technicians should formulate safe and intelligent work habits since these are fully as important as knowledge of the electronic equipment. One primary objective should be to learn to recognize and correct dangerous conditions and avoid unsafe acts. These persons should also know the proper procedures for dealing with fires of an electrical origin, for treating burns, and for giving artificial respiration to persons suffering from electrical shock. In some cases, artificial respiration may have to be accompanied by cardiac massage (CPR) to restore heart beat.

Electrical Shock

Electrical shock may cause burns of varying degree, the stoppage of breathing and unconsciousness, ventricular fibrillation or cardiac arrest, and death. If a 60-hertz alternating current is passed through a person from hand to hand or from head to foot, the effects when current is gradually increased from zero are as follows:

1. At about 1 milliamp (0.001 amp), the shock will be felt.

2. At about 10 milliamps (0.01 amp), the shock will be severe enough to paralyze muscles and a person may be unable to release the conductor.

3. At about 100 milliamps (0.01 amp), the shock is usually fatal if it lasts for 1 second or more. *It is important to remember that, fundamentally, current, rather than voltage, is the criterion of shock intensity.*

It should be clearly understood that the resistance of the human body will vary. That is, if the skin is dry and unbroken, body resistance will be quite high, on the order of 300,000 ohms. However, if the skin becomes moist or broken, body resistance may drop to as low as 300 ohms. Thus, a potential as low as 30 volts could cause a fatal electrical shock. Therefore, any circuit with a potential in excess of this value must be considered dangerous. *Note:* The intentional taking of an electrical shock should be avoided.

Care of Shock Victims

Electrical shock is a jarring, shaking sensation resulting from contact with electrical circuits or from the effects of lightening. The victim usually experiences the sensation of a sudden blow, and if the voltage is sufficiently high, unconsciousness. Severe burns may appear on the skin at the point of contact. Muscular spasms can occur, causing a person to clasp the apparatus or wire which caused the shock and to be unable to let go. Electrical shock can kill its victim by stopping the heart or by stopping breathing, or both. It may sometimes damage nerve tissue and result in a slow wasting away of muscles that may not become apparent until several weeks or months after the shock is received.

The following procedure is recommended for rescue and care of electrical shock victims:

1. Remove the victim from electrical contact at once, *Do not endanger yourself.* This can be done by: (1) throwing the switch, if it is nearby; (2) using a dry stick, rope, leather belt, coat, blanket, or any other nonconductor of electricity; or (3) cutting the cable or wires to the apparatus, using insulated tools while taking care to protect your eyes from the flash when the wires are cut.

2. Determine whether the victim is breathing. If so, keep the victim lying down in a comfortable position. Loosen the clothing about the neck, chest, and abdomen so that the victim can breathe freely. Watch the victim carefully and protect him or her from exposure to cold.

3. Keep the victim from moving about. After shock, the heart is very weak, and any sudden muscular effort or activity on the part of the victim may result in heart failure.

4. Do not give stimulants or opiates. Send for a medical person at once and do not leave the victim until he or she has adequate medical care.

5. If the victim is not breathing, it will be necessary to apply artificial respiration without delay, even though the victim may appear to be lifeless.

WORKING ON ENERGIZED CIRCUITS

Insofar as practical, a technician should not undertake working on energized circuits and equipment. However, as when making operational adjustments, one should carefully observe the following safety precautions:

1. Ensure that you have adequate illumination. You must be able to see clearly, if you are to safely and properly perform the job.

2. Ensure that you are insulated from ground by an approved-type rubber mat or layers of dry canvas and/or wood.

3. Where practical, use only one hand, keeping the other behind you or in your pocket.

4. If the system voltage exceeds 150 volts, rubber gloves should be worn.

5. An assistant should be stationed near the main switch or circuit breaker, so that the equipment may be immediately deenergized in case of an emergency.

6. A person qualified in first aid for electrical shock should be standing by during the entire operation.

7. *Do not work alone.*

8. *Do not* work on any type of electrical apparatus when wearing wet clothing or if the hands are wet.

9. *Do not* wear loose or flapping clothing.

10. The use of thin-soled shoes and shoes with metal plates or hob nails are prohibited.

11. Flammable articles, such as celluloid cap visors, should not be worn.

12. All rings, wristwatches, and similar metal items should be removed before working on the equipment. Also, ensure that clothing does not contain exposed metal fasteners such as zippers, snaps, buttons, and pins.

13. Do not tamper with interlock switches, that is, do not defeat their purpose by shorting them or blocking them open.

14. Ensure that the equipment is properly grounded before energizing.

15. Deenergize the equipment before attaching alligator clips to any circuit.

16. Use only approved meters and other indicating devices to check for the presence of voltage.

17. Observe the following procedures when measuring voltages in excess of 300 volts:
   a. Turn off the equipment power.
   b. Short-circuit or ground the terminals of all components capable of retaining a charge.
   c. Connect the meter leads to the points to be measured.

d. Remove any terminal grounds previously connected.
e. Turn on the power and observe the voltage reading.
f. Turn off the power.
g. Short-circuit or ground all components capable of retaining a charge.
h. Disconnect all meter leads.

18. On all circuits where the voltage is in excess of 30 volts and where metal is present, the worker should be insulated from accidental ground by the use of an approved insulating material. The insulating material should have the following characteristics:

a. It should be dry, without holes, and should not contain conducting materials.
b. The voltage rating for which it is made should be clearly marked on the material, and the proper material should be used so that adequate protection from the voltage can be possible.
c. Dry wood may be used, or as an alternative, several layers of dry canvas, sheets of phenolic insulating material, or suitable rubber mats.
d. Care should be exercised to ensure that moisture, dust, metal chips, and so on, which may collect on insulating materials are removed at once. Small deposits of such materials can become electrical hazards.
e. All insulating materials on machinery and in the area should be kept free of oil, grease, carbon dust, and so on, since such deposits destroy insulation.

REVIEW EXERCISES

1. Why is it especially important for a worker in the appliance field to practice safety?
2. What should be done before operating any piece of equipment?
3. What temperature can the acetylene flame reach?
4. What is the standard shade of eye protection used when welding?
5. What must be done to a container or pipe before using a welding torch on it?
6. What must be done with welding hoses when welding or cutting?
7. Name two uses for a bucket of water in the welding area.
8. What should be done before using new welding equipment?
9. In what position must a fuel gas cylinder be used?
10. Should the fittings and valves on welding equipment be lubricated with oil or grease?
11. What should be done when a cylinder valve becomes dirty?

12. Where should the operator be standing when opening the valve on a welding cylinder?
13. How far should acetylene cylinder valves be opened?
14. How far should oxygen cylinder valves be opened?
15. What is the maximum acetylene pressure that can safely be used on welding equipment?
16. Is it safe to light the welding torch with a match or a cigarette lighter?
17. What must be done to prevent flame backfiring or flashback inside the welding torch?
18. What is the first step in the welding unit shutdown procedure?
19. What is the maximum withdrawal rate of a single welding cylinder?
20. Is it proper procedure to use more than one type of refrigerant in a system?
21. What is the greatest personnel hazard when using refrigerants in a confined space?
22. What could be the result of inhaling refrigerant vapor?
23. Describe the odor of refrigerant vapors when in heavy concentration.
24. What will direct contact of liquid refrigerant with the skin cause other than frostbite?
25. Should hot water be used on refrigerant frostbite?
26. What could possibly happen if the temperature of an overfilled refrigerant cylinder were exposed to temperatures over 125°F?
27. Why should a refrigerant container never be connected to the high side of the refrigeration system?
28. Why is reusing disposable refrigerant cylinders a bad practice?
29. Why should a refrigeration system be left open during a welding operation on tubing?
30. Why are cadmium-oxide fumes dangerous to breathe?
31. What type of equipment is probably the most dangerous with which to work?
32. What complaint is a good indication of a bad heat exchanger in a gas-fired appliance?
33. Is it good practice to leave safety controls bypassed?
34. With what should natural or LP gas be leak-tested?
35. List the possible results of an electrical shock.
36. At what amperage is an electrical shock usually fatal?
37. What is the most dangerous to people, electrical current or voltage?
38. What is the first step to follow when a shock victim is discovered?

2

The Appliance Service Occupation: Problem Solving and Customer Relations

At the present time homeowners are becoming more dependent on machinery to provide them with comfort and the finer things in life. Machines such as automatic washing machines, clothes dryers, refrigerators, freezers, dishwashers, and other such equipment all make our lives easier and happier. These machines all need service and repairs at some point in time. It is the job of the appliance technician to make certain that these machines are installed, maintained, and serviced properly. Most homeowners do not have the time, desire, tools, or perhaps the ability to perform these chores. They, therefore, depend on the appliance technician to provide these services accurately and economically.

CLASSIFICATION OF APPLIANCE PERSONNEL AND THEIR RESPONSIBILITIES

The degree of responsibility of the technicians will depend on their ability and knowledge of the equipment. Therefore, technicians with a wide range of experience would be re-

quired to repair more types of appliances than would technicians with a limited background. This has a tendency to reduce the unnecessary replacement of good parts in an attempt to find the problem.

Service technicians are generally classified in several general categories. Each successive category requires more knowledge and skills. At the same time they have more responsibility and make more money. The major categories are as follows.

Parts Person

This person is responsible for maintaining an adequate inventory of the necessary parts and supplies needed to keep the service and installation departments operating smoothly. The parts person must keep track of the parts and supplies and issue them to the technicians as necessary or he or she may actually sell them to customers, depending on the size of the shop and the nature of the business.

Benchperson

The common names for persons in this category are benchperson, shopperson, and small appliance service technician. The technicians in this group are capable of repairing small appliances such as toasters, mixers, irons, and small heaters that may be brought into the shop for repair. Many of them, while working under the direct supervision of a more experienced technician, can change parts in major appliances. The benchperson has very little, if any, contact with the customer. He or she generally works all day in the shop repairing appliances. In most cases the repair can be performed at a bench or work table.

Installers

These persons are generally more experienced than the benchperson. Installers deliver, install, and demonstrate the operation of the appliance to the customer. In most instances they do little or no troubleshooting or parts replacement. However, they will work under a wide variety of conditions. They are mostly concerned with major appliances, such as washing machines, dryers, refrigerators, freezers, and room air-conditioning units. It is their job to know how these appliances operate and how to get the best performance from them. Most of these technicians' work will occur at the customer's home. They will have very little supervision. They should be familiar with the local building codes and be physically able. They must present themselves in a good manner to the customer.

Major Appliance Technicians

The duties of major appliance technicians are to make service calls and make repairs to major appliances in the customers' home or business. These technicians are usually very knowledgeable about all types of appliances. They will have sufficient experience to handle most problems as they arise. They work with a minimum of supervision. On most calls they will make estimates, make the repairs, and receive payment for the work while they are there. In most instances they will plan their own work schedules. The major appliance technicians must be very good at human relations. They are responsible for a large supply of parts on their trucks as well as the necessary tools and test equipment. They must be well groomed and be physically able to do the required lifting. They must adapt to changing working conditions.

Dispatchers

The dispatchers are responsible for receiving the service calls from the customers. They route the service personnel and establish priority on service calls. They may also make recommendations for special tools and parts that are to be taken to a specific job site.

Service Managers

These persons are in direct charge of the service shop and are usually responsible for the activities there. Their responsibility is to both the customer and the business owner. They should make every effort to keep them both satisfied.

Field Service Technicians (Factory Representatives)

These persons must have a very in-depth understanding of the products that their companies manufacture. They must have a very strong diagnostic ability and be able to teach other technicians about the equipment. Generally, there is a great amount of traveling required in this position. Most of the training courses that they present will be given at night, after normal working hours.

Technical Writers

Technical writers are responsible for writing books of instruction and any other service and installation literature that will help people in the field. They usually work with factory engineers and engineering technicians who have designed and tested the appliance.

The training courses which are considered to be the most beneficial to appliance technicians are physics, math, basic electricity, electronics, English, and some basic-level courses in behavioral sciences and the necessary laboratory exercises which reinforce the theory learned during the lecture period. Even after graduating from a technical school majoring in appliances, they must continue their education by attending manufacturer's product seminars. Many good appliance technicians have failed to succeed in this field because they would not attend these seminars and failed to keep up with the changes in the field.

REQUIREMENTS FOR APPLIANCE TECHNICIANS

Appearance

A neat-appearing technician will impress the customer just about as much as anything. The technician must be neat and clean. Cleanliness is important in both your appearance and your work habits. It is very desirable to keep your tools and parts in order so that they can be easily and quickly found. When you are finished with the job, be sure that the work area is left clean and neat. Wipe off the machine with a clean rag. Do not leave grease, fingerprints or handprints on the machine. A few extra minutes used for this purpose will go a long way toward customer relations in addition to making the machine look like it has been repaired. There will be very few complaints from anyone about this gesture.

Maintaining a Good Attitude

A good attitude toward the customer, the machine, the equipment manufacturer, and your employer is very important. If you display a negative attitude toward any one of these, the customer will also develop a bad attitude toward them. You must always project a positive attitude, confidence, be courteous, and cheerful. Always project a businesslike impression. Be on time for the appointment, if one is made. If you are unable to make the appointment, be courteous enough to phone the customer and set up a new appointment. If you do not, and the customer must phone the company again, the customer will be very displeased with the whole job and will probably be very difficult to satisfy. Always be honest with the customer. If you are not familiar with the piece of equipment that is to be worked on, tell the customer. Many times you have a background that will permit you to call upon past experience to repair the machine. If you cannot fix the machine, inform the customer and give an explanation of what you propose to do. Treat other people the way that you would want them to treat you if you were their customer.

A self-conducted tour through the customer's home or building should not occur. Do not talk about other customers, neighbors, or the product to the customer. In most instances it is better to offer no opinion about the quality of the machine. Even if you think the customer is going to have a lot of trouble with it, do not say so.

REPAIRING THE MACHINE

Any machine that was built by humans can be repaired, if approached with a positive attitude. Sometimes it may be more economical not to repair it, but it can be repaired. When a malfunctioning part is replaced, the machine should function just as well as it did when it was new. It is the job of the technician to find the defective component and repair or replace it. There will be some repair price that you have in your mind that above which the customer must be consulted. Sometimes it is better to give the customer the option of making the repair at any cost rather than just repairing it and submitting a bill. The customer may wish to do something other than repair the machine. The machine may be obsolete and the customer might want to replace it rather than repair it. In any case it is the customer's decision to make. Present the facts as they are. Do not try to sell new equipment at this time. Wait until the customer asks whether it is economically feasible or not, then make your recommendations.

PRESERVICE INSPECTION

It is not wise to make a decision on what the problem may be before you have gotten to the job. Many technicians make this mistake and because of it they have a closed mind when they get to the job site. This makes troubleshooting the machine much harder and takes more time than is necessary. Talk to the customer about what the machine was doing when it malfunctioned. Remember that the customers live with the machine and know its characteristics much better than you. Many times they will tell you exactly what the problem is during their explanation, that is if you listen to what they are saying. Your knowledge of the machine

will pinpoint what components were in operation when the unit failed. These should be the first ones checked.

On any service call it should be your first step to check the installation of the machine. Make certain that everything is as it should be. If it is not be sure to correct anything that is not proper. When you are making an in-warranty service call, explain to the customer what part failed and was replaced or what action was taken, a connection came loose, or what adjustments needed to be made and that there is no charge for the service call. Express your regret for the inconvenience that the failure may have caused, and thank the customer for phoning you promptly so that he or she can realize the full and proper performance of the machine.

CUSTOMER INSTRUCTIONS

When customers make a purchase, there are many things about the new machine that they do not understand or sometimes do not even know are available on the machine. It is the job of the technician to train the customers how to properly use the machine. Usually the salesperson will tell them but may not show them and it slips their memory. This is where you fit in. If there is actually a problem with the machine or its installation, make certain that the customer knows what is required to make the unit operate as it was designed. If the problem is with the customer, explain in a courteous manner how to properly operate the machine. The customer will appreciate this and will feel that he or she has made the right choice. When instructions are properly given, your reputation as a technician will be enhanced as well as that of the equipment.

In most instances you are the only contact that the customer has with the manufacturer. It is your obligation to both the manufacturer and to your employer to make certain that the customer is satisfied. Many times it is your attitude and professionalism that either keeps or loses a customer. When a customer is lost everyone loses.

USE OF YOUR TIME

The effective use of your time affects both you and your employer. A reliable technician is very much in demand. However, one that has the image of being disloyal or dishonest will have a difficult time getting and keeping a job. Being honest and loyal is very much a part of public relations and will affect your success as a technician. Your efforts in these areas will enhance your success as much as your knowledge of the equipment that you are servicing.

Another step in succeeding is the ability to recognize that parts do fail either by abuse, damage, or just plain wearout. It is a part of your job to learn the difference and explain to the customer what was found and how to prevent it from happening again.

Never compare the operation of one piece of equipment with another. Each one has different features and must operate on its own merits. The customer chose the one that was satisfactory at the time of purchase. However, you must be up-to-date on the newer equipment in your field, so that when, and if asked, you can discuss it confidently and freely with the customer. Perhaps the customer will even ask your opinion on which will suit his or her needs the best. But, unless you know the customer, do not offer it.

SERVICE: A NECESSITY

Every machine will need servicing at some time or another. Therefore, a good service technician will always have a good job. The better the technician, the better and more secure is his or her position. Since there is a constant standoff between price and reliability, there will always be some type of service needed. The best service that you can provide is to prevent as many repeat service calls as humanly possible. This will help if you remember that the machine is made up of parts that are assembled to provide a given service and that when one malfunctions the remainder of the machine will not function as it was designed. With this knowledge you can examine the complete machine for the bad part and make the repairs as economically and as accurately as possible.

Customers know that when a machine fails it is going to cost money to have it repaired. However, they are not always aware of the high cost of making these repairs and are due an explanation of the problem and the cost.

It costs money to equip an appliance truck with the necessary tools, instruments, and parts to do the job. This money must come from somewhere and this somewhere is the customer. It is figured into the bill as overhead to operate the business. Then the shop owner must make enough money to pay you and make some for his or her troubles and investment, to which he or she is entitled.

To help this happen is by and large up to the professional appliance technician. A professional is a person who knows what it takes to repair the machine and one who does not "kill" time.

EDUCATION: A NEVER-ENDING TASK

You must continually study to keep up with the constant changes which are taking place on appliances. This is best done by studying manufacturers' literature and attending manufacture-sponsored seminars. Other training programs that are beneficial are usually offered at the local community college or trade school. Basic courses can only provide you with just that, basic information. You will need to find sources of detailed information on any given piece of machinery. This can be accomplished in part by belonging to trade organizations and subscribing to trade publications which offer this type of information. One such publication is Appliance Service News, 5841 West Montrose Avenue, Chicago, IL 60634.

The benefit of staying abreast of the industry is job security and a better chance to stay in business, if you are in business. The appliance industry is a multibillion dollar industry and is still growing. It will, therefore, benefit those who put forth the effort to keep up-to-date of what is happening in the industry.

CUSTOMER RELATIONS

Customer relations is one of the most important aspects of any type of business. A service business is certainly no exception. An appliance service and installation technician should have a good understanding of what makes good customer relations.

As an appliance service and installation technician, your job will involve working directly with the customer and other people who can have an effect on the success of the company, as well as your own personal success. You will also be required to work with equipment and materials. All of these factors are necessary for you to do your job; however, the customers are by far the most important. Your success will depend on them and their attitude toward you. It is sometimes difficult to make friends with difficult customers, but your success depends on how well you handle difficult situations.

Many potential appliance technicians think an ability to use the proper tools is all that is necessary to get the job done. This idea is far from the truth because customers have basic needs that they want satisfied. Therefore, as a technician you must know how to provide these needs and how to get along with customers. Because of this, one of your top priorities must be to continually improve the relationships between you and your customers. It must be remembered that your personal success as a technician is, to a great extent, dependent on how well you get along with the customers.

Many technicians who are not the best at working with equipment and materials are very good at getting along with customers and hold higher-paying positions within the company. This is, for the most part, because the customers call for them to do the needed work.

The Meaning of Human Relations

There are several schools of thought concerning human relations. Some think that the practice of good human relations is applying the golden rule to relationships. Some think that it is applying psychology to make friends and influence people. Others think that human relations makes use of an ethical approach to personal problems. The fact is that human relations is made up of all these factors.

Good human relations means people getting along with other people in a harmonious manner, reaching satisfactory production, and cooperatively achieving an economic as well as some social satisfaction. The key part to any human relations is the motivation of people. Therefore, the practice of good human relations is more than back-slapping, more than being a nice person, and more than glad-handing.

THINGS TO REMEMBER WHEN WORKING WITH OTHERS

As an appliance technician you will be working directly with the customers. Because of this fact, there are several things you should remember about them as individuals:

1. We are all different. We have thousands of differences. Each of our personalities is different, just as our fingerprints are different. We develop differently from the day we are born. We each have our own ideas about life and our wants as an individual.

Due to these individual differences, any study of human relations must start with the individual. It starts, therefore, with you. You are always a part of a team, but you must remember that the team consists of individuals just like yourself. The team is an entity and has a certain amount of power because of its individual members. The team, however, cannot make decisions. It is the individual members who must make the decisions.

2. You must remember that when you are working with a person, you are working with a complete person. You may wish to enjoy a certain part of that person's personality, but this cannot be done. When any part of a person is considered, the whole person must be considered: mind, thoughts, and actions. The person's desires and motivations must also be considered, along with his or her home life, which has an effect on how the person responds to different situations. As a company representative, you must remember that each person is different and in your dealings you must take into consideration the complete person.

3. All the normal behavior of an individual is caused by individual needs and wants. A person is motivated to do things through the belief that doing so will help achieve a goal that he or she feels is worthwhile. Remember, the customer is generally not considering what you think is right but what he or she thinks is right. All of us sometimes think that another person's wants, and the reasons behind them, are foolish; but to him or her they are very important and real needs. These are needs with which you will be dealing when working with customers or fellow employees.

4. Remember that customers are not something that can be programmed for use when we desire and discarded when we are finished with them. We are all human beings and we all need and should be treated with dignity and due respect. This should be done regardless of what we think of a person. Regardless of their stations in life, customers should be shown the proper respect for their personal choices and abilities.

When dealing with people, you should always keep these four aspects clearly in mind. When doing this, you will find that your understanding of other people and your ability to work with them will improve with experience. For example, when you realize that all people are different from each other, you will find that categorizing them is most difficult. You will soon stop trying to handle all people in the same way because you will begin to recognize that you are dealing with the complete person as an individual who is unique.

AFFECTING THE CUSTOMER'S ATTITUDE

Remember, your job is getting things done for people and your effectiveness as a technician will be measured by how the customer relates to you and your company. Your success depends on how the customer sees you. Because of this fact, your customers are a very real asset to you and should be treated with respect and consideration. Your customers, in most cases, want to get along with you and help get the job done as efficiently as is possible. Remember, the work you are doing has an impact on their lives and they hope to gain some personal satisfaction from its successful completion. Whether or not this satisfaction is realized depends on how well you do your job and your style of presenting yourself. A technician who is generally happy and self-confident will usually be seen as a good technician.

UNDERSTANDING YOURSELF

Someone once said that if you want to understand other people, you must first understand yourself. When a person begins to know himself or herself, he or she has taken the first step toward understanding others. For example, when you realize that you have certain attitudes about your dress, how fast you should work, and how you should behave, you should also realize that other people have their ideas about these things and they may not be the same as yours. Therefore, understanding yourself will help you to understand others.

This is especially true of the relationship between the technician and the customer. Any time you become critical about what someone has said or done, you should ask yourself if you have done something to cause the person to react in this manner and what you can do to correct the situation. In most cases, some of the blame is on your shoulders. When none of the blame is yours, change your point of view and look at the situation from the other person's point of view. In most cases this will help you gain a better understanding of the problem. As a result, you will generally feel differently toward the other person and the problem that exists, because you will have greater understanding, compassion, and integrity because you know how you would react under the same circumstances.

Attempting to understand yourself means taking a good look at both your strong points and your weaknesses. Your good points are easy to look at but your weaknesses are a bit more difficult to see. You should realize that you have limitations and hangups just like any other person; you may have a short temper, or you may be prejudiced against certain individuals, and have other such weaknesses. When these shortcomings about yourself are recognized, you are in a better position to control them and do a much better job of human relations.

Earlier we mentioned putting yourself in another person's shoes to help understand his or her behavior. When you do this, you are empathizing; that is, you are trying to see the problem from the other person's point of view. This does not mean that you are going to agree with the person and do things his or her way. It means just the opposite. Once you have an understanding of the problem from the other person's point of view, you are just in a better position to deal with it fairly and effectively. In this manner you are in a better position to appreciate the other person's point of view and feelings without getting involved in his or her personal life. At this point you can truthfully say that you understand why the person feels the way he or she does. When you have truly achieved this ability, you will be in a better position to solve the problem with fairness to all concerned.

We learned earlier that all behavior is the result of something, that there is a reason for that behavior. As a technician, you should attempt to recognize and understand the cause for the behavior of your customers. The specific reason that a customer does something may not have a logical explanation; it may not even be reasonable, and it may be ridiculous to you, but it is important to the customer and it has caused his or her actions. A major task facing you as a technician is to understand why a particular customer or fellow employee behaves the way he or she does, and not turn off the person's actions as absurd or ridiculous. As a matter of fact, when you state that the other person's actions are absurd, you are actually admitting your own inability to see things from the other person's point of view and to understand the reasons for his or her actions. If you are not able to see the reasons, then you are going to have a tough time as a technician.

You should make empathizing so much a part of your actions that it comes as easily as shaking hands with a friend. It should be so easy for you to use that solving disputes will be easily done in most cases.

KNOWING YOUR CUSTOMERS AND FELLOW EMPLOYEES

The better a person knows the people with whom he or she is associated, the easier it will be to understand them, their viewpoints, and their problems. As a technician, therefore, you should make every effort to understand the complete person, whether a customer or a fellow employee. A person's life will influence his or her behavior and actions.

To be good at human relations, it is almost impossible for you to know too much about the people with whom you associate. When the facts about the complete person are known, it is much easier to see his or her point of view and to empathize.

WHAT TO EXPECT FROM GOOD HUMAN RELATIONS

The total concept of human relations is that something takes place between one person and another. When good human relations is practiced, you will be better able to help satisfy some of the needs of those around you. After all, this is what human relations is all about. For example, we all feel a need to belong to something, a group, or someone. Anything that you can do to fill this need will enhance you in that person's eyes and allow you to do a better job.

However, practicing good human relations does not mean that everyone will be all smiles and that happiness will abound in all corners. We all have our own set of

problems, but practicing good human relations will allow us to work around the difficulties that most problems present.

Achieving good human relations will pay off in a big way. The rewards are sometimes difficult to measure, but they are most certainly there. The practice of good human relations often means that a difficult problem can be avoided, and usually stopped before it gets started. Because of this, good human relations will show up in many different ways and will give you the satisfaction of a job well done.

REVIEW EXERCISES

1. The installation, maintenance, and service procedures are accomplished by the _____ _____.
2. What is the first step in servicing appliances?
3. Who is responsible for maintaining an inventory of the necessary parts and supplies?
4. Which appliance service technician has the last contact with the customer?
5. Who is the person who directs the operation of the service shop?
6. With what should the appliance technician be equipped to accomplish his task?
7. In addition to equipment knowledge, what is required of the appliance technician?
8. To be able to succeed in the appliance field, what must a person have?
9. Why must an appliance technician be neat and clean?
10. What is important concerning the customer, machinery, the manufacturer, and your employer?
11. What should never be done on a job?
12. What is the job of the appliance technician?
13. What should be the first step on a service call?
14. What is your obligation to both the manufacturer and your employer?
15. What is one of the most important aspects of any type of business?
16. To be a successful appliance technician, all that is required is the proper use of tools. (True-False)
17. What is the key part to human relations?
18. Where does human relations start?
19. What must be remembered when dealing with people?
20. What causes individual behavior?
21. How should people be treated?
22. How will your effectiveness as a technician be measured?
23. How will a good technician usually be seen?
24. When you want to understand other people, what must first be done?
25. What should be done when attempting to understand yourself?
26. What does it take to be good at human relations?

3

Tools and Instruments

In addition to the commonly used mechanic's hand tools, there are some specialized hand tools used by appliance technicians. The most popular of these tools will be covered at this time, followed by coverage of test instruments that are used in this field.

TUBING CUTTERS

Tubing cutters are used to cut tubing with a square end, enabling the appliance technician to make leak-tight connections. In practice, as shown in Figure 3–1a, the cutter is placed around the tube. The cutter wheel is then tightened on the copper tubing, as shown in Figure 3–1b. Care should be taken not to tighten the cutter wheel too much because the tubing will be flattened and a leak-tight joint will be almost impossible to make. After the cutter wheel touches the tubing, the cutter handle should not be turned more than $\frac{1}{2}$ turn. See Figure 3–2. Make a complete revolution of

the tube with the cutter, then tighten the cutter handle no more than $\frac{1}{2}$ turn again. Repeat this procedure until the tubing has been completely separated. To obtain a square end on the tubing, the tubing must be placed on the rollers so that the cutter wheel will be centered on the tube, as shown in Figure 3–3.

After the tubing has been completely separated, there will be a burr on the inside of the tubing. This burr must be removed or it will cause a restriction to the fluid flowing through it. Most tubing cutters have an attached reaming blade. See Figure 3–4. This blade is rotated outward and then inserted into the tubing end. It is then rotated to remove the burr. Care should be taken to prevent the loose metal chips from falling into the tubing where they will clog the strainers and filters, which will need to be removed for cleaning or replacement.

Tubing cutters are also equipped with a flare cutoff groove in the rollers so that bad flares can be easily cut from

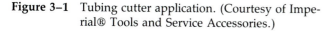

(a) (b)

Figure 3–1   Tubing cutter application. (Courtesy of Imperial® Tools and Service Accessories.)

Figure 3–2   Tubing cutter adjustment. (Courtesy of Imperial® Tools and Service Accessories.)

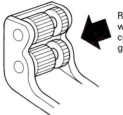

Figure 3–3 Tubing cutter rollers. (Courtesy of Imperial® Tools and Service Accessories.)

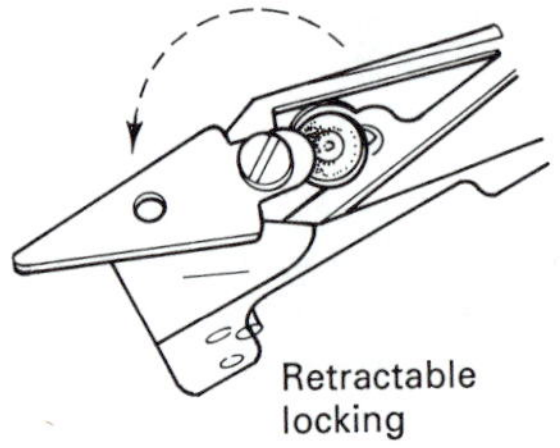

Figure 3–4 Tubing cutter reaming blade. (Courtesy of Imperial® Tools and Service Accessories.)

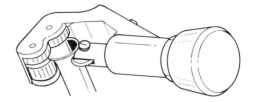

Figure 3–5 Tubing cutter with flare-removing rollers. (Courtesy of Imperial® Tools and Service Accessories.)

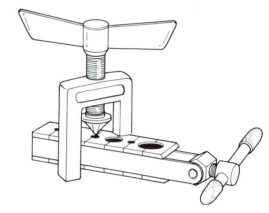

Figure 3–6 Flaring tool. (Courtesy of Imperial® Tools and Service Accessories.)

the tubing. See Figure 3–5. This permits easy removal of the flare with a minimum amount of wasted tube. It also prevents threading of the tube.

FLARING TOOLS

Flaring tools are used to make a flare on the end of tubing so that flare fittings can be used to make leak-tight connections. See Figure 3–6. When a flare is to be made, insert the tubing into the flare block so that approximately $\frac{1}{4}$ inch extends above the flare block. See Figure 3–7a. Swing the clamp into place against the end die and tighten it sufficiently to prevent slippage of the tube in the flare block. See Figure 3–7b. Slide the yoke over the tubing and turn the feed screw 5 or 6 turns after the cone touches the tubing. See Figure 3–7c. When the flare is completed, reverse the flaring process to remove the tube from the flaring block. Backing off of the cone automatically burnishes the flare to a highly polished finish. Flares made in this type of flaring block are stronger because the flare is made above the flaring block die, not against it, and the original wall thickness is maintained at the base of the flare. There is no chance of "washing out" the flare. Flaring tools can be used for re-rounding and sizing bent tubing.

TUBING REAMERS

Tubing reamers are used to remove burrs from the inside of the tubing. See Figure 3–8. After the tubing has been cut, the reamer is placed over the tube and rotated, removing the burrs from either the inside or the outside of the tube. Then the reamer is turned around and placed over the tube again and rotated to remove the burrs. Since these are multisized tools, one tool will fit many sizes of tubing.

REVERSIBLE RATCHETS

These ratchets are used to turn the stem on refrigeration service valves to allow service operations to be completed. The most common size is $\frac{1}{4}$ inch. A $\frac{3}{16}$-inch square is provided on the end with a $\frac{1}{2}$-inch six-point socket. There are a variety of sockets available to fit almost any size valve stem that will enable the technician to connect the refrigeration gauges to the system. Also available are a variety of packing-gland nut sockets that can be used with the reversible ratchet. These are used to tighten packing glands on certain valves to prevent leakage of refrigerant.

GAUGE MANIFOLDS

Probably the most often used tools in the service technician's tool chest are the gauge manifolds. A gauge manifold is comprised of a compound gauge, a pressure gauge, and the valve manifold. See Figure 3–9. All the service operations involving refrigerant, lubricating oil, and evacuation can be accomplished by use of this tool.

Compound Gauges

These gauges are used to read pressures both above and below atmospheric (vacuum). See Figure 3–10. In practice, they are used to determine pressures in the low side of

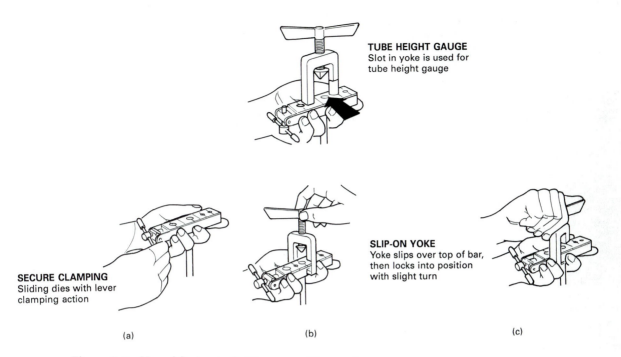

TUBE HEIGHT GAUGE
Slot in yoke is used for tube height gauge

SECURE CLAMPING
Sliding dies with lever clamping action

SLIP-ON YOKE
Yoke slips over top of bar, then locks into position with slight turn

(a)   (b)   (c)

Figure 3–7   Use of flaring tool. (Courtesy of Imperial® Tools and Service Accessories.)

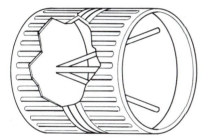

Figure 3–8   Tubing reamer. (Courtesy of Imperial® Tools and Service Accessories.)

a refrigeration system. The outside scale is in pressure (psig). The inside scales are the corresponding temperatures for different refrigerants.

Gauges operate due to the action of a Bourdon tube. See Figure 3–11. When pressure inside the Bourdon tube is increased, the element tends to straighten. As the pressure is decreased, the element tends to curve again. A Bourdon tube is a flattened metal tube sealed at one end, curved, and soldered to the pressure fitting on the other end. The movement of the element will pull a link that is attached to the pointer through a series of gears. This movement will be shown by the gauge hand, or pointer.

Compound retard gauges have a retarder that permits accurate readings within a given range. In refrigeration work this range would be between 0 and 100 psig. These gauges can be recognized by the change in graduations at pressures higher than those usually encountered.

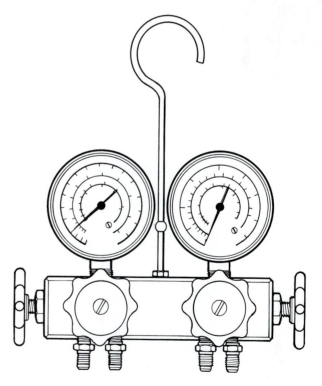

Figure 3–9   Refrigeration gauge manifold. (Courtesy of Robinair Division, SPX Corporation.)

Figure 3–10 Compound gauge. (Courtesy of Marshalltown Instruments.)

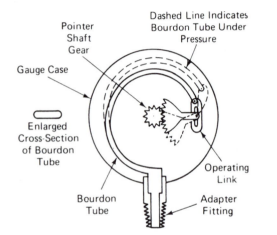

Figure 3–11 Bourdon tube principles. (Courtesy of Billy C. Langley, *Refrigeration and air conditioning,* 3rd ed., 1986, pp. 4–23, 63, 173–178. Reprinted by permission of Prentice Hall, Englewood Cliffs, NJ.)

Figure 3–12 Pressure gauge. (Courtesy of Marshalltown Instruments.)

Pressure Gauges

Pressure gauges are used to determine pressures on the high side of the system. See Figure 3–12. The outside scale is calibrated in psig, and the inside scales indicate the corresponding temperatures of the different types of refrigerants. Some pressure gauges are not designed to operate at pressures below atmospheric. Therefore, caution should be used during evacuation procedures to prevent damage to these gauges.

Valve Manifolds

These manifolds provide openings through which the various service operations are performed on the refrigeration system. See Figure 3–13. The proper manipulation of the hand valves will permit almost any function. When the valves are screwed all the way in, the gauges will indicate the pressure on the corresponding line. The center line is usually connected to a vacuum pump, a refrigerant cylinder, or an oil container.

Charging Hoses

These hoses are flexible and are used to connect the gauge manifold to the system. See Figure 3–14. Charging hoses are equipped with $\frac{1}{4}$-inch flare connections on each end; some automotive units require a different size. One end usually has a valve core depressing attachment for attaching the gauges to Schrader valves. Charging hoses may be purchased in a variety of colors, which facilitate making connections to the unit and are designed for a working pressure of 500 psi and an average bursting pressure of 2000 psi.

POCKET THERMOMETERS

An asset to the appliance service technician is the pocket thermometer. See Figure 3–15. By using pocket thermometers much time can be saved. The operating temperatures of a unit can be checked upon completion of the service work to determine whether or not the unit is operating properly. The two types currently used are the mercury thermometer and the bimetal thermometer. The bimetal type is more popular because of its durability. Pocket thermometers can be purchased with a convenient shock-resistant carrying tube which can be clipped into the shirt pocket for ready use. Pocket thermometers are accurate to within 1 percent.

CHARGING CYLINDERS

These cylinders are used by the service technician to charge the correct amount of refrigerant into a system in a short period of time. In practice, the charging cylinder is loaded with the manufacturer's recommended amount of refrigerant for a particular unit. See Figure 3–16. While the system is under a vacuum, the complete refrigerant charge is dumped

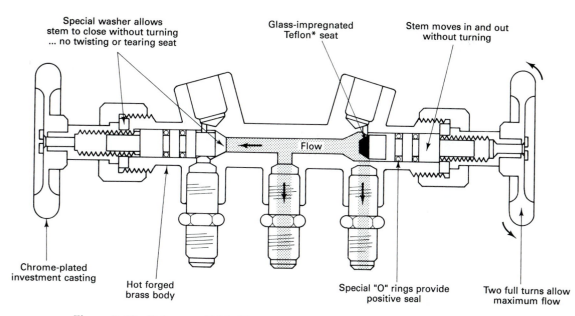

Special washer allows stem to close without turning ... no twisting or tearing seat

Glass-impregnated Teflon* seat

Stem moves in and out without turning

Flow

Chrome-plated investment casting

Hot forged brass body

Special "O" rings provide positive seal

Two full turns allow maximum flow

Figure 3–13 Valve manifold. (Courtesy of Robinair Division, SPX Corporation.)

into it. Some of these units have a built-in electric heater to speed up the dumping process. These cylinders should never have a direct flame applied to them, nor should they be stored in direct sunlight or with liquid refrigerant in them.

FITTING BRUSHES

Fitting brushes are used to clean the inside of fittings that are to be soldered. These brushes can be obtained for each fitting size. In practice, the brush fits tightly into the size fitting for which it has been designed. The brush is then rotated by hand to clean the fitting. If too large a brush is used or if it is rotated in more than one direction, it will be ruined.

MATERIALS

Various materials are peculiar to the appliance industry. Because there are so many in use, we will cover those that are most commonly used.

Figure 3–14 Charging hose. (Courtesy of Robinair Division, SPX Corporation.)

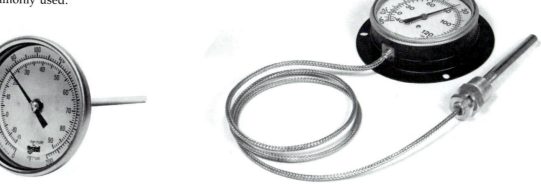

Figure 3–15 Various types of thermometers. (Courtesy of Marshalltown Instruments.)

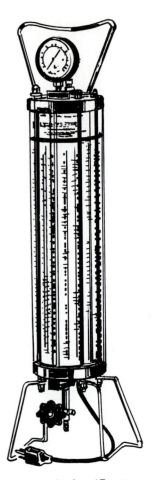

Figure 3–16 Charging cylinder. (Courtesy of Robinair Division, SPX Corporation.)

Silver Solder (Stay-Silv 15)

This solder is the most common agent used in silver brazing joints in modern appliances. It is sometimes referred to as silver brazing. It is available in rods $\frac{1}{8} \times \frac{1}{8} \times 20$ inches, $0.050 \times \frac{1}{8} \times 20$ inches, and 1, 3, 5, and 25 troy ounces and $\frac{1}{16}$ inch in diameter.

To silver-braze copper to steel or to braze dissimilar metals, the recommended product is Stay-Silv 45 alloy, which has a melting point of 1145°F. Also available are the Safety-Silv cadmium-free silver solders 1200 (56 percent silver), Safety-Silv 1370 (45 percent silver), and Safety-Silv 1350 (40 percent silver). Their melting temperatures are according to their designated alloy number as listed. See Figure 3–17. These solders have a strong tensile strength and are not affected by vibration. Some types of solder contain their own flux, eliminating the need for extra flux on the joint. This is especially true on copper tubing.

Almost all alloys used today for brazing operations contain cadmium, a product that gives off poisonous fumes when heated. When using one of these alloys, make certain that there is plenty of ventilation.

In modern times, however, there is no need to use an alloy that contains cadmium for normal silver-brazing operations. Most manufacturers make a cadmium-free silver-brazing alloy. It is recommended that one of these be used when at all possible.

Flux

Flux is used when making sweat joints to aid in making leak-tight connections. It is available in both paste and liquid forms. See Figure 3–18. *Caution:* Flux has an acid content and should not be allowed on clothing, near the eyes, or on open cuts. To remove, wash the area with soap and water or see a physician.

There is a flux designed specifically for each type of solder that cannot be used for another type of solder. Flux should be put on the joint sparingly. Do not overflux a joint or allow flux to enter the piping system because the acid will attack the internal components of the system.

Figure 3–17 Silver solder. (Courtesy of J. W. Harris Co., Inc.)

Figure 3–18 Soldering flux. (Courtesy of J. W. Harris Co., Inc.)

Figure 3–19 Sand cloth. (Courtesy of Billy C. Langley, *Refrigeration and air conditioning*, 3rd ed., 1986, p. 63. Reprinted by permission of Prentice Hall, Englewood Cliffs, NJ.)

Sand Cloth

Sand cloth is mainly used to clean the surfaces to be soldered. It is available in rolls at the supply house. See Figure 3–19. Short strips are torn from the roll and wrapped around the tube. Each end is pulled alternately until the tube is shiny bright. Caution should be used when cleaning tubing to prevent the particles from entering the tube. Perhaps the best way to accomplish this is to turn the end of the tube being cleaned downward so that the particles will fall away from it.

TEST INSTRUMENTS

Test instruments are valuable aids to the service technician. The use of test instruments seems to intimidate many service technicians, however, and they will not make any effort to use these valuable tools of their trade. The use of test instruments is of great benefit to the customer as well as to the service technician. Test instruments should be just as familiar to the service technician as any other tool would be.

Purpose

Test instruments, when used, can help the service technician to perform service operations more accurately and faster, thereby providing a savings to the customer. There are two very strong arguments for the use of test instruments: first, the prevention of the unnecessary changing of parts that usually end up in the scrap box; second, the time saved is money. Customer satisfaction after the sale is not promoted by the parts-changing service technician, who costs the customer and the employer thousands of dollars each year in wasted parts and effort.

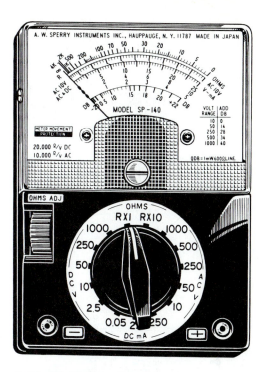

Figure 3–20 Multimeter. (Courtesy of A. W. Sperry Instruments, Inc.)

Application

The proper application of test instruments will provide more accurate trouble diagnosis in the shortest possible time, providing a savings to the customer and effecting better customer relations.

Multimeters The multimeter is probably one of the most frequently used instruments. It is also the easiest instrument to read. See Figure 3–20.

In practice, the selector switch should be set on the proper function for which the meter is to be used. When checking voltage and the correct voltage is not known, start checking the electrical circuit with the meter set on the highest scale possible and work down until one is found that provides a reading at approximately midscale. Midscale readings are the most accurate obtained.

Ohmmeter The ohmmeter is used for measuring resistances, checking continuity, making quick checks on capacitors, and testing some solid-state components. See Figure 3–21. Most ohmmeters are ruined because they are connected to a circuit or a component that has not been disconnected from the source voltage. Ohmmeters have their own power supply and cannot be safely used with any other power source.

In practice, it is better to start with the lowest resistance scale when an unknown resistance is to be checked. Then proceed through the higher ranges until the meter shows a midrange reading. This is the most accurate reading for a meter.

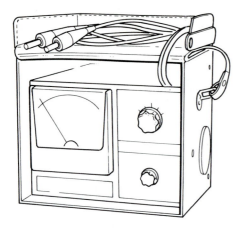

Figure 3–21 Ohmmeter. (Courtesy of Robinair Division, SPX Corporation.)

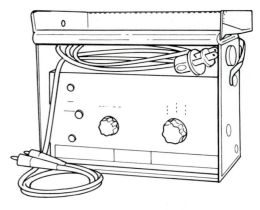

Figure 3–23 Capacitor analyzer. (Courtesy of Robinair Division, SPX Corporation.)

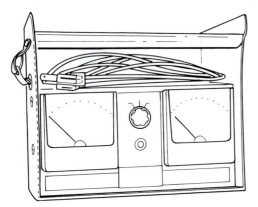

Figure 3–22 Wattmeter. (Courtesy of Robinair Division, SPX Corporation.)

When checking a resistance, first select the proper resistance scale. Install the probes in the proper jacks. Touch the ends of the probes together and turn the adjustment knob until the meter reads zero resistance while the ends of the probes are touching. If the scale is changed, the meter must again be calibrated to zero. If the meter cannot be calibrated to zero, the batteries should be replaced.

When a switch or wire is to be tested for continuity, the needle should deflect to zero, indicating a closed circuit. The part being checked has continuity. If, on the other hand, the needle does not move from the at-rest position, the circuit is open. The part being tested does not have continuity.

If the needle stops between infinity and zero, the meter is measuring the resistance of the part being tested. Infinity means that the resistance is extremely high; probably an open circuit is being tested.

Ammeter The clamp-on ammeter is the most popular current-measuring instrument used for ac circuits because the current flow can be checked without interruption of the electric circuit. All that is necessary to check the current flow is to clamp the meter tongs around one of the wires supplying electric power to the circuit being tested.

Clamp-on ammeters are designed primarily to measure the current flow in ac circuits only. These instruments measure the current flow in a wire by making it serve as the primary side of a transformer. The circuit to the meter movement serves as the secondary side of the transformer. These instruments are most accurate when the current-carrying wire is in the center of the tongs. For best accuracy of the instrument, the mating surfaces of the tongs should be kept clean.

When testing a circuit that has a low current flow, the wire can be wrapped around one tong of the ammeter. The sensitivity of the meter will be multiplied by the number of turns used. To obtain the correct amperage reading, divide the meter reading by the number of turns taken on the tong.

Wattmeter The wattmeter can be used to determine the voltage to the unit and the total wattage draw of the unit. See Figure 3–22. Be sure to set the meter on the highest range possible because the starting wattage of most motors is very high. After the unit is running, the proper scale can be selected so that the needle indicates a midscale reading.

Capacitor analyzer The capacitor analyzer is a very important instrument to the service technician. It can be used to check for open or shorted capacitors, for the microfarad capacity, and for the power factor. See Figure 3–23. Scales are provided for both the voltage at which the instrument will be used and the microfarad rating of the capacitor. It may be used to check both running and starting capacitors.

Temperature tester Most electronic temperature testers have thermistor sensing leads that are used primarily to measure temperatures below 200°F. See Figure 3–24. Do not attempt to alter these leads in any way. If the leads do not function properly, they should be replaced. By using these leads, temperatures at several different locations may be taken without interruption of the normal cycle of the unit.

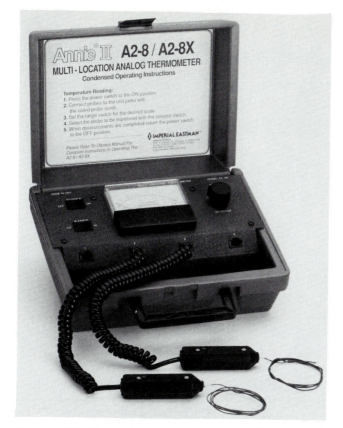

Figure 3–24 Electronic temperature tester. (Courtesy of Mechanical Refrigeration Enterprises.)

Almost any number of leads can be used on instruments equipped with multiple connections.

Temperature testers are also available that will record both the time and the temperature on a piece of paper. These instruments are valuable for helping to locate intermittent temperature-related problems. When this type of instrument is desired, be sure that the chart paper and inking supplies are available.

Micron meter When it is desirable to have an accurate indication of the vacuum in a system, a micron meter is the instrument most often used. These instruments are electrically powered and use a sensing probe to measure the thermal conductivity of the gases in the system being evacuated. These instruments are designed to measure the last part of an inch of vacuum and to indicate this measurement in microns. There are 25,400 microns in 1 inch.

The micron sensing element is not designed to measure above atmospheric pressure. Therefore, some means must be provided to protect the sensor when charging the system with refrigerant.

Electronic leak detector Electronic leak detectors are very sensitive instruments. They operate by moving refrigerant vapor across an ionizing element. There are several ways to warn the user that a leak is present. Some instruments produce a high-pitched screech, some use a blinking light in the probe, and others use a buzzing noise.

These instruments will not function when either ammonia or sulfur dioxide is the refrigerant.

REVIEW EXERCISES

1. What tool is used to cut a square end on tubing?
2. How far should the tubing cutter handle be turned after the cutter wheel touches the tubing?
3. After cutting the tubing what should be done to it?
4. What tool is used to make the fittings on tubing for leak tight connections?
5. What tool is used to remove burrs from the inside of tubing?
6. What tool is used to turn the stem on a refrigeration service valve?
7. What is probably the most used tool in the technician's tool chest?
8. Name the two types of gauges used in refrigeration work.
9. What is the purpose of charging hoses?
10. What tool is used to charge the accurate amount of refrigerant into a system?
11. What type of solders have a strong tensile strength and are not affected by vibration?
12. Why should care be used when using soldering flux?
13. What devices can be used to aid the service technician when performing service operations?
14. What instrument is used to check resistances?
15. What is the most popular ac current measuring instrument?
16. On what scale should a meter be set when first measuring a circuit?
17. For what is a micron meter used?

4

Basic Electricity and Electronics

It is an absolute necessity that the appliance service technician have at least a working knowledge of electricity and electronics. Knowing what electricity can do and the methods used to control it will make your job much easier and you more valuable to the company where you work. The material presented in this chapter is just the beginning of what is to be learned about electricity and electronics. Almost all appliances have solid-state controls on them. They are changing almost on a daily basis; therefore, you should continue to educate yourself and keep abreast of changes in this field.

Most of electricity is based on theories. Therefore, a study of electricity and electronics is largely theoretical.

THE ELECTRON PRINCIPLE

The basis of the electron principle is the fact that all known matter or substance regardless of its form—solid, liquid, or gas—is fundamentally made up of very small particles called atoms. See Figure 4–1. These atoms are made up of even smaller particles called electrons and protons. Both of these particles have an electrical charge. The electrons have a negative charge and the protons have a positive charge. From this it can be concluded that everything on earth is basically electrical in nature, including human beings.

More recent research has uncovered several other important facts concerning the structure of the atom. For example, it was determined that a positively charged proton weighs about 2000 times more than a negatively charged electron. It was also determined that each atom has a core

which is called a nucleus. The protons along with neutrons are located in this center, or nucleus. A neutron is made up of one electron and one proton. The charge of one serves to cancel out the charge of the other. Therefore, the nucleus has no electrical charge. The nucleus is relatively stationary when compared to the electrons, which are rotating around the nucleus at relatively high speeds. The electrons travel around the nucleus in paths called shells or orbits. With the exception of speed, this phenomenon is very similar to the earth and other planets that are rotating around their nucleus, the sun.

This is the very foundation upon which electrical theory is based. Some knowledge of the makeup of matter is necessary to more fully understand electricity. It is the knowledge of matter that enables us to understand why some materials are good electrical conductors and why some are good insulators of electricity and other components used in electrical circuits.

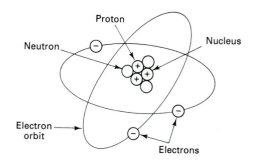

Figure 4–1 The atom.

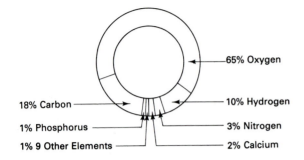

Figure 4-2 Elements of a human body.

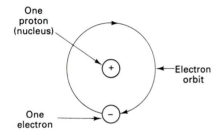

Figure 4-3 A hydrogen atom.

NATURE OF MATTER

It has been determined that everything is made up of matter. Matter can be defined as anything that occupies space or has mass. Not all matter is the same, however, and we cannot learn by simple observation just what type of matter we are looking at. It is, therefore, necessary to complete a chemical analysis of the matter to determine the nature and the behavior of some substances in order to identify them. Almost all materials are made up of some combination of various kinds of matter or mixtures of matter.

When undertaking the study of electron theory, it must be understood that some types of elementary matter may exist that are neither mixtures nor compounds, and that any effort to further break this matter into parts by chemical decomposition will produce no change in any of the characteristics of that matter. When matter has been reduced to its simplest components, these components are known as elements.

Elements Basically, matter is made up of atoms, which are simply a larger collection of elements. An element is considered to be a substance containing only one kind of atom.

There are 106 known elements. Of this number, only 90 are natural elements. The remaining 16 are synthetic elements that have been made in laboratories by atomic research.

Most people do not realize the important role that elements play in our everyday lives. Everything is basically made up of elements, even our bodies, the food we eat, and the clothes that we wear. See Figure 4-2.

In nature, every element is made up of only one kind

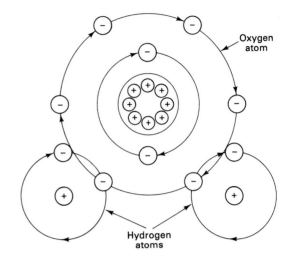

Figure 4-4 A molecule of water.

of atom. From this it can be determined that there are 106 kinds of atoms. However, the electrons and protons in any one element are identical to those in any other element. Because of this, the difference between elements is in the number of electrons and protons that each element contains and how the electrons are arranged in their orbits around the nucleus.

The simplest of all atoms is the hydrogen atom. See Figure 4-3. The hydrogen atom has one electron in orbit around one proton which is the nucleus.

Compounds All known matter, except for the elements, is defined as substances made up of two or more elements. A compound consists of two or more elements chemically combined to form another distinct substance having the same number of elements that are always present in its composition.

When one kind of atom is combined with an atom of another kind, a larger particle which is completely different from either of the other atoms is formed and is known as a molecule. A molecule is defined as the smallest particle of a compound having the same composition as the compound itself.

In high school we learned that the chemical formula for water is H_2O. This formula tells us that there are two elements in this composition, hydrogen and oxygen. There are two atoms of hydrogen and one atom of oxygen. Therefore, every particle of water must contain two hydrogen atoms and one oxygen atom. See Figure 4-4. The atoms are held together because of the crossing orbits of the rapidly rotating electrons. We cannot see a molecule of water with our naked eye. It takes millions of these molecules just to make one drop of water. This is the same with all types of matter. In most cases the compound will be made up of more than two elements.

Electron movement The basis for the well-known facts in electricity that positive charges repel each other, negative charges repel each other, and positive charges attract negative charges is the relationship between electrons

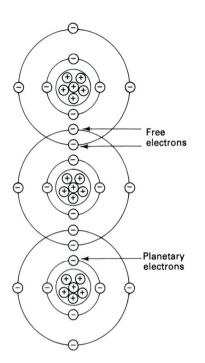

Figure 4–5 The combining of carbon atoms.

and protons. The force that holds the atom together comes from the attraction of the protons in the nucleus for the electrons that surround the nucleus. The strength of this attraction varies with different atoms and the distance from the nucleus to the orbiting electrons. With an increase in the distance between the two, the less is the attraction for each other.

Even though the atoms are locked together by the overlapping orbits of the electrons, some of the electrons are able to move from one atom to another because of the weakness of their attraction to the nucleus. When an electron breaks free of its atom, it is said to be a free electron. The other electrons are closer to the nucleus and, therefore, their attraction to the nucleus is too strong for them to break loose. The atoms left behind are called planetary or bound electrons. The electron principle is that electricity is the movement of free electrons from one atom or molecule to another atom or molecule. Carbon has many free electrons; therefore, it is a good conductor of electricity. See Figure 4–5.

Any material that has a large number of free electrons is considered to be a good conductor of electricity. However, the cost of many of these materials prohibits their practical use as conductors of electricity. Copper is probably the most commonly used material for electrical conductors.

There are many substances that have such strong attractions between the nucleus and the electrons in orbit that they have few free electrons. Therefore, they are poor conductors of electricity and are used as insulators. From this we can see that some materials have a high resistance to the flow of electrons and others offer little resistance to the flow of electrons. Because of this fact, resistance must always be considered as an important characteristic of an electrical circuit.

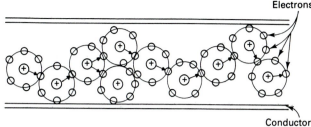

Figure 4–6 Electron flow.

PRESSURE (ELECTROMOTIVE FORCE)

In a battery, the chemical action creates both negatively and positively charged particles. These particles react with the negatively charged zinc electrode and the positively charged copper electrode. When an electrical load is connected across the terminals of the battery, the charge imbalance causes a pressure called electromotive force (emf). It is this force that causes the electrons to flow through the circuit. In reality, this flow of electrons, or current, is many billions of free electrons that repel each other as they travel through the conductor. This direction of flow is from the negative electrode in the battery to the positive electrode.

This flow of electrons is not necessarily one electron that flows completely through the circuit from one end to the other. More likely, one electron will bump the one in front of it, causing that one to bump the one in front of it, and so forth through the length of the conductor. See Figure 4–6.

When a free electron is forced into another atom, it tends to upset the balance of the other electrons with their protons in the receiving atom. This action forces one of the electrons from that atom into another, that atom then becomes negatively charged. At that point one electron is forced from that atom and into another. This action continues through the complete conductor from the negative side of the power source to the positive side of the power source. This movement of electrons from one atom to another is termed dynamic electricity or current in motion. This motion is possible because the free electrons are forced out of a substance with a high negative electrical charge (the zinc electrode) and they are attracted to a substance with a positive electrical charge (the copper electrode).

The greater the difference in electrons between the two materials, the greater the electromotive force that is present. The term voltage is used when referring to electromotive force.

Potential difference (volt) When two substances that have an equal number of electrons are connected by an electrical conductor, there will be no movement of electrons through the conductor. Both substances have the same electrical potential and there is no force to cause the electrons to move. On the other hand, if one of the substances gains a greater number of electrons than the other, a potential difference will exist between them. The greater the difference in the number of electrons in the substances, the

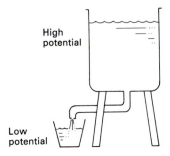

Figure 4–7 Potential difference.

greater the potential difference will be. When the electron principle is considered, free electrons always move from a point of high potential to a point of low potential difference, that is, from a negative electrode to a positive electrode. This principle can be illustrated with the use of water. The potential difference in electricity can be compared with a tank of water in a high tower. The water in the tank has a higher pressure potential than the water in a bucket placed beneath it at the end of a pipe coming from the bottom of the tank See Figure 4–7.

Ampere The ampere (designated amp) is the unit of measurement of current flow or the rate of flow of electrons. The flow of 1 coulomb per second past a given point is equal to 1 amp. The coulomb is a measure of quantity, while the ampere is the measure of strength of the current flow. For example, an ordinary light bulb may require ½ amp and a spotlight may require 150 amps. This indicates that the spotlight requirements are 300 times stronger than the light bulb requirements.

Ampere: A unit of measurement of current or strength or rate of flow of electrons.

$$1 \text{ amp} = 1 \text{ coulomb per second}$$

RESISTANCE

Another very important factor in the understanding of electricity is the resistance to the flow of electrons through a circuit. Almost all metals have a low resistance to this flow and are considered to be good conductors of electricity. They conduct electricity better because of the free electrons in their structure. Copper is one of the most popular metals used in the conduction of electricity. Even copper presents some resistance to the flow of electrons. This resistance is present because of the inherent attempt of the copper atoms to keep their own electrons. In an electrical circuit, the force of the current must overcome the resistance of the conductor. If the applied force is strong or if the resistance is weak, a strong current will flow through the conductor. Also, if the force is weak or the resistance is high, the current flow will be small.

Electrical conductors are the path through which the electrons flow. The resistance of these conductors is dependent on four factors: (1) the diameter of the wire, (2) the

length of the wire, (3) the kind of wire, and (4) the temperature of the wire.

When the wire is long, or small in diameter, the resistance will be greater and the current flow will be smaller. However, if the wire is short, or large in diameter, the resistance will be less and the current flow will be greater. Less current will flow through an iron conductor than through a copper conductor of the same length and diameter. This is because of the difference in the number of free electrons in each substance. With an increase in the temperature of a conductor, there will also be an increase in the resistance of that conductor and less current will flow through it.

In modern appliance control circuits, there is one exception to the temperature-resistance factor that is very important in these control systems. When certain substances, such as oxides of manganese, nickel, and cobalt, are mixed and formed into a solid, a varistor or, as it is more commonly called, a thermistor is formed.

Thermistors are conductors whose resistance to the flow of electricity decreases with an increase in their temperature. Thus, at a low temperature their resistance is very high, allowing only a small current to flow. As their temperature increases, more current flows through them. These thermistors are being used in the control circuits of almost all modern appliances.

Ohm The ohm is the unit of measure of the resistance of an electrical conductor. To establish the unit of measurement, a column of mercury 106.3 centimeters in height and 1 square millimeter in cross section was used. One ohm is the amount of resistance offered by this column of mercury to the flow of 1 amp of electric current having an electromotive force of 1 volt. The temperature was 32°F. This column of mercury is comparable to 1000 feet of number 10 electrical wire, which is $\frac{1}{10}$ inch in diameter, or 2.4 feet of number 36 wire, which is 0.005 inch in diameter.

Ohm: The unit of measurement for the resistance of a conductor to the flow of electrons.

The relationship of volts, amperes, and ohms can now be defined. A volt may be defined as the electromotive force required to push 1 coulomb of electrons in 1 second of time (1 amp) through a resistance of 1 ohm. When all of these are considered, Ohm's law, upon which all electrical measurement is based, becomes very important. Ohm's law states that the current in an electrical circuit is equal to the pressure divided by the resistance.

Ohm's law: The current in an electrical circuit is equal to the pressure (EMF) divided by the resistance.

$$\text{Current} = \text{Pressure/Resistance}$$

This is further simplified by the use of letters which are used to represent the words. *E* represents volts, *I* represents the current flow, and *R* represents the resistance to the current flow in ohms.

$$\frac{I}{E} = R$$

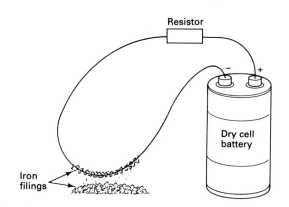

Figure 4–8 Magnetism by current.

ELECTROMAGNETISM

In order to understand the principles of transformers, relays, and other electrical components, some knowledge of electromagnetism is required. Electromagnetism is the production of magnetism by an electric current.

Electromagnets An electromagnet is similar to both the artificial and the natural magnet in its attraction, but it differs in its control. Electromagnets have a strong attraction that can be turned on and off with a flick of a switch. For example, electromagnets are used to open and close the contacts in relays and to open orifices in solenoid valves.

If a wire is connected from the positive to the negative terminals of a battery which has been dipped into a pile of iron filings, some of the filings will cling to the wire. Figure 4–8 shows that a conductor which is carrying an electric current is surrounded by a magnetic field of flux. The current flowing through the conductor produces the magnetic field around the conductor. If the current flow is stopped, the iron filings will drop from the conductor, proving that the field of flux exists only when the current is flowing through the conductor.

Even though the filings are shown to be clinging to only a portion of the conductor, the field of flux is along the complete length of the conductor.

In this example, because the magnetic field of flux has been created by a straight, single conductor, no polarity is apparent. In order to use electromagnetism in practical applications, some polarity must exist. Polarity can be produced simply by forming the straight conductor into a loop. See Figure 4–9.

When a loop is formed, the magnetic field becomes stronger through the loop because the flux lines are more concentrated in this area. The flux lines become more concentrated as they bend toward the center of the loop from a north pole on one side and a south pole on the other side. If more loops of wire are formed into a helix coil, the poles will be made stronger than with a single loop. More turns on the coil produce a stronger magnetic field of flux. This coil is known as a solenoid coil.

When a piece of easily magnetized metal is placed inside the coil of wire, the metal is termed a core or pole piece. When electric power is applied to the coil, the core becomes temporarily magnetized, producing an electromag-

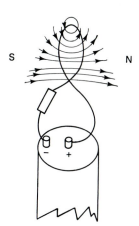

Figure 4–9 Magnetic polarity of a loop.

net. The core of soft iron has a low retentivity and immediately becomes demagnetized when the electric power is turned off.

ELECTRONIC CONTROLS

With the increasing use of solid-state electronic controls on appliances, it is necessary that the service technician learn how they operate. The service technician must learn something about diodes, transistors, and integrated circuit usage in the control systems used on these units. In order to learn how these devices operate, some knowledge of valence bonding and semiconductors is necessary.

An atom that has a valence ring containing only four electrons may be either a conductor or an insulator. The impurities in its composition determine whether it is a conductor or an insulator. These types of materials are considered to be semiconductors.

It is these peculiar properties of semiconductors that have made solid-state devices possible. Silicon is one type of material that is widely used in transistors. Silicon has four electrons in its valence ring. The atoms of pure silicon in the crystalline form have only four electrons in the valence ring and, thus, there is a very tight bonding between the electrons. This bonding is the result of the electrons in the valence ring of the silicon atoms joining together to fill up and share their valence ring. This is commonly known as covalent bonding. See Figure 4–10.

We can see that the covalent-bonded silicon has eight electrons in the valence ring. This makes the valence ring complete. Crystalline silicon is a very good insulator. If certain other materials are added to the silicon, the basic material changes and it becomes a semiconductor because the valence ring is no longer complete.

P and *N* Materials

The process of adding impurities to silicon under carefully controlled conditions is known as doping. Doping affects the silicon in one of two ways. The change depends on whether the doping material has three or five electrons in its valence ring. When an element that has five electrons

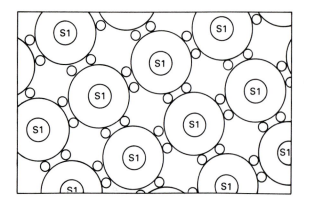

Figure 4–10 Covalent bonding.

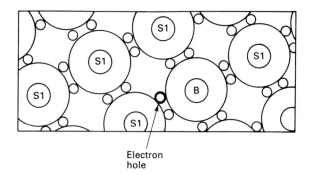

Electron
hole

Figure 4–12 A shortage of electrons.

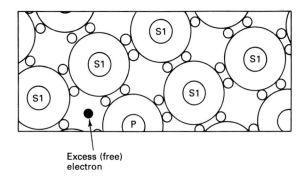

Excess (free)
electron

Figure 4–11 Excess electron.

in its valence ring, such as phosphorous, is combined with silicon, the covalent-bonding process occurs, with one electron extra. See Figure 4–11. Doping in this manner results in a strong-bonded material having a free electron in its structure. The extra electron makes the new material negatively charged, or an *N*-type material.

When a material that has only three electrons in its valence ring, such as boron, is mixed with silicon crystal, a bonded material is produced that has only seven electrons in the valence ring. Thus, there is a hole left in the valence ring. See Figure 4–12. Thus, we now have a material that is positively charged or a *P*-type material. This hole in the valence ring is open to accept any free electrons that may come its way.

Current Carriers

In common theory, the excess electrons are considered to be negative carriers and the holes are considered to be positive carriers. It is believed that the hole can move from one atom to another just as the electrons do. In order for this process to occur, there must be an external voltage applied to the material. Otherwise, no current will flow through the semiconductor.

When electrical power is applied to this semiconductor material, a flow of current will be established. When the *N*-type semiconductor material is charged in this manner, there will be a current flow established. When the *P*-type semiconductor material is charged, the current will flow as

a movement of the holes. Thus, the current flow is caused by the movement of the free electrons into the holes.

Remember that like charges repel and unlike charges attract. The positive side of the power source will attract the electrons and the negative side will repel the electrons. Thus, the flow of electrons is from negative to positive because the electrons from the adjacent atoms will move into the nearest hole in the positive direction, leaving a hole in the last atom. In this manner, there is a flow of electrons from negative to positive as they move from hole to hole. While all this is taking place, the holes are also moving from the positive charge toward the negative charge. This flow is in the opposite direction of the electron flow.

As odd as it may seem, both the holes and the electrons move through a semiconductor material. This movement is a continuous process completely through the material up to the edge of it. At the edge of the material, the positive lead takes the excess electrons and the negative lead supplies electrons to enter the available holes.

Diodes

To form a diode, both the *P*-type and *N*-type materials are joined together. Because of this the diode will allow electric current to flow in only one direction. The current flow occurs when the negative power source is connected to the *N*-type semiconductor and the positive power source is connected to the *P*-type semiconductor. When joined in this manner, the material becomes known as a forward-bias connection. See Figure 4–13. When the positive power source is connected to the *N*-type material and the negative side is connected to the *P*-type material, the connection is known as a reversed-bias condition. The diode will not allow current to flow in this direction. See Figure 4–14.

Because the diode will allow current to flow in one direction and stop it from flowing in the other direction, it is a very useful device. When a diode is connected to an alternating power source, only half of the wave can pass through it in the forward direction, thus changing the power supply from ac to pulsating dc. It can also block a dc current coming into it from the reverse direction.

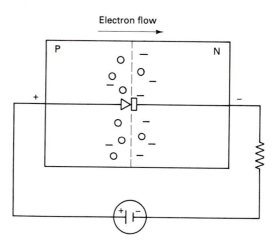

Figure 4–13   Forward-bias connection.

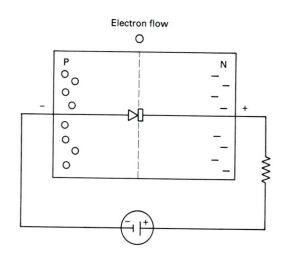

Figure 4–14   Reversed-bias connection.

Transistors

The transistor is a three-layer semiconductor that has a very thin layer of either *P*-type or *N*-type material between two layers of either *P*-type material or *N*-type material. When the center layer is the *N*-type layer between two *P*-type layers, the device is known as a *PNP* transistor. See Figure 4–15. When the *P*-type layer is between two *N*-type layers, the device is known as an *NPN* transistor. See Figure 4–16.

The transistor should only be connected in the forward-bias direction. Do not try to use a *PNP* transistor in place of an *NPN* transistor. Its operation is different and it will not operate when connected incorrectly. The transistor is capable of matching a low-resistance circuit to a high-resistance circuit. Because of this transfer of resistance, the transistor is given its name. The transistor amplifies by power gain.

Integrated Circuits (IC)

This type of circuit is composed of several different types of solid-state devices, such as diodes, transistors, and capacitors. The integrated circuit is manufactured onto a single board with its own circuits; hence, the name IC. The IC board is a complete circuit and many times it is simply called a circuit board.

Circuit Boards

The control components used in electronic control circuits are generally mounted and connected on a printed circuit board or they may be hermetically sealed in a single module. To service this module or circuit board, just remove it and install a new one in its place. None of the devices are field serviceable. Just send the old one to the factory to be repaired in its shop. It is then returned to the field for service.

Circuit boards are easily broken and, therefore, must be handled with care. When they are removed from a unit,

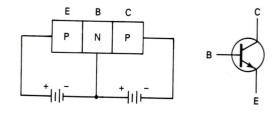

Figure 4–15   *PNP* transistor.

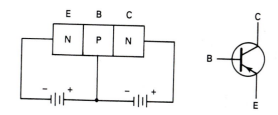

Figure 4–16   *NPN* transistor.

always place a tag, or some other method of identification, on them noting the model number of the unit from which they were taken.

The general assumption is that when the circuit board does not provide the correct voltage at the correct time it is defective. However, care must be taken to make certain that the complete unit is working as it should if it were supplied with the correct voltage before replacing the circuit board. It should not be assumed that the circuit board is bad just because some function of the unit is not being accomplished. Make certain that the component would operate if the board were replaced.

Control Failures

Electronic controls are voltage sensitive. In most units, the voltage is supplied by a built-in, low-voltage dc power supply. The power is then fed to the component through an interconnecting cable to the various control components.

These control components may be either on one control center or they may be on more than one control center, depending on the unit design. Some of these voltages will only be pulses which are used to activate such devices as an SCR, triac, or relay.

Should there be an incorrect signal to any one of the control components, a malfunction will result. Sometimes it is necessary to reset the unit simply by turning off the power supply or just unplugging it from the receptacle. Usually the power should be off for 15 to 20 seconds. Most of the time when the power is restored the unit will operate satisfactorily.

Voltages that cause most of the problems come from the external power source in the form of spikes, voltage surges, or a transient. Transients usually occur when an inductive load such as a motor is turned off. A transient will be more likely to occur if the inductive load is close to the electronic control. Because of this it is recommended that each unit with electronic controls be installed on its own power circuit from the distribution panel. It is also recommended that all electronically controlled appliances be disconnected from the power supply during a thunderstorm. Also, if there is an areawide power outage, these appliances should be disconnected before the power is restored.

REVIEW EXERCISES

1. What is the basis of the electron principle?
2. Of what are atoms made up of?
3. What is the core of an atom called?
4. What is produced when matter has been reduced to its simplest components?
5. What is a substance that contains only one kind of an atom known as?
6. What causes the difference between atoms of different materials?
7. Of what is a molecule made of?
8. What is the smallest particle of a compound known as?
9. What causes the atoms to be held together?
10. What force holds the atom together?
11. What causes a material to be a good conductor of electricity?
12. Name the force that causes the electrons to flow through a circuit.
13. What force (positive, negative) does an atom have when an extra electron moves into one of its orbits?
14. When referring to voltage, what term is used?
15. What does a large difference in the number of electrons between two materials cause?
16. The flow of electricity is in what direction (positive to negative, negative to positive)?
17. Will the resistance of a long small wire be high or low?
18. What happens to the resistance of a conductor when it becomes hot?
19. In what unit is resistance measured?
20. What is the production of magnetism by an electric current known as?
21. What is the process of adding impurities to silicon known as?
22. Is it believed that an electron hole can move just as an electron can?
23. What charge does an electron hole have?
24. In what direction can a diode block dc current?
25. What type of circuit consists of several different types of solid state devices?

5

Machine Lubrication

Anything with moving parts needs lubrication to reduce the amount of friction and heat generated between the parts. The purpose of lubrication is to reduce the amount of friction and wear on the mating parts and to carry away the heat generated by the friction. Fast-moving parts require constant protection with a large and continuous supply of lubricant. The slower-moving parts do not require as much lubrication for proper protection. The speed of the parts and the type of service dictates the type and amount of lubrication needed for proper protection. A contaminated lubricant does not provide adequate lubrication. It also causes corrosion and deterioration of the components.

FRICTION

Friction may be described as the resistance that tends to oppose motion. Friction is not always bad. It is the force that allows bolts and screws to hold machinery together. This type of friction is useful. The friction between bearings and gears in a washing machine transmission, for example, is not good. This type of friction causes wear, excessive operating costs, and high temperatures. This is the type of friction that we want to reduce with lubrication.

Friction can also be caused by poorly aligned components such as pulleys and bearings. Components that are out of alignment will cause excessive wear, higher operating costs, and higher than normal heat generated at the friction points.

EQUIPMENT DESIGN

When engineers design appliances, they take into consideration the normal basic requirements of the types of bearings used, types of lubrication required, and the operating conditions of the equipment. Some of the factors are:

1. The equipment should be able to operate satisfactorily under a heavy load and extreme ambient conditions.
2. The bearing and gear system should be sealed in a shell with a permanent supply of the proper lubricant. The shell is sealed against dust and moisture to prevent contamination of the lubricant.
3. When hermetic sealing is not possible, convenient lubrication points are a part of the design. The manufacturer's lubrication recommendations should be followed for the best performance, operation, and longevity.

However, during the manufacture of these components, some degree of tolerance is allowed because every bearing and shaft obviously cannot be exactly the same dimension. When all of these tolerances are added together some of the bearings may be a little loose or a little closer than desired. Also, the shaft and bearings may not always be in complete alignment. Most manufacturers make limits on the amount of total tolerances allowable. It is not practical or possible to completely eliminate these units from entering the field. These units will sometimes operate satisfactorily for years and others will start causing problems soon after

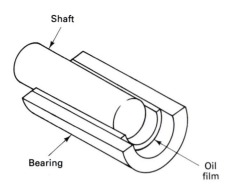

Figure 5–1 Sleeve bearing.

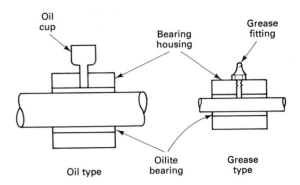

Figure 5–2 Bearing lubrication methods.

being placed into service. The problem will usually appear as abnormal noise or poor operation because of tight, over-heated bearings.

Sometimes the unit is mishandled during shipment and/or installation causing the cabinet frame to warp, which in turn can lead to a misalignment of the bearings, belts, and other moving parts. If the customer uses an excessive amount of washing detergent or cleaning fluids, the bearing lubrication will dry out or be washed away. If the unit is placed in an excessively dirty area, dust may enter the bearings and contaminate the lubricant. The manufacturers are making every effort to eliminate these problems.

BEARINGS

Each different type of bearing receives a special type of lubricant for proper operation and longevity. The type of bearing and its use indicates the type and frequency of lubrication.

Sleeve bearings The sleeve bearing is of a porous, sintered-type construction. Sleeve bearings are usually pre-lubricated by a special process. They do not usually need any further lubrication. They contain about 15 to 20 percent of lubricant by volume. This lubricant is immediately available to the bearing on start-up, thus reducing friction at the time of start-up. See Figure 5–1. When additional lubrication is needed, some bearings have an oil reservoir and a wick system. See Figure 5–2. The oil is conducted from the reservoir to the bearing through the wick. In some designs, the wick actually touches the rotating shaft bearing surface. See Figure 5–3. When the oil reservoir system is used, dust protection is very important. The type of oil most commonly used in this type of bearing is a nondetergent, SAE 20 or 30 weight type.

Needle bearings This type of bearing is usually as-sembled in an outer shell that is packed with the proper type of lubricant (usually oil or grease). The needles are almost in contact with the shaft bearing surface which serves as the inner bearing race. A small amount of lubricant is located between the needle bearings and the shaft bearing.

The shaft bearing is made of hardened steel to permit long life of the shaft. Needle bearings offer less resistance to motion than sleeve bearings.

To accurately check for wear, the bearings must be removed from the shaft. Before the shaft is moved through the bearing, make sure it is clear of rust, burrs, and any other foreign matter that could possibly damage the needle bearing surfaces. Clean the shaft and check it for wear and/or scoring. Check the needle bearings for wear and scoring. If either the shaft or the needles show wear, replace both of them. Be sure to prelubricate the bearings before operating the unit.

Roller (ball) bearings This type of bearing rolls be-tween an inner and an outer race. The bearing is self-con-tained. Most roller bearings are permanently lubricated at the time of manufacture. The entire bearing slips over the shaft and is fastened to the shaft with set screws to prevent movement on the shaft. The bearing has an almost metal-to-metal contact with the races. The lubricant serves as a moving surface between the metal parts and it carries away some of the heat of friction. These bearings have seals that prevent the loss of lubricant, as well as preventing dirt and moisture from entering the bearing races.

When gears or sliding parts are used with the roller bearing inside a housing, an oil-type lubricant is preferred. The oil is usually heavy bodied with additives to obtain the specific properties that are needed for the particular application. When the assemblies are not in an enclosed housing, grease must be used rather than oil.

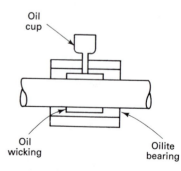

Figure 5–3 Oil wick touching shaft bearing surface.

Plastic bearings Du Pont has designed several types of plastic bearings that are well suited for light-duty self-lubricating applications. These bearings are made from materials such as Teflon, nylon, and Delrin. These materials are very resistant to corrosion, detergent, and water. They are, at present, very popular in small-appliance applications.

LUBRICANT PROPERTIES

The properties of a lubricant can be very technical if studied in detail. However, we will only present a very broad outline of the purposes of a lubricant, without getting too technical.

There are four basic functions that a lubricant must perform:

1. To reduce the friction and wear of mating surfaces.
2. To carry away the heat produced by the friction of the moving parts that are being lubricated.
3. To seal out any contamination from the mating surfaces.
4. To prevent rust and corrosion from forming on the mating surfaces.

The viscosity of an oil is a measure of how fast it will pour at a given temperature. Light-weight oils will run at very low temperatures. High-viscosity oils will not be very fluid at low temperatures. When an oil is too light for the application, it will break down rapidly and not lubricate properly. An oil that is too heavy for the application will develop high temperatures and create a power loss because the film of oil cannot enter the friction area fast enough on a cold start-up.

Oils that are designed to be used in a gear case are usually of the heavy-weight type and have additives such as antifoam, antiwear, extreme pressure protection, corrosion inhibitors, and oxidation inhibitors.

The oil used in refrigeration compressors is a special type that is wax-free, dehydrated, and will flow at the temperatures reached inside the evaporator. This oil must be compatible with the refrigerant used in the system and the compressor components. Never use anything other than refrigeration-type oil in a refrigeration compressor.

Warning: Never overlubricate ball or roller bearings. Too much lubricant inside the bearing can cause excessive churning, friction, and heat. These conditions will cause the lubricant to break down and eventually ruin the bearing. The space inside the bearing should never be more than $\frac{1}{3}$ to $\frac{1}{2}$ full of grease.

Solid or sintered sleeve-type bearings are not generally damaged by overlubrication. The damage that can occur in this instance is ruining what the oil drips onto.

Bearing Failure

A failing bearing will generally produce an unusual noise when the unit is operating. There will usually be play between the shaft and the bearing, especially when sleeve bearings are used. This play can be determined by trying to move the shaft in a sideways direction. In this instance, cleaning or lubricating the bearing will prove a waste of time. The noise may go away for a short period of time, but it will surely return. In the long run, the bearing will sometimes become frozen, causing an overload on the motor or causing the motor to completely stall.

When attempting to make a decision about changing a bearing, the following factors must be considered:

1. The cost of the repair as compared to the cost of a new unit.
2. The general condition of the remaining parts.
3. The cost of the bearing.
4. The amount of time required to complete the job. Will the labor cost overshadow the cost of a new unit?
5. The availability of the proper tools to complete the job.

If the decision to replace the bearing is made, the bearing should be taken to the supply house so that an exact replacement will be obtained.

Bearing Replacement

A bearing that is worn must be replaced in order to maintain its relative position to the remainder of the parts and to prevent overloading of the motor. Do not replace a bearing on a scored or worn shaft or in a housing in which the bearing race has been damaged. This will only delay the replacement of the complete component at an additional cost to the customer. This will usually result in a dissatisfied customer and sometimes a lost customer.

Running Clearance

The running clearance is used in sleeve-type bearings. It is the amount of space between the OD of the shaft and the ID of the bearing. This space is where the lubricant provides a protective film between the moving parts. This fit also allows for some expansion due to the heat build-up during operation.

Press Fit Practices

Without some form of friction, the bearing will turn in the housing and ruin it. The purpose of the shrink fit, press fit, or the interference fit (as they are all known) is to prevent turning of the bearing in the housing. This is accomplished by making the OD of the bearing just a bit larger than the hole in the housing.

SHAFT SEALS

When the housing has a shaft extending through it and the atmosphere is to be kept out and a liquid or a gas is to be kept in, a shaft seal is used. Some examples of shaft seal use are:

1. Where the crankshaft of a refrigeration compressor extends through the compressor housing.

2. The shaft extension where the impeller is connected to the shaft on a disposer or a water pump.

3. The shaft extension from the gearbox of a washing machine.

Locating a leaky seal Any time that there are signs of grease or oil under an appliance, a shaft seal should be suspected. The best and quickest way to determine the location of the leak is to clean the area and watch for an oil or grease drip or find where the grease or oil has been sprayed on adjacent parts from a spinning shaft. All of these conditions are indications of a leaking shaft seal.

Shaft seals are used in washing machines to keep water and detergents out of the transmission. Oil and water dripping out of a relief hole in the gear housing indicate that there is a seal leaking into the transmission housing.

A leaking shaft seal must be replaced. These devices will not wear-in and repair themselves. The only way to solve the problem is to replace the shaft seal. It is good practice to also install a new bearing because the water may have caused the bearing to wear excessively or a worn bearing may have caused the shaft seal to leak.

Types of seals Basically, there are two types of shaft seals used on rotating shafts:

1. Face seal
2. Lip seal

Face seals These are usually two-part seals. One part is pressed into the housing and the other part is pressed onto the shaft. Each part has a mating surface that matches when the seal is properly installed.

Be certain that the shaft and housing are completely clean. A small amount of oil or grease on the shaft and the hole in the housing will help in the installation of this type of seal. Be sure that the seal fits squarely into the housing hole and on the shaft. Any misalignment of these parts will allow the seal to leak.

The seal faces must be kept absolutely clean. Do not touch the sealing faces or a leak will probably occur.

Seating Face-type seals have a very light fit on the face as well as on the shaft and in the housing hole. All of the seats must be clean and free of any foreign material. If there should be a loose fit of the seal to the seat, liquid or gas will leak at this point. In some applications a sealing compound is used to aid in the sealing process. Use the instructions on the compound package. A fit that is too tight may cause damage or distort the seal; sometimes the seal will wear excessively causing it to leak. As always, use an exact replacement part for the job.

When removing the old seal, use caution not to damage any of the sealing surfaces.

Lip seals The inner seal diameter has a smaller-diameter, flexible lip. The inner diameter is smaller than the shaft over which it is placed. The seal is pressed into the housing with the outer shell against the edge of the housing hole. The inner flexible seal will make a pressure contact with the shaft when it is inserted through the seal. When installing this type of seal, make certain that the liquid pressure will be on the underside of the lip, causing a tighter contact with the shaft as the pressure inside the housing is increased.

As always, the shaft must be clean and free of all foreign material before inserting it into the seal. If the shaft surface is not glass smooth, polish it by either sanding or grinding it to the desired finish. When polishing the shaft, be sure that the polishing motion is around the shaft and not along it. Polishing along the shaft will cause the seal to wear much more rapidly.

The lip seal is most satisfactory with either a hardened-steel or a stainless-steel shaft. If the shaft is not hardened, the seal will wear a groove in it.

To be certain that the replacement seal is the proper one, use one with the correct part number for the unit being repaired.

Seal leakage during storage Many times when a unit has been in storage for a long period of time, the seals will leak a bit. This is a normal occurrence. Any unit that has a shaft seal should be operated occasionally to keep the seals lubricated and seating properly.

Shaft seal lubrication Either lip or face seals are used on high-speed shafts to prevent leakage at that point. If they do not receive the proper amount of lubrication, they will deteriorate and start leaking because of the friction heat caused by their contact with the rotating parts.

O-Ring Seals

O-ring seals are used on shafts that rotate relatively slowly or that have a slight sliding motion in their operation. They are also very popular in applications as stationary seals between parts in the place of gaskets.

O-ring seals are designed to fit in a groove either in the shaft or bearing. A part of the seal is also forced into an area in the other part to be sealed. Thus, there is a seal between a shaft and a bearing or between two parts. O-rings are made to very exact tolerances, so that a positive seal can be obtained. They are made from several different compounds which are dictated by their use.

When an O-ring is used as a static seal, no lubricant is needed. However, when an O-ring is used on moving parts, some type of lubrication must be provided. Usually the fluid that is being sealed will lubricate it. There are special applications when the O-ring seal is lubricated by an oil socket next to the O-ring.

Failure O-ring seals are also subject to leaks. The leaks are detected in the same manner as with any other seal. If an O-ring shows signs of leaking, it must be replaced. There is no way to repair a leaking O-ring.

When a leaking O-ring is found, remove the O-ring very carefully. Examine the O-ring, the groove in which it fits, and the mating surface for any rough places, dirt, or foreign material that would cause a leak. The cause of the leak must be repaired or the new O-ring will also leak. Some of the most common causes of O-ring leakage are:

1. The O-ring was pinched during the installation.
2. The O-ring was damaged by sharp edges during the installation.
3. The O-ring was exposed to excessive pressure.
4. The O-ring was exposed to excessive temperature.
5. The O-ring was damaged by a worn bearing.
6. The O-ring groove finish was bad.
7. The O-ring was operating without lubrication on a moving surface.
8. The O-ring was exposed to the corrosive effects of contaminated oil.
9. The O-ring underwent abrasion of the surface because of contaminated oil.

To install an O-ring, use the following procedure: When an O-ring needs to be replaced, be sure to correct the cause of the leak or it will only occur again. As with any other seal, the surfaces must be kept clean of all dirt, corrosion, burrs, and any other foreign material before and during the assembly process. Be sure to use the recommended replacement part.

REVIEW EXERCISES

1. What is the resistance that tends to oppose motion known as?
2. What dictates the type and frequency of bearing lubrication?
3. What percent of lubrication, by volume, do sleeve bearings usually contain?
4. What is the purpose of the wick in a bearing?
5. What weight oil is generally used in sleeve bearings?
6. Does a sleeve or a needle bearing offer less resistance to motion?
7. How are most roller bearings lubricated?
8. Where is the bearing lubricant located to properly lubricate a bearing?
9. What is the measure of how fast a lubricant will pour at a given temperature known as?
10. Is a solid or sintered sleeve-type bearing generally damaged by over lubrication?
11. When a roller bearing is bad will there always be play between the shaft and the bearing?
12. Why must a worn bearing be replaced?
13. Is it good practice to install a new bearing on a shaft that shows a small amount of wear?
14. What is the key to successful bearing replacement?
15. What is the purpose of press fit bearings?
16. What do signs of grease, or oil under an appliance indicate?
17. Will a leaking shaft seal wear in and repair itself?
18. Name the two basic types of shaft seals.
19. When polishing a shaft, what must be the polishing direction?
20. Where is an O-ring seal designed to fit?

6

Refrigeration Principles

The evolution of refrigeration has improved the economy of almost all areas because it is a means of preserving products while they are being shipped to customers. Refrigeration has played a large role in the development of agricultural regions because of the greater demand for products. The dairy and livestock-producing areas have also enjoyed the growth brought about by the use of refrigeration.

We can define refrigeration as the process of removing heat from an enclosed space or material and maintaining that space or material at a temperature lower than its surroundings. As heat is removed, a space or material becomes colder. The more heat is removed, the colder the object becomes. Cold, therefore, is a relative term signifying a condition of lower temperature or less heat.

COMMON ELEMENTS

There are more than 100 basic elements that make up everything around us. There are 92 natural elements; the rest are synthetic. Everything in nature on or around the earth, moon, sun, stars, and the human body is made up of these basic elements. In most cases two or more elements are combined to make a substance.

Atoms

Each element in nature is made up of billions of tiny particles known as atoms. An atom is the smallest particle of which an element can be made up and still maintain the characteristics of that element. An atom is so small that it cannot even be seen with a very powerful microscope. An atom, for our purposes, is considered to be invisible and unchangeable. An atom cannot be divided (broken up) by ordinary means. Atoms of all elements are different. That is, iron is composed of iron atoms, and hydrogen is composed of hydrogen atoms.

Scientists know many things about atoms. How they know these things is outside the scope of this text; we must accept these as true in order to understand the subject being studied.

Molecules

The molecule is the next larger particle of a material. It consists of one or more kinds of atoms. When a molecule contains atoms of only one kind, it is said to be a molecule of that element; two or more elements combined are molecules of a chemical compound. Usually a molecule of an element contains only one atom. However, a molecule can contain several atoms of the same kind. For example, A molecule of iron contains only one iron atom, while a molecule of sulfur usually contains eight sulfur atoms.

A small piece of any element consists of billions of molecules. Each of these molecules consists of one or more atoms of that element.

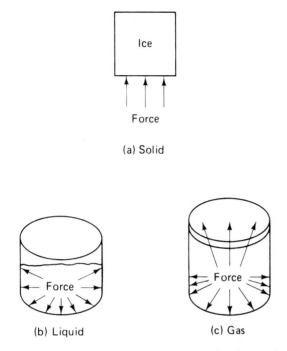

Figure 6–1 Three states of a substance. (Billy C. Langley, *Refrigeration and Air Conditioning,* 3rd ed., 1986, pp. 4–23, 63, 173–178. Reprinted by permission of Prentice Hall, Englewood Cliffs, NJ.)

Chemical compounds The molecule of a chemical compound consists of two or more atoms of different elements. For example, carbon monoxide (CO) is a simple compound with one atom of each element. This combining of different elements causes the material to become entirely different. The new material does not resemble either of the elements that make it up. For example, a molecule of water (H_2O) contains two atoms of hydrogen and one atom of oxygen.

Many of the substances used in our daily living are chemical compounds. Some of these substances are table salt, baking soda, and calcium chloride. Likewise, the refrigerants used in air-conditioning and refrigeration units are chemical compounds.

Example 1 A molecule of Refrigerant-12, a colorless gas, consists of one carbon atom, two chlorine atoms, and two fluorine atoms.

Example 2 A molecule of Refrigerant-22 consists of one carbon atom, one hydrogen atom, one chlorine atom, and two fluorine atoms.

MOLECULAR MOTION

Scientists have found that all matter is made up of small particles called molecules. These molecules may exist in three states: solid, liquid, and gas. Molecules can be broken down into atoms. Atoms are discussed in more detail in Chapter 7.

However, we will study the theory of molecular movement and action because it is involved in air-conditioning and refrigeration systems. A molecule is the smallest particle to which a compound can be reduced before breaking down into its original elements. For instance, water is made up of two elements, hydrogen and oxygen. The movement or vibration of these molecules determines the amount of heat present in a given body. This heat is caused by the friction of the molecules rubbing against each other. The attraction of these molecules to each other is reduced as the temperature increases. When a substance is cooled to absolute zero, all molecular motion stops. At this temperature the substance contains no heat.

Molecules vary in weight, shape, and size. They tend to cling together to form a substance. The substance will assume the character of the combining molecules. Because molecules are capable of moving around, the substance will be, to a degree, dependent on the space between them. The molecules in a solid have less space between them than the molecules in either a liquid or a gas. A liquid has more space between the molecules than a solid and less space than a gas. A gas has more space between the molecules than either a solid or a liquid. Many substances can be made to exist in any of these three forms depending on their temperature and pressure. Water is a very common example of this type of substance.

Solids

In solids, the vibrating rate of the molecules is very slow. Therefore, the attraction of the molecules to each other is very strong and a solid must have support or it will fall. See Figure 6–1a.

Liquids

In liquids, the vibrating motion of the molecules is faster than in solids. Therefore, the attraction of the molecules to each other is less and a liquid must be kept in a container of some type. See Figure 6–1b. The higher the temperature of the molecules, the faster they vibrate. The warmer molecules will move upward in the container toward the surface of the liquid because they are less attracted to each other and require more space. Therefore, they become lighter and rise upward.

The force exerted by a liquid is toward the sides and to the bottom of the container. The force will be greater on the bottom than on the sides because of the weight of the liquid. As the surface of the liquid is approached, the force will decrease because of the reduced weight of the liquid.

Gases

In gases, the vibrating motion of the molecules is even faster than in liquids. Therefore, the attraction of the molecules to each other is very small and a gas must be kept in a closed container or it will escape to the atmosphere. A gas will take the shape of the container on all sides. See Figure 6–1c. The molecules of gas have little or no attraction for each other as well as molecules of other substances.

The force exerted by a gas is equal in all directions. Most gases are lighter than air. Therefore, they tend to float upward, causing a weightless condition.

With the proper regulation of temperature, any substance can be made to remain in any of the three forms: solid, liquid, or gas. Also, any substance can be made to change from one form to another by the proper use of temperature and pressure. This change in form is known as the change of state.

CHANGE OF STATE

The addition of heat to a substance may cause, in addition to a rise in the temperature of that substance, a change of state of that substance. That is, an addition of heat may cause a substance to change from a solid to a liquid, or from a liquid to a gas. There are three states of any substance. The states of water are: ice, water, and steam, that is, solid, liquid, and gas.

HEAT

There are two terms that should not be confused. They are heat and temperature. Heat is considered as the measure of quantity; temperature is the measure of degree or intensity. For instance, if we have a 1-gallon container of water and a 2-gallon container of water and both are boiling, the 2 gallons of water will contain twice as much heat as the 1 gallon, even though they are both 212°F (or the same temperature).

Temperature is measured in degrees with a thermometer. Heat is measured in Btu's (British Thermal Units). A Btu is defined as the amount of heat required to raise the temperature of 1 pound of pure water 1°F.

Methods of Heat Transfer

It is important to know that heat always flows from a warmer object to a cooler object. The rate of heat flow depends on the temperature difference between the two objects. For example, consider two objects lying side by side in an insulated box. One of the objects weighs 1 pound and has a temperature of 400°F, while the second object weighs 1000 pounds and has a temperature of 390°F. The heat content of the larger object will be far greater than that of the smaller object. However, because the temperatures are different, heat will flow from the smaller object to the larger object until their temperatures are the same.

The three ways that heat travels are: (1) conduction, (2) convection, and (3) radiation.

Conduction Conduction is the flow of heat through an object. See Figure 6–2a. For heat transfer to take place between two objects, the objects must touch. See Figure 6–2b. This is a very efficient method of heat transfer. If you have ever heated one end of a piece of metal and then touched the other end, you felt the heat that had been conducted from the heated end of the metal.

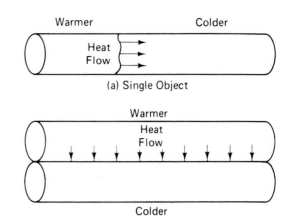

(a) Single Object

(b) Two Objects

Figure 6–2 Heat transfer by conduction. (Billy C. Langley, *Refrigeration and Air Conditioning*, 3rd ed., 1986, pp. 4–23, 63, 173–178. Reprinted by permission of Prentice Hall, Englewood Cliffs, NJ.)

Convection Convection is the flow of heat by the use of a fluid, either gas or liquid. The fluids most commonly used with this method are air and water. The heated fluids are less dense and rise, while the cooler fluids are more dense and fall, thus creating a continuous movement of the fluid. Another example of convection is the heating furnance. The air is heated in the furnace and blown into a room to heat the objects in the room by convection.

Radiation Radiation is the transfer of heat by wave motion. These waves can be light waves or radio-frequency waves. The form of radiation that are most familiar with is the sun's rays. When heat is transferred by radiation, the air between the objects is not heated, as can be noticed when a person steps from the shape into the direct sunlight. The air temperature is about the same in either place. However, you feel warmer in the sunlight. See Figure 6–3. This is because of the heat being conducted by the rays of the sun.

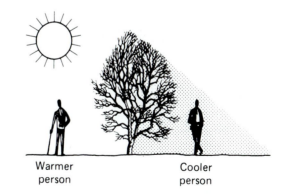

Warmer person   Cooler person

Figure 6–3 Heat transfer by radiation. (Billy C. Langley, *Refrigeration and Air Conditioning*, 3rd ed., 1986, pp. 4–23, 63, 173–178. Reprinted by permission of Prentice Hall, Englewood Cliffs, NJ.)

There is little radiation at low temperatures and at small temperature differences. Therefore, heat transfer by radiation is of little importance in actual refrigeration applications. However, if the refrigerated space is located in the direct rays of the sun, the cabinet will absorb heat. This heat absorption from direct sunlight can be a major factor in the calculation of the heat load of a refrigerated space.

Heat will travel in a combination of these processes in a normal refrigeration application. The ability of a piece of refrigeration equipment to transfer heat is known as the overall rate of heat transfer. As was learned earlier, heat transfer cannot take place without a temperature difference. Different materials have different abilities to conduct heat. Metal is a good conductor of heat.

Sensible heat Heat added to or removed from a substance, resulting in a change in temperature but not change of state, is called sensible heat. The word sensible as applied to heat refers to that which can be sensed with a thermometer. An example of sensible heat is when the temperature of water is raised from 42 to 212°F. There was a change in temperature of 170°F. This change is sensible heat. It can be measured with a thermometer.

Latent heat Heat added to or removed from a substance, resulting in a change of state but no change in temperature, is called latent heat. The types of latent heat are: (1) latent heat of fusion, (2) latent heat of condensation, (3) latent heat of vaporization, and (4) latent heat of sublimation.

Latent heat of fusion Latent heat of fusion is the amount of heat that must be added to a solid to change it to a liquid at a constant temperature. Latent heat of fusion is also equal to the heat that must be removed from a liquid to change this liquid to a solid at a constant temperature.

Latent heat of condensation Latent heat of condensation is the amount of heat that must be removed from a vapor to condense it to a liquid at a constant temperature.

Latent heat of vaporization Latent heat of vaporization is the amount of heat that must be added to a liquid to change it to a vapor at a constant temperature.

Latent heat of sublimation Latent heat of sublimation is the amount of heat that must be added to a substance to change it from a solid to a vapor, with no evidence of going through the liquid state. This process is not possible in all substances. The most common example of this is dry ice. The latent heat of sublimation is equal to the sum of the latent heat of fusion and the latent heat of vaporization.

Specific heat The amount of heat required to raise the temperature of 1 pound of any substance 1°F is called specific heat. Specific heat is also the ratio between the quantity of heat required to change the temperature of a substance 1°F and the amount of heat required to change an equal amount of water 1°F.

Table 6–1 Specific heat of foods.

Food	Specific Heat (Unfrozen)		Specific Heat (Frozen)	
	Btu	cal	Btu	cal
Veal	0.70	176.4	0.39	98.28
Beef	0.68	171.36	0.38	95.76
Pork	0.57	143.64	0.30	75.6
Fish	0.82	206.64	0.43	108.36
Poultry	0.80	201.6	0.42	105.84
Eggs	0.76	191.52	0.40	100.8
Butter	0.55	138.6	0.33	83.16
Cheese	0.64	161.28	0.37	93.24
Whole Milk	0.92	231.84	0.47	118.44

(Billy C. Langley, *Refrigeration and Air Conditioning,* 3rd ed., 1986, pp. 4–23, 63, 173–178. Reprinted by permission of Prentice-Hall, Englewood Cliffs, NJ.)

From the definition of a Btu given earlier, it can be seen that the specific heat of water must be 1 Btu per pound. The specific heat values of some of the more popular foods are given in Table 6–1. If you will note, after the foods are frozen their specific heat values drop considerably. It may be assumed that the specific heat is a little more than one-half of what it was before the foods were frozen.

Superheat Heat added to a vapor after the vapor is no longer in contact with its liquid is called superheat. For example, when enough heat is added to a liquid to cause all the liquid to vaporize, any additional heat added to the vapor is termed superheat. See Figure 6–4.

TEMPERATURE

As stated earlier, temperature is a measure of the degree or intensity of heat. The device used to measure temperature is a thermometer. There are two types of thermometers: Fahrenheit and Centigrade. See Figure 6–5. In the United States the Fahrenheit thermometer is the most often used, while in Europe the Centigrade thermometer is more common.

PRESSURE

Because refrigeration systems depend basically on pressure differences inside the system, a basic understanding of pressure and the laws that govern it is very important to the designer and the technician.

Pressure is defined as the weight of force per unit area and is generally expressed in pounds per square inch (psi) or pounds per square foot. The normal atmospheric pressure at sea level is 14.7 psi.

All substances exert pressure on the materials that support them. A book exerts pressure on the table. A liquid exerts pressure on the bottom and the sides of its container, and a gas exerts pressure on all the surfaces of its container, such as a balloon.

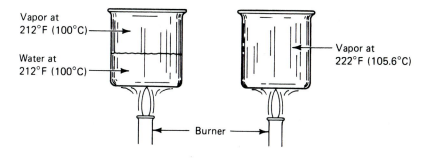

Figure 6–4 Explanation of superheat. (Billy C. Langley, *Refrigeration and Air Conditioning,* 3rd ed., 1986, pp. 4–23, 63, 173–178. Reprinted by permission of Prentice Hall, Englewood Cliffs, NJ.)

If we had a cube of 1 inch in all dimensions that weighed 1 pound, it would exert a pressure of 1 psi on a table top when it is placed on it.

The liquid in a container maintains a greater pressure on the bottom and sides of its container as the liquid level is raised. However, gases do not always exert a constant pressure on the container because the amount of pressure is determined by the temperature and the quantity of gas inside the container.

The air around us exerts pressure, which is called atmospheric pressure. All liquids have a definite boiling temperature at atmospheric pressure. If the pressure over a liquid is increased, the boiling temperature will also be increased. If the pressure over a liquid is lowered, the boiling temperature will also be lowered. All liquids have a definite boiling temperature for each pressure. This is one of the basic principles used in refrigeration work.

Atmospheric Pressure

The pressure exerted on the earth by the atmosphere above us is called atmospheric pressure. At any given point, this atmospheric pressure is relatively constant, except for changes caused by the weather. As a basic reference for comparison, the atmospheric pressure at sea level has been universally accepted as being 14.7 psi. This pressure is equal to that exerted by a column of mercury 29.92 inches high.

The depth of the atmosphere is less at altitudes above sea level; therefore, the atmospheric pressure is less on a mountain. For example, at 5000 feet elevation the atmospheric pressure is only 12.2 psi.

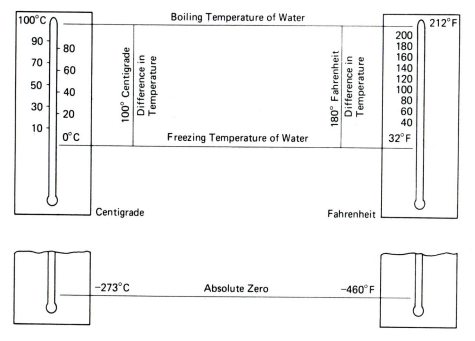

Figure 6–5 Comparison of the Fahrenheit and Centigrade thermometer scales. (Billy C. Langley, *Refrigeration and Air Conditioning,* 3rd ed., 1986, pp. 4–23, 63, 173–178. Reprinted by permission of Prentice-Hall, Englewood Cliffs, NJ.)

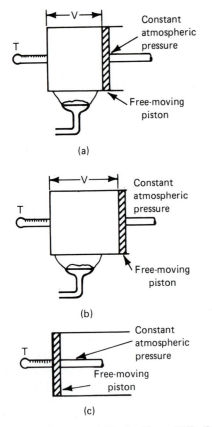

(a)

(b)

(c)

Figure 6–6 Application of Charles' law. (Billy C. Langley, *Refrigeration and Air Conditioning*, 3rd ed., 1986, pp. 4–23, 63, 173–178. Reprinted by permission of Prentice Hall, Englewood Cliffs, NJ.)

Table 6–2 Comparison of Atmospheric and Absolute Pressure at Varying Attributes.

Altitude	psia	Pressure (in. Hg)	Boiling Point of Water (°F)	Refrigerant Boiling Points (°F)		
				R-12	R-22	R-502
0 ft	14.7	29.92	212°F	−22°F	−41°F	−50°F
1000 ft	14.2	28.85	210	−23	−43	−51
2000 ft	13.7	27.82	208	−25	−44	−53
3000 ft	13.2	26.81	206	−26	−45	−54
4000 ft	12.7	25.84	205	−28	−47	−56
5000 ft	12.2	24.89	203	−29	−48	−57
	kg/cm²	cm/Hg	°C	°C		
0 m	1.03	75.9	100	−30	−40.5	−45.6
304 m	.998	73.2	99	−30.5	−41.7	−46
608 m	.96	70.6	97.44	−31.7	−42.2	−47
912 m	.93	68.1	96.3	−32.2	−42.8	−47.8
1216 m	.89	65.6	95.8	−33.3	−43.9	−48.9
1520 m	.857	63.2	94.7	−33.9	−44.4	−49.5

(Billy C. Langley, *Refrigeration and Air Conditioning*, 3rd ed., 1986, pp. 4–23, 63, 173–179. Reprinted by permission of Prentice Hall, Englewood Cliffs, NJ.)

Absolute Pressure

The pressure measured from a perfect vacuum is called absolute pressure. Atmospheric pressure is 14.7 psi. Absolute pressure is normally expressed in terms of pounds per square inch absolute (psia). Absolute pressure is equal to gauge pressure plus atmospheric pressure. To find absolute pressure, add 14.7 to the gauge pressure reading.

Gauge Pressure

Gauge pressure is zero pounds per square inch gauge (psig) when the gauge is not connected to a source of pressure. Pressures below 0 psig are negative gauge readings and are commonly referred to as inches of vacuum. Refrigeration compound gauges are calibrated in inches of mercury (inch Hg) for readings below atmospheric pressure. See Figure 6–6. Since 14.7 psi is equal to 29.92 inches Hg, 1 psi is equal to 2 inches Hg:

$$29.92/14.7 = 2.03 \text{ inches Hg}$$

It should be remembered that gauge pressures are only relative to absolute pressures. See Table 6–2.

Measurement

The measurement of low pressures requires a unit of measurement smaller than the pound or the inch of mercury. The micron is commonly used for measuring low pressures. A micron is a metric measurement of length and is used in measuring the vacuum in a refrigeration system. It is considered as being absolute pressure.

One micron is equal to 1/1000 of a millimeter. There are 25.4 millimeters in 1 inch. Therefore, one micron is equal to 1/25,400 of an inch. A system that has been evacuated to 500 microns would have an absolute pressure of 0.02 inch Hg. At standard conditions this would be equal to a vacuum reading of 29.90 inches Hg, which is impossible to read on a refrigeration compound gauge.

Effects of temperature The effects of temperature on the pressure of gases is of great importance in the refrigeration industry. It must be thoroughly understood before a good knowledge of the refrigeration cycle can be obtained. There are several scientific laws that govern the effects of temperature on the pressure of a vapor within a confined space. These laws are as follows:

Charles' law Charles' law of gases states that "With the pressure constant the volume of a vapor is directly proportional to its absolute temperature." In mathematical form this is stated as $V_1/V_2 = T_1/T_2$ where V_1 is the old volume, V_2 is the new volume, T_1 is the old temperature, and T_2 is the new temperature.

In practical applications this can be proven by use of a properly fitted piston within a cylinder. See Figure 6–6a. In this example the cylinder is fitted with a sliding piston.

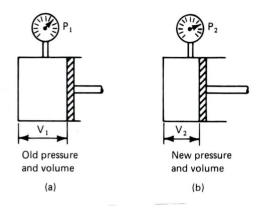

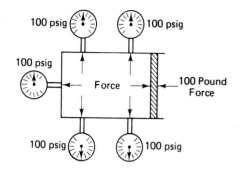

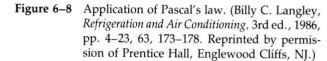

Old pressure
and volume

New pressure
and volume

(a)

(b)

Figure 6–7 Application of Boyle's law. (Billy C. Langley, *Refrigeration and Air Conditioning*, 3rd ed., 1986, pp. 4–23, 63, 173–178. Reprinted by permission of Prentice Hall, Englewood Cliffs, NJ.)

Figure 6–8 Application of Pascal's law. (Billy C. Langley, *Refrigeration and Air Conditioning*, 3rd ed., 1986, pp. 4–23, 63, 173–178. Reprinted by permission of Prentice Hall, Englewood Cliffs, NJ.)

The cylinder is full of vapor at atmospheric pressure. Heat is applied to the cylinder, causing the temperature to rise. Because the piston is easily moved, the volume of vapor increases but the pressure remains constant at atmospheric pressure. See Figure 6–6b. On the other hand, if the gas is cooled, the volume of vapor will become smaller. See Figure 6–6c. If it were possible to cool the vapor to absolute zero temperature, −460°F, the volume would be zero because there would be no molecular movement at this temperature.

• The mechanical equivalent of heat is the heat produced by the expenditure of a given amount of mechanical energy.

• The six components of a basic refrigeration system are: compressor, condenser, flow control device, evaporator, connecting tubing, and refrigerant.

Boyle's law Boyle's law of partial gases states that "With the temperature constant, the volume of a gas is inversely proportional to its absolute pressure." In mathematical form this is stated as $P_1V_1 = P_2V_2$ where P_1 is the old pressure, P_2 is the new pressure, V_1 is the old volume, and V_2 is the new volume.

As before, this can be proven by the use of a cylinder with a properly fitted piston. See Figure 6–7a. By taking simultaneous temperature and pressure readings as the vapor is slowly compressed within the cylinder so that no temperature increase will be experienced, each side of the equation will always be equal. This is possible because a decrease in volume will always be accompanied by an increase in pressure. See Figure 6–7b.

Dalton's law of partial pressures Dalton's law of partial pressures states that "Gases occupying a common volume each fill that volume and behave as though the other gases were not present." This law, along with the combination of Charles' law and Boyle's law, forms the basis for deriving the psychrometric properties of air.

A practical application of Dalton's law is: The total pressure inside a cylinder of compressed air, which is a mixture of oxygen, water vapor, and carbon dioxide, is found by adding together the pressures exerted by each of the individual vapors.

Pascal's law Pascal's law states that "The pressure applied upon a confined fluid is transmitted equally in all directions." A practical application of Pascal's law is shown with a cylinder of liquid and a properly fitted piston. See Figure 6–8. The piston has a cross-sectional area of 1 square inch. With 100 psig pressure applied to the piston, the pressure gauges show that the pressure exerted in all directions is equal.

This is the basic principle used in most hydraulic and pneumatic systems.

The general gas law The general gas law is made by combining Boyle's law and Charles' law of gases. The general gas law is sometimes expressed as

$$P_1V_1/T_1 = P_2V_2/T_2$$

A more useful form of the equation is

$$PV = MRT$$

where P is the absolute pressure of the vapor, in pounds per square feet, V is the volume of the given quantity of vapor, in cubic feet, M is the weight of the given amount of vapor, in pounds, R is the universal gas constant of 1545.3 divided by the molecular weight of the vapor, and T is the absolute temperature of the vapor.

The general gas law may be used to study changes in the conditions of a vapor as long as absolute temperature and absolute pressure are used. Gauge pressure cannot be used. This law is used in calculating the psychrometric properties of air in air-conditioning systems.

Pressure–temperature relationships Of vital importance in refrigeration equipment design and servicing is the relationship between temperature and pressure. The temperature at which a liquid will boil is dependent on the pressure applied to it, and the pressure at which it will boil is dependent on its temperature. From this it can be seen that for each pressure exerted on a liquid there is also a definite temperature at which it will boil, providing an uncontaminated liquid is being measured.

In practice, because all liquids react in the same manner, pressure provides us with a convenient means of regulating the temperature inside a refrigerated space. When an evaporator is part of a refrigeration system that is isolated from the atmosphere, pressure can be applied to the inside of the evaporator which is equal to the boiling pressure of the liquid at the desired cooling temperature. The liquid will boil at that temperature and, as long as heat is absorbed by the liquid, refrigeration is being accomplished.

This process is also reversible. When the pressure over a vapor is increased enough to cause the temperature of the vapor to be higher than the surrounding medium, heat will be given up and condensation of the gas will occur. This is the principle used in the condenser of a refrigeration system.

Cooling

As stated above, cooling is merely the removal of heat from a substance. This cooling can be accomplished in several ways. However, we will discuss only the evaporation and expansion methods at this time.

Evaporation The process that causes the water left in an open container to disappear is called evaporation. Evaporation of water depends on two things: temperature and moisture in the air. When water is left in an open container in the summertime, it dries rapidly because of the high temperature. If the temperature drops, the rate of evaporation will decrease. If the temperature goes up, the rate of evaporation will increase. Even at temperatures of −40°F evaporation takes place.

If the air is nearly saturated with moisture, it will absorb additional moisture very slowly. If the air is dry, it will absorb moisture very rapidly. During a hot spell when the air is muggy or humid, evaporation of the perspiration from the human body is slow because the saturated air absorbs moisture very slowly.

When water evaporates, a sufficient amount of heat will be absorbed in order to supply the latent heat of vaporization. This heat is absorbed from the water itself or from any object (or air) in contact with the water.

The cold, clammy feeling experienced by a person wearing a wet bathing suit is caused by evaporation of the water and absorption of heat from the material of the suit and from the skin of the person.

Expansion The process that causes a cooling effect by compressing a vapor, then rapidly reducing the pressure on the vapor is called expansion. When a vapor is compressed, heat is generated in an amount equivalent to the amount of work done in compressing the vapor. This may be demonstrated by the bicycle pump that gets hot due to the heat developed by compressing air into the tire. The greater the compression, the higher the temperature of the compressed air.

The cooling action that takes place when compressed air is allowed to expand is just the reverse of the compression effect. After a long ride over hot roads, the air in an automobile tire becomes very hot and the pressure within the tire is increased. If the valve stem were opened and the hot air allowed to escape, the air would become cool as it was released due to the expansion and resulting reduction in pressure of the air.

Some of the early refrigeration systems used this principle. The air was compressed and then cooled by passing it through a water-cooled coil. The water was used to absorb and carry away the heat from the compressed air. When the air was allowed to expand, the temperature dropped in relation to the amount of heat that was removed by the water while it was in the compressed state.

There are three steps involved in the expansion refrigeration cycle:

1. Air or gas is compressed to a high pressure.
2. The heat produced by compression is removed.
3. The air or vapor is expanded, causing a reduction in temperature through the absorption of heat by the air.

Subcooling Subcooling occurs when a liquid is at a temperature lower than the saturation temperature corresponding to its pressure. Water at any temperature below its boiling temperature (212°F) at sea level is said to be subcooled.

In refrigeration systems this subcooling may occur while the liquid refrigerant is temporarily stored in the condenser or the receiver. Some of the heat may also be dissipated to the ambient temperature while passing through the liquid line on route to the flow control device. The use of a subcooler will pay for itself through the increased capacity and efficiency of the refrigeration system.

DEHYDRATION

There has been a great amount of money spent on researching the effects of heat and noncondensables in a refrigeration system. Even now, many of the effects are still a mystery. We do know that their presence can result in many forms of damage in a refrigeration system such as sludging, corrosion, oil breakdown, carbon formations, and copper plating. These contaminants are usually the cause of compressor failure. Dehydration, also known as evacuation, is the removal of air, moisture, and noncondensables from a refrigeration system.

Purposes of dehydration Dehydration protects the refrigeration system as much as possible from contaminants, and causes it to operate as efficiently as possible with a minimum amount of equipment failure.

The methods of dehydration are many and varied. As stated above, two methods used to cause a liquid to boil are lowering the pressure exerted on it and applying heat to the liquid. Some of the ways to eliminate moisture from a refrigeration system by the boiling process are:

1. Move the system to a higher elevation where the ambient temperature is high enough to boil water at the existing pressure.
2. Apply heat to the system, causing the moisture inside it to boil.
3. Use a vacuum pump to lower the pressure inside the refrigeration system so that the ambient temperature will boil the moisture.

In practice, the first two choices are impractical. Therefore, the vacuum pump method is the most desirable means of removing moisture and noncondensables from a system. To accomplish effective dehydration, the refrigeration system must be evacuated to at least 500 microns.

Before attempting to charge a system with refrigerant, it must be evacuated with some type of vacuum pump. This is especially true since manufacturers of refrigeration and air-conditioning units have changed most of their equipment to air-cooled condensers. This change has resulted in higher operating head pressures and temperatures and higher condensing pressures, especially when hermetic compressors are used. Also, the greater use of low-temperature refrigeration equipment results in higher compression ratios. When single-stage compressors are used, still higher discharge temperatures are present.

These higher head temperatures make it even more important to remove all the air, as well as the moisture, from the system to a point below the critical point. The process of removing the air from the system can be referred to as removing the noncondensables, or "degassing" the unit. The worst contaminant is the oxygen in the remaining air. Oxygen is one of nature's most chemically active elements, and its rate of reaction with the refrigeration oil in the system increases very rapidly with any increase in temperature above 200°F.

Because of this, the most important factor is removing all of the oxygen from the system, or at least to a very minimum. In the process of removing all of the oxygen, the system will also become adequately dehydrated. Filter–driers are installed in the refrigerant lines to pick up any contaminants, such as soldering flux, any moisture that may exist in the oil and refrigerant charge, and materials of construction.

BASIC REFRIGERATION SYSTEM

The basic components of a refrigeration system are the compressor, condenser, flow control device, evaporator, connecting tubing, and the refrigerant. See Figure 6–9. The compressor is known as the heart of the system. It causes the refrigerant to circulate in the system. The refrigerant is pushed by the compressor to the condenser where both sensible and latent heat are removed. The liquefied refrigerant then goes to the flow control device where the pressure is reduced, allowing the refrigerant to expand and absorb heat from within the refrigerated cabinet. This low-pressure,

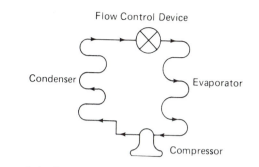

Figure 6–9 Basic refrigeration system. (Billy C. Langley, *Refrigeration and Air Conditioning*, 3rd ed., 1986, pp. 4–23, 63, 173–178. Reprinted by permission of Prentice Hall, Englewood Cliffs, NJ.)

heat-laden refrigerant vapor is then drawn to the compressor where the cycle is repeated.

Refrigerants

The vapors or liquids used in refrigeration systems are known as refrigerants. The evaporation of the refrigerant within the evaporator extracts heat from the surrounding objects. The various parts of the refrigeration unit compress and condense the refrigerant so that it can be used over and over again. Even though there are many different types of refrigerants, only the more common fluorocarbons will be considered here.

For practical purposes, a refrigerant is a fluid that absorbs heat by evaporating at a low temperature and pressure and gives up that heat by condensing at a higher temperature and pressure.

Characteristics of Refrigerants

Most of the commonly used refrigerants exist in a vaporous state under ordinary atmospheric pressures and temperatures. To change these vapors to liquid form, it is necessary to compress and cool them as is done by the condensing unit on a refrigeration system. A fluid is a liquid, gas, or vapor. The words gas and vapor are ordinarily used interchangeably, although to be perfectly technical, perhaps we should explain that a gas near its condensation point is called a vapor. All fluids have both a liquid and a gaseous state. Some fluids have high boiling points, which means that they exist as a gas only when heated to a high temperature or when placed under a vacuum. Fluids that have low boiling points are in the form of vapor at ordinary room temperatures and pressures. Many of the more common refrigerants such as those in the freon group are in this category. If these vapors are to be liquefied, they must be compressed and cooled, or condensed.

Water is a fluid that exists as a liquid at atmospheric pressure and temperature. The boiling point of water under atmospheric pressure at sea level is 212°F. If the water is left in an open basin, it will evaporate very slowly. If heat is applied to the water and its temperature raised to its boiling point, it will then evaporate, or boil, very rapidly. The water will change to the gaseous form of water known as steam or water vapor. If the water is boiled in an open container, its temperature will not rise above 212°F. All the heat supplied by the flame is used to boil off or vaporize the water.

If a liquid refrigerant is similarly placed in an open container, it will immediately begin to boil vigorously and vaporize, but at a very low temperature. Liquid Refrigerant-12, under atmospheric pressure, will boil at −21.6°F. It will absorb sufficient heat from the container and the surrounding air to enable it to boil. It would not be necessary to heat it with a flame as is done with the water.

A vaporizing refrigerant will absorb heat equal to the amount of energy necessary to change the physical form of the refrigerant from the liquid state to the gaseous state. Each refrigerant will absorb an amount of heat per pound of refrigerant equal to its latent heat of vaporization.

Effects of Pressure on Boiling Point

The boiling point of any liquid may be raised or lowered in accordance with the amount of pressure applied to the liquid. The greater the pressure, the higher is the boiling point; the lower the pressure, the lower is the boiling point. Thus, a liquid can be caused to boil at a low temperature by placing it in a partial vacuum.

Standard Conditions

The capacity of any refrigeration unit will vary as the refrigerant temperatures vary on the high- and low-pressure sides of the system. The latent heat of the refrigerant, its condensing pressure, and its vaporizing pressure will also vary as the temperature of the refrigerant varies. For purposes of comparing different refrigerants and refrigeration units, certain standard conditions have been developed. The refrigeration industry has set forth conditions known as standard conditions. These conditions are established with the following temperatures at various locations in the refrigeration cycle: a temperature in the evaporator of 5°F; a temperature in the saturated portion of the condenser of 86°F; a liquid temperature at the flow control device of 77°F; and a suction gas temperature of 14°F.

Now if we use the following factors in comparing any two refrigerants, with these temperatures as a basis, we have a true comparison and can arrive at correct conclusions.

Condensing pressure The condensing pressure will depend on the temperature at which the vapor will liquefy. In refrigeration work, it is desirable to avoid high condensing pressures if at all possible. Therefore, the condensing medium (air or water) must be as cool as possible. Ordinarily, a water-cooled condenser will operate at a lower condensing temperature and pressure than an air-cooled condenser. Because of this, there is some difference in the operating pressures for these two types of condensers.

In general, it may be assumed that the condensing temperature and pressure of an air-cooled unit will be approximately 25 to 35°F higher than ambient temperatures.

Table 6–3 Vaporizing pressures of refrigerants at 5°F.

Refrigerant	Pressure at 5°F	Pressure at −15°C
R-11	−24 in. Hg	41.22 kPa
R-12	11.8	182.05 kPa
R-2	28.1	294.04 kPa
R-500	16.4	213.66 kPa
R-502	36	348.31 kPa

(Billy C. Langley, *Refrigeration and Air Conditioning*, 3rd ed., 1986, pp. 4–23, 63, 173–178. Reprinted by permission of Prentice-Hall, Englewood Cliffs, NJ.)

The actual temperature and pressure will, however, depend on the efficiency of the condenser itself, the location of the condenser, whether or not sufficient air circulation is obtained, and the cleanliness of the condenser surface.

Vaporizing pressure The vaporizing pressure of a refrigerant is important because the refrigerant must evaporate without requiring too low a suction pressure. A temperature of 5°F is considered the temperature of most domestic refrigerator evaporators. This is the same temperature as that set for the standard conditions used for comparison of different refrigerants and refrigeration units. In general, a refrigerant is desired that has an evaporating pressure at or near atmospheric pressure. See Table 6–3. A refrigerant that requires a vacuum to produce evaporation is not practical under ordinary conditions because of the tendency for air to leak into the system. The air will not condense and will cause the condensing pressure to become very high. This high pressure will reduce the efficiency of the refrigeration unit. Refrigerants that have a vaporizing pressure above atmospheric pressure do not allow air to be drawn into the system through a leak.

It should be noted that the pressure in the evaporator and the low side of the system will be the same. Also, the temperature of the evaporating refrigerant will correspond to the temperature shown on the pressure–temperature chart on the gauges, the pressure in the evaporator, or in the low side of the system.

Latent heat of vaporization The amount of heat (in Btu) required to change a liquid to a vapor, the change taking place at a constant temperature, is known as the latent heat of vaporization. This definition can now be developed so as to apply to 1 pound of refrigerant, the vaporization taking place at atmospheric pressure and the liquid to be at a temperature equal to its boiling point when the operation begins. Thus, the latent heat of vaporization of a liquid is the amount of heat (in Btu) required to vaporize 1 pound of the liquid at atmospheric pressure, the liquid to be at its boiling point when the operation begins. To convert 1 pound of water at 212°F into steam at the same temperature and pressure, the water must absorb 970 Btu per pound. This quantity of heat is equal to the total latent heat of 1 pound of water at atmospheric pressure.

Table 6–4 Latent heat of vaporization.

Refrigerant	Latent heat	
	Btu/lb	kcal/kg
R-11	83.459	45.902
R-12	68.204	37.512
R-22	93.206	51.263
R-500	82.45	45.34
R-502 at 40°F	63.1	34.7

(Billy C. Langley, *Refrigeration and Air Conditioning*, 3rd ed., 1986, pp. 4–23, 63, 173–178. Reprinted by permission of Prentice-Hall, Englewood Cliffs, NJ.)

Any refrigerant, when evaporating in the evaporator, must absorb heat from within the cooled space exactly equal to its latent heat of vaporization. When a refrigerant has a high latent heat, it will absorb more heat per pound of liquid than a refrigerant with a lower latent heat of vaporization. Thus, if a refrigerant with a high latent heat of vaporization is used, a smaller compressor, condenser, and evaporator can be used. See Table 6–4.

The latent heat of vaporization of a liquid will vary with the pressure and the corresponding temperature at which the vaporization occurs. When lower temperatures and pressures are encountered, the latent heat of vaporization increases.

Types of Refrigerants

There are many types of commercially available refrigerants. The types that are commonly used in domestic refrigerators and freezers are nontoxic and are not dangerous, for the most part. For normal domestic refrigerators and freezers Refrigerant-12 and Refrigerant-22 are the most popular. Both of these refrigerants are nontoxic and are safe to use in food-handling applications. They operate at relatively low pressures with evaporating pressures above atmospheric pressure.

Handling of Refrigerant Cylinders

The pressure created by a liquid refrigerant in a sealed container is equal to its saturation pressure at that liquid temperature as long as there is space above the liquid for the vapor. However, if the refrigerant cylinder is overfilled, or if the cylinder is gradually and uniformly overheated, the liquid refrigerant will expand until the cylinder becomes full of liquid. When this occurs, hydrostatic pressure builds up rapidly, producing pressures well above saturation pressures. After the cylinder becomes full of the expanded liquid

under gradual and uniform overheating, the true pressure–temperature relationship no longer exists.

The extremely dangerous pressures that can result under such circumstances can cause the rupture of the refrigerant cylinder. Under uniform heating conditions the cylinder can rupture at approximately 1300 psi. If, however, heat is applied with a welding torch, the area of the cylinder wall where the heat is applied may be weakened and the danger of rupture increased.

REVIEW EXERCISES

1. In what three states may a molecule exist?
2. What determines the amount of heat present in a given body?
3. At what temperature does all molecular motion stop?
4. When heat is added to a substance, what other than a rise in temperature happens to that substance?
5. By the regulation of heat, can any substance be caused to remain in any of its three states?
6. To what does the degree or intensity of heat refer?
7. With what is temperature measured?
8. Name the three ways that heat travels.
9. What is the heat added to or removed from a substance resulting in a change in temperature but no change of state known as?
10. What is the amount of heat that must be added to a solid to change it to a liquid called?
11. What is the amount of heat that must be removed from a vapor to condense it to a liquid at a constant temperature known as?
12. What is the amount of heat required to raise the temperature of 1 lb of any substance one degree known as?
13. What is the specific heat of water?
14. What is the heat added to a vapor after the vapor is no longer in contact with its liquid known as?
15. Define pressure.
16. Define atmospheric pressure.
17. What is absolute pressure?
18. What pressure is read when a gauge is not connected to a source or pressure?
19. Name the six components of a refrigeration system.
20. How will an increase in pressure on a liquid affect its boiling point?
21. What is the removal of heat from a substance known as?
22. Define subcooling.
23. What must be done to a refrigeration system before charging it with refrigerant?
24. What is considered to be the evaporating temperature of most domestic refrigerators?

7

Air Control

Air, the substance that completely surrounds us, has many forms and properties. It can be hot, cold, humid, dry, polluted, clear, or it can be found with all of these conditions. We are able to live in just about all of these conditions. However, the most healthful and comfortable form of air occurs in only a very narrow range of these conditions.

There is a comfort zone in which we are the most comfortable. It is generally considered to be between 74 and 80° F with a relative humidity between 40 to 60 percent.

POLLUTANTS

Each of the pollutants in the atmosphere presents a different kind of problem, such as the following:

1. Allergens such as pollen, dust, and molds make breathing difficult for those who are allergic to them.
2. Bacteria and other airborne microorganisms are a health menace.
3. Flammable and toxic gases in large enough quantities can be dangerous.
4. Odors from many different sources are unpleasant, such as those that come from food preparation, evaporating-type chemicals, or unclean bodies.
5. Solid particles, such as soot and fly ash, in addition to irritating the eyes and nose, are also hard on clothes.
6. Radioactive air contaminants are active in very low concentrations.

To reduce the concentration of these pollutants requires an enclosed space with proper air circulation, filters to clean the air, fans to circulate the air, humidifiers to add moisture to the air, and dehumidifiers to remove moisture from the air. Some of these pollutants, such as bacteria, require special equipment to reduce their concentration to acceptable levels. These are controlled through the use of special filters or germicidal lamps. When odors are a problem, charcoal filters (activated carbon) are usually used. It is usually better to dilute the odors with fresh, clean air.

If these pollutants were controlled at the source rather than after they entered the atmosphere, it would be much simpler for all of us. Air conditioning for the removal of the pollutants would not be necessary and we would all be healthier because of it.

AIR MOVEMENT

Air moves naturally from a high-pressure area to a low-pressure area, or from a high-temperature area to a low-temperature area. The greater the difference in either the pressure or the temperature, the faster is the air flow. This is the natural way we are provided with fresh, clean air. However, when an air conditioning system is used, the air is forced with a blower from a high-pressure area—the blower section—to a low-pressure area—the conditioned space.

Air movement in this manner is achieved through the use of two types of fans: the axial-flow type and the squirrel cage (centrifugal) type. The axial-flow fan is the type that is used to set around and blow air directly on a person or something to be cooled. The air flow is in the direction that the fan is pointing, in the direction that the shaft of the motor is pointing.

The centrifugal type is a series of blades that are placed into a circle. The air is drawn into the center of the wheel and is forced out through the housing at some point that will concentrate the volume of air. The air is discharged at a 90° angle from that in which it entered the blower wheel.

These types of fans are used in ducted systems because they can create sufficient pressure to overcome the resistance of the ductwork. Centrifugal fans are generally used in forced-air systems, forced-air combustion furnaces, and forced-air combustion clothes dryers.

The amount of air that can be moved will depend on the size of the motor and the capacity of the fan. Domestic refrigerators use a small 4-inch diameter axial-flow fan with a small $\frac{1}{150}$-hp motor. The fan used in central air-conditioning applications may have a 10- or 12-inch diameter centrifugal blower driven by a $\frac{1}{3}$- or $\frac{1}{2}$-hp motor. The amount of air to be delivered will determine the sizes of the fan and motor.

The velocity of the air movement will determine, to a great extent, the comfort of the occupants in the room. The number of air changes per hour will determine the freshness of the air plus the personal comfort of the occupants. A slow air movement will cause a stagnant, stuffy feeling, while a high volume of air will cause drafts and discomfort. Generally, 15 cubic feet per minute is the minimum and 65 feet per minute is the maximum for human comfort. In most instances velocities around 25 to 35 feet per minute are the most desirable.

The number of air changes per hour in a room will vary depending on the use of the room. If there are a lot of people with heavy smoking, more air changes per hour will be required than if there are only one or two people with little activity and no smoking. Usually $1\frac{1}{2}$ air changes per hour is the minimum. Sometimes a lack of air changes per hour can be temporarily solved by the use of ceiling fans which have several speeds and air volumes.

AIR FILTERING

Air is filtered to remove dust, smoke, and pollen from the air as it passes through the air-conditioning equipment or the air-filtering system. The filter is generally placed in the air stream ahead of the air-conditioning equipment. There are of two types of filters: the filter type and the electronic ionizing type.

1. In the air-filtering type, the air is forced through a filter made of one of several different types of filter media. This medium may be fiberglass, aluminum, sponge rubber, or some other type of material that will collect dust particles as the air flows through it. Some types of media have a coating that helps to collect the dust and pollen. Some of these filters can be cleaned, while others are discarded and new ones installed. These filters should be replaced or cleaned at least once each month, or whenever they become dirty enough to reduce the air flow through them.

2. In the electronic ionizing type, the air passes through an electronically charged field which removes the particles electronically. These filters have positively and negatively charged plates that charge and collect the dust particles as they pass through the ionized area. These plates are charged to about 20,000 volts dc. The ionizing plates are cleaned periodically according to the manufacturer's recommendations. Be sure to follow the manufacturer's recommendations completely when cleaning this type of filter.

GERMICIDAL AND OZONE LAMPS

Airborne bacteria and mold spores are generally controlled through the use of germicidal lamps. Undesirable odors are removed through the use of ozone lamps. In some instances both odors and bacteria can be removed with a single lamp.

Ozone lamps are popular in clothes dryers, range hoods, washrooms, and in smaller air-filtering systems. These lamps must be located so that their output rays will not come into contact with the eyes and skin to prevent burning them. They should be installed in an air flow so that the ozone will be mixed thoroughly with the air.

These are special lamps that operate on 10.5 volts using 350 milliamps of power, ac or dc. The operation of this lamp is due to its ability to arc. In order to arc, it must have a minimum of 28 volts, and it must have a ballast that is either inductive on ac or resistive on ac or dc voltage. The ballast must limit the current flow to the bulb. In some installations a 40-watt, 120-volt incandescent lamp may be used.

Germicidal lamps are sometimes placed in the ductwork of an air-conditioning system to kill any bacteria and other undesirable pathogens. In most installations they are placed after the filter and before the humidification equipment. There should also be a service access door for these lamps.

To kill any microorganisms in the air, germicidal lamps are often used. Their effectiveness is in direct relation to the intensity of the exposure and the length of time exposed. The ultraviolet energy at a 2537 angstrom wavelength will react to the skin and eyes much the same as the direct rays of sunlight. Therefore, precautions must be taken to protect the eyes and skin from excessive radiation. In most cases the light fixture is designed to direct the light rays at the ceiling or high on a wall. When installed in this manner, the rays are allowed to bounce around the room with little or no contact with the occupants.

REVIEW EXERCISES

1. What are pollen, dust, and molds known as?
2. When should pollutants be controlled?
3. Name the two types of fans used to circulate air.
4. In any type of installation, what determines the amount of air that can be moved?
5. In an air conditioned building, what determines the freshness of the air and the personal comfort of the occupants?
6. What process is used to remove the dust, smoke, and pollen from the air passing through an air conditioning unit?
7. How are airborne bacteria and mold spores generally controlled?
8. Where are germicide lamps sometimes used?
9. Name two types of air conditioning filters.
10. What is used to kill microorganisms?

8

Water Chemistry

Water is probably the single most important resource that we have on earth. As it falls from the clouds and flows through the earth, water gathers many impurities because of the pollution and the natural minerals in the earth. Most of these impurities must be removed before the water is useful for human consumption. In addition, some of them must be removed before the water will do the job that we would like it to do.

SYNTHETIC CONTAMINANTS

When rainwater falls to the earth, it passes through a polluted atmosphere and picks up some pollutants, usually evaporative chemicals, before it reaches the earth. These airborne chemicals react with the water to form many types of acids. This then becomes what is known as acid rain. When the rainwater drains through the earth, it absorbs more chemicals which cause the chemicals already in the water to become more active.

When people dump their waste into the ground, the waste flows through the earth and enters the aquifer from which we all obtain our water. This waste only adds more minerals and pollutants to the water supply. The result is that nature is no longer in balance.

A good example is the use of laundry and other detergents that we use to keep ourselves clean. These detergents contain two principle ingredients: a surfactant and a phosphate.

The surfactant (wetting agent) lowers the surface tension of the water to improve its penetration into the fabric being cleaned. At one time the major surfactant was ABS (alkyl benzene sulfonate). This additive was very slow in breaking down in nature. It caused excessive foaming action

in the sewage treatment plants, the lakes, and the streams into which the water containing ABS was discharged. However, this surfactant has been replaced with LAS (linear alkylate sulfonate), which is readily biodegradable in nature. Biodegradation is caused by the useful bacteria (commonly found in sewage, surface water, and soil). These bacteria break down the surfactant and use it for food, converting it to water and carbon dioxide. This step brought one form of pollution under control.

The phosphate used in a laundry detergent serves several important purposes:

It increases the efficiency of the biodegradable LAS.

The soil particles in the wash water are kept in suspension.

It brings the alkalinity level to the necessary level that is needed for the efficient cleaning of laundry.

Oil and greasy soils are emulsified.

It helps in making the water softer by combining with the hardness minerals.

It aids in the reduction of germs left in the clothes.

Phosphates are effective in laundry detergents, but they do not decompose in normal septic tanks or sewage treatment plants. Therefore, phosphates appear as sewage effluent in lakes and streams. When combined with other nutrients that are present in lakes and streams, phosphorous provides an excellent nutrient for plant life. In fact, it may be too good because it will sometimes clog the streams and lakes with algae, killing off the fish population.

Only human efforts can control the waste materials and treat them before they enter the earth and the atmosphere so that reasonably clean uncontaminated water is

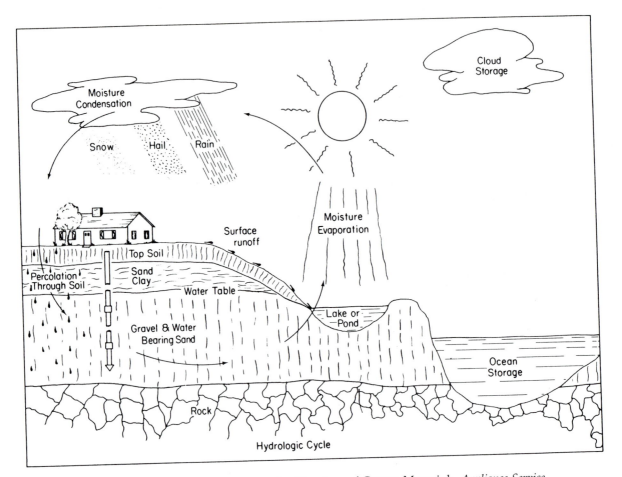

Figure 8–1 Hydrologic or water cycle. (Courtesy of George Meyerink, *Appliance Service Handbook*, 2nd ed., 1988, pp. 40–67, 171–179. Reprinted by permission of Prentice Hall, Englewood Cliffs, NJ.)

available for our use. Because water has a natural affinity for certain kinds of minerals, it is necessary to use some type of water-conditioning equipment. Two of the major problems are hardness and iron.

HARD WATER

When people refer to hard water they mean that the water contains certain minerals that make it less desirable for use in the home without some form of treatment.

Hard water has the characteristics or the quality that results in making an insoluble sticky curd when it is mixed with soap. This hardness is caused by calcium and magnesium, carbonates, and sulfates.

The making of hard water   Water as it is cycled from the clouds in the form of rain to the earth and back to the clouds again is called the hydrologic or water cycle. See Figures 8–1 and 8–2. In this way water is in constant motion, being drawn from the earth in the form of evaporation, stored in the clouds, then released as snow, hail, or rain.

When the water is released from the clouds is just about the only time that the water is in a pure form. This is because during the fall from the clouds to the earth it gathers some impurities in the form of gases, dust, bacteria, smoke, and other forms of pollutants from the air.

Because of this, when the rainwater reaches the earth, it is already contaminated, to some extent. Then when it contacts the earth, it picks up more contaminants.

The water drawn from a well, or ground water, takes on the characteristics of the earth that it travels through, or percolates, as it flows to the aquifer. As the water falls from the clouds through the air, it absorbs some carbon dioxide and other gases, which serve to increase its solvent effect on the soluble minerals in the soil, rocks, and other mineral formations in the earth.

The quality of the water is, therefore, determined by the amount and kind of dissolved minerals and organic matter gathered by it while passing through the earth to the aquifer. The most common ingredients found in the water are dissolved or suspended matter such as carbonates, chlorides, and sulfates of calcium, magnesium and sodium, and copper, iron, and manganese. Organic matter of vegetable or animal origin and gases may also be dissolved in the water. Also present are suspended solids such as iron, silt, refuse, and organic and animal matter.

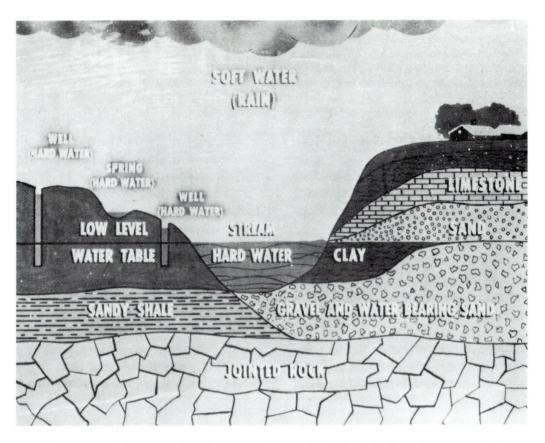

Figure 8–2 Soft water rain. (Courtesy of George Meyerink, *Appliance Service Handbook*, 2nd ed., 1988, pp. 40–67, 171–179. Reprinted by permission of Prentice Hall, Englewood Cliffs, NJ.)

Also, limited amounts of dissolved minerals that are not harmful to the human system may be present. However, in most cases these impurities affect the taste, color, odor, and the general usefulness of the water, usually to such an extent that it must be treated to remove the harmful factors before it can be safely used.

The more common complaints about water are hardness, turbidity, red water, odor, and taste. These may occur individually or in some combination.

The amount of calcium and magnesium in the earth through which the water flows determines just how hard the water will be. The hardness of the water will be different in different parts of the country. See Figure 8–3. In some cases the areas may be smaller than indicated on the map.

Hard-water effects The minerals that cause hard water seriously affect the efficiency of detergents when they are all combined in water. As a result, a grimy scum forms which causes clothes and other things washed in the washing machine to be dingy and gray looking. It also prevents them from being as soft and white as they would normally be. The plumbing system becomes clogged with this scale preventing the proper drainage of the water. See Figure 8–4. Hard water reduces the effect of everything that we use in our homes. Bath soaps do not clean as they should, shampoos do not make the hair soft and manageable, and dishes

and glassware are streaked and dull, just to name a few of the more noticeable effects. All of these conditions are caused by soap that does not dissolve completely, leaving the residues clinging to the glassware, dishes, and the skin. This residue is extremely difficult to remove if it can be removed at all.

Soaps Calcium and magnesium salts have harmful effects in washing, whereas sodium salts do not. All three types are mixed with vegetable oil acids to make soap. The sodium soaps are very popular because they are natural agents for washing. However, the calcium and magnesium salts combined with vegetable oil acids form sticky, noncleaning curds.

When ordinary soaps are used in hard water, an exchange takes place which results in a mixture of hard soaps and soft water. In this manner the initial portion of the soap goes to make the sticky insoluble curd. There is little, or no, suds buildup until all of the minerals that make the water hard are consumed. This is when the curds, or scum, are formed. When clothes are washed in this type of mixture, the curds will not wash off, leaving the clothes dingy and gray looking.

In addition to this, the rinse water is fresh and contains more minerals that cause hard water. When these minerals combine with the soap lather left in the clothes during the

Hard Water 57

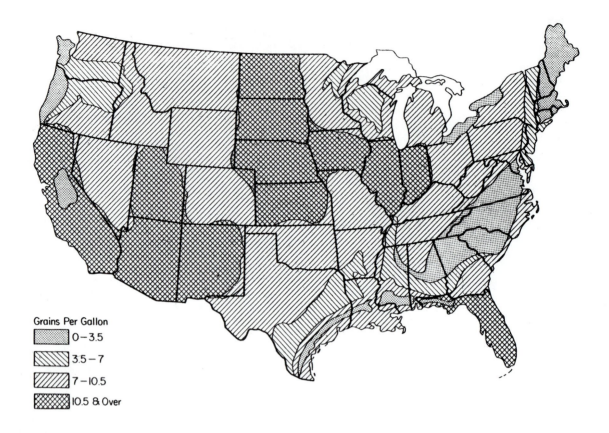

Grains Per Gallon

▦	0 – 3.5
▨	3.5 – 7
▧	7 – 10.5
▩	10.5 & Over

Figure 8–3 Water hardness map of the United States. (Courtesy of George Meyerink, *Appliance Service Handbook,* 2nd ed., 1988, pp. 40–67, 171–179. Reprinted by permission of Prentice Hall, Englewood Cliffs, NJ.)

wash cycle, more curds are formed that will not wash off. The clothes are left with a gray appearance. When soft water is used, there is no adverse reaction with the minerals in the water. The rinsing action merely flushes the lather and dirt down the drain, leaving clean and fresh clothes.

Using soaps in hard water is also expensive. When the hardness of water is increased from 2 to 32 grains of hardness, the soap consumption doubles for the same washing loads. This is in addition to the unpleasant side effects of skin irritation, stiff clothing, poor tasting drinks, and costly damage to the plumbing system.

Testing for hardness Hardness is measured in grains per gallon of water. A grain is a measure of weight of the scale-forming solids in the water. People who have had water softeners installed can often determine whether the water is soft or hard just by feeling it. One of the more common methods used to determine whether or not water is hard is to simply taste it. A more scientific way is to use one of the many test kits that are available for testing the hardness, iron, pH (acidity), and/or other impurities that may be in a specific area. Of major concern to the appliance technician is the amount of hardness in the water. This can be determined by a test known as the green soap test.

One of the methods used for this is to measure an ounce of the water into a clean glass jar. Then add tincture of green soap, shaking the glass of water after adding each drop. Continue to add the drops, one at a time and shaking after each drop, keeping track of the number of drops added to the water. Do this until suds are formed on the water.

Zero soft water requires only one drop of green soap per ounce to form suds. On the other end of the scale, hard water requires one drop of green soap per ounce of water for each grain of hardness before suds are produced. If the water sample should require 15 drops of green soap per ounce, the water would have 15 grains of hardness.

Water needs When a water softener has been installed and the household repeatedly runs out of soft water, the needs of the household should be checked against the capacity of the water softener. There are several factors that could cause this problem. For example, a family may have outgrown the capacity of the softener, or the water in the area may have become more hard since the water softener was installed.

On average, each person in a family uses 50 gallons of water per day. To determine the number of grains that must be removed from the water each day, multiply the number of people in the family times 50. Then multiply the answer by the grains of hardness of the water.

For example, A family of five using water with a 15 grain hardness would require: 5 × 50 × 15 = 3750 grains of hardness to be removed each day just to satisfy the minimum needs of the household. Then dividing this figure into the capacity of the water softener will indicate how long the water softener can go between generations.

In our example the water softener has a capacity of 15,000 grains; then 15,000/3750 = 4 days between generations.

Ion exchange One method of controlling the water hardness for domestic use is with an ion-exchange water conditioner.

Basically, this type of unit consists of a tank containing resin beads that are loaded with sodium ions. These resin beads prefer calcium and magnesium ions. Because of this, as the water flows through this tank of beads, the sodium ions are given up and the beads collect the calcium and magnesium ions from the minerals in the water.

The minerals that are left do not have a negative reaction with the soaps and detergents that the calcium and magnesium ions do. Also, they do not leave any deposits in the water pipes.

As the water is used, the resin beads will have given up all of their sodium ions and the unit will need regeneration. Regeneration is accomplished by flushing a brine solution through the resin beads, removing all of the calcium and magnesium ions. The sodium ions then take their place on the beads. The flushed calcium and magnesium ions are washed down the drain. The resin is then washed with fresh water and the unit will again supply more soft water.

There is no wear or loss of the resin pellets during the regeneration process. The only requirement is that the user replenish the unit periodically with salt for the regeneration process.

Only salt that is recommended specifically for this purpose should be used. This is because all of the impurities settle in the bottom of the salt tank and must be cleaned out periodically.

SOFT WATER

The following list will help clarify some of the questions about water softeners and the use of soft water:

1. Clothes will require several washings before the gray dull color and the curds have disappeared.
2. The first several tanks of hot water will be hard. Then the remaining tanks will be soft.
3. Rinse water from a water softener will not harm a septic tank.
4. Outdoor plants and lawns should not be watered with softened water because: (1) it is expensive and (2) over a long period of use, the excess sodium ions will unbalance the soil composition.
5. House plants can be watered with soft water if care is taken to completely saturate and flush the water

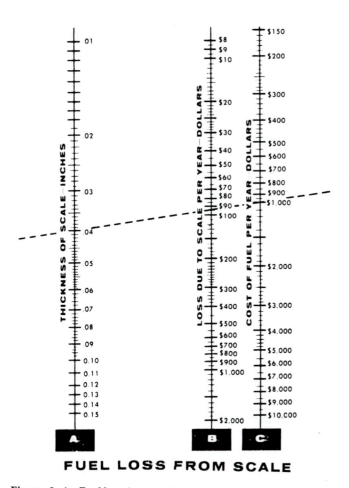

FUEL LOSS FROM SCALE

Figure 8–4 Fuel loss from scale. (Courtesy of George Meyerink, *Appliance Service Handbook*, 2nd ed., 1988, pp. 40–67, 171–179. Reprinted by permission of Prentice Hall, Englewood Cliffs, NJ.)

through the soil each time and not water too frequently. Some types of house plants are not affected; others are affected to some extent.

6. In a well-maintained, balanced aquarium in which water is added only to replenish any loss by evaporation, soft water can be added slowly as required and the fish will not be bothered, because the dilution ratio is so small.

7. In the first washing with softened water, decrease the amount of soap or detergent generally used to one-third or one-quarter of that previously used. On subsequent washings, add or decrease as required to obtain sufficient suds.

8. The rust and scale accumulation in the plumbing will be loosened by the soft water. Be prepared for occasional "slugs" of rusty water until the entire system has been cleared.

9. Soft water should not be used in car batteries. Use only distilled water.

10. Other filters and water-conditioning equipment must be installed in the line between the pressure tank and the softener, as needed.

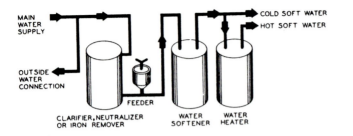

Figure 8-5 Water softener connections for iron removal. (Courtesy of George Meyerink, *Appliance Service Handbook,* 2nd ed., 1988, pp. 40-67, 171-179. Reprinted by permission of Prentice Hall, Englewood Cliffs, NJ.)

11. Some people do not like the taste of softened water and others, under doctor's orders, on a low-sodium diet, must not drink softened water. For these reasons, many installations include one tap from the unsoftened source.

IRON IN WATER

Red water generally contains excessive amounts of iron. There will be red stains on the plumbing fixtures and on the laundry. The taste will generally be changed and the appearance of tea, coffee, and other drinks will often be changed.

The red color is because of the iron hydroxide present in the water. Sometimes the water will appear to be clear when drawn, but it will redden after coming into contact with the air because the soluble iron changes to insoluble iron. The insoluble iron is what causes the problems.

There are two ways to control the amount of iron in the water for domestic use. The type of method used depends on the amount of iron involved. An ion-exchange water softener can handle small amounts of iron very effectively. However, large amounts—more than 3 to 5 parts per million—of iron must be removed by special iron filters.

There are no general recommendations for the treatment of iron in the water because of the complex makeup of this component. Each situation is different and requires that specific consideration be given each application. It must be remembered, however, that the iron filter must be installed between the water source and the water softener. See Figure 8-5. This location is necessary to protect the resin in the water softener from iron fouling. When the resin becomes iron-fouled, it will no longer soften the water.

There a few steps that can be taken to keep the resin from becoming iron-fouled when small quantities of iron are present. One is to prevent the water from coming in contact with air (oxygen) before entering the water softener. This is done by placing an air cell, such as an inflated bag or balloon, in the pressure tank.

Another effective method of accomplishing this is to periodically introduce some sodium hydrosulfite into the brine. This will help to clear the ferric hydroxide from the resin beads, allowing them to perform their function in a normal manner. When the resin beads become badly fouled, it may take several successive treatments to remove all of the iron. When it is finally cleared, the water softener should perform well with only periodic, preventative treatments. A good treatment is 2 to 4 ounces of sodium hydrosulfite for each cubic foot of resin bed. Therefore, a 15,000 grain softener (having 1½ cubic feet of resin) would require only 1 to 2 ounces. To treat the softener, pour the compound into the brine tank every time a 100-pound bag of salt is added.

These iron-reducing chemicals are available under a variety of different brand names: Lykopon, Fer-red, Iron out, and Rover are some of the more common brands available. These products can usually be purchased at the store that sells the water softener salt.

OTHER UNDESIRABLE QUALITIES OF WATER

There are usually other undesirable impurities in natural water. They can be noticed in several different ways, such as taste, color, and odor. These impurities are usually present in a lesser degree than those described above. There is usually someone available who can assist in treating these impurities properly. In most instances the superintendent of the local water company would be the one to start with for help.

REVIEW EXERCISES

1. What is probably the most important resource on earth?
2. What causes acid rain?
3. What is used to lower the surface tension of water?
4. What causes biodegradation?
5. What, in a laundry detergent, aids in the reduction of germs left in the clothes?
6. What do calcium and magnesium, carbonates and sulfate cause?
7. What determines the quality of water?
8. What affect does hard water have on clothes washed in it?
9. Will soap build-up suds in hard water?
10. How is the hardness of water measured?
11. On the average, how much water does each person use daily?
12. What does the user of a water softener need to do periodically?
13. What does red water generally contain in excessive amounts?
14. Where must an iron filter be installed?
15. In most instances, where is the best place to start to solve water problems?

9

Heat Sources and Combustion

When a source of heat is needed for an appliance, both gas and electricity are used. Gas is used for both clothes dryers and incinerators. The remainder of the appliances use electricity for their source of power for operation.

Strictly speaking, fuel is any substance that releases heat when mixed with the proper amount of oxygen. Only those materials that ignite at relatively low temperatures, burn rapidly, and are easily obtained in large quantities at relatively low prices are considered good fuels.

The value of a fuel is derived from the amount of heat released when it is burned and the heat of combustion is measured. This value is obtained when a given amount of fuel is burned under controlled conditions. The apparatus used for this purpose is called a calorimeter. The released heat is absorbed by a definite volume of water and the rise in temperature of the water is measured. The common ratings are given in British thermal units (Btu per pound) or cubic foot of the fuel burned.

When a fuel contains hydrogen, the heat given off during combustion will depend on the state of the water vapor (H_2O) formed when the hydrogen is burned. A heating value known as the higher, or gross, heating value is obtained when this vapor is condensed and the latent heat of condensation is salvaged. If this water vapor is not condensed, the latent heat of vaporization is lost; this is known as the lower, or net, heating value.

The Btu rating of gaseous fuels is given per cubic foot. The gas industry uses the standard conditions of a temperature of 60°, 30 inches of mercury pressure with a saturated condition with water vapor to determine these values.

There are two different types of gaseous fuels used as heat sources for appliances. The most familiar is natural gas. The second one is known as LP (liquefied petroleum) gas.

NATURAL GAS

Natural gas is the lightest of all petroleum products. It is usually found where oil is found, but in some cases it is found elsewhere. Theorists have long argued about the exact origin of natural gas. However, most agree that natural gas was formed during the decomposition of plant and animal remains that were buried in prehistoric times. Because these plants and animals lived during the same period as those that are presently found as fossils, natural gas is sometimes called a fossil fuel.

Both natural gas and petroleum gas are mixtures of hydrocarbons. Both are considered fossil fuels and both are composed of various chemicals obtained from the hydrogen and carbon contained in prehistoric plants and animals.

The gas industry may be broken down into the various areas of exploration, production, transmission, and distribution.

The exploration section of the industry performs the function that its name implies. The people who are employed in exploration simply explore new areas, determine the location of the gas or petroleum field, and complete the necessary reports, purchases, and other essential duties prior to the actual drilling.

When all the exploration functions are completed, the production department accomplishes the actual drilling of the well. The gas, or crude oil, is brought to the earth's surface and is blocked at this point. The well remains in this state until the gas or oil is needed.

When the need arises the transmission department receives the gas from the well at pressures ranging from 500 to 3000 pounds per square inch gauge (psig). Even with these high pressures, the resistance of the pipe and the distance covered require that booster pumps be used to transfer the gas from the well to the refinery. At the refinery, the gas passes through a drying process that removes moisture, propane, and butane. During this process most of the odor is also removed from the raw gas and an odorant is added to aid in leak detection.

After the refining processes are completed, the distribution department receives the gas through measuring gates where the number of cubic feet of gas is recorded. The gas is then passed through a series of regulators that reduce the pressure in steps. The steps are necessary to prevent the regulators from freezing and becoming inoperative.

The gas pressure is reduced to correspond with the requirements of one of two distribution systems, either the intermediate or the low-pressure system. In most cases, the low-pressure system is taken from the intermediate system. The intermediate distribution system maintains pressures ranging from 18 to 20 psig, while the low-pressure system has a pressure of approximately 8 ounces. The low-pressure system is generally used when cast-iron pipe is used for distribution. Cast-iron pipe does not seem to hold the gas as well at higher pressures as does the copper or polyethylene pipe used in the intermediate systems.

When the intermediate system is used, the gas pressure is reduced from 18 to 20 psig to $4\frac{1}{2}$ ounces where it enters the house. Both systems use gas meters at this point, but the low-pressure system does not require a regulator. Even though the 8 ounces of pressure in the low-pressure distribution system is higher than that required in the meter loop (the meter loop consists of all the components from the main line through the meter and regulator, if used), no regulator is needed because of the pressure drop through the meter loop. This pressure drop occurs because gas does not travel in a straight line but rolls instead, reducing the flow of gas by about 10 cubic feet for each turn. This rolling action also brings about the need for straightening vanes in the line directly ahead of the gas meter. If these vanes were omitted, the meter would not measure the flow of gas correctly.

On leaving the meter loop, the gas flows into the house piping. Most of the appliances used are manufactured to operate on $4\frac{1}{2}$ ounces of gas pressure; however, natural-gas furnaces are built to operate on $3\frac{1}{2}$ inches of water column of gas pressure in the furnace manifold. This requires an additional regulator at the furnace.

The measurement of small pressures requires a manometer, a U-tube, or a manifold gas pressure gauge. A pressure indicated by a water column is a very small amount. A pressure of 1 psi will support a column of water 2.31 feet high, or 27.7 inches. A pressure as low as 0.05 psi

will support a column of water 1.39 inches high (27.7×0.05).

When the pressure is measured inside a pipe that is slightly higher than atmospheric pressure, the water column in the U-tube will be depressed in one leg and pushed up in the other. See Figure 9–1. The height of the water column in the U-tube is 3.2 inches, which corresponds to a pressure of 0.116 psi inside the pipe to which the tube is connected. If there were no pressure difference between the inside and the outside of the pipe, the water in both columns would stand at exactly the same level. The pressures which are measured in gas appliance manifolds are always measured with a water gauge, which indicates the pressure in inches of water column. The use of inches of water column when measuring low pressures eliminates the need for converting measurements made with a water gauge to pounds per square inch.

Natural gas has a specific gravity of 0.65, an ignition temperature of 1100°F, and a burning temperature of 3500°F. One cubic foot of natural gas will emit from 900 to 1400 Btu per cubic foot, with the greater amount of natural gas used emitting approximately 1100 Btu per cubic foot. The Btu content of natural gas will vary from area to area. The local gas company should be consulted when the exact Btu content is desired.

Natural gas is made up of 55 to 98 percent methane (CH_4), 0.1 to 14 percent ethane (C_3H_8), and 0.5 percent carbon dioxide (CO_2). It requires 15 cubic feet of air per cubic foot of gas for proper combustion. It is lighter than air. Because methane and ethane have such low boiling points (methane −258.7°F and ethane −127.5°F, natural gas remains a gas under the pressures and temperatures encountered during its distribution. Because of the varying amounts of methane and ethane, the boiling point of natural gas will vary according to the mixture.

LIQUEFIED PETROLEUM (LP) GAS

Liquefied petroleum is both butane and propane, and in some cases it is a mixture of the two. These two fuels are refined natural gases and were developed for use in rural areas. They are transported by truck and stored in containers specifically made for LP gas installations.

Liquefied petroleum is a liquid until the vapor is drawn off. When liquefied petroleum is extracted from raw gas at the refinery, it is in the liquid state, under pressure, and remains in this state during storage and transportation. Only after the pressure is reduced, does liquefied petroleum become a gaseous fuel.

LP gas has at least one definite advantage in that it is stored in the liquid state and thus the heat is concentrated. This concentration of heat makes it economically feasible to provide service anywhere that portable cylinders can be used. The fact that 1 gallon of liquid propane becomes 36.31 cubic feet of gas when evaporated illustrates its feasibility.

When LP gas is stored in a container, it is both in the liquid and gaseous state. See Figure 9–2. To make the action of LP gas more easily understood, let us review briefly the

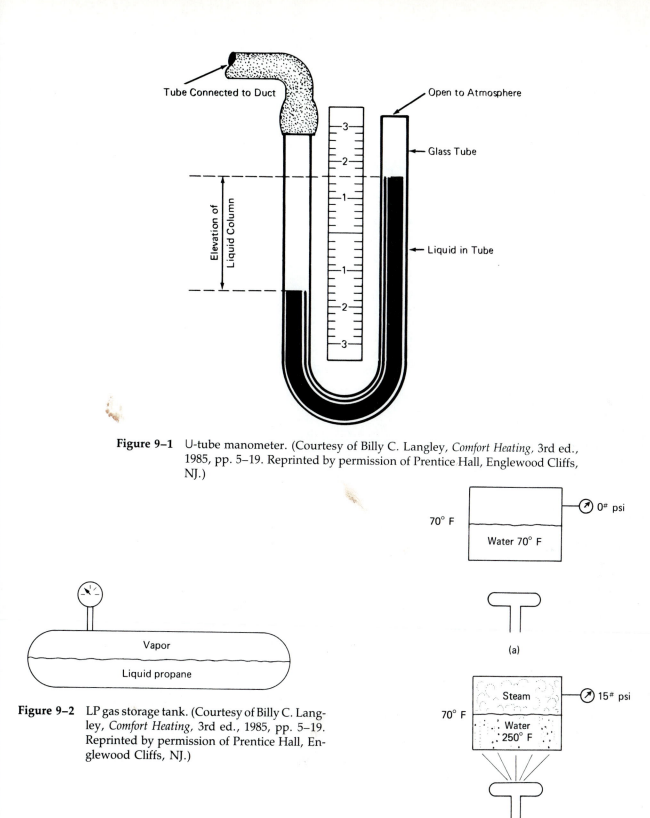

Figure 9–1 U-tube manometer. (Courtesy of Billy C. Langley, *Comfort Heating*, 3rd ed., 1985, pp. 5–19. Reprinted by permission of Prentice Hall, Englewood Cliffs, NJ.)

Figure 9–2 LP gas storage tank. (Courtesy of Billy C. Langley, *Comfort Heating*, 3rd ed., 1985, pp. 5–19. Reprinted by permission of Prentice Hall, Englewood Cliffs, NJ.)

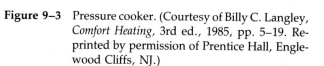

Figure 9–3 Pressure cooker. (Courtesy of Billy C. Langley, *Comfort Heating*, 3rd ed., 1985, pp. 5–19. Reprinted by permission of Prentice Hall, Englewood Cliffs, NJ.)

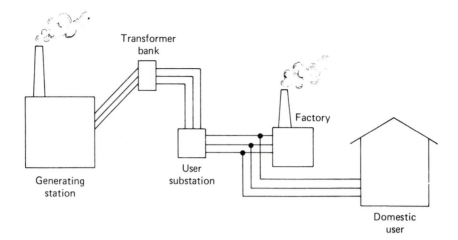

Figure 9–4 Electrical distribution. (Courtesy of Billy C. Langley, *Comfort Heating*, 3rd ed., 1985, pp. 5–19. Reprinted by permission of Prentice Hall, Englewood Cliffs, NJ.)

boiling point and pressures of water. See Figure 9–3. When the pressure cooker is first filled with water with no heat applied and the top remaining off, there is no pressure on the surface of the water. See Figure 9–4a. However, when the top is put securely in place and heat is applied, as in Figure 9–4b, the pressure will begin to rise after the boiling point of the water is reached. As more heat is applied, more pressure will be created above the water by the evaporating liquid. When a constant temperature is maintained, a corresponding pressure will also be maintained. Likewise, when the pressure is reduced, the boiling point is reduced.

If we apply this principle to LP gas in a storage tank, it can be readily seen that as vapor is withdrawn from the tank, more liquid will evaporate to replace that which was withdrawn. We must remember that each liquid had its own boiling point and pressure.

Butane (C_4H_{10}), like propane (C_3H_8) and natural gas, is placed in the hydrocarbon series of gaseous fuels, because it is composed of hydrogen and carbon. Butane has a boiling point of 31.1°F, a specific gravity of 2, and a heating value of 3267 Btu per cubic foot of vapor. It requires 30.97 cubic feet of air per cubic foot of vapor for proper combustion. At sea level it has a gauge pressure of 36.9 pounds at 100°F. Butane expands to 31.75 cubic feet of vapor per gallon of liquid. It is heavier than air. The ignition temperature is approximately 1100°F and the burning temperature is 3300°F.

Propane has a boiling point of −43.8°F, a specific gravity of 1.52, and a heating value of 2521 Btu per cubic foot of vapor. It requires 23.82 cubic feet of air per cubic foot of vapor for proper combustion. At sea level it has a gauge pressure of 175.3 pounds at 100°F. Propane expands to 36.35 cubic feet of vapor per gallon of liquid. It is also heavier than air. It has an ignition temperature of 1100°F and a burning temperature of 2975°F.

When we study the physical properties of LP gases, we can see that each has both good and bad properties. These properties should be given a great deal of consider-

Table 9–1 LP gas vapor pressures.

Temp. °F	Propane	Butane	Temp. °F	Propane	Butane
0	38.2 psig		70	124 psig	16.9 psig
10	46		80	142.8	22.9
20	55.5		90	164	29.8
30	66.3		100	187	37.5
40	78	3 psig	110	212	46.1
50	91.8	6.9	120	240	56.1
60	107.1	11.6	130	272	66.1

(Courtesy of Billy C. Langley, *Comfort Heating*, 3rd ed., 1985, pp. 5–19. Reprinted by permission of Prentice-Hall, Englewood Cliffs, NJ.)

ation when determining which fuel to use for any given application. The two characteristics deserving the most consideration are the Btu content and the vapor pressure. See Table 9–1. When considering these fuels, the pressure is the major limiting factor, especially in colder climates. As we study the table, we can see that when the temperature of liquid butane reaches 30°F or lower, there is no pressure in the tank. Therefore, butane would not be suitable as a fuel at these lower temperatures without some source of heat for the storage tank. This source of heat may be steam or hot-water pipes around the tank, electrical heaters around the tank, or even having the tank buried in the ground. However, these all add to the initial cost of the equipment.

If we look at the pressures of propane, we see that they are suitable throughout a wide range of temperatures. Therefore, from the pressure standpoint, propane would be the ideal fuel. On the other hand, the lower Btu rating makes it less desirable than butane. To overcome this dilemma, the two fuels may be mixed to obtain some of the desirable characteristics of each gas. An example of this may be a mixture of 60 percent butane and 40 percent propane. At a temperature of 30°F, the mixture will have a vapor pressure of approximately 24 psig and a heat content of 2950 Btu per cubic foot. Since these fuels are usually mixed before delivery to the local distributor, it is difficult to know exactly what the tank pressure and Btu content are. As long as there is enough pressure in the storage tank to allow 11 inches of water column of pressure to enter the house piping, there is little or nothing a service technician can do.

Before becoming too deeply involved in working with LP gases, state and local authorities should be consulted. Some states maintain strict control over the personnel working with these fuels.

ELECTRICITY

Electricity is not a new phenomenon; its existence has been known for centuries. The early applications of the heat-producing ability of electricity were limited. It was used only in a few specialized areas, mainly industrial processes and as portable heaters to help supplement inefficient heating systems. Today, electricity is used in almost every appliance and at a cost that most people can afford.

Electrical power is generated at the utility company's generating station. As it leaves the generating station, it passes through a bank of transformers to increase the voltage. From the transformer bank, the electricity is distributed to user substations. The voltage is reduced at the substation by another bank of transformers to a voltage that can be used by commercial manufacturing plants. The voltage is again reduced by the building's current transformer and is carried through the meter loop to the disconnect switch. From the disconnect switch, the electricity is distributed through the house wiring to the various appliances and electric heating units. See Figure 9–4.

The major users of electricity are electric ovens, ranges, countertop cooking units, electric motors, electric clothes dryers, and other types of resistance heaters. There are three types of resistance heating elements: (1) the open wire, (2) the open ribbon, and (3) the tubular-cased wire. See Figure 9–5. An electric resistance heating element can be defined as: An assembly consisting of a resistance wire, insulated supports, and terminals for connecting the electrical supply wire to the resistance wire. Resistance heating will convert electrical energy into heat energy at the rate of 3412 Btu per kilowatt (1000 watts). Theoretically, electric heating elements are 100 percent efficient; that is, for each Btu input to the heating element, 1 Btu in usable heat is recovered.

The open-wire heating elements are usually made of nichrome wire, which is wire made from nickel and chromium, but without iron, wound in a springlike shape and mounted in ceramic insulators to prevent electrical shorting to the metal frame. The open-wire elements have a longer life than the others because they release all the heat directly into the air stream and, thus, operate cooler.

The tubular-cased element is the one that is used in electric cooking surfaces and ovens. The nichrome wire is placed inside a tube and insulated from it by magnesium-oxide powder. Thus, a tubular-cased heating element will not require external insulation as do the other two resistance

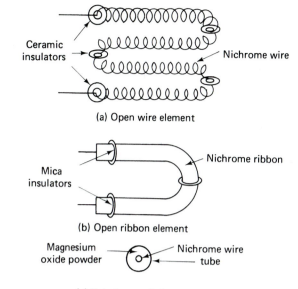

(a) Open wire element

(b) Open ribbon element

(c) Tubular cased element

Figure 9–5 Resistance elements. (Courtesy of Billy C. Langley, *Comfort Heating*, 3rd ed., 1985, pp. 5–19. Reprinted by permission of Prentice Hall, Englewood Cliffs, NJ.)

elements. It is also less efficient because of the energy loss caused by the extra material that the heat must pass through before reaching the air or the utensil being heated. It is, however, safer to use because of the interior insulation used in its manufacturing process. The tubular-cased elements have a shorter life than either of the other two types of elements because of their higher operating temperature. The control of these elements is more difficult than the others because of the extra material involved.

COMBUSTION

The available energy contained in a fuel is converted to heat energy by a process known as *combustion*. Combustion may be defined as the chemical reaction of a substance with oxygen resulting in the evolution of heat and some light.

There are three basic requirements for combustion: sufficiently high temperatures, oxygen, and fuel. See Figure 9–6. When the air-fuel mixture is admitted to the combustion chamber, some means must be provided to bring the

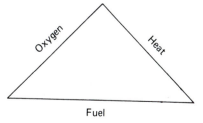

Figure 9–6 Basic combustion requirements. (Courtesy of Billy C. Langley, *Comfort Heating*, 3rd ed., 1985, pp. 5–19. Reprinted by permission of Prentice Hall, Englewood Cliffs, NJ.)

Table 9–2 Fuel gas limits of flammability.

Gas	Upper Limit	Lower Limit
Methane	14	5.3
Ethane	12.5	3.2
Natural	14	3
Propane	9.5	2.4
Butane	8.5	1.9
Manufactured	29	4

(Courtesy of Billy C. Langley, *Comfort Heating*, 3rd ed., 1985, pp. 5–19. Reprinted by permission of Prentice-Hall, Englewood Cliffs, NJ.)

mixture to its flash point. This is usually done by some type of ignition device. If, for any reason, the temperature of the gas–air mixture is reduced below its flash point, the flame will automatically go out. For example, if the temperature of a mixture of natural gas and air is reduced below its flash point of 1000°F, there will be no flame.

Also, an ample supply of properly distributed oxygen must be supplied. The oxygen requirements governing the combustion process will vary with each different fuel. They also depend on whether or not the fuel and air are properly mixed in the correct proportions.

The third requirement for combustion is the fuel. The properties of fuel were discussed earlier in this chapter. The physical properties of each fuel must be considered when determining its requirements for combustion. All of the basic requirements for combustion must be met or there will be no combustion.

An important factor to keep in mind when making adjustments involving gaseous fuels is the limits of flammability, which are stated in percentages of the gas in the air of a mixture that would allow combustion to take place. To simplify this, if there is too much gas in the air, the mixture will be too rich to burn. If there is too little gas in the air, the mixture will be too lean to burn. The upper and lower limits are shown for the more common gaseous fuels in Table 9–2.

Complete combustion can be obtained only when all of the combustible elements are oxidized by all of the oxygen with which they will combine. The products of combustion are harmless when all of the fuel is completely burned. These products are carbon dioxide (CO_2) and water vapor (H_2O).

The rate of combustion, or burning, depends on three factors:

1. The rate of reaction of the substance with the oxygen.
2. The rate at which the oxygen is supplied.
3. The temperature due to the surrounding conditions.

All the oxygen supplied to the flame is not generally used. This is commonly called *excess oxygen,* or *excess air.* This excess oxygen is expressed as a percentage, usually 50 percent, of the air required for the complete combustion of a fuel. For example, natural gas requires 10 cubic feet of air for each cubic foot of gas. When 50 percent excess air is added to this figure, the quantity of air supplied is calculated to be 15 cubic feet of air for each cubic foot of natural gas. There are several factors governing the excess-air requirements: the uniformity of air distribution and mixing, the direction of gas flow from the burner, and the height and temperature of the combustion area. Excess air constitutes a loss and should be kept to a minimum. However, it cannot usually be less than 25 to 35 percent of the air required for complete combustion.

Excess air has both good and bad effects in the combustion process. It is added as a safety factor in case the 10 cubic feet of required air is reduced for some reason, such as dirty burners, improper primary air adjustments, or a decrease in the supply of primary air. The adverse effect is that the nitrogen in the air does not change chemically and tends to reduce the burning temperature and the flue-gas temperature, thereby reducing the efficiency of the heating equipment. The air supplied for combustion contains about 79 percent nitrogen and 21 percent oxygen.

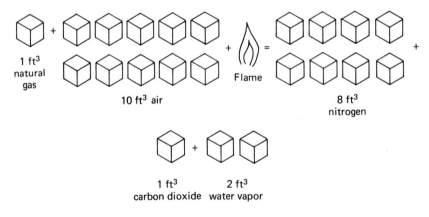

Figure 9–7 Elements of combustion. (Courtesy of Billy C. Langley, *Comfort Heating*, 3rd ed., 1985, pp. 5–19. Reprinted by permission of Prentice Hall, Englewood Cliffs, NJ.)

The products of combustion created when 1 cubic foot of natural gas is completely burned are 8 cubic feet of nitrogen, 1 cubic foot of carbon dioxide, and 2 cubic feet of water vapor. See Figure 9–7. These products are harmless to human beings. In fact, carbon dioxide is the ingredient added to water that makes soft drinks fizz.

The by-products of combustion are carbon monoxide—a deadly product; aldehyde—a colorless, inflammable, volatile liquid with a strong pungent odor and an irritant to the eyes, nose, and throat; keytones—used as paint removers; oxygen acids; glycols; and phenols. These by-products are harmful and must be safeguarded against by the proper cleaning and adjustment of the heating equipment.

Probably the most important step in maintaining good combustion is the proper adjustment of the ratio of primary air to secondary air.

REVIEW EXERCISES

1. Define a fuel.
2. What gas, that is used for a heating fuel, is the lightest?
3. Of what are both natural gas and petroleum gases mixtures?
4. On most natural gas distribution systems, what is the pressure of the gas entering the residence?
5. How high will 1 psi support a column of water?
6. Name the instruments that are used to measure gas pressure.
7. What is the ignition temperature of natural gas?
8. Name the major methods of using electricity.
9. Define combustion.
10. Name the three things that determine the rate of combustion.

10

Room Air Conditioning Units

The appliance technician must be familiar with the procedures used to install and service room air-conditioning units. The installation of these types of units is varied, depending on the make, model, and the type of structure in which the unit is to be installed. However, the servicing of room units is not as varied. Standard service procedures can be used on almost any unit with great success. Because of this we will not include the installation procedures in this book. It is recommended that the service technician refer to the installation instructions packaged with the unit being installed.

CONSTRUCTION AND OPERATION

The following is a description of the construction and operation of room air-conditioning units.

Construction

Referring to Figure 10–1, the refrigeration components of a typical room air-conditioning unit are mounted on a heavy steel base pan that is protected from the elements by a durable enamel finish. The evaporator and condenser are positioned so that they can be served by a single-fan motor.

The evaporator sits on an insulated condensate drain tray. The compressor is mounted to the base pan by welded studs that pass through the compressor base brackets. Rubber grommets are used to cushion the compressor and reduce sound transmission. Retaining nuts and washers secure the compressor to the studs.

An insulated bulkhead separates the room-side components (evaporator, evaporator blower, and electrical controls) from the outside components (compressor, condenser, condenser fan, and fan motor). An enamel-finished cabinet shell houses the entire assembly, protecting the components. The cabinet leaves the back side of the condenser exposed. It is louvered either on both sides or in the rear to provide inlet air for the condenser fan. Some models have a removable rear grille that protects the condenser without restricting the air flow over the condenser. The front of the cabinet is open, with the edges formed to accept the decorative front panel. The cabinet of larger units is partially enclosed on the bottom, providing support for the air-conditioning unit. The unit can be pulled out of the cabinet for service.

Air flow through the evaporator and the condenser is provided by a double-shaft fan motor mounted to the back side of the center partition. The condenser fan operates within a shroud enclosing the front side of the condenser. The condenser fan blows outside air through the condenser

Base pan

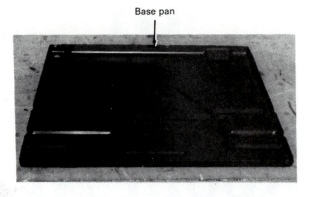

Condenser   Evaporator

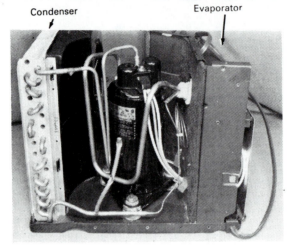

Figure 10–1 Refrigeration components of a room air-conditioning unit. (Courtesy of WCI Major Appliance Group.)

fins. On almost all models, the room air is drawn through the evaporator by a squirrel cage blower wheel or a propeller-type fan. The air is then discharged into a full-width plenum located directly above the evaporator. Occasionally, the evaporator will be located in front of the plenum. In this design, the air is blown through the evaporator. In either case, the unit is designed to provide cooling at the rate specified on the nameplate. All the exterior surfaces of the blower or fan housing and the plenum that are exposed to the outside air are insulated to prevent any loss of cooling capacity. The surfaces of the cabinet that are exposed to refrigerated air are insulated to prevent sweating (condensation).

The electric controls are mounted to a metal plate that is attached to a metal box located adjacent to the evaporator, thus providing a safety enclosure for the controls. An escutcheon, with operating instructions and control knobs, completes the control assembly.

There are several different cabinet designs to accommodate the various size, capacity, and installation conditions. Models are available with an electric heater coil mounted in the discharge plenum. The installation kits are varied for the different installations that may be encountered.

Operation

The room air conditioner is designed to circulate, ventilate, filter, cool, dehumidify, and, with some models, heat rooms in homes, offices, and other closed areas where a comfortable ambient is desired. Most models can be operated through the use of two- or three-speed fans to provide the amount and degree of air conditioning needed for maximum comfort.

Adjustable louvers allow the user to direct the conditioned air to maximize the unit's effectiveness. To rid the air of unwanted odors or to maintain a freshness in the air, some models are equipped with an exhaust door installed in the discharge air plenum. When opened, a small percentage of the recirculated air is discharged to the outside. Fresh air infiltrates the building structure to replace the exhausted air. Some louver designs provide the ability to close off a portion or all of the cold-air outlet, thereby increasing the exhaust.

Filtering is accomplished through the use of a cleanable, plastic foam filter sheet that is placed in the return air stream. This filter is either attached directly to the face of the evaporator by pressing it onto bayonet-type retainers (solid-front models), or it is held against the back side of the

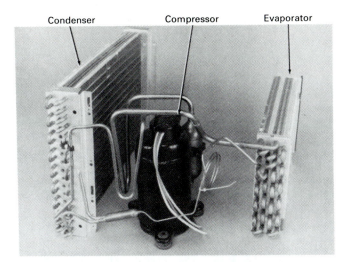

Figure 10–2 Typical sealed refrigeration system. (Courtesy of WCI Major Appliance Group.)

air intake grille of the decorative cabinet front by plastic tabs (grille-front models). Pull-and-clean filters come with products that have a tip-out front grille.

Cooling is provided by operating the compressor along with the fan. On most models, room air is drawn through the front panel, the filter, and evaporator by the evaporator fan, and discharged into the plenum and back into the room. As the air passes over the evaporator fins, it is cooled. Excessive moisture is also condensed from it. The condensate runs off the evaporator into a plastic drain tray under the evaporator. It then flows through a tube past the cabinet bulkhead and into a trough stamped into the base pan. This water is collected in a sump under the condenser fan. The condensate water is picked up by the fan and thrown onto the condenser where it evaporates, significantly increasing the efficiency and capacity of the unit. A variable rate of cooling is provided by operating the unit at the various fan speeds available.

The operation of the air conditioner is controlled by a thermostat that senses the temperature of the room air as it enters the unit. When the room air temperature drops to the cutoff setting of the thermostat, it turns off the compressor. The fan motor remains running to keep the room air circulating over the thermostat. This causes the air conditioner to respond quickly to changes in the overall room temperature. Some models feature an energy-saving switch that causes the fan to cycle off and on with the compressor. Most people find the constant air movement more comfortable.

FUNCTIONAL SYSTEMS

The following is a discussion of the functional systems of a room air-conditioning unit.

Sealed Refrigeration System

Referring to Figure 10–2, a typical sealed refrigeration system consists of a cooling coil (evaporator) located on the room side, a heat rejection coil (condenser) located outside, and an electric self-contained, motor/compressor to circulate a refrigerant through the system and to develop the necessary pressure differential to make the system work. Copper tubing, sized to carry the volume of refrigerant to be circulated, connects these components in a continuous loop—from the evaporator to the compressor to the condenser and back to the evaporator. The tube connecting the condenser to the evaporator also serves as a refrigerant flow control. This tube (restrictor tube) has a very small inside diameter. The combination of this small diameter and the extra length of this tube restricts the flow of liquid refrigerant, maintaining the pressure differential necessary for the refrigeration system to function as designed. A cone-shaped strainer is placed in the condenser outlet tube to prevent any foreign material from clogging the restrictor tube. On some large units, an automatic (expansion) valve, located at the evaporator inlet, is used for refrigerant flow control.

Operation of the sealed refrigeration system Operation of the refrigeration system involves three basic physical laws:

1. The physical state of a substance (solid, liquid, or gas) is directly related to the heat contained in the substance.
2. A large amount of heat is required to change the state of a substance (solid to liquid, liquid to gas, and vice versa) with no change in its temperature.
3. The temperature at which a substance changes from a liquid to a gas and a gas to a liquid depends on the pressure on the substance.

Figure 10–3 illustrates a typical refrigeration system, indicating the pressure, temperature, and physical state of the refrigerant throughout the system during the refrigeration cycle as described below.

In operation, the compressor lowers the pressure on the liquid refrigerant in the evaporator to the point at which

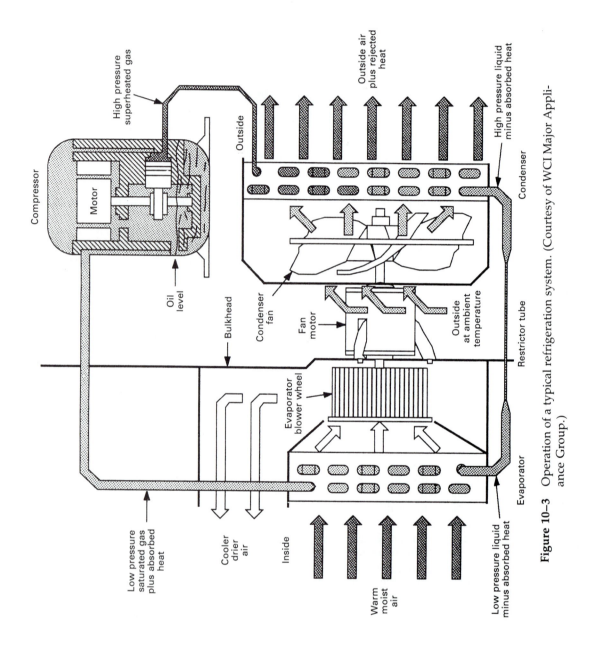

Figure 10–3 Operation of a typical refrigeration system. (Courtesy of WCI Major Appliance Group.)

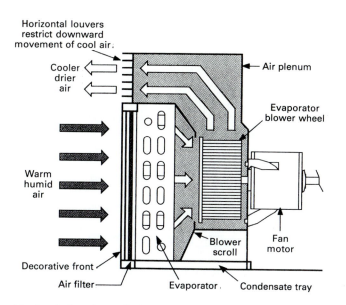

Figure 10–4 Air circulation through the evaporator. (Courtesy of WCI Major Appliance Group.)

the refrigerant will change from a liquid to a gas at temperatures ranging between 35 and 50°F. The heat required for this change is obtained from the room air circulated through the evaporator fins. Heat is absorbed by the refrigerant not only from the air, but from the moisture in the air as well. The air is cooled and the moisture condenses on the evaporator fins.

As the refrigerant liquid continues to vaporize, the pressure in the evaporator tends to rise. However, the compressor removes the vapor at a rate which maintains the desired pressure and temperature in the evaporator.

The compressor discharges the refrigerant gas into the condenser where the pressure builds up against the restrictor tube. As the pressure of the gas increases, so does its temperature. When the temperature of the gas in the condenser exceeds the temperature of the air passing through the condenser, heat is transferred from the gas to the outside air. This loss of heat causes the refrigerant gas to condense back to a liquid.

The small-diameter restrictor tube between the condenser and the evaporator maintains the pressure differential created by the compressor. It also meters the liquid refrigerant back into the evaporator. The refrigeration cycle is now complete and will continue until the compressor is turned off by the thermostat.

Air-Handling Systems

The following is a discussion of the air handling systems used on room air-conditioning units.

Room air circulation A blower wheel, or propeller fan, pulls the warm humid air through the air filter and evaporator where it is cleaned, cooled, and dehumidified. The conditioned air is then discharged through the plenum and a louvered grille into the room. See Figure 10–4. Most

models have both horizontal and vertical louvers. The downward movement of the horizontal louvers is restricted to prevent recirculating the cold air directly over the thermostat. To do so would cause short cycling of the unit. The vertical louvers, on some models, can be closed completely on one or both sides of the air discharge grille to increase the velocity of the discharge air or to maximize the exhaust feature for clearing smoke and/or odors from the room.

Most models of room air conditioners have a small exhaust door located in the discharge air plenum. This door is cable operated from the control panel.

The amount of air circulated can be varied by selecting from the fan speeds available (up to three speeds on some models). It should be noted that the position of the blower wheel or fan with the blower wheel or fan scroll is critical as to the amount of air delivered into the room.

On blow-through models, the warm humid air is pulled through the filter and discharged through the evaporator where it is cooled and dehumidified before reentering the room.

Condenser air A propeller fan pulls outside air through the cabinet-side louvers and discharges that air through the condenser, carrying away the heat removed from the room by the refrigeration system. A slinger ring (molded onto the outside of the fan blades; see Figure 10–5) or a vortex chamber (molded into the bottom part of the condenser fan shroud) enables the fan to pick up condensate water that has drained away from the evaporator and to throw it onto the condenser. This helps to cool the condenser significantly. Driven by the same motor, the condenser fan speed varies with the evaporator fan speed. On some models with solid side cabinets, the condenser does not extend completely across the cabinet. An open area adjacent to the condenser is provided through which fresh air is drawn. The air is then blown through the condenser.

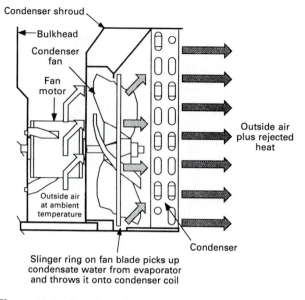

Figure 10–5 Location of the slinger ring. (Courtesy of WCI Major Appliance Group.)

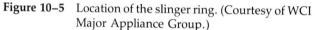

Figure 10–6 Fresh air intake. (Courtesy of WCI Major Appliance Group.)

The fins on the half of the condenser nearest to the fresh air intake are slanted away from the air intake. This directs the warm air coming from the condenser away from the intake. See Figure 10–6.

It should be noted that the position of the fan within the fan shroud is critical as to the amount of air delivered through the condenser.

Condensate Water Disposal System

The following is a description of the condensate water disposal system components and their operation:

1. *Slinger ring design* The slinger ring is attached to the outer circumference of the condenser fan. It picks up condensate water that has drained from the cooling coil and throws it onto the condenser. This disposes of the condensate water and increases the efficiency of the unit through the evaporative cooling effect on the condenser. See Figure 10–7.

2. *Vortex chamber design* A vortex chamber is molded into the bottom of part of the condenser fan shroud. Condensate water from the evaporator flows into a sump below the chamber. See Figure 10–8. The condenser fan blades create a vacuum inside the chamber forming a vortex which pulls water up from the base pan and onto the fan blades. A small amount of water is put into the air stream and onto the condenser fins, while the remainder is recycled.

Slinger ring

Figure 10–7 Operation of the slinger ring. (Courtesy of WCI Major Appliance Group.)

Note: On through-the-wall heat/cool models, a thermostatic drain valve empties the base pan of all water when the outside temperature drops below 40°F. See Figure 10–9. This prevents the fan motor from stalling should the condensate water freeze around the blades during winter operation.

FUNCTIONAL COMPONENTS

Functional components are those that cause the system to operate as designed.

Refrigeration Components

These are the components that allow the refrigerant to either circulate or cause it to be circulated to cool and dehumidify the room air as it passes through the unit.

Compressor/compressor motor A direct-drive motor/compressor assembly is spring-mounted within a welded steel shell. The motor is a permanent split capacitor (PSC) type with either an external or an internal motor overload protector. (See the Electrical Controls section presented later in this chapter.) The electrical connections to the compressor motor are attached to a glass-insulated terminal assembly that is fused into an opening in the steel shell. The compressor is equipped with suction and discharge mufflers to reduce the sound level. The entire assembly is bolted to the air-conditioner base where it is cushioned by rubber grommets.

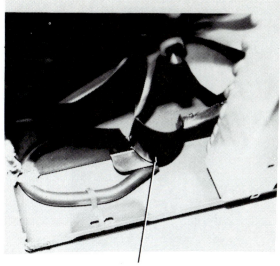

Vortex chamber

Figure 10–8 Location and operation of the vortex chamber. (Courtesy of WCI Major Appliance Group.)

Drain valve

Figure 10–9 Location and operation of the drain valve. (Courtesy of WCI Major Appliance Group.)

There are two types of compressors used in modern room air conditioners. Some models are equipped with low-side reciprocating piston-type compressors. Other models are equipped with high-side rotary compressors. An important difference to remember when servicing these two types of compressors is the operating refrigerant pressures in the compressor shell.

Reciprocating compressors operate with the low side of the suction pressure within the compressor shell. See Figure 10–10. Refrigerant vapor from the evaporator is drawn into the shell to cool the motor windings before being compressed. The high-pressure vapor is discharged directly into the condenser.

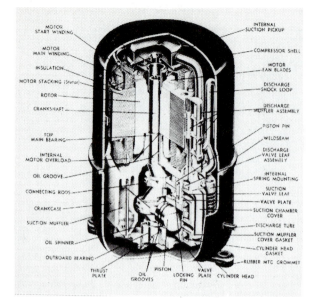

Figure 10–10 Reciprocating compressor. (Courtesy of WCI Major Appliance Group.)

Figure 10–11 Rotary compressor. (Courtesy of WCI Major Appliance Group.)

In a rotary compressor, the low-pressure refrigerant vapor from the evaporator is drawn directly into the compression chamber. See Figure 10–11. The high-pressure vapor is discharged into the compressor shell where it picks up excessive motor heat before passing into the condenser.

After an extended down time, rotary compressors take longer to start the refrigeration process than reciprocating compressors. The reason is that while the temperature of the evaporator and condenser changes with the fluctuation of ambient temperature, neither type of compressor can provide a temperature-stable mass. Since all of the refrigerant in the system condenses in the compressor shell during extended idle periods, the high-side rotary compressor depends on the heat generated by the motor and compressor to vaporize the refrigerant and pump it through the system. On the other hand, the low-side reciprocating compressor quickly lowers the pressure on the liquid, causing it to vaporize faster. The compressor immediately starts to pump it through the system.

The motors of both the reciprocating and the rotary compressors are PSC (permanent split capacitor) motors with both starting and running windings. The starting device is the running capacitor connected between the starting winding S terminal and the running winding R terminal. (See the Electrical Diagrams presented at the end of this chapter.) The capacitance (in microfarads) and voltage of the running capacitor must be rated as indicated by the manufacturer. The run capacitor must be sized according to the manufacturer's specifications, unless a starting problem is encountered. In such cases see the Start Assist Device section presented later in this chapter. Too high a capacitance rating will cause overheating of the compressor motor. It may also cause damage to the start windings because of the excessive winding current and temperature. The motor starting torque, while not as high as with a relay and a starting capacitor combination, is more than adequate for

air-conditioner operation. Also, the maintenance of a relay and starting capacitor is eliminated.

These permanent split capacitor (PSC) compressors must have the rated voltage applied for starting. The applied voltage must be within ±5 percent of the rated voltage at the moment the compressor is attempting to start.

To test the compressor, use the following procedure: The compressor can be tested by conducting a capacity test. Do not attempt to pull a vacuum to test the compressor valves. The compressor motor could be damaged.

To test the compressor motor, use the following procedure: Caution: Use extreme care in removing and replacing electrical connections to the compressor terminals. Damage to the terminal assembly can result in an expulsion of a terminal, hot refrigerant, and oil under pressure.

Note: Since the compressor is driven by a permanent split capacitor motor, the running capacitor should be tested first. Refer to the capacitor test procedure in the Electrical Controls section presented later in this chapter.

1. Disconnect the air conditioner from the electrical power.

2. Remove the electrical leads from the compressor terminals.

3. Check the compressor for grounded windings by checking from each terminal to any copper tubing connected directly to the compressor. If continuity is indicated, the motor windings are grounded and the compressor will have to be replaced.

4. Check for open windings by checking from the start terminal to the run terminal. If no continuity is indicated, one of the windings is open and the compressor will have to be replaced.

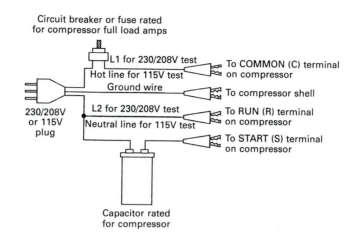

Figure 10-12 Compressor test cord. (Courtesy of WCI Major Appliance Group.)

5. For compressors with an internal overload protector: If continuity is indicated in step 4, then check from COMMON to START and COMMON to RUN. Check the resistance of the start and running windings with the information concerning the specific compressor being tested. If no continuity is indicated, the internal motor protector is open, and the compressor must be given time for the windings to cool below 172°F and then be rechecked. This may take quite some time.

After the compressor has cooled and the continuity from COMMON to START and COMMON to RUN is restored, the cause of the overloading and overheating of the compressor must be determined. Some possible causes are:

1. *Shorted winding* Recheck the resistance of the start and run windings and compare this with the information for the compressor being tested.
2. *Low voltage* Check the voltage at the electrical outlet with the air conditioner disconnected, and again at the compressor (COMMON TO RUN) when the compressor starts and while it is running. Excessive voltage drop during start-up or while it is running may be the cause of the motor protector opening.
3. *Defective run capacitor* If the compressor will not start (or starts slowly) or the compressor runs but draws high current, then test the capacitor.
4. *Short cycling on the thermostat* Note: This problem has been eliminated on models with electronic controls. The compressors are automatically prevented from restarting for 3 to 4 minutes after a shutoff by an electronic timing switch and relay.
5. Dirty condenser.
6. Insufficient air flow over the condenser from other causes.

If an internal motor protector does not reset after the motor windings have cooled below 172°F, a defective protector is indicated. The compressor will have to be replaced.

The compressor can also be operated on a test cord when connected as illustrated in Figure 10–12. *Caution:* Use only a known good electrolytic capacitor rated for the compressor being tested. The capacitor is a running capacitor. It must be in the circuit while the compressor is running.

Evaporator The evaporators used on all air-conditioner models are of the tube-and-fin type. The aluminum fins are punched to fit over staggered rows of copper tubing.

The evaporators used on energy-efficient units have refinements to improve the rate of heat transfer from the air to the fins and from the tubing to the refrigerant. The fins are embossed and louvered to present a greater surface to the passing air. See Figure 10–13. The tubes are also rifled, increasing the surface in contact with the refrigerant. This rifling causes the refrigerant to swirl as it passes through the evaporator, thus ensuring 100 percent contact between the refrigerant liquid and the interior of the evaporator tubes. See Figure 10–14.

Condenser The construction of the condenser is similar to that of the evaporator. The physical size of the condenser must be sufficient to dissipate all of the heat of compression and that of the motor as the refrigerant passes through the motor/compressor. It must do this without allowing the temperature and pressure of the refrigerant to exceed acceptable limits. The condenser is one part of the air conditioner that requires periodic maintenance. Accumulated lint and other airborne materials must be cleaned from its surface and from the surface of the fins.

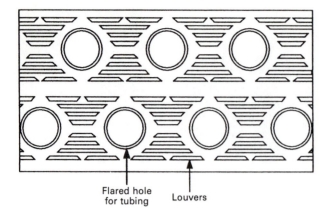

Flared hole
for tubing   Louvers

Figure 10–13 Energy-efficient evaporator tubing fin. (Courtesy of WCI Major Appliance Group.)

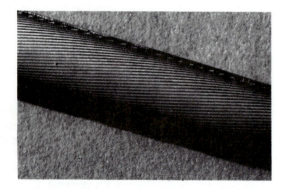

Figure 10–14 Cross section of copper tubing showing rifling. (Courtesy of WCI Major Appliance Group.)

Refrigerant filter screen On some room air-conditioning units, a filter screen is used in place of the filter drier. The refrigerant filter is installed in a T-shaped charging connector just ahead of the cap tube. See Figure 10–15.

Restrictor tube (capillary tube) The restrictor tube is a refrigerant metering device that determines the amount of high-pressure liquid refrigerant that flows from the condenser to the evaporator. This flow is dependent on the length of the tube as well as its diameter. At normal operating pressures, the feed of refrigerant into the evaporator is fairly constant. Under a heavy heat load, the condenser temperature and pressure increases causing more refrigerant to be forced through the restrictor. Under these conditions, the evaporator pressure and temperature will also increase somewhat. Under low-load conditions, the pressures and temperatures drop. The evaporator may frost heavily or ice up.

Automatic expansion valve The automatic expansion valve is a variable refrigerant metering device. See Figure 10–16. Currently there are two types used on room air conditioners. The automatic expansion valve varies the

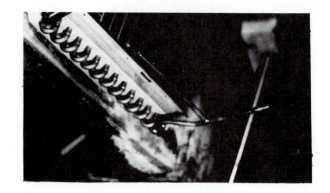

Figure 10–15 Refrigerant filter location. (Courtesy of WCI Major Appliance Group.)

Figure 10–16 Expansion valve used in 1989 air conditioner. (Courtesy of WCI Major Appliance Group.)

amount of refrigerant flow to the evaporator maintaining a constant pressure within the evaporator. The constant pressure permits the air conditioner to operate at its designed evaporator temperature under all operating conditions. This refrigerant control also provides freeze-up protection. As the evaporator pressure and temperature fall, the valve opens, permitting more refrigerant to flow into the evaporator. The increased refrigerant flow raises the suction pressure and temperature to its designed level, eliminating evaporator freeze-up. When the compressor stops, the increase in evaporator pressure causes the valve to close.

To test the automatic expansion valve, use the following procedure:

1. Disconnect the unit from the electrical power supply.
2. Remove the cabinet shell.
3. Install an adapter or tool so that the compound gauge can be attached to the suction line.

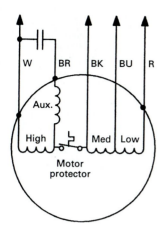

Figure 10–17 Schematic diagram of a typical multispeed fan motor. (Courtesy of WCI Major Appliance Group.)

Figure 10–18 Evaporator fans. (Courtesy of WCI Major Appliance Group.)

4. Operate the unit, observing the suction pressure. The operating pressure will vary from about 55 to 85 psig, depending on the inside and outside ambient temperatures. After a few minutes of operation with a normal charge of refrigerant, the operating pressure should remain constant as long as the ambient temperatures remain constant.

Fan motor Two types of fan motors are used. These are the shaded pole (SP) and the permanent split capacitor (PSC). The motor sizes vary from $\frac{1}{20}$ hp to $\frac{1}{3}$ hp and are rated for 120-volt ac or 240-volt ac. All motors have a drive shaft at each end for attachment of the evaporator fan and the condenser fan. All fan motors are designed for multispeed operation, either high/low or high/medium/low. Operating speeds range from 780 rpm to 1850 rpm.

Fan motors do not have start windings, as such, or starting switches. Shaded-pole motors employ a shading coil in the stator to produce the starting torque in the proper direction. When the motor is up to speed, the shading coil has little effect on the operation of the motor. Permanent-split-capacitor motors utilize an auxiliary winding and a low-microfarad electrolytic capacitor to produce the starting torque and rotational direction. When the motor is up to speed, the auxiliary winding and capacitor remain in the circuit. The capacitor is wired in parallel to the main winding, boosting the operating efficiency of the motor. The fan capacitor is often combined with the compressor running capacitor in a single-metal container. The center terminal is common to both capacitors and is shared by both the fan motor lead and the compressor motor lead.

Fan speed change is produced in the motor by changing the power connection to the motor at the selector switch. All motors are four-pole motors with a synchronous speed of 1800 rpm. The stator winding is tapped in one or two places to produce the different speeds. See Figure 10–17. The first tap is the main winding connection. Under load, the fan speed will be 1000 to 1500 rpm depending on the model. The second tap and third taps provide medium and low speed, respectively. As more turns of the winding are used, the voltage per turn decreases, reducing both the torque and the speed.

All fan motors are mounted to the bulkhead with three bolts. The evaporator fan shaft extends through the bulkhead. Rubber grommets or flexible mounting legs cushion the motor and reduce sound transmission to the bulkhead.

Fan motors are permanently lubricated and need no maintenance.

To test the fan motor, use the following procedure:

1. Disconnect the service cord from the electrical power.
2. Follow the control panel removal procedure.
3. Check the capacitor using an ohmmeter. See the capacitor test in the Electrical Controls section presented later in this chapter.
4. If the capacitor checks OK, remove the fan motor leads from the selector switch. Use an ohmmeter to check the motor windings for proper resistance and possible grounds.
5. If the windings check OK, inspect the motor for tight bearings or blockage.

Fans and Blowers There are several different types of fans and blower wheels used to move air across the evaporator and condenser of the various models. Evaporator fans operate within a scroll housing. Centrifugal (squirrel cage) blower wheels or radial-flow propeller fans draw air into the center of the scroll and discharge it from the outlet of the scroll into the plenum. See Figure 10–18.

Condenser fans are mostly axial-flow propeller fans that draw air through the cabinet-side louvers and blow it through the condenser. Axial-flow fans operate within a plastic shroud. Some have slinger rings around the blade tips to pick up the condensate water. See Figure 10–19. Air conditioners without the side louvers are equipped with radial-flow propeller-type fans that operate in a scroll. See Figure 10–20.

The efficiency of any fan is directly related to its position within its housing or shroud.

Figure 10–19   Side louver intake condenser fans. (Courtesy of WCI Major Appliance Group.)

Figure 10–20   Scroll condenser fan. (Courtesy of WCI Major Appliance Group.)

The heater on compact models extends across the width of the air outlet grill.

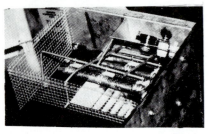

On intermediate models, the heater is concentrated in a smaller area and mounted over the evaporator blower outlet.

Figure 10–21   Heating element. (Courtesy of WCI Major Appliance Group.)

Compact Heat/Cool units are additionally protected by a solid state thermal fuse.

Intermediate models have the thermal fuse incorporated into the thermal cut-out.

Figure 10–22   Thermal cutout and fuse. (Courtesy of WCI Major Appliance Group.)

Heating Element, Thermal Cutout, and Thermal Fuse (Heat/Cool Models Only)

An open-coil heating element is used on all heat/cool air conditioners. The heating coil is strung through ring-type porcelain insulators that are supported by a steel frame. The assembly is mounted in the discharge plenum. On compact models, the heater extends across the width of the air outlet grille. On some intermediate models, the heater is concentrated in a smaller area and mounted over the evaporator blower outlet. A metal screen prevents user contact with the heating element. On heavy-duty and newer intermediate models, the heater is mounted behind the evaporator. Reflective insulation is cemented to the cabinet shell directly above the element. See Figures 10–21, 10–22, and 10–23.

The heating element is protected by a thermal cutout connected in series with the heating element. The thermal cutout is mounted on the heater frame. If the fan motor should fail or the air volume become appreciably reduced during heater operation, the thermal cutout will cycle the heater. The contacts of the thermal cutout open at 120°F or 136 ± 5°F and close at 90°F or 98 ± 6°F, depending on the model.

Note: The temperature ratings for these devices are given for information only. The actual cutout temperature cannot be accurately tested in the field. If a thermal cutout is suspected of being faulty, it should be replaced.

To test the heating element and thermal fuse, use the following procedure:

1. Check the wattage and compare the values against the values for that unit.

2. Disconnect the electrical power cord from the electrical outlet.

3. Check the resistance of the heating element.

4. Check the continuity of the thermal fuse.

To test the thermal cutout, use the following procedure:

1. Disconnect the electrical power cord from the electrical outlet.

2. Check for continuity. The contacts should be closed at room temperature.

ELECTRICAL CONTROLS

The following is a description of the electrical controls commonly used on room air-conditioning units.

Electronic Controls

Figures 10–24 through 10–29 illustrate the various electronic controls used on some models of room air-conditioning units.

A typical solid-state electronic control system is available on one or more models in most capacities. See Figure 10–24. The system features the reliability of solid-state engineering and push-button controls with indicator lights. It incorporates the temperature control selector switch, energy-saving switch, fan speed, and all other feature controls found on conventionally controlled models, plus additional controls and refinements.

Figure 10–23 Heating element on heavy-duty and modular intermediate models. (Courtesy of WCI Major Appliance Group.)

Figure 10–24 Single-sensor type I electronic control (two fan speeds, no filter check, and no frost protection). (Courtesy of WCI Major Appliance Group.)

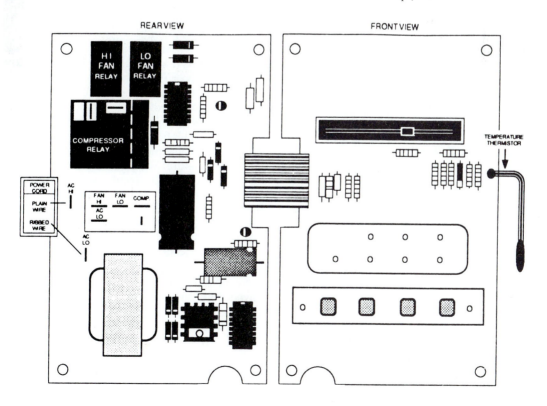

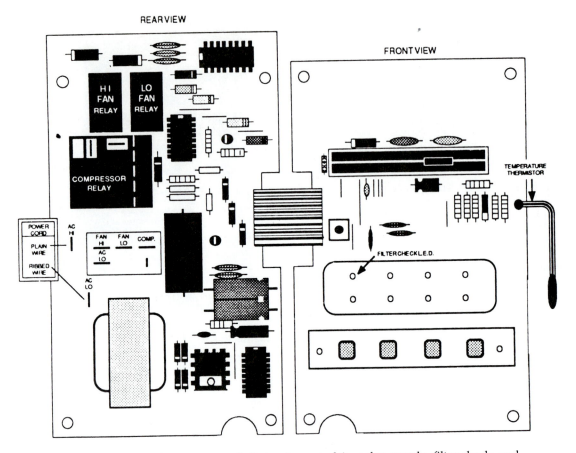

Figure 10–25 Single-sensor type I electronic control (two fan speeds, filter check, and no frost protection). (Courtesy of WCI Major Appliance Group.)

A slide temperature control lets the owner select an actual temperature setting in degrees Fahrenheit. A thermistor sensor is used that makes this control 50 percent more accurate than the thermostat used on conventional models. This reduces the amount of room temperature swing between cycles and provides a more constant level of cooling comfort. The thermistor is positioned on the lower right face of the evaporator by a plastic retainer/spacer. It senses the incoming room air at this point.

The power for the compressor and fan motor is controlled through relays mounted on the circuit boards of the control. An improved solid-state energy saver is programmed for an additional 3 minutes of fan run time after the compressor cycles off. This delivers additional cool air to the room, while warming the coil to near room temperature. The warm coil permits the temperature control to more accurately sense the true room temperature.

An automatic compressor delay circuit eliminates excessive cycling on the overload protector. The compressor will not operate for 3 to 4 minutes if the unit is turned off and then immediately turned back on. This time delay allows the pressures inside the compressor to equalize. The fan will run during this period, but there will be no cooling.

A filter check feature appears on some models. This device lights an indicator lamp after 200 hours of operation. See Figure 10–25. A filter check reset button, when pressed, restarts the electronic counter at zero hours. The owner is

instructed to reset the clock after cleaning the filter. (*Note:* The electronic counter automatically goes to zero hours after a power interruption.)

A two-sensor control is used on models when icing of the evaporator is a problem. See Figures 10–26 and 10–27. The extra sensor is attached to either the upper or lower left face of the evaporator. It will sense freeze-up due to operation when the outside temperature is below 70°F or when the air flow is reduced by a clogged filter or other obstruction. This freeze-up control is much more responsive than conventional thermostats sensing both air and evaporator temperature. See Figure 10–31.

Further improvements were made on the energy-saving feature. See Figures 10–28 and 10–29. The control now monitors room air temperature by cycling the fan on for 3 minutes every 15 minutes during the time the compressor is cycled off by the temperature control. This provides improved sensing of changing temperature conditions in the room.

A three-speed fan control and a delayed-start/delayed-stop feature are also available on some models. The start-up of the air conditioner can be delayed for up to 14 hours. It can also be operated up to 14 hours and then shut off automatically by an electronic timer.

To test the electronic control, use the following procedure: Except for a jumper test of the compressor relay, the electronic control cannot be field-tested by the technician. All

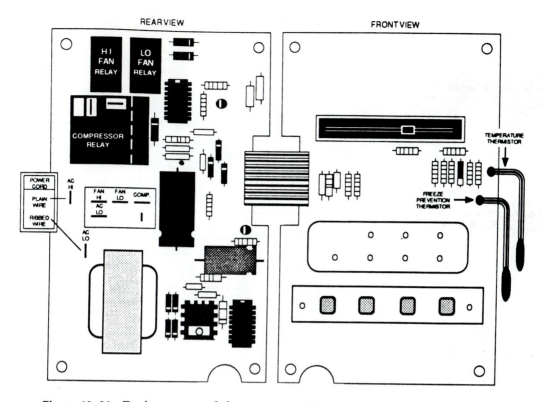

Figure 10–26 Dual-sensor type I electronic control (two fan speeds, frost protection, and no filter check). (Courtesy of WCI Major Appliance Group.)

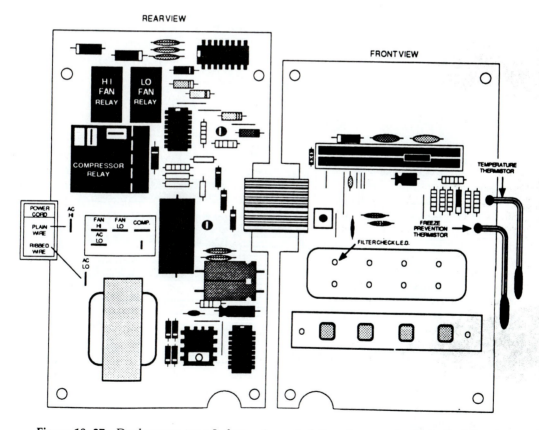

Figure 10–27 Dual-sensor type I electronic control (two fan speeds, filter check, and frost protection.) (Courtesy of WCI Major Appliance Group.)

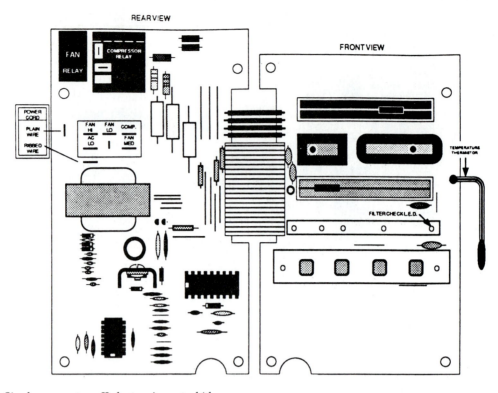

Figure 10–28 Single-sensor type II electronic control (three fan speeds, filter check, and no frost protection). (Courtesy of WCI Major Appliance Group.)

Figure 10–29 Dual-sensor type II electronic control (three fan speeds, filter check, and frost protection). (Courtesy of WCI Major Appliance Group.)

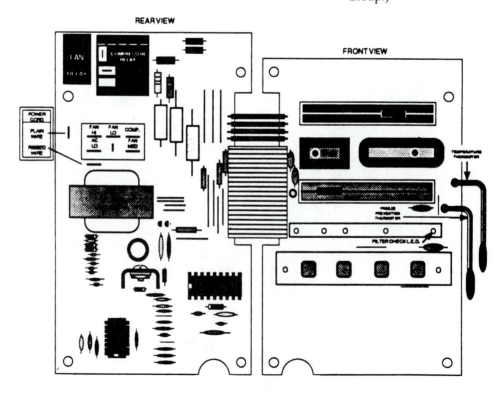

Figure 10–30 Temperature thermistor. (Courtesy of WCI Major Appliance Group.)

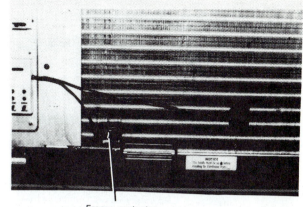

Freeze protector

Figure 10–31 Freeze protection and temperature thermistor. (Courtesy of WCI Major Appliance Group.)

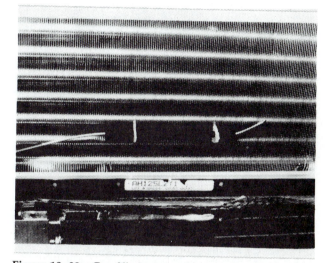

Figure 10–32 Gas-filled power element. (Courtesy of WCI Major Appliance Group.)

components operated by the electronic control, including the power cord wiring, must be tested and proven OK before the electronic control is replaced.

To test the compressor relay, use the following procedure: Refer to Figures 10–24 through 10–29.

1. Disconnect the power cord from the electrical supply.

2. Connect a jumper wire with ¼-inch female spade connectors to the relay between the COMM and N.O. terminals.

3. Plug in the unit, set the control for HI COOL and the energy saving switch to OFF. Set the temperature control to the coldest setting. If the compressor starts and operates with the jumper, replace the control.

4. If the compressor does not start, remove the jumper and check the compressor capacitor, overload, and wiring.

Be aware of the following conditions that could mislead your diagnosis: The automatic compressor restart delay prevents the compressor from restarting for 3 to 4 minutes to allow the pressure in the sealed system to equalize. When the energy-saving switch is on, the fan will cycle off with the compressor. Models with the filter check counter automatically restart at zero hours after a power interruption.

Control Thermostat

Three types of thermostats are used to control the room temperature on conventionally controlled air-conditioning units. Each thermostat consists of a switch to control the compressor, fan, and a temperature-sensitive power element, or actuator, to operate the switch. See Figures 10–30 and 10–31.

Gas-filled power element This type of thermostat is sensitive to temperature changes along the entire length of the sensing tube and at the bellows of the power element. The coldest point on the power element becomes the control point for the thermostat. The sensing tube touching the evaporator, or a cold temperature chilling the bellows, will result in short cycling of the compressor. On some low-ampere models, the capillary sensor is coiled next to the thermostat body and is located in the control box. Air from the room is drawn through the control box to activate the thermostat. See Figure 10–32.

Liquid-filled power element This type of thermostat has a liquid-filled sensing bulb at the end of the sensing tube. This makes the thermostat sensitive to temperature changes only at the bulb.

Figure 10–33 Liquid-filled power element. (Courtesy of WCI Major Appliance Group.)

The power element bellows at the switch and the capillary tube (between the bellows and the bulb) are filled with vapor that is boiled out of the liquid in the bulb by the warm room air.

The bulb is positioned on the face of the evaporator where it senses the temperature of the incoming room air. It is mounted in a plastic bulb clamp that clips into the evaporator fins. The clamp also insulates the bulb from the cold evaporator. The thermostat bulb should not touch the evaporator or other cold parts of the air conditioner.

Where liquid-filled thermostats are used on heat/cool units, a small heater is attached to the power element at the switch body. This prevents the liquid in the bulb from totally vaporizing under the influence of warm room temperatures, then condensing in the bellows at the switch body when chilled below the outside air temperatures. See Figure 10–33.

Coiled capillary This type of thermostat has a gas-filled power element coiled behind the body of the thermostat and mounted inside the control housing. Air from the room is drawn through the control housing by the evaporator fan, causing the power element to operate the bellows in the thermostat body.

Frost control Under certain operating conditions, a room air conditioner can freeze up. This means that ice has formed on the cooling coil—even though the unit is set for cooling and is operating, little air is delivered to the room.

Freeze-up may occur when the unit is set for cooling and the outside temperature drops below 70°F (particularly when the temperature control is set to the maximum cooling position).

Selected model room air conditioners have the thermostat located directly on the cooling coil. At this point it senses frost buildup and turns off the compressor to prevent further frost formation. The fan continues to operate, melting the frost accumulation. The compressor is recycled when the evaporator coil is defrosted.

Cool ambient operation Occasionally, an air conditioner will not run when the outside temperature is less than 60°F even though the room is warm. The metal in the air conditioner and the body of the thermostat can become so cold that the vapor-actuating medium in the thermostat power element condenses and cannot close the thermostat contacts. The volume of vapor and liquid in the power element bellows is so great, as compared to the volume of liquid remaining in the bulb, that thermostat operation will be controlled by the temperature of the thermostat body and not by the room-side thermostat bulb. Even heating the bulb will not expand the power element charge enough to operate the switch.

Note: If the compressor is forced to run in a very low outside ambient temperature, little cooling, if any, will occur in the room because all the refrigerant will be condensed in the compressor and condenser. The condenser pressure will build very slowly and little refrigerant will circulate through the evaporator.

To test the thermostat (cool-only models), use the following procedure:

1. Disconnect the service cord from the electrical power.
2. Remove the cabinet front.
3. Remove the control panel to gain access.
4. Remove the wires from the thermostat.
5. Attach an ohmmeter to the thermostat terminals.
6. Warm the sensor—the contacts should close (0 ohms).
7. Cool the sensor (use ice or cold water)—the contacts should open (infinite ohms).

To test the control thermostat (heat/cool models), use the following procedure:

1. Disconnect the service cord from the electrical supply.
2. Turn the thermostat to the coldest position and check the continuity across terminals 1 and 2. Continuity should exist.
3. Check the continuity across terminals 2 and 3. No continuity should exist.
4. Turn the thermostat to the warmest position and cool the thermostat bulb. Check the continuity across terminals 2 and 3. Continuity should now exist.
5. With the thermostat bulb still cold, check the continuity across terminals 1 and 2. No continuity should exist.
6. If the thermostat does not check as outlined above, replace the thermostat. (*Note:* The thermostat bulb must be mounted with the tip up.)

Control Thermostat Calibration

The following is a description of the procedure used to calibrate the control thermostat.

Operating range of the thermostat The operating range is the difference between the cutout temperature at the coldest thermostat setting and the cutout temperature at the warmest thermostat setting. This difference usually represents an approximately 20 to 21°F temperature change in a 180° rotational movement of the thermostat shaft, or a change equal to approximately 5°F for every 45° rotation. For example, if the thermostat has a cutout temperature between 58 to 64°F at its coldest setting, then the cutout temperature at its warmest setting should be 20 to 21°F warmer, or 78 to 84°F to 79 to 85°F.

Operating differential of the thermostat Determine the cut-in and the cutout temperature of the thermostat by sensing the return-air dry-bulb temperature with an accurate temperature tester. The temperature should be sensed ¼ to ½ inch away from the decorative front, along the lower right half of the air inlet. Average the readings.

The difference between the cut-in and the cutout temperature is the operating differential of the thermostat. If this difference is less than or greater than the manufacturer's recommended differential, the control should be replaced. For example, if the temperature sensed at the cutout on the thermostat is 79°F, then the cut-in should occur when the inlet air temperature increases 2.5 to 5°F.

For proper performance, it is important that the sensing bulb of the thermostat is positioned correctly. On most models, the bulb is positioned ¼ to ⅜ inch in front of the evaporator. If the thermostat capillary line or bulb is too close or touching the evaporator, it will sense the coil temperature directly instead of the air temperature, causing the compressor to short cycle.

To test the thermostat using a water bath, use the following procedure: Submerge 6 to 8 inches of the capillary line, or the bulb, in a water bath. Set the thermostat at the coldest position. Starting at approximately 75°F, slowly cool the bath and check the cutoff temperature. Raise the bath temperature slowly and check the differential. Repeat the test to confirm the results.

Selector Switch

The selector switch is a multicontact switch that is used to complete selected circuits to the fan motor and the compressor (and to the heater on heat/cool models).

Several combinations of fan speed and compressor operation provide for HI and LO cooling, HI and LO fan, Energy Saving, and LO heat (heat/cool models). Medium fan speed is available on some models. Rotary, push-button, and slider switches are used. The rotary and push-button switches have gauged contacts, whereas the slide switches are single pole with one, two, or three positions.

To test the selector switch, use the following procedure:

1. Disconnect the service cord from the electrical power.
2. Remove the cabinet front.
3. Remove the control panel and turn it outward to gain access to the selector switch.
4. Remove the wires from the switch.
5. With an ohmmeter, check the continuity of each position.

Figure 10–34 Example of an externally mounted motor protector. (Courtesy of WCI Major Appliance Group.)

Refer to the unit electrical schematic to identify the terminals involved with each position.

Motor Protectors

Two types of motor protectors are used to protect the motors of an air conditioner against damage from electrical and mechanical overload. They are the external and the internal overload.

External overload Refer to Figure 10–34. The externally mounted motor protector is a bimetal, self-resetting switch. It is mounted on the compressor shell. It is actuated by a heater through which all compressor motor current flows. Shell and ambient temperatures have some influence on its operation. Excessive motor current and excessive shell temperature will cause the protector to open. Defective external overload protectors can be replaced. Be sure to use an exact replacement for proper protection of the motor.

Internal motor protector An internal motor protector is used to protect all fan motors and some compressor motors. This device is actuated by excessive motor current and/or temperature. It is embedded in the motor windings. It is connected in series with the common wire to the motor. This application makes it sensitive to the current draw and temperature of both the main and phase windings. When the protector opens from excessive current draw, it will reset automatically after a short time. After repeated operation, caused by prolonged overload or repeated attempts to start, the temperature-sensitive portion of the protector will extend the time required to reset.

Cycling on the motor protector When the compressor or fan motor becomes overheated and/or draws too much current, the motor protector trips open and turns off the motor. If this happens repeatedly, the motor is said to be cycling on the motor protector or the overload.

Cycling on the motor protector may be caused by

1. Insufficient air circulation through the condenser. Be certain that the condenser is clean. Shine a light through the fins. Check the entire condenser.

2. Low line voltage. Check the voltage at the compressor terminals while the unit is running and while it is trying to start.

3. Compressor stalling due to the system pressure not unloading.

If, while the air conditioner is operating, the compressor is turned off, then quickly back on, the overload protector contacts will open, breaking the circuit to the compressor common wire. This occurs because the high-side and the low-side refrigerant pressures of the system have not had time to equalize. Approximately 3 minutes are required for the pressures to equalize enough for the compressor to start. The previous condition can also be caused if the room temperature thermostat is short cycling because of too narrow a temperature differential. This condition can also occur because the sensing tube of the thermostat is being precooled by touching the evaporator.

To test the compressor external motor overload protector, use the following procedure:

1. Disconnect the service cord from the electrical power.

2. Remove the compressor terminal cover.

3. Remove the wire connected to the motor protector from the control panel.

4. Check for continuity across the terminals of the overload protector. Continuity should exist at ambient temperature.

5. If the motor protector is open at ambient temperature, replace it.

To test the compressor internal motor protector, use the following procedure: Refer to the compressor motor test procedure in the Functional Components section presented later in this chapter.

To test the fan motor internal motor protector, use the following procedure: If the fan motor will not run, make the following test with an ohmmeter to determine if the motor is defective or the internal protector is open.

1. Disconnect the service cord from the electrical power.

2. Remove the control panel cover and rotate it to gain access to the fan motor leads.

3. Disconnect the red and white fan motor leads. Check the motor windings for continuity.

4. If no continuity is found, the protector, or a winding, is open.

5. Allow the fan motor to cool and recheck it. If the circuit through the protector and windings remains open, replace the motor.

6. If continuity is found, check each winding in the motor and compare it to the motor manufacturer's specifications.

Note: If the motor is warm, the resistance of the windings will increase somewhat. A shorted winding will read lower resistance.

Running Capacitors

Running capacitors are used on some compressors and some fan motors to increase the efficiency of the motor by improving the power factor. They are connected in series with the phase or auxiliary winding of the motor that operates in parallel with the main or run winding. The running capacitor provides enough electrical phase shift in the phase winding to develop a starting torque. During the initial acceleration of the motor, the phase winding and the running capacitor perform the same function as the start winding on ordinary electric motors. However, when the motor is up to speed, they may remain in the circuit for efficiency and lower cost of operation.

Note: The names phase winding, auxiliary winding, and start winding are used interchangeably throughout the trade, as are running winding and main winding.

Running capacitors differ from starting capacitors in several ways. They are constructed with a metal case and contain a liquid electrolyte. The microfarad rating is relatively low, but the rated voltage is substantially higher than the line voltage. When both the compressor and the fan motor require running capacitors, both capacitors may be packaged in the same container. These dual capacitors have three terminals.

Always replace running capacitors with like type and microfarad rating. Particular attention must be paid to the voltage rating. (*Note:* A 5-microfarad increase is permissible on compressor run capacitors where periodic or marginal voltage is experienced.)

To test the running capacitor, use the following procedure: Caution: An internally shorted capacitor may explode if energized with line voltage. Check it only with an ohmmeter or a capacitor tester. After testing, always discharge the capacitor with a 20,000-ohm, 2-watt resistor placed across the terminals.

1. Disconnect the service cord from the electrical power.

2. Remove the front panel (cabinet shell on slider models), then remove the control panel.

3. Discharge the capacitor. (*Note:* The best method of discharging a capacitor is with an insulated copper wire in

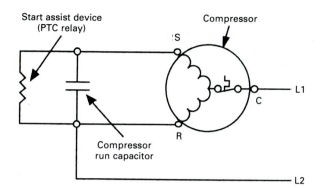

Figure 10–35 Schematic diagram of PTC relay installation. (Courtesy of WCI Major Appliance Group.)

series with the 20,000-ohm, 2-watt resistor. Place this high-resistance jumper across the capacitor terminals.)

4. Disconnect the capacitor wiring.

5. Connect an ohmmeter across the terminals of the capacitor to be checked:

   a. If the capacitor is good, the needle should jump toward 0 ohms and quickly drop back to infinity.

   b. If the needle does not move, the capacitor is open.

   c. If the needle reads a constant value at or near 0 ohms, the capacitor is shorted.

   d. If the needle jumps toward zero and falls back to a constant high-resistance value (not infinity), the capacitor has a high-resistance leak.

   e. Check for a ground from each terminal of the capacitor to the bare metal of the capacitor case. The resistance should be infinite.

   f. If the resistance is less than infinite, the capacitor is grounded.

Replace the capacitor if it is open, shorted, grounded, or has a high-resistance leak.

Start Assist Device: PTC Relay

The start assist kit is to be used under conditions of low voltage supply or tight compressor assemblies. The device is essentially a solid-state relay (PTC thermistor) that has a very steep electrical resistance curve relative to its temperature. At room temperature, its resistance is low. At a higher temperature, its resistance increases sharply. This device is connected across the terminals of the compressor run capacitor as shown in Figure 10–35.

At room temperature, its resistance is about 50 ohms. When the electrical power is supplied to the compressor, the capacitor is effectively shorted out of the circuit and the phase winding receives a large current that develops the required starting torque. After approximately $\frac{1}{5}$ second, the current through the thermistor heats the device and raises its resistance to about 80,000 ohms. At this point, the thermistor has no effect on the circuit and the compressor runs normally with the phase winding operating in series with the run capacitor. Since the device is across the terminals of the capacitor, a slight current trickles through the thermis-

tor. It, therefore, stays at a high temperature and a high resistance. As long as the compressor runs normally, it has no further effect on the circuit. When the compressor circuit is opened, the thermistor cools down, reverts back to its low resistance, and is ready for the next start.

Installing the PTC relay This relay can be mounted in any position. No additional capacitor is needed.

To install the PTC relay, use the following procedure:

1. Disconnect the service cord from the electrical power.

2. Fasten the mounting bracket to any existing hole in the unit control compartment.

3. Snap the relay into the mounting bracket.

Caution: Maintain a 5-inch clearance between the terminal spade connectors and the unit metal parts.

4. Connect one lead between one end of the relay and one side of the run capacitor. (*Note:* If the unit has a dual capacitor, connect the PTC relay to the high-capacity side.)

5. Connect the other lead between the other end of the relay and the other side of the compressor run capacitor. (*Note:* The connecting leads are equipped with piggyback spade connectors to ensure available terminals at the compressor capacitor.)

PARTS REPLACEMENT OPERATIONS

Caution: Review the safe-servicing procedures in the front of this book before attempting these parts replacement procedures and repairs.

Control Replacement (All Models)

To replace the control on all models, use the following procedure:

1. Disconnect the service cord from the electrical power.

2. Follow the control panel removal procedure to gain access to the selector switch, thermostat, solid-state control module, and the exhaust air control.

3. Remove the mounting screws, the electrical and grounding lead connections. Detach the thermostat sensor from the evaporator.

4. Reverse the above procedure to reassemble the unit. Check all wiring connections for tightness. Check all grounding provisions.

Solid-State Control Module

To remove the solid-state control module, use the following procedure:

1. Disconnect the service cord from the electrical power.

2. Remove the control panel.

3. Disconnect the electrical connector from the module. See Figure 10–36.

Figure 10–36 Solid-state control panel removal. (Courtesy of WCI Major Appliance Group.)

Figure 10–37 Auto-sweep motor. (Courtesy of WCI Major Appliance Group.)

4. Disconnect the line cord to the module (note the location of the ribbed wire).

5. Disconnect the sensor from its retainer on the evaporator.

6. Remove the four hex-head screws that secure the module to the control panel.

7. Reverse the above procedure to reassemble the unit.

Auto-Sweep Motor (Some Models)

To service the auto-sweep motor, use the following procedure:

1. Disconnect the service cord from the electrical power.

2. Follow the Control Panel Removal procedure to gain access to the auto-sweep motor.

3. Remove the mounting screws, and the electrical and ground lead connections.

4. Reverse the above procedure to reassemble the unit. Check all wiring connections for tightness. Check the safety grounding provisions. See Figure 10–37.

Fan Motor (All Models)

Servicing the fan motor requires several steps, depending on the type of chassis involved. A general procedure will be presented here followed by specific requirements for other models.

General procedure Slight variations will be encountered on some models. Refer to the specific procedure at such points.

1. Disconnect the service cord from the electrical power.

2. Remove the cabinet shell.

3. Remove the control panel. Disconnect the fan motor electrical leads.

4. Remove the screws securing the condenser shroud, or release the tabs holding the shroud. On heavy-duty models, remove the upper half of the shroud. On models with solid side cabinets, leave the shroud attached to the condenser.

Figure 10–38 Positioning condenser for access to fan motor. (Courtesy of WCI Major Appliance Group.)

Figure 10–39 Positioning evaporator for access to fan motor. (Courtesy of WCI Major Appliance Group.)

Compact and standard intermediate models
To service these units, use the following procedure:

1. Carefully lift the condenser over the base pan flange. Do not overstress or kink the refrigeration tubing. See Figure 10–38.
2. Loosen the screw or clamp on the condenser fan hub. Remove the fan shroud.
3. If a blower wheel is used on the evaporator side, proceed to step 11. If a propeller-type evaporator fan is used, proceed to step 4.
4. Remove both the evaporator top cover and the vertical partition between the control box and the evaporator.

5. Remove the screws and brace (top left) securing the evaporator. Carefully lift the evaporator to clear the drain tray and move it to the left to free the right end. Pull the evaporator forward over the edge of the base pan. See Figure 10–39. If possible, work the refrigerant tubing through the bulkhead to avoid bending it.
6. Loosen the clamp on the evaporator fan hub. Proceed to step 8.
7. Loosen the set screw of the clamp on the blower wheel hub. See Figure 10–40.
8. Remove the fan motor ground wire. Free the motor leads from the clips and the control box.
9. Remove the fan motor mounting bolts from the condenser side of the bulkhead. Remove the fan motor (the fan or blower wheel will remain in the fan housing). *Note:* If nuts are found on the evaporator side, it may be necessary to move the evaporator—see steps 4 and 5 above.

Figure 10–40 Evaporator blower wheel removal. (Courtesy of WCI Major Appliance Group.)

10. Reverse the above procedure to reassemble the unit.

11. Rewire the fan motor per the instructions included with the new fan motor. Make sure that the green wire is attached to the new motor. Reseal the wiring and tubing at the bulkhead.

Intermediate modular chassis models *To service this type of unit, use the following procedure:*

1. Remove the screws holding the bulkhead to the base. Lift out the bulkhead with the condenser shroud and the fan motor attached to it.

2. Remove the blower wheel and the condenser fan. Remove the motor.

3. Reverse the above procedure to reassemble the unit.

Heavy-duty models *To service this type of unit, use the following procedure:*

1. Remove the screws holding the bulkhead to the base. Lift out the bulkhead with the fan motor attached to it.

2. Remove the blower wheel and the condenser fan. Remove the motor.

3. Reverse the above procedure to reassemble the unit.

Evaporator Fan (Blower Wheel)

This is a general procedure. Slight variations will be encountered on some models. The procedure at such points will be obvious.

1. Disconnect the service cord from the electrical power.

2. Remove the cabinet shell.

3. Remove the control panel and the evaporator top cover.

4. On intermediate models, remove the screws holding the bulkhead to the base. Release the clips holding the condenser shroud. Lift out the bulkhead, fan motor, and shroud. Remove the blower wheel clamp and the blower wheel.

5. On compact models remove the screws and the brace (top left) securing the evaporator. Remove the vertical partition between the control box and the evaporator.

6. Carefully lift the evaporator to clear the drain tray. Move it to the left to free the right end. Pull the evaporator forward over the edge of the base pan. See Figure 10–39. If possible, work the refrigerant tubing through the bulkhead to avoid bending it.

7. Loosen the clamp on the fan (or blower wheel) hub. Remove the fan housing and the fan.

8. On heavy-duty models, remove the screws holding the bulkhead to the base. Remove the top half of the condenser shroud. Lift out the bulkhead with the motor attached. Remove the clamp and the blower wheel.

9. Reverse the above procedure to reassemble the unit.

Condenser Fan

To service the condenser fan, use the following procedures:

1. Disconnect the service cord from the electrical power.

2. Remove the cabinet shell.

3. Remove the screws securing the condenser shroud. Move the shroud toward the bulkhead and pull the locking tabs from the condenser end plate.

Note: On models with solid side cabinets, leave the fan shroud attached to the condenser.

4. Carefully lift the condenser over the base pan flange. Position the condenser outside the base. Do not overstress or kink the refrigerant tubing. See Figures 10–39 and 10–40.

5. On heavy-duty models remove the top half of the condenser shroud.

6. Loosen the set screw or clamp on the condenser fan hub. Remove the fan.

7. Reverse the above procedure to reassemble the unit.

Evaporator Tray (All Models)

This is a general procedure. Slight variations will be encountered on some models.

1. Disconnect the service cord from the electrical power.

2. Remove the cabinet shell.

3. Remove the control panel and the evaporator top cover.

4. Remove the screws and the brace (top left) securing the evaporator. Remove the vertical partition between the control box and the evaporator.

5. Lift the evaporator to clear the tray and move it to the left to free the right end. Pull the evaporator forward over the edge of the base pan. See Figure 10–39. If possible, work the refrigerant tubing through the bulkhead to avoid bending it. *Note:* For heavy-duty chassis, proceed to step 7.

6. Support the evaporator with a wood block so that the tray can be removed. Slowly pull the tray forward to disconnect it from the drain tube. Proceed to step 8.

7. On heavy-duty chassis:
   a. Remove the evaporator fan housing.
   b. Disconnect the drain tube between the evaporator tray and the condenser tray.
   c. Remove the screws holding the right side of the bulkhead to the base pan. Lift the bulkhead to clear the drain tray.
   d. Lift the front of the tray, slide it to the left and lift it out.

8. Reverse the above procedure to reassemble the unit. Make certain that the blower wheel does not rub on the evaporator fan housing. Reseal the bulkhead to the base pan and the refrigerant tubing at the bulkhead.

Figure 10–41 Drain valve, base pan replacement. (Courtesy of WCI Major Appliance Group.)

Thermal cut-out

Thermal fuse

Thermal cut-out

Figure 10–42 Heat/cool model, thermal fuse, thermal cut-out. (Courtesy of WCI Major Appliance Group.)

Condenser Tray (All Models)

To service these units, use the following procedure:

1. Disconnect the service cord from the electrical power.
2. Remove the unit from the cabinet shell.
3. Remove the screws securing the condenser shroud and the condenser.
4. Carefully lift the condenser over the base pan flange. Position the condenser outside the base. Do not overstress or kink the refrigerant tubing.
5. Loosen the set screw or the clamp on the condenser fan hub. Remove the fan and the fan shroud.
6. Disconnect the evaporator tray drain hose from the condenser tray. Remove the tray.
7. Reverse the above procedure to reassemble the unit.

Base Pan Drain Valve (Heat/Cool Models)

To service these components, use the following procedure:

1. Disconnect the service cord from the electrical power.
2. Remove the cabinet shell.
3. Remove the valve mounting screws to the base pan. See Figure 10–41.
4. Clean the trough area of the base pan.
5. Install the new valve assembly.
6. Reverse the above procedure to reassemble the unit.

Thermal Fuse, Thermal Cut-out, and Heating Element (Heat/Cool Models)

See Figure 10–42. *Note:* An external thermal fuse is used on compact models.

1. Disconnect the service cord from the electrical power.
2. Remove the cabinet shell.
3. Remove the protective screen.
4. Remove the heater lead wires.
5. Remove the heater.
6. Use an ohmmeter and check the thermal fuse and/or thermal cutout for continuity.
7. Visually check the heater element for an open coil.
8. Replace the parts as needed. Make sure that all the connections are tight. Make sure that the thermal fuse is positioned away from the heater lead.
9. Reverse the above procedure to reassemble the unit.

Exhaust Air Door

Note: On compact and heavy-duty cabinets, the shell must be removed to make repairs to the exhaust air door.

The actuating cable and/or the control lever replacement or adjustment can be performed through the discharge

Figure 10–43 Side-mounted type (reciprocating compressor). (Courtesy of WCI Major Appliance Group.)

Figure 10–44 Top-mounted type (rotary compressor). (Courtesy of WCI Major Appliance Group.)

air plenum, except on blow-through models. On blow-through models, the exhaust door is located behind the evaporator coil.

The exhaust air door control knob and/or cable can be serviced by removing the control panel.

Compressor Motor Protector (External Protector Only)

To service these units, use the following procedure (See Figure 10–43):

1. Disconnect the service cord from the electrical power.
2. Remove the cabinet shell.
3. Remove the compressor retainer and terminal cover.
4. Remove the protector lead from the compressor terminal.
5. Remove the remaining electrical lead from the protector terminal.
6. Remove the protector.
7. Reverse the above procedure to reassemble the unit.

Top-mounted type (rotary compressor)
To service these components, use the following procedure See Figure 10–44.:

1. Disconnect the service cord from the electrical power.
2. Remove the chassis from the cabinet shell.
3. Remove the compressor retainer and terminal cover.
4. Remove the protector lead from the compressor terminal.
5. Remove the remaining lead from the protector terminal.
6. Remove the protector.
7. Reverse the above procedure to reassemble the unit. Make certain that all the electrical connections are tight.

Adjusting the Condenser Fan Blade (Intermediate Modular Chassis)

To adjust the condenser fan blade on intermediate modular chassis, remove the unit from the cabinet shell. Locate a round plug on top of the condenser shroud. See Figure 10–45. Cut the membrane on each side of the plug and push on one side to roll the plug out of the opening. Insert a screwdriver blade through the opening and loosen the clamp on the condenser fan blade. After adjusting the blade, tighten the clamp and cover the hole with a piece of two-inch-wide ducttape.

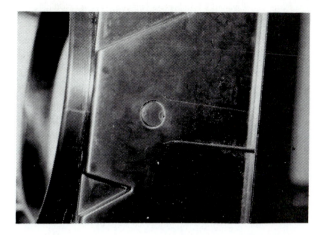

Figure 10–45 Adjusting the condenser fan blade. (Courtesy of WCI Major Appliance Group.)

TROUBLESHOOTING

Caution: Review the safe-servicing procedures in the front of this book before attempting these diagnosis and repair procedures.

Air-Conditioner Voltage Limits

Nameplate Rating	Minimum	Maximum
120 volts ac	103.5 volts ac	126.5 volts ac
240 volts ac	207 volts ac	253 volts ac
208/240 volts ac	197.5 volts ac	253 volts ac

Low Voltage

Low voltage is a common cause of trouble in the operation of any room air conditioner. It becomes doubly important, because of the motor size, that the service technician checks the voltages when servicing room air conditioners.

Improper voltage may result in one or more of the following complaints:

1. Unit will not start.
2. Compressor motor cycling on the motor protector.
3. Premature failure of the motor protector.

4. Blown fuses.
5. Premature failure of the compressor or the fan motor.
6. Noticeable dimming of lights when the air conditioner is running.
7. Evaporator icing. Low voltage may reduce the fan speed resulting in an inadequate air flow over the evaporator, thereby allowing it to ice up.

Often, low voltage can be attributed to the use of extension cords or an inadequately wired circuit. However, low voltage into the building and loose fuses or connections in the power supply should not be overlooked. Low voltage may also be a general condition in the area (a responsibility of the power company).

All units will start and run on the minimum voltage stated on their nameplate and they will perform satisfactorily if the voltage remains constant. Low voltage caused by defective wiring will not remain constant under a load.

Testing for low voltage should be done with a reliable voltmeter with a capacity to measure the required voltage. Measurements should be taken at the electrical service entrance and at the electrical outlet serving the air conditioner. Readings should be taken with the unit off, while the unit is starting, and again while the unit is running. The lowest reading should not drop below the lowest value listed on the unit nameplate.

High Voltage

High voltage can be as equally troublesome by causing motors to overheat, cycle on their protectors, or break down electrically. This problem can be solved by the power company.

Electronic Control

This type of control is not repairable. If any component on the control is defective, the entire control must be replaced.

Important note: Repair or replace any malfunctioning line voltage component before testing or replacing the electronic control. Do not assume a service problem is directly caused by the electronic control system. A line voltage component (including the power cord and wiring) that has opened, shorted, grounded, or otherwise malfunctioned may have created the service problem.

11

Refrigerators

In most modern refrigerators, the refrigerator and freezer are combined into one single cabinet. The cabinet may be either one single door or there may be two or more doors, depending on the cabinet design and the use for which it was intended. The cabinets are designed to be either free-standing or they may be built into a wall.

The freezer section of the unit may be either above, below, or beside the refrigerator section, depending on the cabinet design. Usually when the freezer is either above or below the refrigerator section, it is very small and will only hold a small amount of frozen food. Most of the side-by-side cabinets have a much larger freezer section which will hold much more frozen food and perhaps an automatic ice maker.

REFRIGERATOR CABINET

A domestic refrigerator is made up of a specially designed and insulated storage box. The shell of the cabinet is constructed of steel which is finished on the exterior to match modern residential interior-design colors. The inside of the refrigerator is known as the liner, and it is usually made from a type of material that will not impart foul tastes to the stored foods. Between the shell and the liner is an insulation that retards the flow of heat from the exterior to the interior of the refrigerator. The insulation is made from a material that has a low heat transfer coefficient, reducing the amount of energy required to keep the interior of the refrigerator at the desired temperature.

The refrigerator cabinet is equipped with fittings that allow the door to close properly and reduce the amount of heat that will leak into the refrigerator. There are also door-activated switches that will start or stop a fan motor when the door is opened or closed, and there are types that will turn on the interior light when the door is opened and turn off the light when the door is closed. There may be other features that will operate from the door switches depending on the unit design and functions.

We will use the Frigidaire F line models of refrigerators for a more detailed description of the refrigerator cabinet and components, as well as service and repair procedures.

INSTALLATION

The refrigerator may be installed wherever practical in the home, but several factors should be considered, as follows:

1. The floor should be firm, level, and capable of supporting the loaded weight of the cabinet without vibration. If the floor is not strong enough, it should be reinforced with ¾-inch plywood placed under the refrigerator for added support.

2. Refrigerators are designed to operate efficiently at room temperatures between 60 and 110°F. If the refrigerator is installed where the temperature will be below 60°F, the operating efficiency is affected. This may result in user complaints. Therefore, installations in ambient temperatures below 60°F are not recommended.

3. If possible, avoid locating the refrigerator next to a stove, radiator, or hot-air register. Similarly, an installation that exposes the refrigerator to the direct rays of the sun is undesirable. The user should be advised that although the refrigerator is capable of compensating for these handicaps, it will run longer and consequently use more electricity.

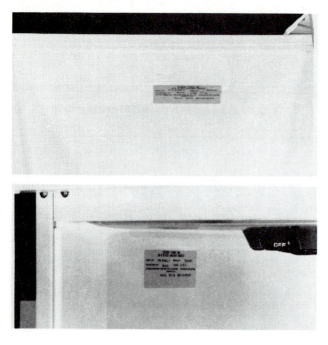

Figure 11–1 Serial plate location. (Courtesy of WCI Major Appliance Group.)

Model and Serial Number

The model, serial number, and the amount of refrigerant charge are located on a nameplate located on the refrigerator. The model and serial number must be written down for future reference concerning any warranty questions that might arise. The amount of refrigerant charge is necessary for properly charging the unit should refrigerant cycle repairs be necessary. See Figure 11–1. Each manufacturer has different locations for these plates, but they must be located and the information on them noted.

Leveling Legs

Some manufacturers provide screw-type levelers which are assembled to the front corners of the cabinet. The rear cabinet base is extruded to form rear sliders. The screw-type levelers may be adjusted up or down by turning them either to the left or to the right. These adjustments should be made at the time of the installation to prevent an unnecessary callback.

Rollers

Some manufacturers equip their refrigerators with four rollers. To adjust the front rollers, first remove the kickplate grille if the refrigerator is equipped with one. Using an open-ended wrench, adjust the rollers up or down as required to level the refrigerator.

Installing the Refrigerator in Place

Note: Refrigerators should not be installed where the ambient temperature will fall below 60°F, because the refrigerator compressor will not run enough to maintain the desired temperature inside the cabinet.

The cabinet dimensions are important when planning the installation of a refrigerator. Be sure to allow ½ inch on each side of the cabinet to allow the refrigerator to be slid into place.

Level the refrigerator from front to rear and then from side to side. Accurate leveling is especially important on those models equipped with ice makers and ice and water dispensers.

Air Circulation

Proper air circulation around the cabinet is a must if efficient refrigerator operation is to be maintained. The owners guide for the particular manufacturer and model number must be referred to for the recommended clearance. Avoid locating the refrigerator next to a heat source or in an area which is exposed to the direct sunlight for long periods of time.

Electrical Circuit

Connect the refrigerator power supply cord to a properly grounded electrical wall outlet. A separate circuit from a fuse or circuit breaker is preferred so that interference from other appliances will not affect operation of the refrigerator. The circuit should be capable of currying 20 or 25 amps. The circuit should include the properly sized circuit breaker or time-delay fuse.

The voltage measured at the wall receptacle where the refrigerator is connected must not vary more than ±10 percent from the nominal 120-volt rating.

All refrigerators are equipped with a power supply cord equipped with a three-prong grounded plug and a ground wire that is attached to the refrigerator cabinet for user protection against shocks and other electrical hazards.

In installations having only a two-prong outlet, a qualified electrician must be contacted to provide the correct type of electrical connection that complies with the National Electrical Code. *Caution:* Do not under any circumstances cut or remove the grounding prong from the power supply cord.

A two-prong adaptor is not an approved device for connecting a refrigerator to an electrical wall outlet.

ELECTRICAL COMPONENTS

Use the following procedures when working with the electrical systems on refrigerators.

Electrical Grounding

All refrigerators are equipped with an electrical power supply cord equipped with a three-prong grounding plug

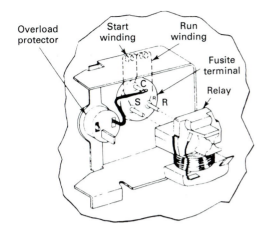

Figure 11–2 Terminal pins (relay and overload protector). (Courtesy of WCI Major Appliance Group.)

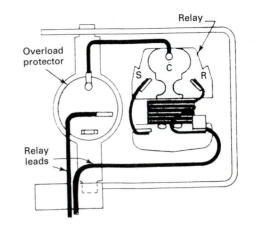

Figure 11–3 Starting relay and overload protector. (Courtesy of WCI Major Appliance Group.)

and a ground wire attached to the refrigerator cabinet to protect the user against electrical shock. Each electrical component is either mounted, or connected through, a ground wire to the cabinet to complete the ground. Certain components, such as defrost timers, may be "double insulated" and, therefore, do not require a ground wire.

Be sure that the wall electrical receptacle is of the three-prong type and is properly grounded in accordance with the National Electrical Code and/or local codes.

Relay and Overload Protector

The compressor starting relay is the push-on type that mounts on the start (S) and run (R) terminals of the compressor. See Figure 11–2.

The relay coil carries the main winding current. The relay armature holds the start winding contacts in the open position except during the starting period.

At the moment when the thermostat closes the electrical circuit and the compressor starts, a surge of electric current passes through the main motor winding and through the relay coil. This current energizes the relay coil and pulls up the relay armature, causing the start winding contacts to close.

The current flow through the compressor start winding introduces a second, out-of-phase, magnetic field in the stator and starts the motor turning. As the motor speed increases, the main winding current is reduced. At a predetermined condition the main winding current, which is also the current through the relay coil, drops to a value below that necessary to hold up the relay armature. The armature drops out and opens the start winding contacts and takes the start winding out of the circuit.

In series with the compressor motor winding is a separate bimetallic overload protector, held in place on the compressor by a spring clip. The short wire lead on the overload protector connects to the common (C) terminal on the compressor. See Figure 11–3.

Should the current in the motor windings increase to a dangerous point, heat is produced by the high current passing through the bimetallic disc. This heat will cause the disc to deflect and open the contacts. This breaks the electric circuit to the compressor motor windings, stopping the motor before any damage can occur to it.

The dome-mounted overload protector provides added protection for the compressor motor. This protection is in addition to the protection against excessive current draw. It also protects against any excessive rise in temperature in the compressor housing.

After an overload or a temperature rise has caused the overload protector to break the electrical circuit, the bimetallic disc cools and returns the contacts to the closed position. The time required for the overload switch to reset varies with room temperature and the compressor dome temperature.

The overload protector is specifically designed with the proper electrical characteristics for the compressor motor and its application. Any replacement must be made with an exact replacement, that is, the same part number. *Never substitute an overload protector with another overload protector having an unauthorized part number.* The wrong protector can result in a burned-out motor. If the relay is found to be inoperative, change both the relay and the overload protector. If the overload protector is inoperative, change only the overload protector.

When the thermostat cuts off after a normal cycle, or when the service cord is pulled from the electrical outlet during a running cycle, about 8 minutes time is required for "unloading" (longer if the compressor stopped during a pull-down). Unloading is the reduction of the refrigerant pressure differential between the high side and the low side of the system. During this unloading period, the overload will trip if the service cord is plugged into the electrical outlet.

To check for an open overload protector, short across its terminals. See Figure 11–3. If the compressor starts, replace the overload protector. If the compressor does not start, look for trouble elsewhere (line voltage during the starting period, inoperative relay, inoperative compressor, etc.).

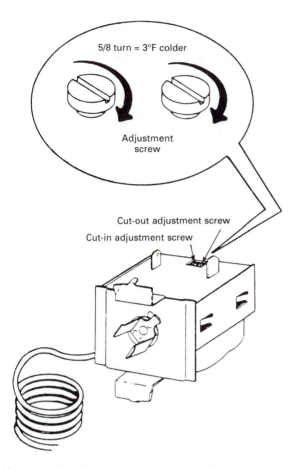

5/8 turn = 3°F colder

Adjustment screw

Cut-out adjustment screw

Cut-in adjustment screw

Figure 11–4 Cutler/Hammer control thermostat. (Courtesy of WCI Major Appliance Group.)

Since the relay is current operated and designed for a specific compressor and motor current value, the correct size relay, represented by the part number, is an absolute necessity. *Never substitute a starting relay with another starting relay having an unauthorized part number.*

The starting relay cannot be properly repaired or adjusted in the field; it must always be replaced.

If the compressor repeatedly starts and runs for a few seconds and then cycles on the overload protector, the starting relay contacts may be stuck closed. In this case the excessive current is causing the overload protector to trip.

Compressor Motor Electrical Check

When checking for electrical trouble, always be sure that there is a live electrical circuit to the cabinet and that the temperature selector dial is not in the OFF position.

When the sealed unit will not start and the cabinet temperature is warm, the trouble may be in the relay, overload, thermostat, the wiring, or it may be in the compressor motor itself.

When the compressor will not run, make a voltage check across the power lead terminals on both the relay and

the overload protector. Check the wiring diagram for the refrigerator being serviced.

The voltmeter should show line voltage if the thermostat knob is in the normal operating position and not in the OFF position. If this check does not show a live circuit, the thermostat and wiring should be checked for an open circuit. Pay particular attention to all terminal connections for a loose, corroded, or overheated wire condition.

A control thermostat check can be made by using a piece of wire as a temporary bridge across the two thermostat terminals. If the compressor now starts and runs with the bridge, the control thermostat is faulty and should be changed.

If the voltage check shows power supply at the relay terminals, check the compressor by means of a manual test set.

If the compressor motor does not start and run with either the test set or the regular electrical accessories, check the line voltage to see that there is no more than 10 percent variation from the rated voltage. If the voltage is correct and the compressor will not start and run, change the compressor.

If the compressor starts and runs with the test set in place, replace the starting relay.

Control Thermostat

The control thermostat is a variable cut-in type. When the thermostat knob is changed from one setting to another, both the cut-in and the cutout temperatures change. The degree of temperature change is determined by the knob setting.

The Cutler/Hammer control thermostat has two adjustment screws. Both the cut-in and the cutout screws must be adjusted. See Figure 11–4.

The GE control thermostat has only one adjustment screw. See Figure 11–5.

All control thermostats are located in the refrigerator compartment and mounted in a control housing.

The thermostat sensing element for the different models of refrigerators are located as listed below:

Single-door manual-defrost models The sensing element is attached to the right front corner of the evaporator and senses the evaporator temperature.

Top-freezer cycle-defrost models The sensing bulb is attached to the bottom of the cold plate and senses the plate temperature. See Figure 11–6.

Top-freezer frost-proof models The sensing element for these models is located in the control housing and senses the air temperature. See Figures 11–7 and 11–8. The medium-profile housing has the sensing element formed in a "figure 8" configuration around the mounting projections within the housing.

The low-profile housing has the sensing element inserted into a plastic tube-type sensing well. The element must be inserted to a depth of 7 inches. If the element cannot

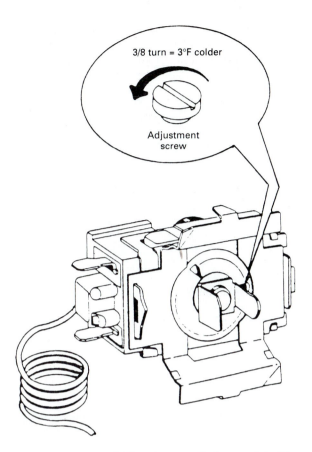

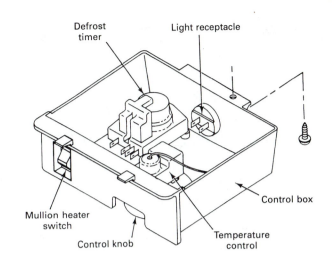

Figure 11–7 Top-freezer frost-proof control housing (medium profile). (Courtesy of WCI Major Appliance Group.)

Figure 11–5 General Electric control thermostat. (Courtesy of WCI Major Appliance Group.)

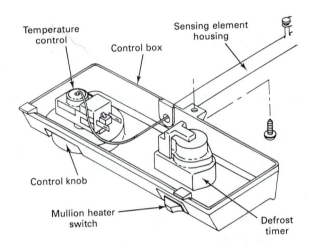

Figure 11–8 Top-freezer frost-proof control housing (low profile). (Courtesy of WCI Major Appliance Group.)

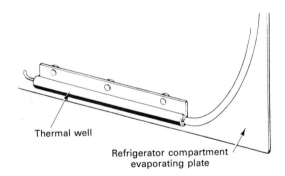

Figure 11–6 Top-freezer cycle defrost thermal well. (Courtesy of WCI Major Appliance Group.)

be inserted the full 7 inches, the sensing well should be replaced.

Single-door defrost models The control thermostat is mounted in a housing, located behind the light shield, at the top of the refrigerator. The sensing element is inserted into a plastic tube-type sensing well. See Figure 11–9.

Side-by-side control thermostat mounting The control thermostat is mounted in a one-piece plastic control housing. The sensing element is attached to the projection, on the bottom of the housing, with a single bend. See Figure 11–10.

The control housing is attached to the refrigerator liner with two screws and contains the control thermostat, defrost timer, and lamp receptacle.

Light Switch

The single-pole, single-throw refrigerator compartment door-activated light switch is illustrated in Figure 11–11.

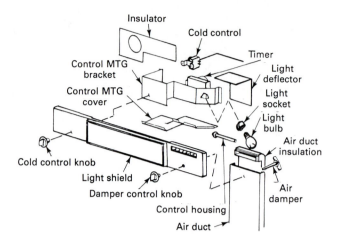

Figure 11–9 Single-door frost-proof control housing. (Courtesy of WCI Major Appliance Group.)

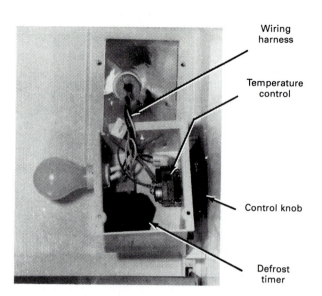

Figure 11–10 Side-by-side control thermostat mounting. (Courtesy of WCI Major Appliance Group.)

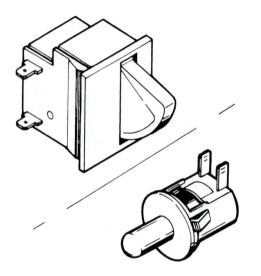

Figure 11–11 Single-pole, single-throw refrigerator compartment door switches. (Courtesy of WCI Major Appliance Group.)

When the door is closed, the light circuit is open (off). Refer to the wiring diagram that accompanies the refrigerator. Push-on-type terminal connectors secure the wiring harness leads to the switch terminals.

To replace the switch, pry it out of the mounting, disconnect the wire leads, and replace the switch.

Power-Saver Switch

A single-pole, single-throw switch is wired in series with the mullion heater. The switch has two positions, ON and OFF. When the switch is in the ON, Reduces Exterior Moisture, or the High Humidity position, the mullion heater is energized. When the switch is in the OFF, Saves Power, or Low Humidity position, the circuit to the mullion heater is open. This feature allows a reduction in operating cost by turning the mullion heater off in dry weather.

Perimeter and Mullion Heaters

To reduce the possibility of condensation forming on the exterior of the refrigerator cabinet in high-humidity areas, top-freezer and side-by-side models are equipped with perimeter and mullion heaters.

Top-freezer models These models have a perimeter hot tube which is part of the refrigerant system in place of an electric perimeter heater.

In addition to the perimeter hot tube, these models have a low-wattage mullion heater that is attached to the aluminum foil. The foil is adhered to the back side of the removable crossrail.

Side-by-side models These models have a one-piece perimeter and mullion hot tube that is a part of the refrigeration system. No electric heaters are used.

Freezer Compartment Fan and Motor Assembly: All Frost-Proof Models

The fan and motor assembly is located behind the freezer compartment air duct directly above the freezer compartment evaporator.

The fan is a suction type. It pulls the air up through the evaporator and blows it through the refrigerator compartment air duct and freezer compartment fan grille.

It is wired in series with the control thermostat and the defrost timer contacts. The freezer compartment fan motor operates when the control thermostat contacts are closed, except during the defrost cycle.

The location of the fan blade on the motor shaft is most important. Mounting the fan blade too far back or too far forward on the motor shaft in relation to the orifice ring

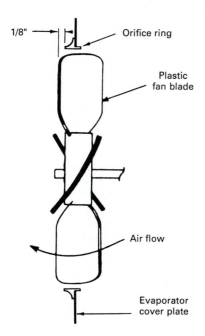

Figure 11–12 Position of fan in orifice. (Courtesy of WCI Major Appliance Group.)

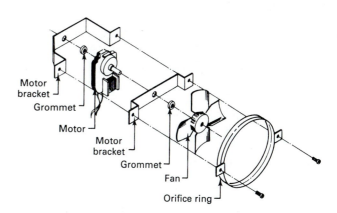

Figure 11–13 Fan (motor and bracket assembly). (Courtesy of WCI Major Appliance Group.)

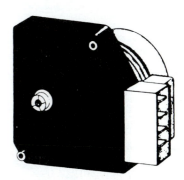

Figure 11–14 Defrost timer. (Courtesy of WCI Major Appliance Group.)

will result in improper air circulation. The freezer compartment fan must be positioned with ⅛ inch of the fan blade forward from the orifice ring. See Figure 11–12.

To remove the fan and motor assembly, use the following procedure:

1. Remove the fan cover.
2. Pull the plastic orifice ring out of the air duct fan opening.
3. Remove the two fan motor mounting screws and pull the fan and motor out through the opening.
4. Then disconnect the wire harness connector.
5. Reassemble the fan and motor in the reverse order. See Figure 11–13.

Defrost Timer

In addition to the control thermostat, the refrigerator is equipped with an automatic defrost timer. See Figure 11–14.

The automatic defrost timer is driven by a self-starting electric motor. The timer motor is wired in series with the control thermostat and runs only when the compressor is energized.

The defrost timer is located in the refrigerator compressor compartment within the control housing. An opening in the housing allows access to the timer shaft so that it may be advanced manually for test purposes.

Defrost Thermostat

The defrost thermostat is a temperature-sensing device. It is wired in series with the defrost timer and evapora-

tor defrost heater. Refer to the wiring diagram that accompanies the refrigerator. It senses a rise in the evaporator temperature during the defrost cycle and cycles the defrost heater off after all the frost has melted. It is calibrated to permit a defrost cycle only when the temperature is below a preset temperature. See the specifications for the refrigerator being serviced.

Defrost Heater

The defrost heater is a radiant U-shaped resistance heater. The defrost heater is wired in series with the defrost thermostat and contacts 2 and 1 of the defrost timer. The defrost heater is energized during that period of the cycle when the defrost thermostat contacts are closed.

The length of time that the heater is energized depends on the amount of frost accumulation on the evaporator.

Ice and Water Dispensers: Side-by-Side Models

The following is a description of the different systems used on these types of refrigerators.

Water systems Water for the ice maker and water dispenser is controlled by a dual water valve located at the

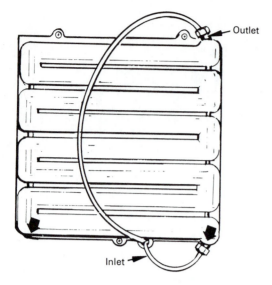

Figure 11–15 Water storage tank. (Courtesy of WCI Major Appliance Group.)

bottom back of the refrigerator. Plastic tubing connects the water to the ice maker fill tube.

Plastic tubing connects the water valve to the inlet (bottom) of the storage tank in the refrigerator, the outlet (top) of the storage tank to the elbow below the lower freezer door hinge, and from the elbow to the water dispenser in the freezer door. See Figure 11–15.

All sections of the water tubing may be replaced if necessary.

Water and ice dispensing The water and ice dispensers are located in the freezer door. Both are activated by push buttons. The water dispenser button, when pushed, operates a microswitch that closes the electrical circuit to the water valve. The ice dispenser button, when pushed, opens the ice door and operates a microswitch that closes the circuit to the dispenser motor. See Figure 11–16.

The trim plate, surrounding the push buttons, is removed by pulling it straight out.

Figure 11–16 Dispenser module assembly. (Courtesy of WCI Major Appliance Group.)

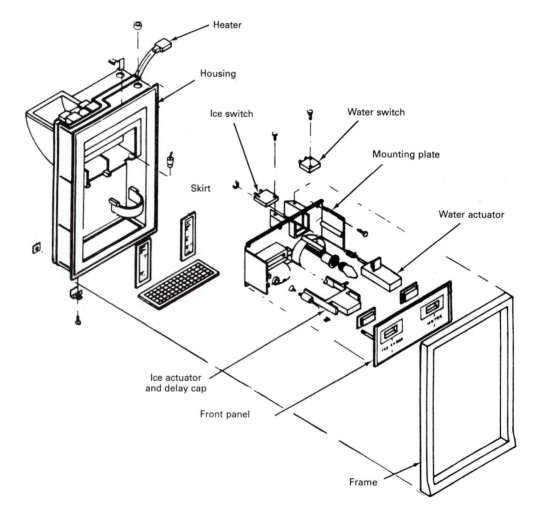

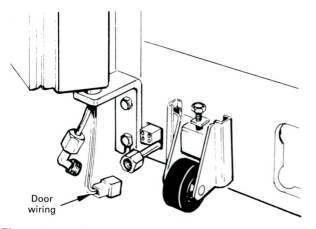

Figure 11–17 Dispenser electrical checks. (Courtesy of WCI Major Appliance Group.)

Switch plate assembly With the trim plate removed, the module light and the adjustment for the ice door time delay are accessible for service or replacement. Use only original equipment 10-watt lamp for replacement.

The ice door delay is adjusted with a $\frac{1}{16}$-inch Allen wrench. Turn the adjustment screw clockwise to increase, and counterclockwise to decrease, the time delay. The delay should be adjusted for no more than 3 to 5 seconds.

Ice chute door The ice chute door is located on the inside of the freezer door. Located above the ice chute door is a spring-loaded flipper. The purpose of the flipper is to prevent ice from dropping on top of the ice chute door and possibly interfering with its operation. The ice chute door closes primarily by its own weight and, therefore, must operate freely. The flipper spring adds only a small amount of closing force to the door.

To remove the ice chute door, first tape the flipper in its full open position, then pull the ice chute door straight out to "snap" it off its pivot shaft. The flipper is removed by first removing the small coil spring, then pulling the flipper to one side and disengaging the opposite side hinge pin.

Dispenser internal components The only functional components within the door that may require service are:

1. The dispenser module heater.
2. Wiring to the dispenser module.
3. The water dispenser water tubing.

All ice and water dispenser electrical components may be checked for continuity at the electrical connector located beneath the lower freezer door hinge. See Figures 11–17 and 11–18.

1. Disconnect the refrigerator service cord from the wall receptacle.
2. Remove the base grille and freezer door lower hinge cover.
3. Disconnect the electrical connector and make continuity checks, with an ohmmeter, as follows:

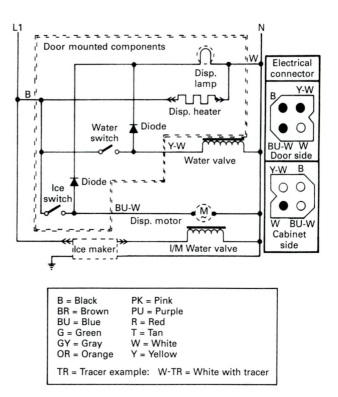

B = Black	PK = Pink
BR = Brown	PU = Purple
BU = Blue	R = Red
G = Green	T = Tan
GY = Gray	W = White
OR = Orange	Y = Yellow
TR = Tracer example: W-TR = White with tracer	

Figure 11–18 Schematic diagram (dispenser assembly). (Courtesy of WCI Major Appliance Group.)

Door-Mounted Components

Continuity Check	Connector, Door Side
Dispenser heater	B to W (black to white)
Diode (ice)*	W to BU-W (white to blue-white)
Diode (water)*	W to Y-W (white to yellow-white)
Switch (ice)†	B to BU-W (black to blue-white)
Switch (water)†	B to Y-W (black to yellow-white)

* Must have a good light bulb in the receptacle.
† Must remove the light bulb before making this check.

Cabinet-Mounted Components

Dispenser motor	W to BU-W (white to blue-white)
Water valve	W to Y-W (white to yellow-white)

Diodes The freezer door wiring harness has two diodes in the circuits to the dispenser light. See Figure 11–19. The only function of the diodes is to prevent feedback to the water-dispensing valve or the ice-dispensing motor when the opposite dispensing button is depressed.

A good diode will show continuity in one direction only. The continuity must be in the same direction for both diodes. An open diode will show no continuity in either direction. An open or shorted diode must be replaced. It is not necessary to replace the entire door wiring harness if a diode fails. Be sure to use the recommended diode for the refrigerator being serviced.

Diode replacement (see Figures 11–18 and 11–19) The following are the recommended steps to use when replacing a diode:

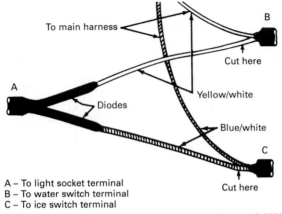

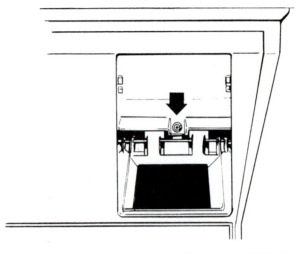

A – To light socket terminal
B – To water switch terminal
C – To ice switch terminal

A-3398

Figure 11–19 Diode and terminal locations. (Courtesy of WCI Major Appliance Group.)

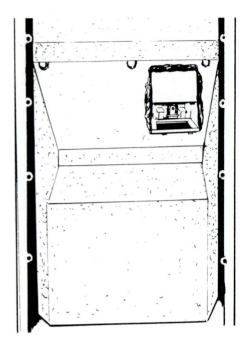

Figure 11–21 Styrofoam insulation. (Courtesy of WCI Major Appliance Group.)

Figure 11–20 Inner panel removal. (Courtesy of WCI Major Appliance Group.)

1. Disconnect the refrigerator electrical cord from the wall receptacle.

2. Remove the control panel by pulling it straight out and off the two pins.

3. Remove the screws and pull the switch mounting plate forward.

4. Disconnect the short wires from a wye (Y) between the lamp receptacle and each of the two dispenser switches. See Figure 11–48.

5. Cut both short wires close to terminals B and C, so that the terminals stay with the main wiring harness.

6. Discard the wye (Y) formed by the two diodes and short wires.

7. Using a good quality electrical tape, insulate the cut end of the wires.

8. Install the piggyback terminals of the replacement wye (Y) onto the open terminals of the dispenser switches.

9. Install the main harness wires on the proper dispenser switches at the open piggyback terminals.

10. Install the other end of the new wye (Y) onto the open terminal of the lamp receptacle.

11. Reassemble the dispenser housing.

12. Check the operation of the ice and water dispensers and the light.

The dispenser components within the freezer door are accessible for service by removing the freezer inner door panel and gasket. Remove one screw in the ice chute opening and all inner panel screws. See Figure 11–20.

The dispenser module is encased in styrofoam insulation.

To remove the styrofoam insulation, use the following procedure:

1. Carefully loosen the section of heater adhered to the ice chute opening.

2. Remove the fiberglass insulation from the freezer door and pull the styrofoam straight off the dispenser. See Figure 11–21.

The dispenser wiring and the water tube enter the freezer door through the bottom hinge and pass through a plastic sleeve. The sleeve is shaped to allow room for the wiring and water tube to flex and move as the freezer door is opened and closed.

The wiring and water tube are held in place along the side of the dispenser module with a plastic shield. See Figure 11–22.

The door wiring and water tube pass through a rubber bushing within the hollow hinge pin of the lower freezer door hinge. See Figure 11–23. A bronze bearing is attached to the freezer door and turns on the hollow hinge pin. The wiring and water tube remain stationary.

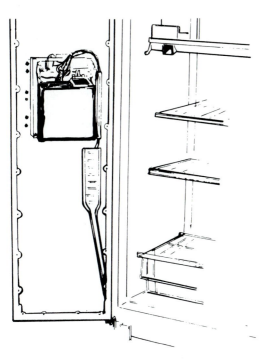

Figure 11–22 Wiring and water tube routing. (Courtesy of WCI Major Appliance Group.)

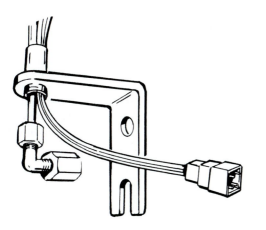

Figure 11–23 Lower freezer door hinge. (Courtesy of WCI Major Appliance Group.)

Figure 11–24 Dispenser light insulation. (Courtesy of WCI Major Appliance Group.)

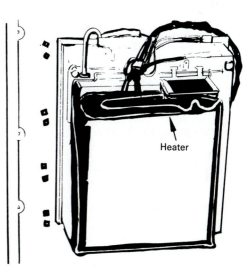

Figure 11–25 Dispenser module heater. (Courtesy of WCI Major Appliance Group.)

A piece of ½-inch fiberglass insulation is taped in place on top of the dispenser module. This insulation prevents heat caused by the dispenser light from overheating the styrofoam insulation. It must, therefore, be in place on reassembly of the unit. See Figure 11–24.

The one-piece, 10-watt dispenser heater is connected to the wire harness with a two-wire disconnect. Carefully note the positioning and routing of the heater before attempting to install a replacement. See Figure 11–25.

The dispenser module is set into the freezer door from the front outer panel. It is held in place with push nuts on the plastic pins along each side. A foam gasket is used to form a seal between the module front flange and the outer door panel. Additional, silicone rubber sealer may be used on the inside to assure a positive seal.

Dispenser drive motor The dispenser drive motor is located directly behind the ice container. The front of the motor assembly is formed with a guide to assist in properly positioning the ice container. The motor mounting screws and wiring disconnects are located behind the plastic cover at the top of the motor assembly. See Figure 11–26. *Note:* It is not necessary to remove the ice maker to service the motor.

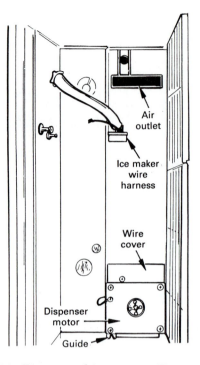

Figure 11–26 Dispenser drive motor. (Courtesy of WCI Major Appliance Group.)

To remove the motor assembly, use the following procedure:

1. Remove the cover by removing one screw.
2. Unplug the wire connector and remove the two motor mounting screws. See Figure 11–27.

The drive motor can then be lifted up and out of the freezer.

A plastic drive coupling is threaded to the motor shaft.

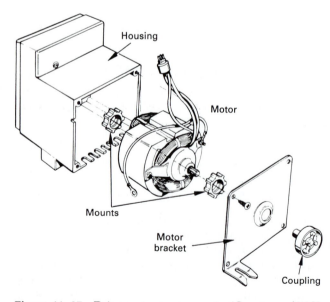

Figure 11–27 Drive motor components. (Courtesy of WCI Major Appliance Group.)

Use a straight screwdriver through the back of the housing to hold the motor shaft while unscrewing the coupling.

To remove the motor from the plastic housing, remove the ground wire screw and the four mounting screws from the metal front plate. The motor is supported by two rubber bushings and is held in place between the plastic housing and the metal front plate. See Figure 11–28.

When reinstalling the motor assembly into the freezer, be sure that the recessed area at the back of the motor housing is in place over the metal tab on the back of the freezer.

SC and SF compressor electrical circuit (see Figures 11–29 through 11–31) The SG and SF series of very high efficiency compressors are equipped with all new electrical components. The new electrical components consist of a solid-state PTCR* relay, a thermally operated overload protector, and a run capacitor. See Figure 11–29.

SG and SF compressor start circuit (see Figure 11–30) When the compressor is first energized, the solid-state relay has low resistance (3 to 12 ohms), and both the run and start windings are energized to start the compressor. The run capacitor is being shunted (bypassed) by the PTC relay, and it has little function during the compressor starting process.

SG and SF compressor run circuit (see Figure 11–31) When the self-heating solid-state relay has reached sufficient temperature, it will abruptly change from low (3 to 12 ohms) resistance to very high resistance (10 to 20 ohms) and, in effect, "switches off" the start windings. The relay no longer shunts the run capacitor. The run capacitor is now in series with the start windings. The only purpose of the run capacitor is to improve the compressor operating efficiency, which it does by "correcting" the power factor of the compressor motor.

---

* PTCR (positive temperature coefficient resistor) defines a resistor which increases in resistance as its temperature is increased. PTCR is commonly abbreviated as just PTC.

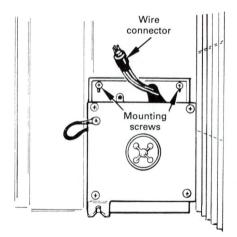

Figure 11–28 Drive motor removal. (Courtesy of WCI Major Appliance Group.)

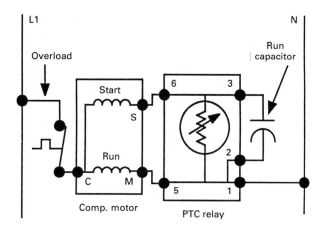

Figure 11–29 SG and SF compressor electrical circuit. (Courtesy of WCI Major Appliance Group.)

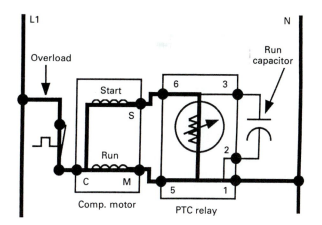

Figure 11–30 SG and SF compressor start circuit. (Courtesy of WCI Major Appliance Group.)

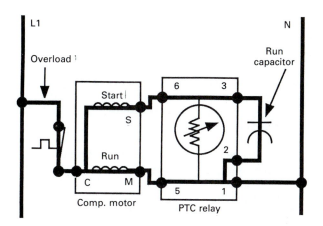

Figure 11–31 SG and SF compressor run circuit. (Courtesy of WCI Major Appliance Group.)

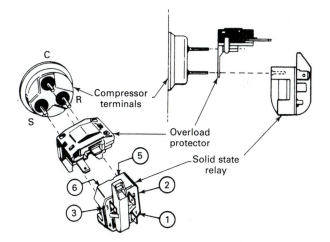
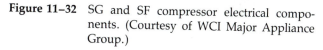

Figure 11–32 SG and SF compressor electrical components. (Courtesy of WCI Major Appliance Group.)

SG and SF compressor operating characteristics The following are the operating characteristics of these compressors:

1. When the compressor electrical circuit is energized, the start winding current causes the PTC relay to heat, and after an appropriate amount of starting time, "switch off" the start winding circuit. *Note:* The PTC relay will "switch off" the start winding circuit even though the compressor has not started, such as when attempting to restart after a momentary power interruption.

2. Because the PTC opens the compressor start circuit whether or not the compressor has started, the overload protector is designed and calibrated to open the compressor electrical circuit with locked rotor run winding current only. *Caution: Use only the correctly specified overload protector for service replacement.*

3. With an open PTC relay, the compressor will not start because there is little or no current to the start windings. The overload protector will open due to the high locked rotor run winding current.

4. With a "shorted" PTC relay or capacitor, the compressor will start, and the overload protector will quickly open due to the high current of the combined run and start windings.

5. With an open or weak capacitor, the compressor will start and run, apparently as normal. The compressor, however, will be operating at a reduced efficiency.

Solid-state relay, SG and SF compressors (see figures 11–29 and 11–32) The solid-state relay has no moving parts. It consists of a PTC resistor mounted in a plastic case with the appropriate terminals.

PTC (positive temperature coefficient) simply denotes a resistor that increases in resistance as its temperature is increased. The self-heating PTC resistor used in the solid-state relay has the unique characteristic of changing from low to very high resistance abruptly, thus serving as an on/off switch.

The solid-state relay plugs directly onto the compressor start and run terminals. Relay terminals 1, 2, and 5 are connected together within the relay, as are terminals 3 and 6. The run capacitor is connected to relay terminals 2 and 3, so it is connected in electrical parallel with the PTC resistor.

One side of the 120-volt power supply is connected to relay terminal 1.

To check the relay, use the following procedure:

1. Disconnect the refrigerator electrical cord from the wall receptacle.

2. Remove the relay cover, disconnect the three (3) wires, and pull the relay off the compressor terminals.

3. With an ohmmeter, check the resistance between relay terminals 2 and 3. The resistance should range from 3 to 12 ohms at normal room temperature. A shorted relay will read "0" resistance. An open relay will read very high or infinite resistance.

Overload protectors SG and SF compressors (see figures 11–29 and 11–32) The overload protector is a thermally operated type; it can be opened by either excessive heat or current. Unlike prior overloads, the internal bimetal is not self-heating and is not a part of the electrical circuit. The overload has a small built-in coil heater in electrical series with the compressor start and run windings.

The overload plugs directly onto the compressor common terminal.

To check the overload protector, use the following procedure:

1. Disconnect the refrigerator electrical cord from the wall receptacle.

2. Remove the relay cover, pull the relay off the compressor, disconnect one wire from the overload protector, and pull it off the compressor.

3. With an ohmmeter, check the resistance between the tab terminal and the female pin terminal. At normal room temperature the overload protector should have less than 1-ohm resistance. An open overload protector will have infinite resistance.

Run capacitor, SG and SF compressors (see figure 11–32) The run capacitor has permanently attached wires that are connected to relay terminals 2 and 3. The capacitor does not have an identified terminal and can be wired without regard to polarity.

To check the run capacitor, use the following procedure:

1. Disconnect the electrical service cord from the wall receptacle.

2. Remove the relay cover and disconnect the capacitor wires from the relay.

3. Discharge the capacitor by shorting across the terminals with a 500-K (1 watt) resistor for 1 minute.

4. With an ohmmeter, check the resistance across the capacitor wire terminals, with the meter set on the "ohms times 1000" scale.

   a. The needle should jump toward 0 ohms and quickly move back to infinity.

   b. If the needle does not move, the capacitor is open.

   c. If the needle reads a constant value at or near 0 ohms, the capacitor is shorted.

   d. If the needle jumps toward 0 ohms and then moves back to a constant high resistance (not infinity), the capacitor has a high-resistance leak.

   e. Check the resistance from each capacitor terminal to its bare metal case. If the resistance is less than infinity, the capacitor is grounded.

AIR CIRCULATION

In order for a refrigerator to operate as it was designed, there must be proper air circulation throughout the inside of the cabinet. Also, forced-air circulation is required for frost-free operation of the unit.

Principle of Frost-Free Operation

Frost-free refrigerators operate on the principle that moisture or frost transfers or migrates to the coldest surfaces (the evaporator in the refrigerator) in the freezer compartment. For example, a small amount of water spilled from an ice cube tray in the freezer compartment will freeze immediately; however, this ice in time will evaporate and transfer to the colder surfaces of the freezer evaporator coil.

Air Circulation Pattern

The following is a description of the air flow patterns for different refrigerator models:

Top-freezer frost-free models Top-freezer frost-proof models with a single evaporator in the freezer compartment have forced-air cooling in the freezer and refrigerator compartments.

The fin-and-tube-type evaporator is located on the back wall of the freezer compartment. A circulating fan (suction type) pulls air from the freezer and refrigerator compartments across the evaporator surfaces. The cold air is forced into a fan cover and is then discharged down through a duct in the back wall and is discharged into the refrigerator compartment. See Figures 11–33 and 11–34.

Air from the freezer and refrigerator compartments flows to the evaporator through air return ducts in the divider between the freezer compartment and the refrigerator compartment as illustrated.

The air-circulating fan in the freezer compartment operates only when the compressor is running. During the defrost period, however, the compressor and the air-circulating fan do not operate. The automatic defrost timer opens the electrical circuit to both the fan motor and the compressor.

Side-by-side frost-free models All side-by-side models have forced-air cooling in the freezer and refrigerator compartments.

The fin-and-tube evaporator is located on the back wall of the freezer compartment. A circulating fan (suction type) pulls the air from the freezer and refrigerator compart-

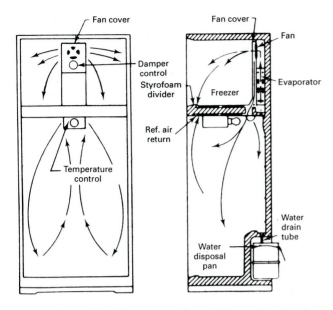

Figure 11–33 Top-freezer air flow (12- to 14-cubic-foot models). (Courtesy of WCI Major Appliance Group.)

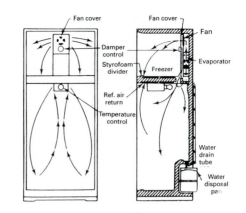

Figure 11–34 Top-freezer air flow (16.6- to 20.6-cubic-foot models). (Courtesy of WCI Major Appliance Group.)

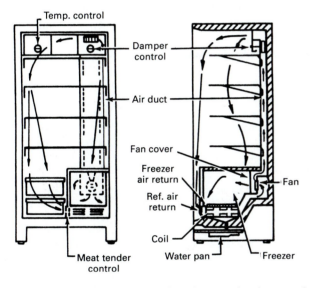

Figure 11–35 Single-door air flow (17.5-cubic-foot models). (Courtesy of WCI Major Appliance Group.)

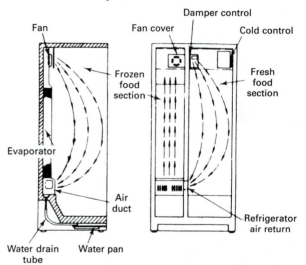

Figure 11–36 Side-by-side, without dispensers, air flow. (Courtesy of WCI Major Appliance Group.)

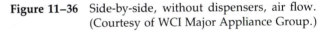

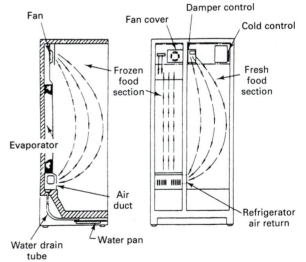

Figure 11–37 Side-by-side, with ice and water dispensers, air flow. (Courtesy of WCI Major Appliance Group.)

ments, across the evaporator on the back wall and is discharged into the freezer and refrigerator compartments. See Figures 11–35 and 11–36. Air from the refrigerator compartment flows to the evaporator through an air duct that is located in the separating wall.

The air-circulating fan operates only when the compressor is running. During the defrost period, however, the compressor and the circulating fan do not operate. The automatic defrost timer opens the electrical circuit to both the fan motor and the compressor.

For an illustration of the air flow in side-by-side models with ice and water dispensers, refer to Figure 11–37.

Table 11–1 Typical perimeter hot tube package.

Perimeter Hot Tube Package P/N	Models
9960003	12–14 cu. ft Top Freezer Models
9960001	16–20.6 cu. ft. Top Freezer Models
9960005	19–24 cu. ft. Side By Side Models

(Courtesy of WCI Major Appliance Group.)

REFRIGERATION SYSTEMS AND SERVICE

The following is a description of refrigeration systems and their service.

Refrigeration Systems

The basic components of a refrigerator are the compressor, condenser, evaporator, capillary tube, suction line, and dryer. In addition, some models may have a defrost water evaporating plate assembly and/or a perimeter hot tube.

Refrigerant cycle The refrigerant cycle is a continuous cycle that occurs whenever the compressor is operating. Liquid refrigerant is evaporated in the evaporator by heat which enters the cabinet through the insulated walls and heat introduced by the product load and door openings. The refrigerant vapor passes from the evaporator, through the suction line, to the compressor. The pressure and temperature of the vapor is raised in the compressor by compression. The vapor is then forced through the discharge valve into the discharge line and into the condenser. Air passing over the condenser surface removes the heat from the high-pressure vapor, which then flows from the condenser to the evaporator, through the small-diameter liquid line (capillary tube). Before it enters the evaporator, it is subcooled in the heat exchanger by the low-temperature suction vapor in the suction line.

Perimeter hot tube To reduce the possibility of condensation forming on the exterior of the cabinet in high-humidity areas, a refrigerant tube (perimeter tube), part of the refrigerating system, extends across the top and down both sides of the cabinet. When the compressor operates, warm refrigerant flows through the perimeter tube warming the cabinet front exterior.

The perimeter tube is not replaceable. If a leak occurs in this section of the tubing, an electrical heater wire must be installed within the tubing. Be sure to use the proper repair components. Refer to Table 11–1 for a typical perimeter hot tube repair package.

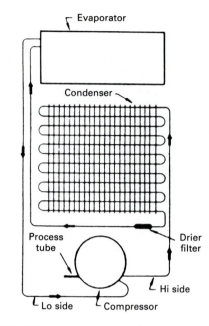

Figure 11–38 Single-door manual-defrost models. (Courtesy of WCI Major Appliance Group.)

Refrigerant Cycle Charts

The following charts are representative of the refrigerant cycle used for various types of refrigerator cabinets. See Figures 11–38 through 11–42.

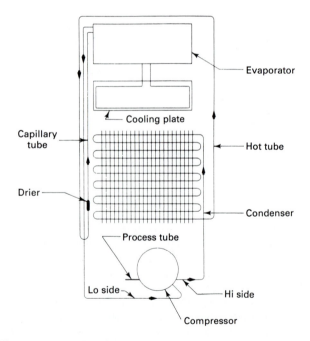

Figure 11–39 Top-freezer cycle defrost models. (Courtesy of WCI Major Appliance Group.)

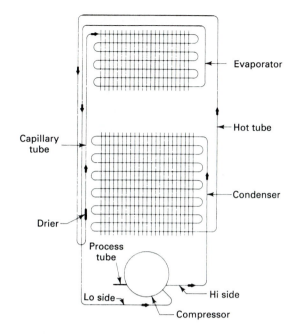

Figure 11–40 Twelve- to 20.6-cubic-foot freezer frost-proof models. (Courtesy of WCI Major Appliance Group.)

Refrigerator service All models are serviced by changing component parts: the compressor, evaporator, heat exchanger, condenser, water-evaporating plate assembly, and filter drier.

Caution: Remove any component part from the interior of the cabinet before attempting to solder. The excessive heat from soldering will warp the plastic liner. Do not heat the exterior surface of the cabinet above 175°F, because overheating could cause distortion of the insulation. Also, do not use a torch directly on the insulation, because it will char at approximately 300°F and will flash ignite (burn) at 500°F.

Sealed-System Repair

Caution: Any time that the sealed system is opened and the refrigerant charge is removed, the proper procedures should be used to capture the refrigerant and prevent it from entering the atmosphere. Protect the flooring and walls from torch burns and oil stains.

Low- or high-side leak or undercharge A loss of refrigerant results in excessive to continuous compressor operation, above-normal refrigerator compartment temperature, a partially frosted evaporator (depending on the amount of refrigerant loss), below-normal freezer compartment temperature, low suction pressure (vacuum), and low wattage. The condenser will be "warm to cool," again, depending on the amount of refrigerant loss.

When refrigerant is added to the system, the frost pattern will improve, the suction and discharge pressures will rise, the condenser will become hot, and the wattage will increase.

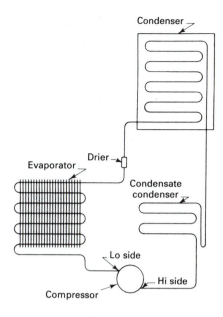

Figure 11–41 Single-door frost-proof models. (Courtesy of WCI Major Appliance Group.)

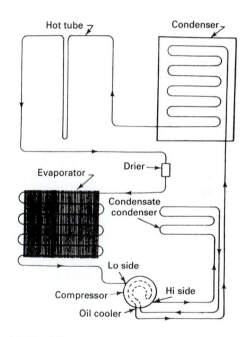

Figure 11–42 Nineteen- to 24-cubic-foot side-by-side models. (Courtesy of WCI Major Appliance Group.)

In the case of a low-side refrigerant leak, resulting in a complete loss of refrigerant, the compressor will run, but will produce no refrigeration. The suction pressure will drop below atmospheric pressure, and air and moisture will be drawn into the system, saturating the drier filter.

If a slight undercharge of refrigerant is indicated, without a leak being found, the charge can be corrected without changing the compressor.

If there is reason to believe that the system has operated for a considerable length of time with no refrigerant and the leak occurred in the evaporator, excessive amounts of moisture may have been drawn into the system. In such cases the compressor may need to be replaced to prevent repetitive service.

If a high-side leak is located and some refrigerant remains in the system, it is not necessary to change the compressor.

Testing for refrigerant leaks If the system is diagnosed as short of refrigerant and the system has not been recently opened, there is probably a leak in the system. Adding refrigerant without first locating and repairing the leak, or replacing the component would not permanently correct the difficulty. *The leak must be found and repaired.* Sufficient refrigerant may have escaped to make it impossible to leak-test effectively. In such cases, add a ¼-inch line-piercing valve to the compressor process tube. Add sufficient refrigerant to increase the pressure to 75 psi. Through this procedure minute leaks are more easily detected before removing the refrigerant from the system. *Note:* The line-piercing valve (clamp-on type) should be used for adding refrigerant and test purposes only. It must be removed from the system after it has served its purpose.

The various types of leak detectors available are liquid detector (bubbles), halide torch, and halogen sensing electronic.

A leak can sometimes be found by the presence of oil around it. To be conclusive, however, use a leak detector.

Liquid detector (bubbles) can be used to detect small leaks in the following manner: Brush the liquid detector over the suspected area and watch for the formation of bubbles as the gas escapes. If the leak is small, it may be several minutes before a bubble will appear. *Caution:* Use the bubble method only when you are sure that the system has a positive pressure. Using it where a vacuum is present could pull the liquid leak detector into the system.

When testing with the *halide torch,* be sure that the room is free from refrigerant vapors. Watch the flame for the slightest change in color. A very faint green indicates a small leak. The flame will be unmistakably green to purple when large leaks are encountered. To simplify leak detection, keep the system pressurized to a minimum of 75 psi.

For more sensitive testing use an electronic-type leak detector. *Halogen-sensing electronic-type* leak detectors are capable of detecting minute refrigerant leaks, even though the surrounding air may contain small amounts of refrigerant.

Cleaning the refrigerant system Remove any inoperative system components. Flush the high side and the low side of the system with R-12 liquid refrigerant (invert the drum). *Caution:* Any flushing operation must be done in such a manner as to prevent pollution of the atmosphere.

Connect the high side and the low side of the system together. Also, connect the oil cooler tubes together. Then evacuate the system using a vacuum pump. Never use the new compressor for this purpose because it will quickly become contaminated. Break the vacuum with refrigerant.

Repeat the process. Then, and this is extremely important, repeat the process a third time for a total of three purges and three evacuations.

System repair procedure When a diagnosis has definitely established the need for entering the refrigeration system, remove the refrigerant from the system in the proper manner to prevent the refrigerant entering the atmosphere.

1. Cut the low-side process tube as close to the pinch-off point as possible and install a process tube adapter with a hand valve.
2. Attach the low-side manifold hose to this adapter.
3. Install a new service drier and install a process adapter and high-side manifold hose to this adapter.
4. Attach the remaining (center) manifold hose to the refrigerant cylinder.
5. Repair or replace the system component which has failed because of leaks, restriction, or mechanical failure.

Caution: Any time the sealed system is opened and the refrigerant charge removed, a new service drier must be installed and the system thoroughly evacuated.

Silver soldering (brazing) The following steps are recommended for this procedure:

1. The proper cleaning of each piece of metal is extremely important in silver soldering.
2. Apply the proper flux to each joint. Do not allow the flux to get inside the tube. Apply the flux to the male portion of the joint only.
3. Apply heat to the joint. The tip of the torch flame should be held close to the tubing. This is the hottest part of the flame. Apply the heat evenly over the joint, keeping in mind the fact that copper carries away heat faster than steel if you are making a copper to steel joint.
4. The flux changes three times as it is warmed. The first change is to a liquid. The liquid appears to dry and become a white powder. The powder then changes back to a watery liquid appearance while the metal becomes cherry red.
5. You should now apply the silver solder to the joint. Capillary action will pull the silver solder into and around the entire joint; however, do not use so much that it sags under the joint.
6. After the joint has cooled, wash the joint thoroughly with a wet cloth to remove any flux.
7. Leak-test the joint.
8. Properly remove the refrigerant from the system.

Attach the center manifold hose to the vacuum pump.

Pump a minimum of 29.5 inches of vacuum for at least 30 minutes.

Pinch off the dryer process tube.

With the pinch-off tool on the drier process tube, remove the process tube adapter.

Flatten the end of the drier process tube and seal by silver soldering.

Charging procedure   Use the following recommended steps for the charging procedure:

1. Turn the hand valve off and attach the center manifold hose to the charging cylinder.
2. Open the valve on the charging cylinder and purge the manifold hose until the liquid refrigerant is at the hand valve connection.
3. Check the unit serial plate for the correct charge of R-12.
4. Check the pressure reading on the gauge at the top of the charging cylinder. Turn the plastic shroud on the charging cylinder to the point where the pressure reading for the refrigerant being used is over the sight glass tube and corresponds with the gauge pressure on the cylinder.
5. Observe the refrigerant level in the sight glass. Subtract the amount of charge and note the shutoff point.
6. Open the hand valve on the process tube very slowly and allow the refrigerant to enter the system.
7. Leak-test the high and low side.
8. Pinch off the process tube. With the pinch-off tool on the process tube, remove the process tube adapter.
9. Flatten the end of the process tube and seal by silver soldering.
10. Leak-test the end of the process tube.

SERVICE DIAGNOSIS

A prime requisite on the initial contact is: Always allow the customer to explain the problem. Many times the trouble can be diagnosed more quickly, based on the customer's explanation. Most of all, do not jump to conclusions until the full story has been heard and the information obtained from the customer has been evaluated. Then proceed with the diagnosis.

Before starting a test procedure, connect the refrigerator electrical service cord to the power source, through a wattmeter, combined with a voltmeter. Then make a visual inspection and an operational check of the refrigerator to determine the following:

1. Is the refrigerator properly leveled?
2. Is the refrigerator located for the proper dissipation of heat from the condenser? Check the recommended spacing from the rear wall and clearance above the cabinet.
3. Feel the condenser. With the compressor in operation, the condenser should be hot, with a gradual reduction in temperature from the top to the bottom of the condenser.
4. Are the door gaskets sealing properly?
5. Does the door, refrigerator or freezer, activate the light switch?
6. Is the freezer compartment fan cover in place?
7. Is the freezer compartment fan properly located on the motor shaft?
8. Is the thermostat thermal element properly positioned?
9. Observe the frost pattern on the evaporator.
10. Check the thermostat knob setting.
11. Check the air-damper control knob setting.

12. Make a mark on the bracket opposite the slotted shaft of the defrost timer to determine if the timer advances.
13. Are the air ducts free of obstructions?

The service technician should inquire as to the number of people in the family to determine the service load and daily door openings. In addition, the service technician should know the room temperature.

After this phase of diagnosis is completed, a thorough operational check should be made of the refrigeration system and any components not previously checked in the following order.

Freezer and refrigerator compartment air temperatures   The freezer and refrigerator compartment temperatures are affected by an improper door seal, frost accumulation on the evaporator, the service load, the ambient temperature, percentage of relative humidity, thermostat calibration (cut-in and cutout), the location of the freezer compartment fan blade on the motor shaft, and the compressor efficiency.

From this, it is evident that the temperatures are not always the same in every refrigerator, even under identical conditions.

Line voltage   It is essential to know the line voltage at the refrigerator. A voltage reading should be taken at the instant the compressor starts, and also while the compressor is running. Line voltage fluctuation should not exceed ±10 percent from the nominal rating. Low voltage will cause overheating of the compressor motor windings, resulting in compressor cycling on the thermal overload, or the compressor may fail to start.

Inadequate line wire size and overloaded lines are the most common reasons for low voltage at the refrigerator.

AUTOMATIC ICE MAKERS

The following is a description of the typical ice makers used in refrigerators.

Top-freezer models, optional accessory ice maker   The automatic ice maker is mounted in the freezer compartment. It is attached to the freezer compartment wall with screws and may be removed for servicing.

The ice maker is designed to produce ice cubes automatically. The length of time between harvest cycles will vary, depending on the load condition, door openings, ambient temperature, and freezer temperature. These factors must be taken into consideration when checking the ice production rate.

The ice maker is wired across the line and will harvest in either the refrigeration or defrost cycles.

The water valve and solenoid assembly is mounted in the compressor compartment.

A $\frac{3}{16}$-inch polyethylene tube extends from the water valve up the rear wall of the refrigerator to a water inlet spout which directs the water into the fill trough. A bead of sealer around the inlet water tube prevents the migration of air and moisture into the freezer compartment.

Figure 11–43   Ice maker. (Courtesy of WCI Major Appliance Group.)

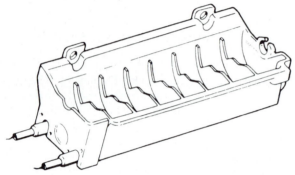

Figure 11–44   Ice mold. (Courtesy of WCI Major Appliance Group.)

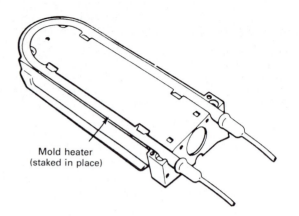

Figure 11–45   Mold heater (staked in place). (Courtesy of WCI Major Appliance Group.)

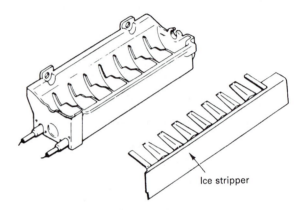

Figure 11–46   Ice stripper. (Courtesy of WCI Major Appliance Group.)

Front Cover

A decorative front cover, made of molded plastic, encloses the operating mechanism of the ice maker and protects it from moisture. It is essential that the cover be in place on an operating ice maker to prevent accidental contact by the user. See Figure 11–43.

Component Parts of the Ice Maker

The following is a description of the component parts of an ice maker.

Ice mold   The ice mold is a die-cast aluminum with the ice maker thermostat bonded to its front surface. The mold has a semicircular interior partitioned into equal-sized compartments. Water enters at the rear of the mold through a fill trough. Openings in each compartment permit all compartments to fill with water. A film of silicone grease on the top of each edge of the mold prevents siphoning of water by capillary attraction. See Figure 11–44.

Mold heater   A mold heater, rated at 165 watts and covered with an aluminum sheath, is embedded in the grooved section on the underside of the mold. When the mold heater is energized, heat melts the ice contact surface within the mold, allowing the harvest of the ice pieces. See Figure 11–45.

The mold heater is wired in series with the ice maker thermostat. The ice maker thermostat acts as a safety device.

The original heater is staked in place, but it can be removed for replacement. The replacement heater is secured to the mold by four (4) flat-head retaining screws that thread into holes in the mold.

A thermal plastic sealer is placed between the heater and the mold to assure a good thermal contact.

Ice stripper   An ice stripper is attached to the mold to prevent ice pieces from falling back into the mold. It also serves as a decorative side cover. See Figure 11–46.

Ice ejector   The ejector blades are molded from Delrin and extend from a central shaft, which turns in nylon bearings at both the front and rear of the shaft. Each blade sweeps an ice section out of the mold. The drive end of the ice ejector is D shaped. Silicone grease is used to lubricate the bearing surfaces. See Figure 11–47.

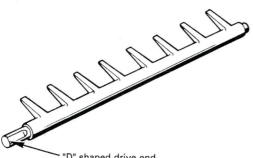

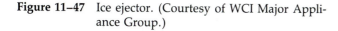

Figure 11–47   Ice ejector. (Courtesy of WCI Major Appliance Group.)

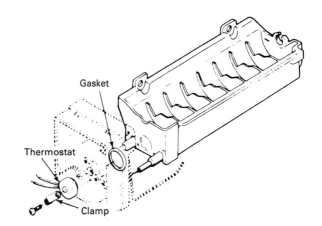

Figure 11–49   Thermostat and gasket. (Courtesy of WCI Major Appliance Group.)

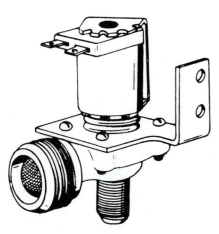

Figure 11–48   Water valve. (Courtesy of WCI Major Appliance Group.)

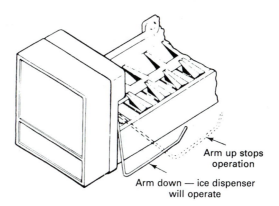

Figure 11–50   Signal arm. (Courtesy of WCI Major Appliance Group.)

Water valve assembly   The water valve is solenoid operated, and when energized releases water from the supply line into the ice mold. The amount of water is directly proportional to the length of time that the water valve is energized. A flow washer inside the water valve maintains a constant rate of water flow over a supply line pressure range of 15 to 100 psi. It will not compensate for pressures less than 15 psi or greater than 100 psi. A 60-mesh screen, placed ahead of the flow washer, filters our foreign materials. See Figure 11–48. The solenoid coil draws 10 to 15 watts of electrical power. The coil is wired in electrical series with the mold heater, across the supply voltage.

Thermostat   The thermostat is a single-pole, single-throw, bimetallic disc-type thermal switch. It automatically starts the harvest cycle when the ice is frozen. The thermostat closes at a temperature of $18 \pm 6°F$. It is wired in electrical series with the mold heater. The ice maker thermostat acts as a safety device against overheating in case of a mechanical failure. A thermal mastic bond is provided where the thermostat is mounted against the mold. A gasket prevents water from leaking into the support housing. See Figure 11–49.

Signal arm and linkage   The signal arm is cam driven and operates a switch to control the quantity of ice produced. In the harvest cycle, the arm is raised and lowered during each of the two revolutions of the timing cam. If the signal arm comes to rest on the top of a piece of ice in the storage container during either revolution, the switch will remain open and stop the ice maker at the end of that revolution. When sufficient ice is removed from the container to lower the arm, ice production will resume. To manually stop the ice maker, raise the signal arm until it locks in the upper position. Operation is resumed when the arm is manually lowered. See Figure 11–50.

Timing switches   The three timing switches used are of the single-pole, double-throw type. They are identical except for their function and they can be used interchangeably. Their main functions are as follows:

1. *Holding switch*   The holding switch assures completion of a revolution once ice maker operation has started. See Figure 11–51.

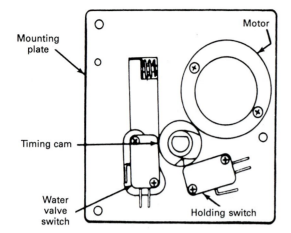

Figure 11–51 Holding and water valve switch. (Courtesy of WCI Major Appliance Group.)

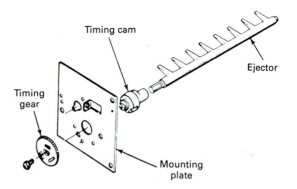

Figure 11–53 Timing gear and cam. (Courtesy of WCI Major Appliance Group.)

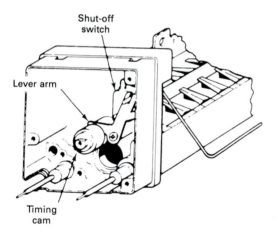

Figure 11–52 Shutoff switch. (Courtesy of WCI Major Appliance Group.)

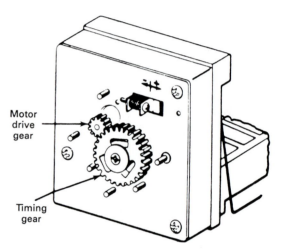

Figure 11–54 Motor drive and timing gears. (Courtesy of WCI Major Appliance Group.)

2. *Water valve switch* The water valve switch opens the water valve during the fill cycle. It is the only adjustable component in the ice maker.

3. *Shutoff switch* The shutoff switch stops operation of the ice maker when the storage container is full of ice.

Timing cam and coupler Three separate cams are combined in one molded Delrin part. One end is attached to a large timing gear. The other end is coupled to the ejector. The functions of the cams are:

1. The inner cam operates the shutoff switch lever arm. See Figure 11–52.
2. The center cam operates the holding switch.
3. The outer cam operates the water valve switch.

Timing gear The large molded plastic timing gear is driven by the timing motor. The gear, in turn, rotates the cam and ejector. A D-shaped hole in the gear fits over the

timing cam hub. Spacer tabs on the back side of the gear prevent them from binding on the mounting plate. See Figures 11–53 and 11–54.

Motor A low-wattage stall-type motor drives the timing gear. This gear turns the timing cam and the ejector blades approximately 1 revolution every 3 minutes. See Figure 11–55.

Fill trough The fill trough is molded nylon. It supports the inlet tube and directs the water into the mold. It also forms a bearing for one end of the ejector blades.

Wiring A four-prong plug connects the ice maker wiring to the cabinet wiring harness. The ice maker assembly is wired across the line and will harvest in either the refrigeration or the defrost cycle. A wiring diagram is located inside the front cover of the ice maker.

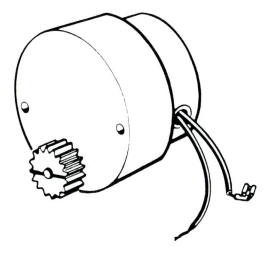

Figure 11–55 Motor. (Courtesy of WCI Major Appliance Group.)

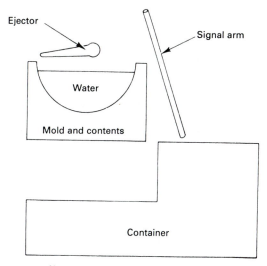

Figure 11–56 Position of the ice maker components. (Courtesy of WCI Major Appliance Group.)

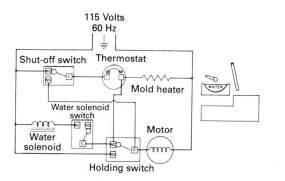

Figure 11–57 Freeze cycle. (Courtesy of WCI Major Appliance Group.)

Operating Cycle

The ice maker operation, water refilling, and controlled storage require proper functioning and timing of all components. Consider the following:

1. The refrigerator has been properly installed and connected to a source of electrical power and water.
2. The freezer compartment evaporator has pulled down to temperature.
3. Several ice-making cycles have been completed and the ice maker is in the freezing cycle.
4. The ice maker thermostat is a single-throw switch wired in electrical series with the mold heater.
5. The ejector blades make 2 revolutions per cycle. Ice is not stored on the blades after a harvest cycle.
6. The water solenoid valve is wired in electrical series with the mold heater.

See Figures 11–56 through 11–66 for a normal operating sequence of the ice maker.

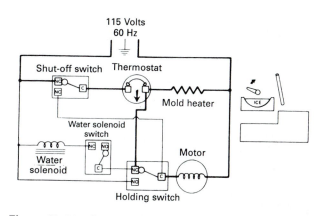

Figure 11–58 Continuing the harvest cycle. (Courtesy of WCI Major Appliance Group.)

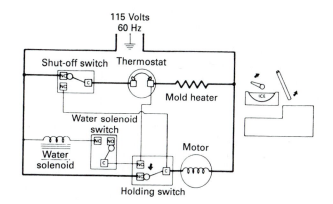

Figure 11–59 Continuing the harvest cycle. (Courtesy of WCI Major Appliance Group.)

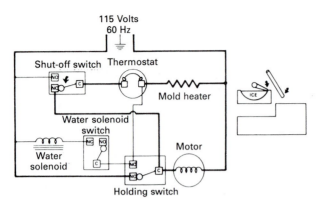

Figure 11–60 Ejector blades reach the ice in the mold. (Courtesy of WCI Major Appliance Group.)

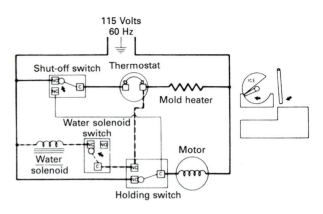

Figure 11–61 Near the completion of the first revolution. (Courtesy of WCI Major Appliance Group.)

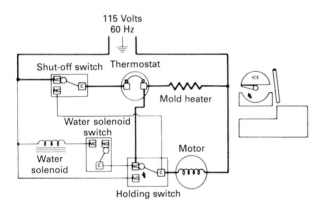

Figure 11–62 End of the first revolution. (Courtesy of WCI Major Appliance Group.)

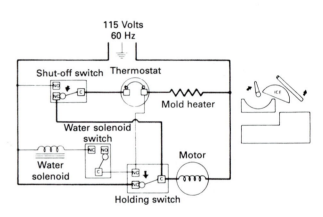

Figure 11–63 Again after a few degrees of revolution. (Courtesy of WCI Major Appliance Group.)

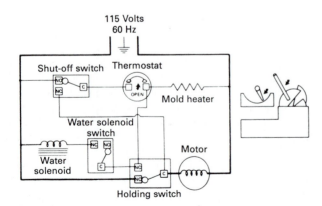

Figure 11–64 Sometime during the second revolution. (Courtesy of WCI Major Appliance Group.)

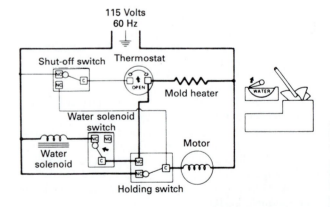

Figure 11–65 Near the completion of the second revolution. (Courtesy of WCI Major Appliance Group.)

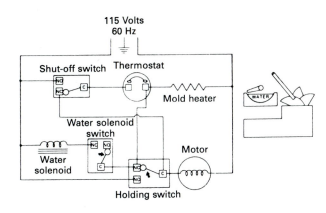

Figure 11–66 Ejection of the ice. (Courtesy of WCI Major Appliance Group.)

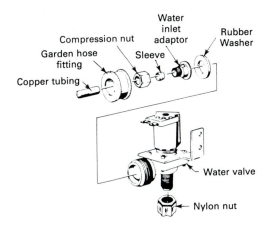

Figure 11–67 Water valve connector. (Courtesy of WCI Major Appliance Group.)

Installing the Water Supply Line to the Ice Maker

Make sure that the installation complies with all applicable plumbing codes.

The ice maker kit includes an adaptor to connect ¼-inch tubing to the ice maker water valve. See Figure 11–67.

The ice maker should be connected to a frequently used cold-water line to ensure a fresh supply of water.

Note: A vertical cold-water line should be selected for the water supply. If a vertical water line is not available, a horizontal line may be used, providing the connection is on the side or top of the pipe and not on the bottom. Scale and foreign material in the pipe could cause stoppage of water flow if the connection is on the bottom.

Be sure to leak-test all connections after the water supply has been turned on.

Test-Cycling the Ice Maker

It may be necessary, on some occasions, to test-cycle an ice maker to ensure its operation. This can be done either on the repair bench or while it is mounted in the refrigerator.

If the ice maker is in an operating refrigerator, take precaution against the formation of condensate. This is done by allowing the cold metal surfaces to warm up before removing the front cover. This process can be sped up by cycling the assembly with the cover in place and the water supply valve closed.

To manually cycle the ice maker, slowly turn the ejector blades clockwise until the holding switch circuit is completed to the motor. When the motor starts, all the components except the ice maker thermostat should perform normally to complete the cycle. Then remove the front cover by prying it loose with the blade of a screwdriver placed at the bottom of the support housing.

If further test-cycling is necessary, place the blade of a screwdriver in the slot located in the motor drive gear. Turn the shaft counterclockwise until the holding switch circuit is completed to the ice maker motor. See Figure 11–68.

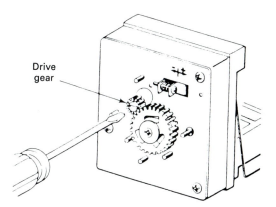

Figure 11–68 Test-cycling the ice maker. (Courtesy of WCI Major Appliance Group.)

Water Valve Switch: Water Fill Volume

The amount of water fill is directly proportional to the length of time that terminals C to NC of the water valve switch are closed. This occurs when the switch plunger drops into a cavity formed in the cam.

Different water valves have different flow rates. For this reason any time a water valve is replaced the water valve switch must be adjusted. See Figure 11–69.

The correct water fill volume is 145 cubic centimeters, or about 5 ounces. To measure, test-cycle the ice maker and collect all the water. Measure the volume of water in a container that is calibrated in cubic centimeters or ounces.

The fill volume is adjusted by increasing or decreasing the length of time that the water valve switch remains closed.

To adjust the valve switch, first determine how much more or less water is needed. The adjusting screw is calibrated so that one complete turn of the screw changes the water fill about 18 cubic centimeters. Turning the screw clockwise decreases the fill; turning the screw counterclockwise increases the fill.

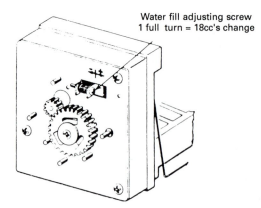

Water fill adjusting screw
1 full turn = 18cc's change

Example An ice maker is test-cycled and the water fill sample is 158 cubic centimeters. Subtracting 145 cubic centimeters from 158, the adjustment needed is 13 cubic centimeters. Since one turn of the adjusting screw changes the fill 18 cubic centimeters, a ¾-turn clockwise would reduce the fill about 13 cubic centimeters, the desired amount.

Figure 11–69 Adjusting the water fill. (Courtesy of WCI Major Appliance Group.)

12

Freezers

INSTALLATION INSTRUCTIONS

The following discussion provides the recommended installation instructions for freezers. The appliance technician should be familiar with these instructions, so that the best possible installation can be provided to the customer and the equipment will operate as it was designed. Each of these characteristics has a special importance and each one should be accomplished as completely as possible.

Model and Serial Number Label

The model number, serial number, refrigerant type, and the refrigerant charge are printed on a self-adhesive label. This label is located on the cabinet back, near the left-side lid hinge, on almost all chest freezer models, and on the front of the cabinet base, near the bottom door hinge, on almost all upright freezer models.

Leveling Legs

All upright freezers have screw-type levelers located at each of the four corners of the cabinet. Chest model freezers are not equipped with levelers.

Installation

All freezers should be installed in a cool, dry area away from heat sources such as a range, water heater, furnace, or heat register. Allow at least 4 inches of space on all sides of chest freezers, and all sides and the top of upright freezers, for the internal condenser to work efficiently.

Upright freezers must be level from side to side, and they must have a slight tilt toward the rear so that the door will swing toward the cabinet from a 45° open position, to assist in maintaining a proper door gasket seal. If the cabinet is allowed to tilt forward, the weight of the door plus the weight of the food stored in the door will result in a poor gasket seal.

Chest freezers are leveled by placing wood or metal shims under the cabinet as required. The weight of the freezer must be equally supported at each of the four cabinet corners to maintain proper lid alignment and lid gasket seal.

The floor must be sufficiently strong to support the weight of the freezer, when fully loaded.

Note: Automatic-defrost upright model freezers should not be installed where the ambient air temperature will be lower than 32°F, because the defrost water drain may freeze.

Electrical Circuit

Connect the freezer power cord to a properly grounded electrical wall outlet. A separate circuit from the fuse or circuit breaker box is preferred. The circuit should be protected by a 15- or 20-amp circuit breaker or time-delay fuse.

The voltage measured at the electrical outlet, as the compressor starts, must not vary more than ±10 percent from the 120-volt rating of the appliance.

All freezers are equipped with a power supply cord having a three-prong grounded plug and a ground wire that is attached to the freezer cabinet for protection against shock hazards. See Figure 12–1.

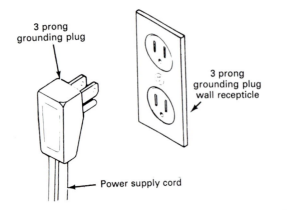

Figure 12–1 Three-prong power plug. (Courtesy of WCI Major Appliance Group.)

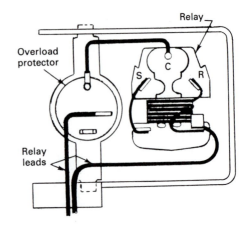

Figure 12–2 Separate mount (magnetic relay and overload). (Courtesy of WCI Major Appliance Group.)

When only a two-prong electrical wall plug is available, it is the responsibility and obligation of the customer to contact a qualified electrician and have it replaced with a properly grounded three-prong electrical wall outlet in accordance with the National Electrical Code. *Caution:* Do not under any circumstances cut or remove the grounding prong from the power supply cord.

The use of a two-prong adaptor is not an approved device for connecting a freezer with a three-prong plug to an electrical wall outlet.

ELECTRICAL COMPONENTS

The following is a description of the electrical components of a freezer. The appliance service and installation technician must be familiar with these components and their operation.

Electrical Grounding

All freezers are equipped with a power supply cord incorporating a three-prong grounding plug and a ground wire. The ground wire is attached to the freezer cabinet for protection against electrical shock hazard. Each electrical component is either mounted, or connected through a ground wire, to the cabinet to complete the ground. Certain components, such as defrost timers, may be "double insulated" and may not require a ground wire.

Be sure that the electrical wall receptacle is of the three-prong type and is properly grounded in accordance with the National Electrical Code and/or local codes.

Magnetic Relay and Overload Protector

There are two (2) types of magnetic relays and thermal-disc overloads used. Figure 12–2 shows a separately mounted relay and overload. Figure 12–3 shows a combination relay and overload on a mounting bracket. The two (2) types differ only in their mounting. They are electrically and functionally the same.

The starting relays are the push-on type that mount on the start (S) and run (R) terminals of the compressor, as shown in Figures 12–2 and 12–3. The overload protector is connected to the common (C) terminal of the compressor.

The relay coil carries the main winding current. The relay armature holds the start winding contacts in the open position, except during the starting period.

At the moment of starting, when the thermostat closes the electrical circuit, a surge of electrical current passes through the main motor winding and through the relay coil. This energizes the relay coil and pulls up the relay armature, causing the relay start winding contacts to close.

The current through the start winding introduces a second, out-of-phase, magnetic field in the stator and starts the motor. As the motor speed increases, the main winding current is reduced. At a predetermined condition, the main winding current, which is also the current through the relay coil, drops to a value below that necessary to hold up the armature. The armature drops and opens the start contacts and takes the starting winding out of the circuit.

In series with the motor windings is a bimetallic overload protector. The protector is held in place on the compressor by either a spring clip or the relay. The overload protector connects to the common (C) terminal on the compressor. See Figures 12–2 and 12–3.

Should the current to the motor windings increase to a dangerous point, the heat developed by the passage of current through the bimetallic disc will cause it to deflect and open the contacts. This breaks the circuit to the motor windings, stopping the motor before any damage can occur.

Excessive heat radiating from the compressor can also "trip" the overload protector. This is the reason that the overload protector is mounted against the compressor shell.

After the current overload or the excessive heat has caused the overload protector to break the circuit, the bimetallic disc cools down and returns the contacts to the closed position. The amount of time required for the overload switch to reset varies with the room temperature and the compressor shell temperature.

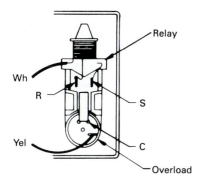

Figure 12–3 Combination mount (magnetic relay and overload). (Courtesy of WCI Major Appliance Group.)

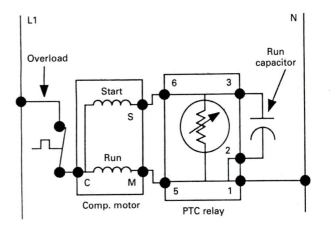

Figure 12–4 Electrical circuits. (Courtesy of WCI Major Appliance Group.)

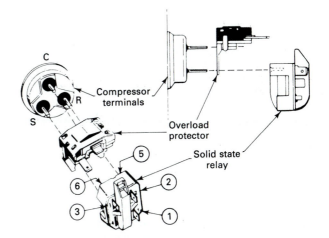

Figure 12–5 Electrical components. (Courtesy of WCI Major Appliance Group.)

The overload protector is specifically designed with the proper electrical and heat characteristics for the compressor motor and its application. Any replacement must be made with an exact replacement, that is, the same part number. Never sibstitute an overload protector with another unauthorized part number. The wrong protector can result in a burned-out motor if the relay is defective. If the relay is found to be inoperative, change both the relay and the overload protector. If the protector is inoperative, change only the overload protector.

When the thermostat cuts off after a normal cycle, or when the service cord is pulled from the electrical outlet during a running cycle, about 8 minutes (longer if it occurred during a pull-down) is required for "unloading" (the reduction of the pressure differential between the high side and the low side of the refrigeration system). During this unloading period, the overload will trip if the service cord is plugged into the electrical outlet.

To check for an open overload protector, short across its terminals. See Figures 12–2 and 12–3. If the compressor starts, replace the overload. If the compressor does not start, look for other trouble (low line voltage during the starting interval, inoperative relay, or inoperative compressor).

If the compressor repeatedly starts and runs for a few seconds and then cycles on the overload protector, the starting relay contacts may be stuck closed. The excessive current is tripping the overload.

To check the continuity of a magnetic relay, use the following procedure:

1. Remove the relay from the compressor.
2. Check the continuity of the relay coil, by connecting an ohmmeter to the relay wiring terminal and the run (R or M) pin contact. The coil should have continuity.
3. Check the continuity of the relay switch by connecting an ohmmeter to the relay wiring terminal and the start (S) pin contact. With the relay in its normal upright position, there should be no continuity. With the relay "upside down," there should be continuity.

Very High Efficiency Compressor: Electrical Components and Circuits

The new series of very high efficiency compressors are equipped with all new electrical components consisting of a solid-state PTCR* relay, a thermally operated overload protector, and a run capacitor.

Solid-state relay See Figures 12–4 and 12–5. The solid-state relay has no moving parts. It consists of a PTC resistor mounted in a plastic case with appropriate terminals.

PTC (positive temperature coefficient) simply denotes a resistor that increases in resistance as its temperature is increased. The self-heating PTC resistor used in the solid-state relay has the unique characteristic of changing from

* PTCR (positive temperature coefficient resistor) defines a resistor that increases in resistance as its temperature is increased. PTCR is commonly abbreviated as just PTC.

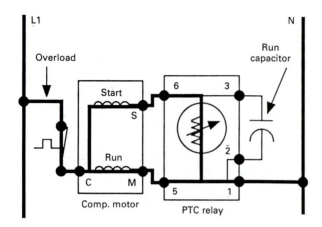

Figure 12–6 Compressor start circuit. (Courtesy of WCI Major Appliance Group.)

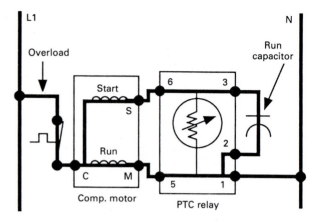

Figure 12–7 Compressor run circuit. (Courtesy of WCI Major Appliance Group.)

very low to very high resistance very abruptly, thus serving as an on/off switch.

The solid-state relay plugs directly onto the compressor start (S) and run (R) terminals. Relay terminals 1, 2, and 5 are connected within the relay, as are terminals 2 and 3. It is connected in electrical parallel with the PTC resistor. See Figure 12–4.

One side of the 120-volt ac electrical power is connected to relay terminal 1.

To check the relay, use the following procedure:

1. Disconnect the freezer service cord from the wall receptacle.
2. Remove the relay cover, disconnect the three (3) wires, and pull the relay off the compressor terminals.
3. With an ohmmeter, check the resistance between terminals 2 and 3. The resistance should be from 3 to 12 ohms, at normal room temperature. A shorted relay will read 0 resistance. An open relay will read very high or infinite resistance. Replace a relay that has an ohm reading of anything other than 3 to 12 ohms at normal room temperature.

Overload protector See Figures 12–4 and 12–5. The overload protector is completely thermally operated. It will open from excessive heat or current. Unlike prior overloads, the internal bimetal is not self-heating and is not a part of the electrical circuit. The overload has a small built-in coil heater that is in electrical series with the compressor start and run windings.

The overload protector plugs directly onto the compressor common terminal.

To check the overload protector, use the following procedure:

1. Disconnect the freezer service cord from the wall receptacle.
2. Remove the relay cover and pull the relay off the compressor terminals. Disconnect one (1) wire from the overload protector and pull the overload off the compressor.

3. With an ohmmeter, check the resistance between the tab terminal and the female pin terminal. At normal room temperature the overload protector should have less than 1 ohm of resistance. An open overload protector will have infinite resistance. Replace an overload found to have high resistance.

Run capacitor See Figure 12–5. The run capacitor has permanently attached wires that are connected to relay terminals 2 and 3. The capacitor does not have an identified terminal and can be wired without regard to polarity.

To check the run capacitor, use the following procedure:

1. Disconnect the freezer service cord from the wall receptacle.
2. Remove the relay cover and disconnect the capacitor wires from the relay.
3. Discharge the capacitor by shorting across the terminals with a 500-K (1 watt) resistor for 1 minute.
4. With an ohmmeter, check the resistance across the capacitor wire terminals with the meter set on the "ohms times 1000" scale.
   a. The needle should jump toward 0 ohms and quickly move back to infinity.
   b. If the needle does not move, the capacitor is open. It must be replaced.
   c. If the needle reads a constant value at or near 0 ohms, the capacitor is shorted. It must be replaced.
   d. If the needle jumps toward 0 ohms and then moves back to a constant high resistance (not infinity), the capacitor has a high-resistance leak. It must be replaced.

Compressor start circuit See Figure 12–6. When the compressor circuit is first energized, the solid-state relay has low resistance (3 to 12 ohms), and both the run and the start windings are energized to start the compressor. The run capacitor is being shunted (bypassed) by the PTC relay. It has little function during the compressor starting phase.

Compressor run circuit See Figure 12–7. When the self-heating solid-state relay has reached sufficient temperature, it will abruptly change from low resistance (3 to 12 ohms) to a very high resistance (10 to 20 kiloohms) and, in effect, "switches" off the start windings. The run capacitor is now in electrical series with the start winding. The only purpose of the run capacitor is to improve compressor operating efficiency, which it does by "correcting" the power factor of the compressor motor.

Compressor operating characteristics The following is a description of the operating characteristics of a compressor motor:

1. When the compressor electrical circuit is energized, the start winding current causes the PTC relay to heat. After an appropriate amount of starting time, the relay "switches off" the start winding circuit. *Note:* The PTC relay will "switch off" the start winding circuit even though the compressor has not started (such as when attempting to restart after a momentary power interruption).

2. Because the PTC relay opens the compressor start circuit whether or not the compressor has started, the overload protector is designed and calibrated to open the compressor electrical circuit with locked rotor run winding current only. *Caution:* Use only the correctly specified overload protector for service replacement.

3. With an "open" PTC relay, the compressor will not start because there is little or no current to the start winding. The overload protector will open because of the high locked rotor run winding current.

4. With a "shorted" PTC relay or capacitor, the compressor will start, but the overload protector will quickly open because of the high combined current of the run and start windings.

5. With an open or weak capacitor, the compressor will start and apparently run normally. The compressor, however, will be operating at reduced efficiency.

Compressor motor electrical check When checking the freezer for electrical trouble, always be sure that there is a "live" electrical circuit to the freezer cabinet, and that the temperature selector dial is not in the OFF position.

When the compressor will not start and the cabinet temperature is warm, the trouble may be in the relay, overload, thermostat, wiring, or in the compressor motor itself.

If the compressor will not run, make a voltage check across the winding terminals on the relay and the overload protector. The voltmeter should show line voltage if the thermostat knob is in the normal operating position. If this check does not show a "live" circuit, the thermostat and wiring should be checked for an open circuit. Pay particular attention to all terminal connections.

A control thermostat check can be made by using a piece of wire as a temporary bridge across the two thermostat terminals. If the compressor now starts and runs with the bridge, the control thermostat is at fault and should be replaced.

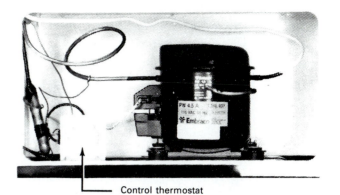

Figure 12–8 Compact chest control thermostat. (Courtesy of WCI Major Appliance Group.)

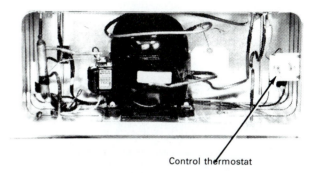

Figure 12–9 Standard chest control thermostat. (Courtesy of WCI Major Appliance Group.)

If the voltage check shows power supply at the relay terminals, check the compressor by means of an external compressor motor test cord. *Note:* Follow the operating instructions included with the compressor test cord.

If the compressor does not start and run with either the test cord or the regular accessories, check the line voltage. There should be no more than a ±10 percent variation from the normal 120 volts. If the voltage is correct and the compressor will not start and run, replace the compressor.

If the compressor starts and runs with the test cord, replace the relay.

Control thermostats The control thermostats on all freezer models are of the variable cut-in and cutoff type, having gas-charged capillary tube sensing elements.

All Chest Models

The control thermostat is mounted to a bracket inside the compressor compartment. The sensing element is inserted about 18.5 to 19.5 inches into a control well located in the top of the compartment. See Figures 12–8 and 12–9.

To replace the control thermostat, use the following procedure:

1. Disconnect the freezer power cord from the electrical outlet.

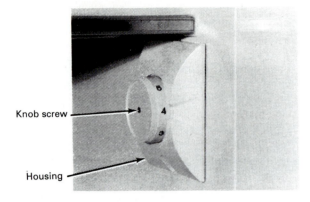

Figure 12–10 Control thermostat housing. (Courtesy of WCI Major Appliance Group.)

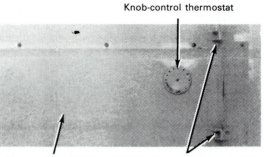

Figure 12–12 Control thermostat location. (Courtesy of WCI Major Appliance Group.)

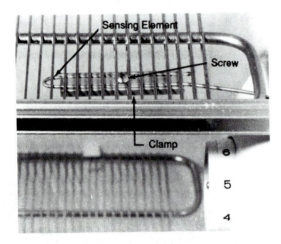

Figure 12–11 Control thermostat sensing element. (Courtesy of WCI Major Appliance Group.)

2. Remove the compressor compartment cover. Mark the sensing element at the point where it enters the control well in the top of the compressor compartment, either with tape or a marking pen.

3. Disconnect the wiring, including the ground wire, from the control thermostat.

4. Remove the control knob and the thermostat mounting screws. Gently pull the sensing element from the control well.

5. Carefully straighten the sensing element of the replacement control thermostat. Using the original thermostat as a pattern, mark the sensing element for the correct depth of insertion into the control well.

6. Push the new control thermostat sensing element into the control well to the proper depth, about 18.5 to 19.5 inches. Seal the end of the well with a mastic sealer, such as Permagum.

7. Reverse the above procedure to reassemble the unit.

Upright manual defrost models The control thermostat is mounted in a plastic housing on the right side of the freezer liner. Two (2) passes of the sensing element are clamped to the bottom of a shelf. See Figures 12–10 and 12–11.

To replace the control thermostat, use the following procedure:

1. Disconnect the freezer power cord from the electrical outlet.

2. Remove the thermostat knob and unclamp the sensing element from the bottom of the shelf.

3. Remove the screw from the bottom of the control housing. Pull the bottom of the housing out and down to disengage the housing tabs from the slots in the liner.

4. Disconnect the wiring, incuding the ground wire, from the control thermostat.

5. Remove the control thermostat mounting screws and the thermostat from the freezer.

6. Reverse the above procedure to reassemble the unit.

Upright automatic-defrost models The control thermostat is mounted to the upper right corner of the evaporator cover. Two passes of the sensing element are clamped to a bracket located above the right side of the evaporator. See Figures 12–12 and 12–13.

To replace the control thermostat, use the following procedure:

1. Disconnect the freezer power cord from the electrical outlet.

2. Remove enough of the freezer shelves to allow full access to the evaporator cover.

3. Remove the control thermostat knob and unsnap the thermostat from the evaporator cover.

4. Tip the thermostat forward and push it back through the opening in the evaporator cover.

5. Remove the evaporator cover mounting screws and the rear shelf supports for the lower two (2) or (3) shelves.

6. Pull the top of the evaporator cover forward. Disconnect the wiring from the thermostat and the sensing

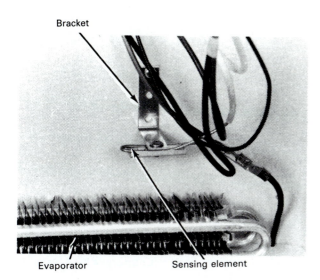

Figure 12–13 Control thermostat sensing element. (Courtesy of WCI Major Appliance Group.)

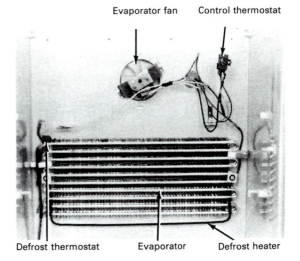

Figure 12–14 Automatic-defrost components. (Courtesy of WCI Major Appliance Group.)

element from the bracket. Remove the thermostat from the freezer.

7. Reverse the above procedure to reassemble the unit.

Automatic-Defrost Components

Components that are unique to automatic-defrosting models include a fin-and-tube-type evaporator, evaporator fan assembly, defrost thermostat, defrost heater, and defrost limiter.

Evaporator assembly The fin-and-tube evaporator is located at the lower back of the freezer compartment. See Figure 12–14. Note the styrofoam air barriers located at each side of the evaporator. These air barriers must be in place for proper operation of the freezer.

Evaporator fan and motor The evaporator fan assembly is located behind the evaporator cover, just above the evaporator. The propeller-type fan draws into the duct (located behind the freezer liner) and forces it up to the top of the freezer compartment. See Figures 12–14 and 12–15.

To remove the evaporator fan assembly, use the following procedure:

1. Disconnect the freezer power cord from the electrical outlet.

2. Remove enough of the freezer shelves to allow full access to the evaporator cover.

3. Remove the control thermostat knob and unsnap the thermostat from the evaporator cover.

4. Tip the thermostat forward and push it back through the opening in the evaporator cover.

5. Remove the evaporator cover mounting screws and the rear shelf supports for the lower two (2) or three (3) shelves. Remove the evaporator cover.

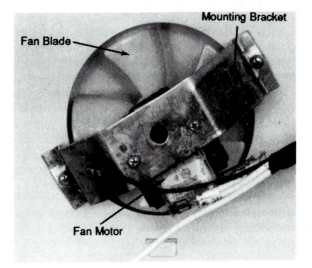

Figure 12–15 Evaporator fan assembly. (Courtesy of WCI Major Appliance Group.)

6. Disconnect the wiring, including the ground wire. Remove the screws attaching the fan assembly to the liner.

7. With the fan assembly removed from the freezer, the fan blade is removed by simply pulling it off the motor shaft. The motor is removed by removing the two (2) mounting screws.

8. Reverse the above procedure to reassemble the unit. *Note:* When installing the fan blade on the motor shaft, be sure that the hub is flush with the end of the motor shaft.

Defrost thermostat The disc-type defrost thermostat is located at the upper left corner of the evaporator. It is held in place on the evaporator tubing with a built-in clip.

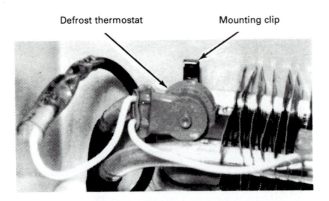

Figure 12–16 Defrost thermostat location. (Courtesy of WCI Major Appliance Group.)

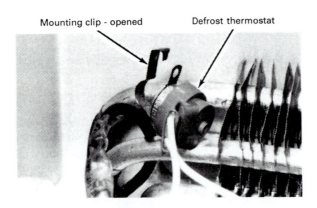

Figure 12–17 Defrost thermostat mounting clip. (Courtesy of WCI Major Appliance Group.)

See Figures 12–14, 12–16, and 12–17. The defrost thermostat contacts may be checked for continuity with an ohmmeter. At a normal room temperature of 65°F or higher, the contacts should be open.

To remove the defrost thermostat, use the following procedure:

1. Disconnect the freezer power cord from the electrical outlet.
2. Remove enough of the freezer shelves to allow full access to the evaporator cover.
3. Remove the control thermostat knob and unsnap the thermostat from the evaporator cover.
4. Tip the thermostat forward and push it back through the opening in the evaporator cover.
5. Remove the evaporator cover mounting screws and the rear shelf supports for the lower two (2) or three (3) shelves. Remove the evaporator cover.
6. With a small screwdriver, open the defrost thermostat mounting clip and remove the thermostat from the evaporator. See Figures 12–16 and 12–17.
7. Disconnect the defrost thermostat wiring.

Note: The defrost thermostat has insulated in-line tab-type terminals. After assembly, the terminals are sealed against moisture infiltration with a hot-melt-type sealer. To

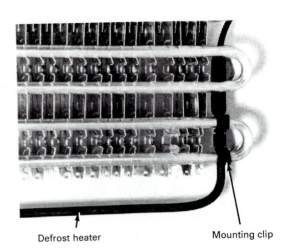

Figure 12–18 Defrost heater. (Courtesy of WCI Major Appliance Group.)

disconnect the sealed terminals, slit the plastic sleeve lengthwise and heat the hot-melt sealer just enough to pull the wire free. Then disconnect the terminal. If the original insulated terminal is used for reassembly, connect the terminal and push the wire into the reheated hot-melt sealer. Reseal the plastic sleeve with additional hot-melt sealer or a silastic-type sealer. *Do not* allow the hot-melt sealer to contaminate the terminals while disconnecting or connecting the wiring.

8. Reverse the above procedure to reassemble the unit.

Defrost heater The radiant-type defrost heater is mounted at each lower corner of the evaporator with spring steel clips. See Figures 12–14 and 12–18. The heater is centered between the front and rear passes of the evaporator. It has $\frac{1}{2}$-inch clearance from the bottom of the evaporator fins. The defrost heater may be checked for continuity and resistance with an ohmmeter by disconnecting the wire leads.

To remove the defrost heater, use the following procedure:

1. Disconnect the freezer power cord from the electrical outlet.
2. Remove enough of the freezer shelves to allow full access to the evaporator cover.
3. Remove the control thermostat knob and unsnap the thermostat from the evaporator cover.
4. Tip the thermostat forward and push it back through the opening in the evaporator cover.
5. Remove the evaporator cover mounting screws and the rear shelf supports for the lower two (2) or three (3) shelves. Remove the evaporator cover.
6. Disconnect the defrost heater wiring.
7. Remove the styrofoam air barriers from each side of the evaporator. Remove the spring steel heater mounting clips by pulling them forward off the evaporator tubes.
8. Pull the heater down and away from the evaporator.

9. Reverse the above procedure to reassemble the unit. Be sure that the defrost heater is spaced ½ inch down from the evaporator fins.

Defrost timer The defrost timer is mounted to a bracket in the compressor compartment. All wiring connections are made with a single four-conductor connector. The timer will initiate a defrost cycle after 12 hours of compressor run time. It will defrost for up to a maximum of 30 minutes. The timer contacts and motor may be checked with an ohmmeter for continuity or resistance.

To remove the defrost timer, use the following procedure:

1. Disconnect the freezer power cord from the electrical outlet.
2. Disconnect the timer wiring by pulling the four-conductor connector straight off the timer terminals.
3. Remove the timer mounting screws and remove the timer from the compressor compartment.
4. Reverse the above procedure to reassemble the unit.

To check the timer, use the following procedure:

1. Remove the timer from the freezer as instructed above.
2. Check the timer motor by reading the resistance between terminals 1 to 3. The motor coil should have about 8000 to 9000 ohms of resistance.
3. Check the defrost timer contacts by advancing the timer at least ¼ turn and continuing until it "clicks" just once. The defrost heater terminals 1 to 2 should be closed. The compressor terminals 1 to 4 should be open.
4. Advance the timer slowly until it "clicks" a second time. The defrost heater terminals 1 to 2 should be open. The compressor terminals 1 to 4 should be closed.

Interior Lights

The following is a description of the interior lights used on freezer cabinets.

Chest models The interior light is located on the inner lid panel. The light has a 25-watt, candelabra base, "showcase"-type bulb. The combination lamp socket/mercury switch is attached to the lid with a retaining clip. The 25-watt bulb is replaced by simply opening the freezer lid and unscrewing the bulb.

To remove the lamp socket assembly, use the following procedure:

1. Disconnect the freezer power cord from the electrical outlet.
2. Remove and disassemble the lid. See the Lid Removal Section presented earlier in this chapter.
3. Disconnect the lamp socket wiring and remove the socket retaining clip.

4. Reverse the above procedure to reassemble the unit.

Upright models The interior light is located at the top of the freezer liner. The light is a 25-watt, candelabra base, "showcase"-type bulb. The lamp socket "snaps" into a rectangular opening in the freezer liner. The door-activated light switch is mounted in the right-side breaker trim. Wiring to the lamp socket passes through the foam insulation and is not replaceable. A lamp shield, in front of the bulb, prevents contact with food packages.

To remove the lamp socket assembly, use the following procedure:

1. Disconnect the freezer power cord from the electrical outlet.
2. Remove the lamp shield and unscrew the lamp bulb.
3. With a small screwdriver, carefully push in on the socket retaining tabs and work the socket out of the liner opening. Disconnect the socket wiring.
4. Reverse the above procedure to reassemble the unit.

To remove the light switch, use the following procedure:

1. Disconnect the freezer power cord from the electrical outlet.
2. Remove the right-side breaker trim. See the Breaker Trim Removal section presented earlier in this chapter.
3. Disconnect the switch wiring. Then depress the switch retaining tabs and carefully work the switch out of the breaker trim.
4. Reverse the above procedure to reassemble the unit.

Temperature Warning Alarm

The temperature warning alarm system consists of an amber "power-on" signal light, a red temperature alarm signal light, a temperature alarm buzzer, a buzzer on/off switch, and a temperature sensor. Wiring between the sensor and the alarm assembly passes within the foam insulation and is not replaceable. See Figures 12–19 and 12–20.

The amber "power-on" light will glow anytime that the freezer is connected to a "live" electrical outlet.

The red alarm light and alarm buzzer will be energized by the sensor if the freezer temperature rises to about 22°F. The light will continue to glow and the alarm buzzer will sound continuously, until the freezer temperature has been reduced to a safe food storage level. However, the alarm buzzer may be turned off by the buzzer on/off switch until the red alarm light goes out. The buzzer can then be turned back on.

When the freezer is initially placed in operation, the buzzer switch should be turned to the OFF position until the red warning light has gone off.

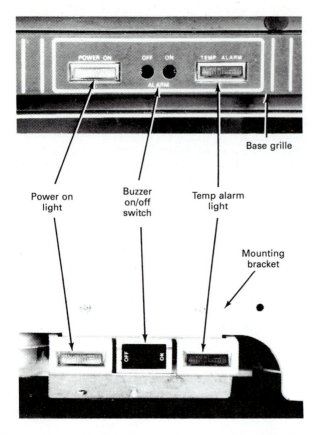

Figure 12–19 Upright freezer alarm assembly. (Courtesy of WCI Major Appliance Group.)

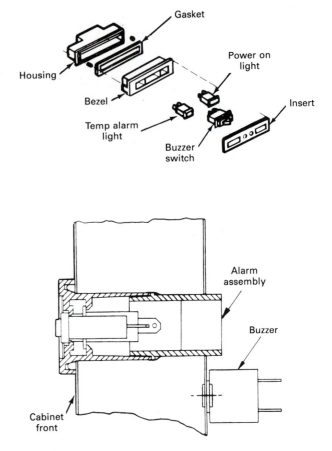

Figure 12–20 Chest freezer alarm assembly. (Courtesy of WCI Major Appliance Group.)

The temperature alarm system operates on the standard 120-volt power supply to the freezer.

Upright models All alarm system components, except the sensor, are mounted to a bracket located behind the right side of the base grille. See Figure 12–19. The alarm assembly components are accessible for service by removing the base grille and the bracket mounting screws.

The alarm sensor is located on the upper right side of the freezer liner. See Figure 12–21. The housing is removed by squeezing the top and bottom together to release the mounting tabs. The sensor can then be unplugged.

Chest models The alarm signal lights and the buzzer on/off switch are contained within a housing assembly. The entire assembly is mounted through a hole in the front of the cabinet. It is held in place by two (2) pressure fit studs on the bezel. The alarm buzzer is mounted to the inside wall of the compressor compartment, just below the alarm assembly. See Figure 12–20.

The alarm sensor is located on the freezer liner. See Figure 12–21. The housing is removed by squeezing the top and bottom together to release the mounting tabs. The sensor can then be unplugged.

AIR CIRCULATION PATTERN

Automatic-defrost freezers have a single evaporator. They have forced-air cooling throughout the freezer.

The fin-and-tube-type evaporator is located on the back wall of the freezer. An air-circulating fan (suction type) pulls the air from the freezer across the evaporator surfaces. The cold dry air is forced into an air duct and is then discharged into the top of the freezer. See Figure 12–22.

The air-circulating fan in the freezer operates only when the compressor is running. During the defrost period, however, the compressor and the circulating fan do not operate. The automatic-defrost timer opens the electrical circuit to the fan motor and the compressor motor.

Principle of Automatic-Defrost Operation

Automatic-defrost freezers operate on the principle that moisture or frost transfers or migrates to the coldest surfaces (evaporator) in the freezer. For example, a small amount of water spilled from an ice cube tray in the freezer will freeze immediately; however, this ice in time will evaporate and transfer to the colder surfaces of the freezer evaporator coil.

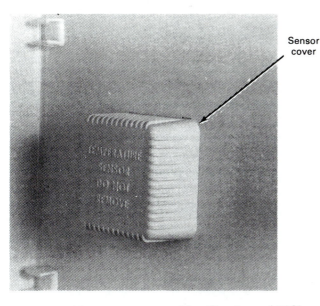

Sensor cover

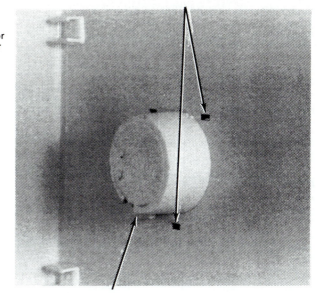

Cover mounting holes

Sensor assembly

Figure 12–21 Alarm sensor assembly. (Courtesy of WCI Major Appliance Group.)

REFRIGERATION SYSTEMS AND SERVICE

Warning: The instructions given here are furnished only as a guide. Persons attempting the use of these instructions to make repairs to the sealed refrigeration system should have a working knowledge of refrigeration and previous training on sealed-system repair.

Safety

Caution: Compressor testing Whenever testing a compressor, extreme caution should be used to prevent damaging the terminals. A compressor with a damaged terminal or a grounded winding can expel a terminal from its insulated housing when the compressor is energized. If this happens, a mixture of refrigerant and oil will be released which could be ignited by an external heat source (open flame, heater, etc.). Also, if there is air in the system when this happens, a spark at the compressor shell could ignite the refrigerant and oil mixture.

Warning: Charging sealed systems Overcharging a freezer system with refrigerant can be dangerous. If the overcharge is sufficient to immerse the major parts of the motor and compressor in liquid refrigerant, a situation has been created which, when followed by a sequence of circumstances, can lead to the compressor shell seam separating.

A hydraulic block occurs preventing the compressor from starting. This condition is known as locked rotor current. Electrical current continues to flow through the compressor motor windings which become, in effect, electrical resistance heaters. The heat produced begins to vaporize the excess refrigerant liquid, causing a rapid increase in system pressure. If the compressor protective devices fail, the pressure within the system may rise to extremes far in

AIR FLOW

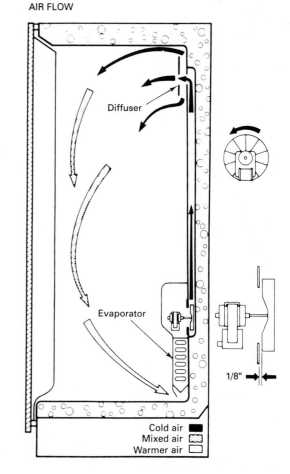

Diffuser

Evaporator

1/8"

Cold air
Mixed air
Warmer air

Figure 12–22 Automatic-defrost freezer air flow. (Courtesy of WCI Major Appliance Group.)

excess of the design limits. Under these conditions, the weld seam around the compressor shell can separate with explosive force, spewing oil and refrigerant vapor, which could ignite.

To eliminate this exceedingly rare but potential hazard, never add refrigerant to a sealed system. If refrigerant is required, evacuate the existing charge and recharge the system with the correct amount of refrigerant as specified for the system.

Soldering

Caution: Wear the proper and approved safety glasses when working with, or on, any pressurized system or equipment. Have an approved dry-type fire extinguisher handy when using any type of gas-operated torch.

1. All joints to be soldered must have a proper fit. The clearance between the tubes to be soldered should range from 0.001 to 0.006 inch. It is not practical to actually measure this clearance; however, a dry or loose fit is undesirable. The tubing joints should overlap about the distance of their diameter except for the restrictor tubes which should be inserted 1.25 inches.

2. Clean all joint areas with steel wool or preferably an abrasive cloth, such as grit cloth No. 23 or Scotch-Brite.

3. Apply a thin film of liquid flux that is recommended for silver soldering the surfaces to be joined and the surfaces immediately adjacent to the joint.

4. Align the tubing so that there is no stress on the joint. Do not move the tubing while the solder is solidifying or leaks will result.

Caution: During the application of heat, use wet cloths to prevent the heat from conducting to areas other than the soldered joint. Use a sheet of metal as a heat deflector to keep the flame away from flammable materials and painted surfaces.

5. Use a torch of adequate capacity, so that the joint can be quickly heated with a minimum of heat travel to other points. Use a good grade of silver solder.

6. Solder the connections. If the tubing is properly cleaned and fluxed, the solder will readily flow. Do not use an excessive amount of solder, just enough to make a good bond.

7. Allow the joint to cool, then wash the exterior with water to remove excess flux.

Refrigeration Systems

The basic components of a refrigerator are the compressor, condenser, evaporator, capillary tube, suction line, and drier. In addition, some models may have a defrost water-evaporating plate assembly and/or a perimeter hot tube.

Refrigerant cycle The refrigerant cycle is a continuous cycle that occurs whenever the compressor is in operation. Liquid refrigerant is evaporated in the evaporator by the heat entering the freezer cabinet through the insulated

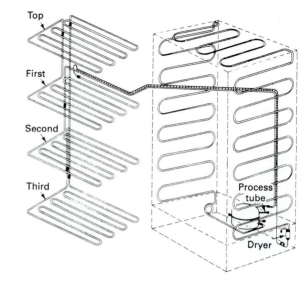

Figure 12–23 Upright, manual-defrost system layout. (Courtesy of WCI Major Appliance Group.)

walls and that introduced by the product load and door openings. The refrigerant vapor is then drawn from the evaporator, through the suction line, to the compressor. The pressure and temperature of the vapor are raised in the compressor by compression. The vapor is then forced through the discharge valve into the discharge line and condenser. Air passing over the condenser surface removes heat from the high-pressure vapor, condensing it to a liquid. The liquid refrigerant flows from the condenser to the evaporator through a small-diameter liquid line (capillary tube). Before it enters the evaporator, it is subcooled in the heat exchanger by the low-temperature suction vapor in the suction line.

Refrigerant system layout and refrigerant flow Figures 12–23 through 12–30 show the system layout and refrigerant flow in the various types of refrigerating systems.

Low- or high-side leak or undercharge A loss of refrigerant results in excessive or continuous compressor operation, above-normal freezer compartment temperature, a partially frosted evaporator (depending on the amount of refrigerant loss), below-normal freezer compartment temperature, low suction pressure (vacuum), and low compressor wattage. The condenser will be "warm to cool," again, depending on the amount of refrigerant loss.

When refrigerant is added, the frost pattern will improve, the suction and discharge pressures will rise, the condenser will become hot, and the compressor wattage will increase.

In the case of a low-side refrigerant leak, resulting in a complete loss of refrigerant, the compressor will run, but with no refrigeration. The suction pressure will drop below atmospheric pressure. Air and moisture will be drawn into the system, saturating the drier.

If a slight undercharge of refrigerant is indicated and no leak is found after a thorough leak test, the charge can

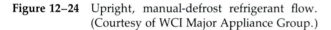

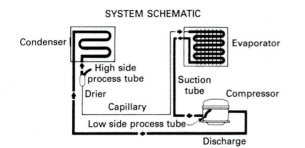

Figure 12–24 Upright, manual-defrost refrigerant flow. (Courtesy of WCI Major Appliance Group.)

SYSTEM SCHEMATIC

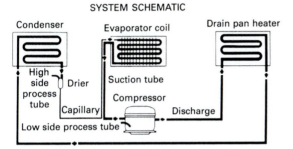

Figure 12–26 Upright, auto-defrost refrigerant flow. (Courtesy of WCI Major Appliance Group.)

be corrected without changing the compressor.

Even though there is no reason to believe that the system has operated for a considerable length of time with no refrigerant and a leak is found in the evaporator, excessive amounts of moisture may have entered the system. In such cases the compressor may need to be replaced to prevent repetitive service.

If a high-side leak is located and some refrigerant remains in the system, it is not necessary to change the compressor.

Testing for refrigerant leaks If the system is diagnosed as short of refrigerant and the system has not been opened, there is probably a leak in the system. Adding refrigerant without first locating and repairing the leak or replacing the compressor would not permanently correct the difficulty. *The leak must be found.* In some cases, sufficient refrigerant may have escaped to make it impossible to leak-

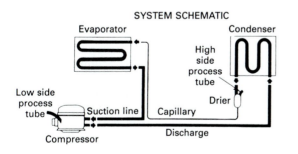

Figure 12–28 Compact chest refrigerant flow. (Courtesy of WCI Major Appliance Group.)

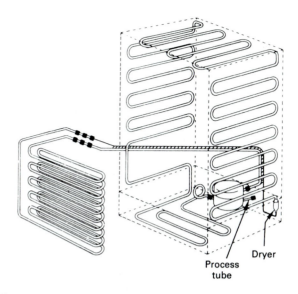

Figure 12–25 Upright, auto-defrost system layout. (Courtesy of WCI Major Appliance Group.)

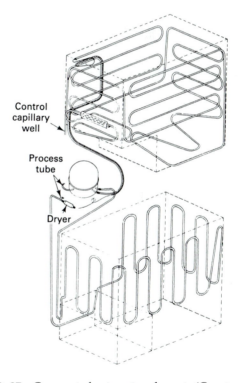

Figure 12–27 Compact chest system layout. (Courtesy of WCI Major Appliance Group.)

test effectively. In such cases, add a ¼-inch line-piercing valve to the compressor process tube. Add sufficient refrigerant to increase the pressure to 75 psig. Through this procedure, minute leaks are more easily detected before discharging the system and possibly contaminating the surrounding air. *Note:* The line-piercing valve (clamp-on type) should be used for adding refrigerant and for test purposes only. It must be removed from the system after serving its purpose.

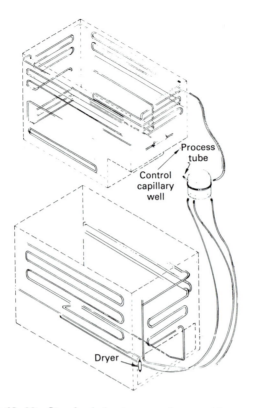

Figure 12–29 Standard chest system layout. (Courtesy of WCI Major Appliance Group.)

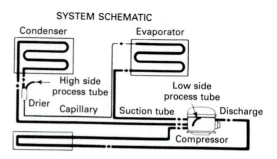

Figure 12–30 Standard chest refrigerant flow. (Courtesy of WCI Major Appliance Group.)

Procedure for checking internal leaks Before checking for internal leaks, check all accessible system components and joints for leaks.

If an internal system leak is suspected, it must be determined if the leak is in the condenser or evaporator. Use the following procedure:

1. Remove the charge from the system by using refrigerant recovery equipment.*

2. Disconnect the condenser tube from the drier and pinch off and solder both the drier and condenser tube closed.

3. Remove the suction and discharge tubes from the compressor. Oil cooler lines, if so equipped, may remain connected to the compressor. Solder a ¼-inch charging hose fitting to both tubes.

4. Connect pressure gauges and access valves to both tubes. Pressurize the system to 250 psig using dry nitrogen or carbon dioxide.

Warning: Never pressurize with oxygen. Never open a high-pressure tank unless it is equipped with a pressure regulator. Never put high pressure on the dome of the compressor. It might explode. Make sure that the gauge fittings are in good condition and do not leak.

5. Leave the pressure on each side of the system for 24 hours. Any drop in pressure is an indication of a leak.

Compressor replacement Before installing the new compressor, check for possible system contamination by obtaining an oil sample from the old compressor. If the oil has a burned odor, but no color change or residue, a normal compressor change according to the instructions below may be made. If the oil has the odor of burned sugar and shows contamination (dark color), the system should be cleaned, using the proper procedures to prevent contaminating the atmosphere. Install a new drier with each new compressor installation.

Replacement of the compressor and installation of the filter drier must be done in a continuous sequence so that the system is not exposed to the atmosphere any longer than is necessary. Also, avoid opening the system when any of the components or lines are cold.

All replacement compressors are shipped with rubber plugs in the suction, discharge, and process tubes. They will also contain a holding charge of inert gas. Compressors have a low-side process tube attached to the compressor shell.

A high-side process tube is attached to the filter drier.
Warning: Do not operate a reciprocating compressor when charging liquid refrigerant into the system through the process tube.

All models To replace the compressor in a freezer, use the following procedure for all models:

1. Disconnect the unit from the source of electrical power.

2. Remove the compressor cover (chest models) from the freezer to gain access to the compressor.

3. Install the required equipment and remove the refrigerant using the proper refrigerant recovery methods.

4. Remove the leads from the compressor motor terminals.

5. Remove the mounting nuts and bolts.

6. After the refrigerant has been completely removed, cut the suction and discharge lines as close to the compressor as possible. Leave only enough tubing to pinch

---

* The federal government has restrictions governing the disposal of chlorofluorocarbons (CFCs) such as R-12 and R-22. Therefore, when removing the refrigerant from a system, use the proper refrigerant recovery equipment.

off and seal the defective compressor. Plug or tape the open-system tubing to avoid the entrance of moisture and air into the system. Remove the inoperable compressor and transfer the mounting parts to the new compressor.

7. Release the holding charge from the new compressor. (Release it slowly to avoid oil discharge.)

8. Install the new compressor in exactly the same manner as the original compressor.

9. Reform both the suction and discharge lines to align with the new compressor. If they are too short, use additional lengths of tubing. The joints should overlap ½ inch to provide sufficient area for a good solder joint. Clean and mark the area where the tubing should be cut. Cut the tubing with a tubing cutter. Work as quickly as possible to avoid letting moisture and air into the system. (*Note:* If the new low-side process tube is too short, silver solder a 4-inch piece of tubing onto the process tube at this time.)

10. Solder all connections according to the soldering procedure outlined above.

11. Remove the original and install a new filter drier at the condenser outlet. Evacuate and charge the system using the recommended procedure described in the Evacuating and Recharging section presented later in this chapter.

12. Reconnect the compressor terminal leads in accordance with the freezer wiring diagram.

Evaporator replacement Chest model evaporators are formed in place and are not accessible for repair or replacement. Contact the local distributor of the freezer in question to determine if the evaporator is still under warranty. On the other hand, upright model evaporators can be replaced using the following procedures:

Automatic-defrost models Use the following procedure when replacing the evaporator in these units:

1. Disconnect the unit from the source of electrical power.

2. Remove sufficient assemblies from the freezer to gain access to the evaporator.

3. Remove the refrigerant charge in accordance with proper procedures to avoid polluting the atmosphere.

4. Remove the evaporator assembly from the freezer. Cut the suction and discharge lines as close to the evaporator as possible.

5. Clean the suction and capillary lines with abrasive cloth.

6. Connect the lines to the replacement evaporator and solder the joints.

7. Remove the original and install a new filter drier at the condenser outlet. Evacuate and recharge the system using the recommended procedure described in the Evacuating and Recharging section presented later in this chapter.

8. Reverse the above procedure to reassemble the unit. All sealing materials must be replaced where the lines pass through the cabinet.

Manual-defrost models Use the following procedure when replacing the evaporator in this model unit:

1. Disconnect the unit from the source of electrical power.

2. To prevent damage to the painted steel liner, do not attempt to solder inside the cabinet. Pull the evaporator shelves out of the cabinet by disconnecting the thermostat sensor from the shelf and removing the shelf fronts, the screws attaching the top shelf to the liner, and the plastic ties from the refrigerant lines.

3. Remove the refrigerant, using the proper recovery methods.

4. Remove the defective shelf by unsoldering the joints.

5. Prepare the joints by cleaning them with an abrasive cloth.

6. Connect the replacement shelf to the evaporator assembly and silver solder the joints. Clean the joints and coat them with an aluminum paint.

7. Reassemble the evaporator assembly into the cabinet and align the refrigerant lines into the left corner of the liner.

8. Remove the original and install a new filter drier at the condenser outlet. Evacuate and recharge the system using the recommended procedure described in the Evacuating and Recharging section presented later in this chapter.

9. Reverse the above procedure to reassemble the unit. All sealing materials must be replaced where the lines pass through the cabinet.

Condenser replacement: all models The condensers are formed in place and are not accessible for repair. However, repairs can be made by installing a service replacement condenser kit.

Each service replacement condenser kit consists of a condenser assembly to be installed on the back of the cabinet, mounting hardware, replacement filter drier, and complete instructions.

To install the condenser kit, use the following procedure:

1. Disconnect the unit from the source of electrical power.

2. Remove the charge of refrigerant from the system using the proper recovery techniques.

3. Install the replacement condenser in accordance with the instructions included with each kit.

4. Evacuate and charge the system using the recommended procedure described in the Evacuating and Recharging section presented later in this chapter.

Suction line and capillary tube replacement (upright models only) Chest model suction line and capillary tube assemblies are formed in place and are not accessible for repair or replacement. For upright models, proceed as follows:

1. Disconnect the unit from the source of electrical power.

2. Remove the refrigerant using the proper removal and reclaiming procedures.

3. Remove and replace the suction line and capillary

3. Feel the condenser. With the compressor in operation, the condenser should be hot, with a gradual reduction in temperature from the refrigerant entry to the refrigerant exit of the condenser.

4. Are the lid and door gaskets sealing properly?

5. Does the door or lid activate the light switch?

6. Is the evaporator fan properly located on the motor shaft? (automatic-defrost models)

7. Is the thermostat sensing element properly positioned?

8. Observe the frost pattern on the evaporator.

9. Check the thermostat knob setting.

10. Inscribe the bracket opposite the slotted shaft of the defrost timer to determine if the timer advances (automatic-defrost models).

The service technician should inquire as to the number of people in the family to determine the service load and the daily door openings. In addition, the service technician should know the room temperature.

After this diagnosis is completed, a thorough operational check should be made of the refrigeration system.

Freezer air temperatures Freezer temperatures are affected by improper door or lid seal, frost accumulation on the evaporator, service load, ambient temperature, percentage of relative humidity, thermostat calibration (cut-in and cutout), location of the evaporator fan blade on the motor shaft, and the compressor efficiency.

From this, it is evident that the temperatures are not always the same in every freezer, even under identical conditions.

Line voltage It is essential to know the line voltage at the freezer. A voltage reading should be taken at the instant that the compressor starts and while the compressor is running. The line voltage fluctuation should not exceed ±10 percent from the nominal rating. Low voltage will cause overheating of the compressor motor windings, resulting in the compressor cycling on the thermal overload, or the compressor may fail to start.

Inadequate line wire size and overloaded lines are the most common reasons for low voltage at the freezer.

13

Dehumidifiers and Humidification

DEHUMIDIFIERS

The purpose of a dehumidifier is to remove moisture from the air being circulated through the unit. This reduced relative humidity helps in preventing rust, mildew, and rot on the surfaces inside the room or any other enclosed space where the dehumidifier is located.

The component parts of a dehumidifier are a motor-compressor unit, a refrigerant condenser, an air-circulating fan, a refrigerated coil (evaporator), some means of collecting and disposing of the condensed moisture, and a cabinet to house these various components. See Figure 13–1.

In operation, the fan draws the moisture-laden air through the cooling coil where it is cooled below its dew point, causing the moisture to condense out. The moisture either drains off into a water receptacle or it may pass through a drain and into the sewer system. The cold air then passes through the hot condenser where it is reheated. The air is then heated further by other unit-radiated heat and is discharged into the room at a temperature somewhat higher and with a lower relative humidity. The continuous circulation of the room air gradually reduces the humidity inside the room.

Design and Construction

Hermetic-type motor-compressors of the size and type that produce the rated output of the overall unit are used.

The refrigerant condenser, in most cases, is a conventional finned-tube-type coil. The control of refrigerant flow is generally done with a capillary tube device. However, some units in the higher-capacity range may use an expansion valve for this purpose.

A direct-driven propeller-type fan is generally driven by a shaded-pole motor, creating the required air flow through the unit. The air flow rate through these units is from 125 to 250 cubic feet per minute, depending on the moisture removal capacity of the unit. Generally, the dehumidifier output is increased with an increase in the air flow. Very high rates of air flow may, however, create an objectionable noise. Residential dehumidifiers in most cases will maintain a satisfactory humidity level within the space when the air flow rate and the unit placement permit the entire volume of air inside the space to be passed through the dehumidifier once per hour. For example, an air flow rate of 200 cubic feet per minute would be sufficient to provide for one air change per hour in a room that has a volume of 12,000 cubic feet. The AHAM Dehumidification Selection Guide should be consulted for more detailed information.

The evaporator is generally of the bare-tube construction, although finned-tube coils are sometimes used if the fins are spaced wide enough to allow for rapid runoff of the water droplets. Bare-tube coils mounted vertically tend to collect smaller droplets of water, allow faster runoff, and cause less water reevaporation than the finned-tube or the horizontally mounted bare-tube coils. Bare-tube coils that

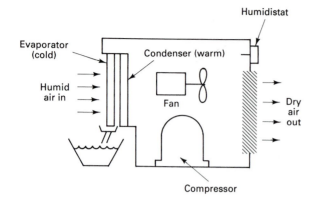

Humidistat

Evaporator (cold)

Condenser (warm)

Humid air in

Fan

Dry air out

Compressor

Figure 13–1 Typical dehumidifier unit.

are continuously wound in the form of a flat circular spiral (sometimes consisting of two coil layers) and mounted with the flat dimensions in a vertical plane are considered a good design compromise because they have most of the advantages of the vertical-tube coil.

The dehumidifier evaporator is protected against corrosion by several methods such as painting, waxing, and anodizing (on aluminum). None of these finishes produce a considerable loss in capacity of the unit.

Most dehumidifiers are provided with removable water receptacles that will hold approximately 16 to 24 pints of water. Generally they are made from plastic, which withstands corrosion much better than metal. They are made so that they can be removed and emptied with a minimum amount of trouble. Most dehumidifiers are equipped with a connection for connecting a flexible hose to either the water receptacle or some other means of connecting the drain hose. The flexible hose permits direct gravity drainage to the sewer system and eliminates the need for emptying the condensate by hand.

There are several different types of cabinet design, so that the interior design of the building can be matched as closely as possible with the cabinet. The more expensive the unit, the more features and higher output capacity along with a more expensive cabinet design. Some are equipped with a humidity-sensing control for cycling the unit and maintaining the desired humidity inside the space. Normally, humidistats are adjustable from 30 to 80 percent relative humidity. Some humidistats are equipped with a continuous-run setting. Most are equipped with an on and off point on the humidistat, while others will include an additional sensing and switching device that will automatically turn off the unit when the water receptacle is full and needs to be emptied. Some are equipped with a warning light indicating that the receptacle is full and that the unit is off.

These units are designed to provide the maximum performance at the standard rating conditions of 80°F dry-bulb temperature and 60 percent relative humidity. When the room temperature drops to a point that the system is not loaded, as it should be at a condition of 65°F dry-bulb temperature and 60 percent relative humidity, the refrigerant pressure and the corresponding evaporator temperature

usually drop to a point where frost will occur on the cooling coil. This is especially noticeable on units that use a capillary tube as the refrigerant flow control device.

Some dehumidifiers are equipped with special defrost controls that will turn the compressor off when frost occurs on the evaporator. This control is generally a bimetal thermostat, strategically mounted on the evaporator tubing, that allows the dehumidifying process to continue at a reduced rate when frosting occurs. Under some frosting conditions, the humidistat can be set for a higher relative humidity setting, which will reduce the number and duration of the running cycles, permitting satisfactory operation at low-load conditions. During late fall and early spring operation of the dehumidifier, supplementary heat must be provided, so that the space can be maintained within the desired conditions and frosting of the evaporator can be prevented.

Dehumidifier units are generally equipped with rollers or casters so that they can be easily moved to the desired place of operation.

HUMIDIFICATION

One of the most important aspects of air conditioning, humidification, is unfortunately one of the least understood. This is undoubtedly because it is intangible; you cannot see it, you cannot touch it, it has no odor, no color, and no sound.

To be sure that we are starting out on common ground, we will define some of the words that we are going to use in the discussion of humidity and humidification.

Humidity   The water vapor within a given space.
Absolute humidity   The weight of water vapor per unit volume.
Percentage humidity   The ratio of the weight of water vapor per pound of dry air to the weight of water vapor per pound of dry air saturated at the same temperature.
Relative humidity   The amount of water vapor (percent) actually in the air compared to the maximum amount that the air could hold under the same conditions. The warmer the air, the more moisture it can hold. Air in a building heated to 70°F can hold about 8 grains of moisture per cubic foot. This is 100 percent relative humidity. If there are only 2 grains per cubic foot in the building, this is one-fourth of the amount of moisture that the air can hold. Therefore, the relative humidity is one-fourth or 25 percent; the air could hold four times as much water vapor.

The important thing to remember is that when air is heated it can hold more moisture. This reduction of relative humidity is taking place in every unhumidified or underhumidified building where heating is used.

To solve this problem, we add moisture, so there is more water available for the air to absorb. We humidify because there are benefits that are as important as heating to overall indoor comfort and well-being during the heating season.

These benefits can be grouped into three general classifications:

1. Comfort
2. Preservation
3. Health

Comfort

Have you ever stepped out of the shower and noticed how warm the bathroom is? It is probably about 75°F in the bathroom, and the relative humidity is probably about 70 to 80 percent because of the water vapor that was added to the air while showering. Now, if the phone rings, and you step out into the hall to answer it, you nearly freeze. Yet the temperature there is probably about 70°F—just 5°F cooler than in the bathroom—and you are shivering. This shivering is because you just became an evaporative cooler. The air out in the hall is dry, the relative humidity is possibly about 10 to 15 percent. You are wet and this thirsty air absorbs moisture from your skin. The water evaporates, and as it does, your skin is cooled. This same type of thing happens day after day, every winter, in millions of buildings. People are turning thermostats up to 75°F and higher in order to feel warm because it feels drafty and chilly when the evaporative cooling process is taking place. The proper relative humidity level makes 70°F feel more like 75°F.

The chilling effect is not the only discomfort caused by too dry air. Static electricity is usually an indication of low relative humidity and a condition that is consistently annoying. Proper relative humidity will alleviate this discomfort.

Preservation

The addition or reduction of moisture drastically affects the qualities, dimensions, and the weight of hygroscopic materials. Wood, leather, paper, and cloth, even though they feel dry to the touch, contain water—not a fixed amount, but an amount that will vary greatly with the relative humidity level of the surrounding air. For example, 1 cubic foot of wood with a bone-dry weight of 30 pounds at 60 percent relative humidity will hold over 3 pints of water. If the relative humidity is reduced to 10 percent, the water held by the wood will not even fill a 1-pint bottle. We have, in effect, withdrawn 2½ pints of water from the wood by lowering the relative humidity from 60 to 10 percent.

This type of action goes on, not only with wood, but with every single material in a building that has the capability of absorbing and releasing moisture. Paper, plaster, fibers, leather, glue, hair, skin, and so on, will shrink as they lose water and swell as they take on water. If the water loss is rapid, warping or cracking takes place. As the relative humidity changes, the condition and dimensions of the materials change as constantly as the weather. This is why proper relative humidity is important.

Too much moisture also can be damaging. Everyone has seen windows fog during the winter, maybe a little fog on the lower corners, maybe a whole window fogging or completely frosting over. This latter condition is an indication that the indoor relative humidity is too high.

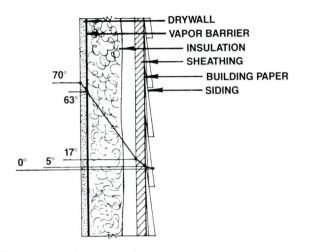

Figure 13–2 Outside wall construction. (Courtesy of Research Products Corp.)

The formation of condensation is due to the effect of vapor pressure. Dalton's law explains vapor pressure. It states, "In a gaseous mixture, the molecules of each gas are evenly dispersed throughout the entire volume of the mixture." Taking a building as the volume involved, water vapor molecules move throughout the entire building. Because of this tendency of these molecules to disperse evenly, or to mix, the moisture in the humidified air moves toward drier air. In other words, in a building the moist indoor air attempts to reach the drier outside air. It moves toward the windows, where there is a lower temperature, and therefore, an increase in relative humidity to a point at which the water vapor will condense out of the air and onto the cold surface of the window. This is the dew point and it occurs at various temperatures, depending on the type of windows in the building.

Usually, condensation on inside windows is a type of measurement of the allowable relative humidity inside a building. We can further assume that if this condensation is taking place on windows, it is also taking place within the walls if there is no vapor barrier. See Figure 13–2.

The typical outside wall has a drywall (or plaster), a vapor barrier (on the warm side of the insulation), the insulation, air space, sheathing, building paper, and siding. Given an indoor temperature of 70°F, a relative humidity of 35 percent, and an outside temperature of 0°F, the temperature through the wall drops to about 63°F at the vapor barrier, down to 17°F at the sheathing, and on down to 0°F outside. If a psychrometric chart was checked, we would discover that 70°F indoor air at 35 percent relative humidity has a dew point temperature of 41°F. That temperature occurs right in the middle of the insulation. This is where condensation occurs—and where the trouble is—without a vapor barrier and a humidifier that can be controlled.

Properly controlled relative humidity is the important factor in avoiding the damaging effects of too dry air and too high relative humidity.

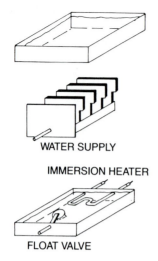

PAN-TYPE HUMIDIFIERS
Evap. Rate .0036 gals./hr./ft.^{2}
at Room Temperature and still air.
.36 gals./hr. Requires 100 ft.^{2} Pan

Pan with Plates or Discs
Mounted in Warm Air
Capacity Rating .12 to .36 gals./hr.

Capacity Rating with a:
1,000-Watt Element = .36 to .48 gals./hr.
Hot Water Coil (160°F) = .48 to .74 gals./hr.
Steam Coil (2 psig) = 1.2 to 2.4 gals./hr.

WATER SUPPLY

IMMERSION HEATER

FLOAT VALVE

Figure 13–3 Pan-type humidifier. (Courtesy of Research Products Corp.)

Health

In the struggle between the nose and the air-conditioning equipment, sometimes the heater wins and sometimes the cooler; but seldom does the nose win. Nasal mucus contains some 96 percent water. It is more viscous than mucus elsewhere in the body and even a slight drying increases the viscosity enough to interfere with the work of the cilia. Demands on the nasal glands are great even under excellent conditions and they cannot cope with extreme dryness indoors in winter.

Experience has shown that with approaching winter, the first wave of patients with dry noses appears in the doctor's office when the relative humidity indoors falls to 25 percent. It would seem, therefore, that 35 percent would be regarded as a passing grade, but 40 percent is something to strive for. It boils down to this: A pint of water is a lot of water for a small nose to turn out. In disease or old age, the nose simply does not deliver enough moisture; drainage stops and germs take over.

Dr. Joseph Lubart, an expert on the common cold, states in the *New York State Journal of Medicine:*

Prevention of the common cold at the present is our nearest approach to a cure. The most important prevention measure would appear to be proper regulation of the humidity especially during the heating season with its distressing drying of the indoor air and the creation of an environment favorable to the cold bug.

Types of Humidifiers

There are many types of humidifiers available. They vary in price, capacity, and in principle of operation. For classification purposes, it is simpler, and more logical, to consider humidifiers in three general types:

1. Pan type
2. Atomizing type
3. Wetted-element type

Pan type   The pan type is the simplest type of humidifier. It has a low capacity. On a hot radiator, it may evaporate 0.07 pounds of water per hour. In the warm air plenum of a heating furnace, it would evaporate approximately 1.5 pounds of water per hour.* To increase the capacity, the air-to-water surfaces must be increased by placing water wicking plates in the pan. The capacity goes up as the air temperature in the furnace plenum increases. Greater capacity is also possible through the use of steam, hot water, or an electric heating element immersed in the water. See Figure 13–3. A 1200-watt heating element, for example, in a container with water supplied by a float valve could produce 4 pounds of moisture per hour.

Atomizing type   The atomizing type atomizes the water by throwing it from the surface of a rapidly revolving disc. It is generally a portable or console-type unit. However, it can also be installed so that the water particles will be directed into a ducted central system. See Figure 13–4.

Wetted-element type   The wetted-element type, in its simplest form, operates in the manner of an evaporative cooler. Here air is either pushed or pulled through a wetted element or filter and evaporative cooling takes place. By increasing the air flow or by supplying additional heat, the evaporation rate of the humidifier can be increased. The heat source for evaporation can be from an increase in water temperature or an increase in air temperature. See Figure 13–5.

The air for evaporation can be taken from the heated air of the furnace plenum and directed through the humidifier by the humidifier fan. See Figure 13–6. It can also be drawn through the wetted element by the air pressure differential of the furnace blower system. See Figure 13–7.

---

* A plenum is a pressurized chamber with one or more distributing ducts connected to it.

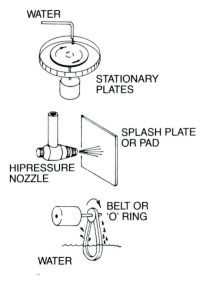

ATOMIZING HUMIDIFIERS
Centrifugal Wheel or Cone Type
Capacity Rating
.12 to .24 gals./hr.

Spray or Mist
Capacity Rating
.12 to 1.2 gals./hr.

Splashing Ring Plenum-Mounted
Capacity Rating
.12 to .36 gals./hr.

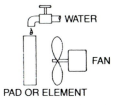

Figure 13–4 Atomizing-type humidifier. (Courtesy of Research Products Corp.)

Figure 13–5 Wetted-element-type humidifier. (Courtesy of Research Products Corp.)

Furnace-mounted humidifiers, usually of the wetted-element type, can be constructed so that they produce 10 or more pounds of moisture per hour. Because of their higher capacity, this type usually has a humidistat or a control that will activate a relay or water valve and start a fan that operates until the control is satisfied. Normally more water is supplied to the unit than is evaporated, and this flushing action washes a large portion of the hardness salts from the evaporative element to the floor drain, eliminating them from the humidifying system.

Checking the Humidifier

In most instances when a lack of humidity is experienced, a few visual checks and a test of the relative humidity will usually indicate if the humidifier is working. Test the air with a sling psychrometer to determine the relative humidity. Not many humidifiers will satisfy the job during extreme conditions.

Check to make certain that the humidistat is working properly. Check to be sure that sufficient water is being supplied to the humidifier for proper operation. Check the element to make certain that it is not clogged or dirty and will not allow the water to enter the air stream.

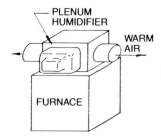

Figure 13–6 Humidifier installation. (Courtesy of Research Products Corp.)

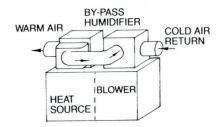

Figure 13–7 Bypass-type humidifier. (Courtesy of Research Products Corp.)

14

Dishwashers

INSTALLATION INSTRUCTIONS

Note: All electrical and plumbing work must be done by qualified personnel. Be sure that all applicable codes for electrical and plumbing connections are followed.

Note: All service expenses that occur as the result of an improper installation are the responsibility of the installer and are not covered by the warranty.

Installer

Follow these instructions carefully. It is recommended that dishwashers be installed by a person with experience in the electrical and plumbing trades.

The following are some of the things that must be checked for the installation to be successful:

1. The enclosures for the sides and back must be provided at the time of installation.

2. On models with reversible door and service panels, reverse the panels prior to the installation. Refer to the owner's manual and later in this chapter for instructions.

3. The cabinet enclosure must be of the correct height, width, and depth.

4. The rear wall must be free of obstructions.

5. The cabinet enclosure must not be more than 10 feet from a waste-water connection.

Step 1   Refer to Figure 14–1 to determine the rough in dimensions for the electrical, drain, and water supply.

Drain hose   The drain hose must be at least 2 inches in diameter to provide for adequate clearance. Do not leave excess drain hose behind or under the washer. *Be sure that the drain hose is not kinked.* A high loop is necessary for all models.

Water and electrical supply   *Caution:* Do not use a hose of any type for the hot-water supply.

The electrical and water pipe must enter the shaded area and run straight from the back of the enclosure to the fill valve or the electrical junction box. Do not run across in front of the pump and motor assembly.

A hot-water shutoff valve in the supply line is recommended.

The electrical installation must be in accordance with the local codes and the National Electrical Code.

Grounding instructions   This appliance must be connected to a grounded metal, permanent, wiring system, or an equipment-grounding connector must be run with the circuit conductors and connected to the equipment-grounding terminal, or lead of the appliance.

Step 2: Position, align, and level the dishwasher

1. Remove the lower service access panel, by removing the two (2) lower front screws. Open the dishwasher door and remove the two (2) screws securing the top of the service access panel. See Figure 14–2.

2. Close the dishwasher door and remove the service access panel by pulling it straight forward. See Figure 14–3.

3. Remove the four (4) screws securing the cabinet toe plate. See Figure 14–4.

4. Adjust the leveling legs so that the top of the dishwasher is $\frac{1}{4}$ inch below the countertop prior to moving the dishwasher into the enclosure.

5. Adjust the rear leveling legs to level the dishwasher from front to back.

6. Slide the dishwasher into the enclosure and make the final leveling adjustments, front to back and side to side. The top of the dishwasher door should be level and aligned

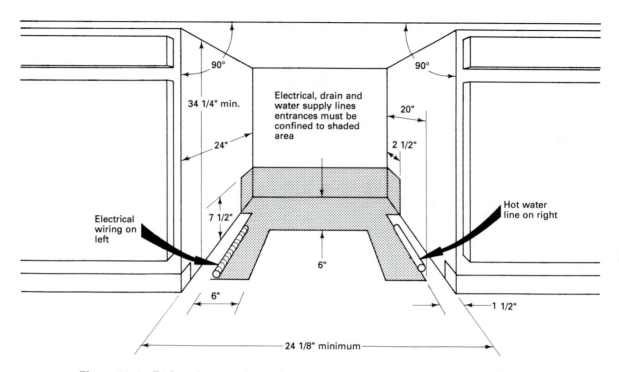

Figure 14–1 Dishwasher rough in dimensions. (Courtesy of WCI Major Appliance Group.)

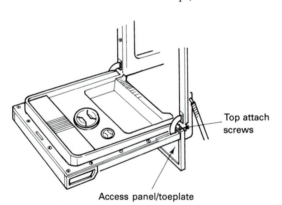

Top attach screws

Access panel/toeplate

Figure 14–2 Service access panel screws removal. (Courtesy of WCI Major Appliance Group.)

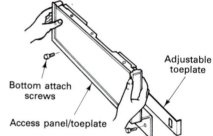

Bottom attach screws

Adjustable toeplate

Access panel/toeplate

Figure 14–3 Service access panel removal. (Courtesy of WCI Major Appliance Group.)

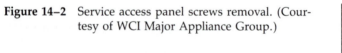

Figure 14–4 Cabinet toe plate removal. (Courtesy of WCI Major Appliance Group.)

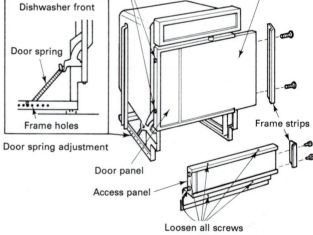

Figure 14–5 Door spring adjustment. (Courtesy of WCI Major Appliance Group.)

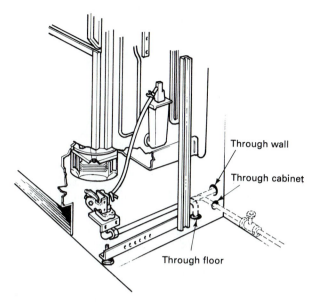

Through wall

Through cabinet

Through floor

Figure 14–6 Connecting the water supply to the valve. (Courtesy of WCI Major Appliance Group.)

with the countertop. Be sure that all four (4) leveling legs are firm against the floor.

7. Fasten the brackets at the top front of the dishwasher to the countertop using wood screws. If the brackets extend beyond the counter, cut the excess off and smooth any rough edges.

8. Open and close the door. There should be no interference or noise during this operation. If there is a problem, repeat the leveling procedure.

9. Adjustment of the door springs may be necessary for correct operation of the door. The door should not slam when closing and it should stay open with its own weight. It is allowable for a door to move $\frac{1}{2}$ inch from the full open position.

There are four (4) adjustment positions for the door springs. Adjustment can be made from the pump compartment. Both springs must be in the same position on each side. See Figure 14–5.

Decorator door and service panel inserts Refer to Figure 14–5.

10. Remove the two (2) screws securing the dishwasher door side trim and remove. Loosen the opposite side trim screws.

11. Slide the insert out of the frame. Place the chosen panel color to the outside, replace the panel and trim frame, replace the side trim, screws, and tighten the opposite trim screws.

12. Remove the two (2) screws securing the service access panel side trim and remove. Loosen the opposite side trim screws.

13. Slide the insert out of the trim frame. Place the chosen panel color to the outside, replace the panel into the trim frame, replace the side trim screws, and tighten the opposite trim screws.

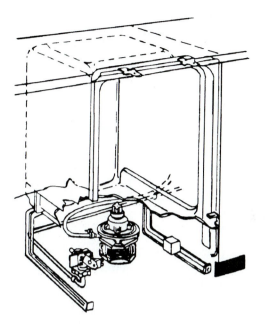

Figure 14–7 Attaching the drain hose to the dishwasher. (Courtesy of WCI Major Appliance Group.)

Step 3: Connect the water supply Refer to Figure 14–6.

1. Install approved tubing or pipe of $\frac{3}{8}$ inch minimum ID to the water valve. Connection to the valve is $\frac{3}{8}$-inch pipe thread.

2. Use thread-sealing compound or Teflon tape on all threaded connections. The water line must be free of all scale, burrs, chips, and lubricant before connecting it to the valve.

3. Keep the valve from twisting when making the water pipe connection by grasping the brass insert of the water valve with water pump pliers.

Caution:

1. Do not overheat the inlet valve when soldering connections. Overheating will damage the internal parts.

2. Use care to see that no thread compound, thread chips, or foreign material can enter the valve.

3. If the valve clogs with foreign material, remove the four (4) screws at the inlet end of the valve for access to the filter screen.

4. The valve inlet connection is $\frac{3}{8}$-inch pipe thread.

Step 4: Connect the drain Connect the drain hose to the house drainage system. See Figures 14–7 through 14–9. Attach the hose to the dishwasher as shown in Figure 14–7.

The discharge end of the drain hose supplied with the dishwasher can be used for a 1-, $\frac{3}{4}$-, or $\frac{5}{8}$-inch opening. Cut the drain hose at the right ridge on the coupler for $\frac{3}{4}$ and $\frac{5}{8}$ inch. Do not cut this hose to shorten it. Coil any excess under the cabinet or in the shaded area. See Figures 14–1 and 14–8.

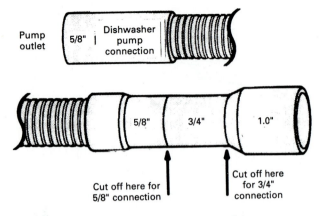

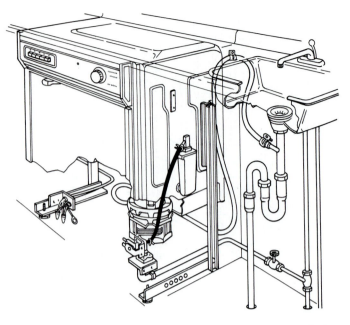

Large end on hose fits disposer inlet fitting.

Figure 14–8 Cutting the drain hose for proper-size pipe connection. (Courtesy of WCI Major Appliance Group.)

Figure 14–9 Connecting the drain hose to the house drain system. (Courtesy of WCI Major Appliance Group.)

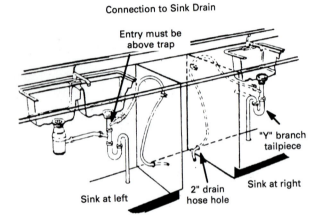

Figure 14–10 Drain connections. (Courtesy of WCI Major Appliance Group.)

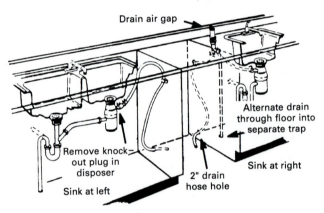

Figure 14–11 Drain connections with an air gap. (Courtesy of WCI Major Appliance Group.)

Reverse the ends of the drain hose when connecting it to an air gap. The small end ($\frac{1}{2}$ inch) goes on the air gap, the $\frac{5}{8}$ inch will connect to the dishwasher. See Figure 14–10. *Important note:* If the drain line goes to a separate trap beneath the floor, make sure that the trap is vented to prevent siphoning. Install an air gap at the countertop level. The air gap is available at plumbing supply houses and hardware stores. See Figure 14–11.

Step 5: Connect the electricity

1. Armored or BX cable must not be allowed to come in contact with any electrical connections, such as the heater or relay.

2. Provide a separate 15-amp, 120-volt ac, 60-hertz circuit using a 15-amp time-delay fuse or a 15-amp circuit breaker.

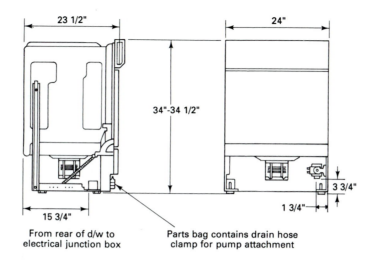

Figure 14–12   Wiring to the dishwasher. (Courtesy of WCI Major Appliance Group.)

3. A 20-amp circuit is required if the dishwasher is to be installed on the same branch circuit as the garbage disposer.

4. Use only UL approved wiring connections and wire nuts.

5. Connect the white wire from the dishwasher to the incoming white wire. See Figures 14–12 and 14–13.

6. Connect the black wire from the dishwasher to the incoming black wire. Connect the incoming ground wire (bare or green conductor) to the green grounding screw on the unit.

Caution: The joining of aluminum building wire to stranded copper wire involves special procedures, and should be done by qualified personnel using materials recognized by UL (or CSA) as suitable for this purpose.

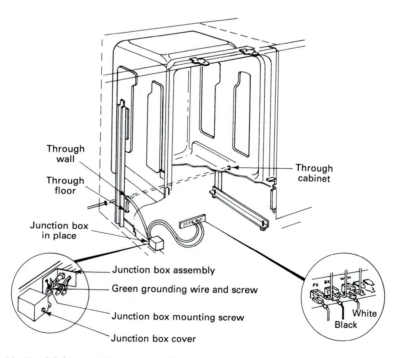

Figure 14–13   Making wiring connections to the dishwasher. (Courtesy of WCI Major Appliance Group.)

Figure 14–14 Checking for water leaks. (Courtesy of WCI Major Appliance Group.)

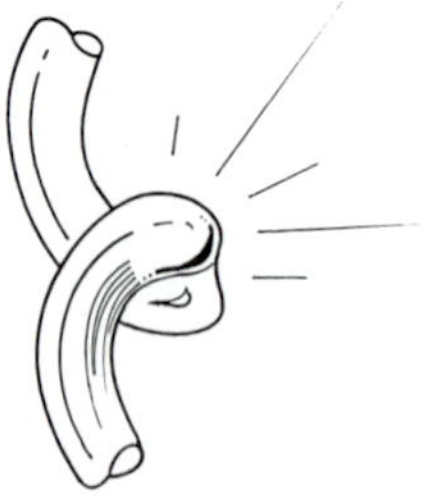

Figure 14–15 Checking for kinks in the drain hose. (Courtesy of WCI Major Appliance Group.)

Step 6: Final check

1. Is the dishwasher securely fastened in place?
2. Is the water supply turned on?
3. Are the water connections tight? Check for leaks. See Figure 14–14.
4. Is the water inlet valve wiring in place?
5. Make certain that the electricity is turned on.
6. Make certain that there are no kinks in the drain line. See Figure 14–15.
7. Remove the packing material and the customer literature package from the inside of the dishwasher.
8. Spin the spray arm to check for movement.
9. Start the dishwasher, allowing at least one complete fill and drain.
10. Check the water level. The water should fill to the heater element. Operate the drain cycle.
11. Check to see that the electrical supply lines are not touching any parts in the machine compartment.
12. Check for unit leaks, especially around the door, pump mounting gasket, at the hose connections, and the plumbing connection at the water valve.
13. Make certain that the leveling feet are in contact with the floor. Attach the service panel and toe plate with the four (4) screws removed earlier.

Note: Installation errors are not covered by the warranty.

Operation

All of the dishwasher operations are controlled by the timer and selector switch. Several dishwasher cycles and options are available: Heavy Soil, Normal Soil, Light Soil, and a Rinse-Hold cycle.

Energy options include a heated dry cycle.

The door latch holds the door closed and activates the door latch switch to complete the electrical circuit for dishwasher operation. If the door is opened during the cycle, the latch switch opens and all operations cease.

After the cycle selection is made and the door latch is closed, the timer energizes the water fill valve allowing water to enter the dishwasher. The timed fill period allows 1.3 to 1.5 gallons of water to enter the dishwasher for either the wash or the rinse cycles. Should the timer fill contacts fail to close during the fill cycle, a float switch assembly

(located in the left front corner of the tub) will open the electrical circuit to the fill valve at a preset "overfill" level.

A water temperature above 14°F is required to activate the dishwasher detergent. The tub and dishes act together as a large heat sink which may lower the temperature of the incoming hot water below the detergent activation temperature. Proper water temperature is maintained in the dishwasher by an 800-watt heater, located in the bottom of the tub. The heater operates during the wash and rinse cycles. The detergent is dispensed automatically into the wash water, and a rinse agent is dispensed into the final rinse water.

The motor and pump assembly has two (2) impellers, one for wash and one for drain. During the wash and rinse cycles (motor operates counterclockwise), the upper impeller recirculates the water at a rate of 50 gallons per minute.

The water is distributed at two levels: through the wash arm under the lower dishrack and through the center wash tower under the upper rack. The holes in the wash arm provide water to the lower dishrack, while the slots in the center wash tower provide water to the upper dishrack.

During the drain cycle (motor rotates clockwise), the lower drain impeller pumps the water out the drain hose.

The heat for drying is furnished by the 800-watt heating element.

SYSTEMS OPERATION

Detailed information on the testing and replacement of individual components will be given later in this chapter.

Electrical Control System

All electrical components and controls operate on 120-volt ac, 60-hertz power. The dishwasher requires a separate 15-amp circuit (20 amps if a garbage disposer is to be installed on the same branch circuit). The dishwasher must be protected by a 15-amp time-delay fuse or circuit breaker.

The timer is the control center of the dishwasher. The multiplug connections provide easy service. The timer is manually advanced to the desired dishwashing cycle. The user is instructed to advance the timer with the door unlatched (door switch open) and to start operation by closing and latching the door. This prevents timer contact damage.

After advancing the timer and making the desired cycle selections, latch the door. From that point, the timer operates automatically, driven by a small synchronous motor. A signal lamp lights on the control panel indicating operation. After completion of the selected dishwashing cycle, the signal lamp turns off.

On some models, control of the electrical components is shared by the timer and a selector switch. This switch is wired in series or in parallel with certain timer contacts. By opening or closing these selector switches, the control circuits built into the timer can be cut off or bypassed adding variation to the dishwasher operation. Cycle selector switches allow the user to choose the desired dishwashing cycles available. An energy selector switch permits the user to select additional heat to raise the temperature of the wash and rinse water in the tub and/or to select the heat for drying. A float-operated water level safety switch and a door latch safety switch complete the electrical control system.

Timer

The mechanical timer used on some dishwashers features electrical terminal blocks for easy connection of the electrical leads. The timer contacts are arranged in pairs or threes providing single-pole, single-throw and single-pole, double-throw switching. These sets of contacts are arranged in rows on the side of the timer. Extensions of the contact arms through the terminal blocks provide the male terminals for connection to the dishwasher wiring harness connector. Each set of contacts (pairs or threes) has one contact connected to a common, internal buss bar.

The timer contacts are opened and closed by a slowly rotating plastic drum on which ridges or tracks of varying height are molded. The drum is turned by a small synchronous motor, crank gear, and pawl. One end of the timer drum has rachet-type teeth molded on its circumference. The pawl is moved back and forth by the crank gear every 90 seconds, turning the drum forward approximately 5°.

The timer can be manually advanced by turning the drum shaft clockwise. However, this should never be done while power is supplied to the timer. This can be accomplished by unlatching the door, since the door safety switch is connected in the timer power lead. The user is instructed to leave the door unlatched when advancing the timer to the desired dishwashing cycle. Cycle selections are sometimes made on a separate push-button-operated, multicontact selector switch.

Water Fill System

The water fill system consists of a solenoid-operated valve, a fill funnel, and an iner connecting hose.

The timer gives 1 minute of fill time for each wash or rinse fill. The valve flow rate is 2.5 gallons per minute with an incoming water pressure of 15 to 120 psi.

A fill safety switch is wired in series with the fill valve solenoid.

Should the timer motor fail during the fill cycle, or the timer contacts fail to open, water could continue to enter the dishwasher. Should this occur, the float assembly located in the left front corner of the tub will rise, allowing the fill safety switch to open and stop the water flow. *Note:* The float switch will not protect against a mechanical failure of the valve.

Water Circulation System

The water pump housing and motor assembly is clamped to the bottom of the tub with six (6) screws and a plastic collar. The bottom of the tub forms a sump. A rubber mounting gasket, surrounding the pump housing, isolates the pump housing vibration from the tub. An impeller shroud in the bottom of the recirculating housing prevents foreign objects or large food particles from entering the pump.

Wash and rinse cycle The upper impeller, diffuser, and impeller housing act as a "two way" centrifugal pump. During the wash and rinse cycles, when the motor is rotating counterclockwise, the curved open vanes of the upper impeller pump the water upward into the impeller housing where the diffuser captures and directs the turbulating water into the wash arm and the center wash tower. The pulverizer blade, mounted under the upper impeller, chops up any soft food particles that enter the pump during water circulation.

While the upper impeller is providing the pumping action necessary to circulate the water, the lower impeller (under the pump-plate) "windmills" or rotates without a purpose. *Note:* Future models may be equipped with a center wash arm, in which case the lower impeller would provide water to this wash arm during the wash and rinse cycles.

Two-level wash is provided as follows:

1. *Wash arm* The wash arm is made of stainless steel. The discharge ports along the top are set at an angle to provide a propulsion thrust to rotate the arm during the wash and rinse cycles. This provides a constant rotation speed resulting in an even distribution of water over any dishes in the lower rack. A large opening in the center provides water to the center wash tower. See Figure 14–16.

2. *Center wash tower* A center wash tower is attached to the wash arm hub. A vertical spray tube extension tip located within the tower is free to turn with the wash arm. The force of the water delivered to the wash arm lifts the extension tower tip through a hole in the lower dishrack. The top end of the tube is formed to deliver separate jets of water which cascade over the dishes in both racks during the wash and rinse cycles.

Drain cycle When the timer advances to a drain cycle, the motor stops momentarily and then rotates clockwise (changes the direction of rotation). With the lower impeller rotating clockwise, water is gravity fed through the impeller shroud and into the impeller's pump chamber. The straight vanes of the impeller now force the water out the drain port.

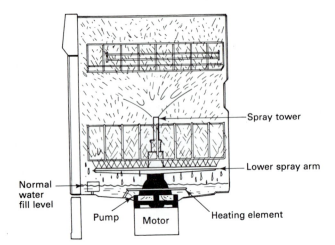

Figure 14–16 Water distribution system. (Courtesy of WCI Major Appliance Group.)

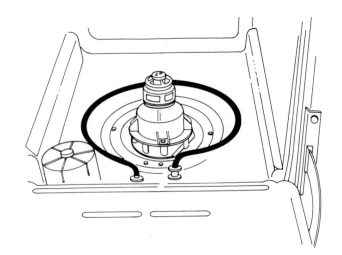

Figure 14–17 Heating element. (Courtesy of WCI Major Appliance Group.)

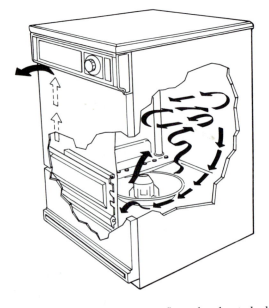

Figure 14–18 Dishwasher air flow for heated drying. (Courtesy of WCI Major Appliance Group.)

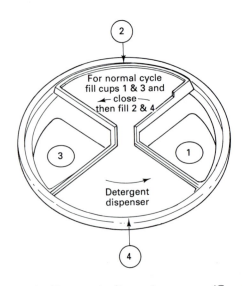

Figure 14–19 Detergent dispensing cups. (Courtesy of WCI Major Appliance Group.)

While the lower impeller is pumping the water out the drain, the upper impeller "windmills" or rotates without a purpose.

Drying system

The drying system consists of the timer, an option selector switch, and an 800-watt tubular heater. See Figure 14–17. The heater is located in the bottom of the tub. Operation of the heater is controlled by both the timer and the selector switch. Drying can be accomplished in two (2) ways, depending on the user's choice. The user, by setting the energy selector switch, may elect to allow the dishes to dry

without the aid of additional heat. This method may be satisfactory if the wash and rinse water is very hot, the water is very soft, and a rinse agent is used. When drying without additional heat, it is suggested that the dishwasher door be unlocked and left slightly open after the completion of the drying cycle.

During heated drying, the 800-watt heater is energized by the timer during the last 26 minutes of dishwasher operation following the last drain cycle. Water remaining on the dishes is vaporized by the addition of heat. The moist hot air is drawn into the vent at the bottom of the inner door panel, and it is discharged just beneath the control panel and left of the door handle. See Figure 14–18.

DETERGENT DISPENSING SYSTEM

The detergent dispenser is mounted on the tub side of the panel. It has four (4) cups in which the user places the specified amount of detergent. See Figure 14–19. Two (2) cups are open and dispense their contents into the tub when the door is closed. The other two (2) pie-shaped cups are covered by a rotating cover. The cover rotates counterclockwise to enclose the dispenser cups and protect them from water spray. The dispensing cover shaft rotates in a bearing that passes through the door panel. It is attached to a spring-operated cam.

Power to operate the dispenser is provided by winding up the cam spring when rotating the cover into its closed position. The detergent dispenser is controlled by the timer and activated by a trip lever and bimetal arm. When energized, the bimetal arm releases the spring-loaded cam, rotating the cover and exposing the detergent to the water spray.

FUNCTIONAL COMPONENTS AND PARTS

The following is a discussion of the functional components and parts of a dishwasher.

Safety Precautions

Always turn off the electrical power supply before servicing any electrical component, making ohmmeter checks, or replacing any parts.

All voltage checks should be made with a voltmeter having a full-scale range of 130 volts or higher.

After service is completed, be sure that all safety grounding circuits are complete, all electrical connections are secure, and all access panels are in place.

Timer

The electromechanical timer features an electrical terminal box for easy connection of the electrical leads. Timer contacts are arranged in pairs or threes providing single-pole, single-throw and single-pole, double-throw switching. These sets of contacts are arranged in rows on the side of the timer. Extensions of the contact arms through the terminal block provide the male terminals for connection to the dishwasher wiring harness connector. Each set of contacts (pairs or threes) has one contact connected to a common, internal buss bar.

Timer contacts are opened and closed by a slowly rotating plastic drum on which ridges or tracks of varying height are molded. The drum is turned by a small synchronous motor, a crank gear, and a pawl. One end of the timer drum has ratchet-type gear teeth molded on its circumference. The pawl is moved back and forth by the crank gear every 75 seconds, turning the drum forward approximately 5° to 7°.

The timer can be manually advanced by turning the knob ckockwise. However, this should never be done while electrical power is being supplied to the timer. This can

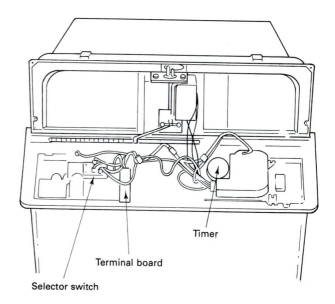

Figure 14–20 Timer mounting detail. (Courtesy of WCI Major Appliance Group.)

be accomplished by unlatching the door, because the door safety switch controls the power lead to the timer. The user is instructed to leave the door unlatched when advancing the timer to the desired dishwashing cycle. On some models, additional cycle selections are made on a separate push-button, multicontact selector switch.

Testing the timer Proceed as follows:

1. Index the timer to the first increment of operation, which is a drain period.
2. If the pump motor fails to operate during the first increment, check for electrical power at the motor lead block. If there is no power, check the door latch switch and the selector switch. If there is power, check the motor as described later in this section.
3. If the pump does operate, let the timer motor advance the timer through the drain increment to determine if the timer motor and drive train are fully operative.
4. Let the timer advance or index it forward to the portion of the cycle in question.
5. If a component controlled by the timer fails to function as the timer advances through the cycle, check for voltage at the timer terminal. If voltage is supplied to the component, check the component as described in this chapter.

Continuity through the timer contacts, other controls, and wires can also be checked with an ohmmeter (with the power disconnected).

If the timer contacts fail to close in the sequence shown in the timer chart, or are burned (have resistance measurable with an ohmmeter), or if the timer does not advance automatically, refer to Figure 14–20 and replace the timer.

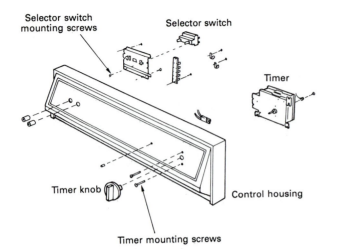

Selector switch mounting screws

Selector switch

Timer

Timer knob

Control housing

Timer mounting screws

Figure 14–21   Selector switch. (Courtesy of WCI Major Appliance Group.)

To remove or replace the timer, use the following procedure:

1. Disconnect the dishwasher from the electrical supply.
2. Remove the hex-head screw securing the door latch knob to the door latch arm.
3. Remove the timer control knob by pulling it straight off.
4. Open the door and remove the two (2) screws across the top of the door and the two (2) screws along both sides of the control panel. This will release the control panel from the door.
5. Carefully pull the control panel forward to gain access to the controls.
6. Remove the two (2) screws securing the timer.
7. Remove the timer from the control panel.
8. Remove the harness connector block from the timer.
9. Disconnect the timer wiring to the door safety switch, one brown lead and one black lead.
10. Disconnect the light-blue lead with the tan tracer from the timer.
11. Disconnect the black lead connector block.
12. Install the new timer and check the operation.
13. Reverse the above procedure to reassemble the unit.

Selector Switch (Push-Button)

On certain models, a multifunction, push-button selector switch provides the user with a selection of dishwashing cycles and heating options. See Figure 14–21.

Only one cycle and one heat option selection is possible. When a push button is pressed, it releases another push button that may have been previously selected. The internal contacts are operated by a slide bar arrangement.

The selector switch is mounted to the dishwasher control panel. It is easily accessible for service, testing, and/or replacement.

The selector switch can be tested using an ohmmeter or continuity tester, along with the wiring schematic and the selector switch chart.

To test the push-button selector switch, use the following procedure:

1. Disconnect the dishwasher from the electrical supply.
2. Remove the $\frac{1}{4}$-inch hex-head screw securing the door latch knob to the door latch arm.
3. Open the door and remove the two (2) screws across the top of the door and the two (2) screws along both sides of the control panel. This will release the control panel from the door.
4. Carefully pull the control panel forward to gain access to the controls.
5. Remove the electrical leads from the selector switch.
6. Place the leads of an ohmmeter across the switch terminal contacts. (Refer to the wiring diagram and selector switch chart for the proper switch closing.)

To remove or replace the selector switch, use the following procedure:

1. Disconnect the dishwasher from the electrical supply.
2. Remove the $\frac{1}{4}$-inch hex-head screw securing the door latch knob to the door latch arm.
3. Open the door and remove the two (2) screws across the top of the door and the two (2) screws along both sides of the control panel. This will release the control panel from the door.
4. Carefully pull the control panel forward to gain access to the controls.
5. Remove the one (1) screw securing the exhaust vent shield.
6. Remove the push-button selector knobs, pull them straight off.
7. Remove the electrical leads from the selector switch.
8. Remove the two (2) screws securing the selector switch assembly.
9. Remove the two (2) screws securing the selector switch to its mounting bracket.
10. Install the new selector switch.
11. Reverse the above procedure to reassemble the unit.

Door Latch Assembly and Safety Switch

The latch and safety switch are located in the door, with the strike located at the top of the tub. See Figure 14–22. The dishwasher will not operate until the door is closed, the latch handle is rotated to hold the door firmly against the tub seal, and the normally open contacts of the single-pole, single-throw door safety switch are closed.

To remove or replace the door latch assembly, use the following procedure:

1. Disconnect the dishwasher from the electrical supply.
2. Remove the ¼-inch hex-head screw securing the door latch knob to the door latch arm.
3. Open the door and remove the two (2) screws across the top of the door and the two (2) screws along both sides of the control panel. This will release the control panel from the door.
4. Carefully pull the control panel forward to gain access to the latch and controls.
5. Remove the electrical leads from the door latch safety switch.
6. Remove the two (2) screws securing the door latch safety switch and the plastic shield.
7. Remove the four (4) Phillips-head screws securing the door latch assembly to the inner door panel.
8. Install the new door latch.
9. Reverse the above procedure to reassemble the unit.

To remove or replace the door strike, use the following procedure: The door strike is mounted to the upper front tub liner. The door latching arm rotates in behind the door strike to hold the dishwasher door secure. The strike may be adjusted by adding or removing shims, to ensure a proper door seal.

1. Disconnect the dishwasher from the electrical supply.
2. Open the dishwasher door and remove the two (2) screws securing the strike to the tub liner.
3. The strike tapping plate is held in place by a screw behind the strike. See Figure 14–22.
4. Install the new strike.
5. Reverse the above procedure to reassemble the unit.

To test the door safety switch, use the following procedure:

1. Disconnect the dishwasher from the electrical supply.
2. Remove the ¼-inch hex-head screw securing the door latch knob to the door latch arm.
3. Open the door and remove the two (2) screws across the top of the door and the two (2) screws along both sides of the control panel. This will release the control panel from the door.
4. Carefully pull the control panel forward to gain access to the controls.
5. Remove the electrical leads from the door latch safety switch.
6. Use an ohmmeter and check for continuity.
7. If the switch checks OK, adjust the latch arm assembly (the mounting plate holes are elongated so that the latch assembly will close the switch when the door is closed and locked).

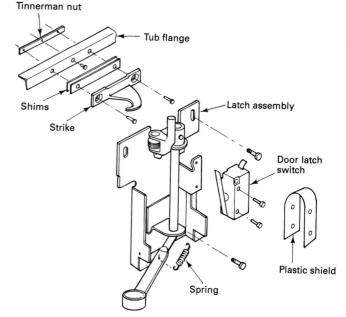

Figure 14–22 Door latch and strike assembly. (Courtesy of WCI Major Appliance Group.)

8. Replace the switch if it is defective.
9. Reverse the above procedure to reassemble the unit.

To remove or replace the door safety switch, use the following procedure:

1. Disconnect the dishwasher from the electrical supply.
2. Remove the ¼-inch hex-head screw securing the door latch knob to the door latch arm.
3. Open the door and remove the two (2) screws across the top of the door and the two (2) screws along both sides of the control panel. This will release the control panel from the door.
4. Carefully pull the control panel forward to gain access to the controls.
5. Remove the electrical leads from the door latch safety switch.
6. Remove the two (2) screws securing the door latch safety switch.
7. Remove the plastic shield and switch assembly.
8. Install the new door safety switch.
9. Reverse the above procedure to reassemble the unit.

Water Fill Valve

The water valve is solenoid operated. The solenoid is activated by the timer. The flow of water is controlled by a rubber flow washer capable of maintaining a flow rate of 2.5 gallons per minute with an incoming water pressure of 15 to 120 psi. See Figure 14–23.

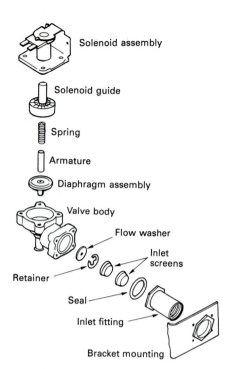

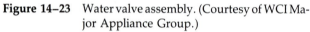

Figure 14–23 Water valve assembly. (Courtesy of WCI Major Appliance Group.)

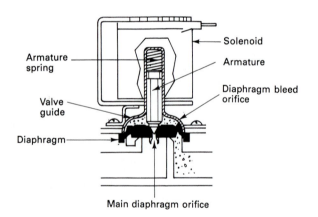

Figure 14–24 Water valve closed. (Courtesy of WCI Major Appliance Group.)

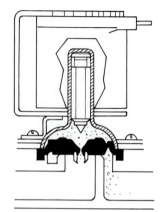

Figure 14–25 Water valve open. (Courtesy of WCI Major Appliance Group.)

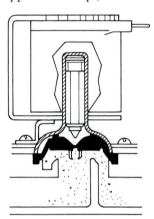

Figure 14–26 Water valve (diaphragm up). (Courtesy of WCI Major Appliance Group.)

The solenoid valve provides a water inlet connection and passage with a large orifice and seat where the water flow can be stopped. A movable rubber diaphragm, operated by water pressure, rests against the valve seat to start and stop the flow of water. It has a small bleed orifice outside the seat contact area, with a larger main orifice at its center. The armature of the solenoid serves to open and close the main orifice at the center. The armature operates within a closed plastic tube (valve guide) that is sealed by the outer edge of the diaphragm to the valve body. A coil spring holds the armature down against the diaphragm main orifice when the solenoid is not energized.

When a valve is in the closed position, the solenoid is not energized. Water is bled through the diaphragm bleed orifice placing the incoming line pressure on top of the diaphragm. The bottom of the diaphragm is essentially at atmospheric pressure (open to the outlet). The pressure differential will hold the valve shut. See Figure 14–24.

When the solenoid is energized, the resulting magnetic field pulls the armature up into the valve guide. See Figure 14–25. The armature spring is compressed by this action. When the armature moves up, it allows the water on top of the diaphragm to drain out through the main orifice.

The diaphragm bleed orifice is much smaller than the main orifice and will not admit enough water to maintain the pressure on the top side of the diaphragm. Thus, as the pressure on the top of the diaphragm is reduced to almost zero, the pressure on the bottom lifts the diaphragm off the valve seat allowing a full flow of water. See Figure 14–26.

When the solenoid is deenergized, the armature spring forces the armature down, closing the diaphragm main orifice. See Figure 14–27. Water continues to flow through the diaphragm bleed orifice, building up pressure and forcing the diaphragm down against the valve seat. See Figure 14–24.

To test the water valve, use the following procedure:

1. Disconnect the dishwasher from the electrical supply.
2. Remove the two (2) ¼-inch hex-head (black) screws securing the bottom of the service panel.
3. Open the dishwasher door and remove the two (2) ¼-inch hex-head (black) screws, one (1) on each rear corner, securing the top of the service panel.
4. Using an ohmmeter, check the resistance of the solenoid valve coil.
5. To check for the proper mechanical operation of the water valve, remove the connector from the solenoid terminals and attach a separate 120-volt, grounded power source to the terminals (attach a ground lead to the water valve body). Turn on the electrical power for a few seconds and then turn the power off. If the water flow does not shut off within 2 seconds, replace the valve.

To remove or replace the water fill valve, use the following procedure. The water fill valve is mounted to the right hinge support of the dishwasher.

1. Disconnect the dishwasher from the electrical supply and shut off the water supply.
2. Remove the two (2) hex-head (black) screws securing the bottom of the service panel. (See the Float Switch Assembly section presented later in this chapter.)
3. Open the dishwasher door and remove the two (2) ¼-inch hex-head screws, one (1) in each rear corner, securing the top of the service panel.
4. Remove the lower service panel by pulling it straight off.
5. Remove the four (4) screws securing the toe plate to the cabinet.
6. Remove the incoming water line (union) connector and the outlet hose.
7. Remove the two (2) mounting screws.
8. Remove the wiring harness connector.
9. Install the new water inlet valve. Check it for proper operation.
10. Reverse the above procedure to reassemble the unit.

Troubleshooting the Water Valve

Use the following procedures when problems occur with a dishwasher. The different problems and their solutions are presented here.

No water to the tub

1. Be sure that the main water supply is turned on.
2. Check the float assembly for free movement up

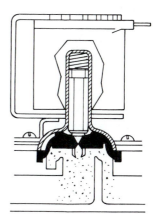

Figure 14–27 Water valve closing. (Courtesy of WCI Major Appliance Group.)

and down. (See the Float Switch Assembly section presented later in this chapter.)
3. Remove the two (2) ¼-inch hex-head (black) screws securing the bottom of the service panel.
4. Open the dishwasher door and remove the two (2) hex-head screws (black), one (1) in each rear corner, securing the top of the service panel.
5. Remove the lower service panel by pulling it straight off.
6. Advance the timer to the fill position and latch the door. Start the dishwasher.
7. Check for voltage at the fill valve solenoid. If voltage is present, disconnect the electrical power and check the resistance of the solenoid coil. Replace the valve if it is found to be defective. If the solenoid checks OK, shut off the water supply and the electrical supply. Remove the water supply line from the valve and check the inlet filter screen for debris. See Figure 14–23. Clean the screen and reassemble the valve.
8. If there is no voltage present at the solenoid, check the continuity through the latch switch, timer contacts, float switch, and the wiring (with the power off).

Water level too low
The water level should just cover the heating element.

1. Check the incoming water pressure; 15 psi is the minimum pressure for an adequate fill.
2. Check the main water supply valve to see that it is "full open."
3. Check for a clogged screen in the inlet valve.

Water will not shut off

1. Turn off the electrical supply. If water continues to flow, close off the main water supply valve and replace the fill valve.
2. If the water shuts off when the power is disconnected, check for welded timer contacts or a timer motor that fails to advance. Replace the timer if either is defective.

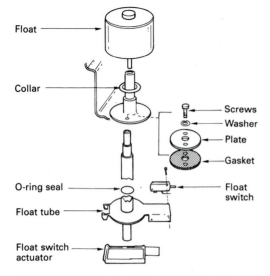

Figure 14–28 Float (water fill safety). (Courtesy of WCI Major Appliance Group.)

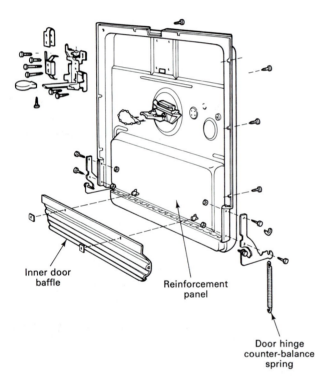

Figure 14–29 Inner door panel and reinforcement. (Courtesy of WCI Major Appliance Group.)

Float Switch Assembly

The water float assembly is located in the left front corner of the tub. See Figure 14–28. Its function is to shut off the electrical power to the water fill solenoid should too much water enter the dishwasher. The float switch (seated directly under the float beneath the tub) is a normally open switch. The weight of the float assembly holds the switch closed. If the switch fails to close, check the following:

1. Loose connection at the switch terminal.
2. Switch not installed properly.
3. Bent or binding switch actuator.
4. Warped stem on the float, not contacting the actuator blade.

To remove or replace the float switch, use the following procedure:

1. Disconnect the dishwasher from the electrical supply.
2. Remove the two (2) $\frac{1}{4}$-inch hex-head (black) screws securing the bottom of the service panel.
3. Open the dishwasher door and remove the two (2) $\frac{1}{4}$-inch hex-head (black) screws, one (1) in each rear corner, securing the top of the service panel.
4. Remove the lower service panel by pulling it straight off.
5. Open the dishwasher door, remove the lower dishrack.
6. Remove the float, pulling it straight up.
7. Remove the two (2) $\frac{5}{16}$-inch hex-head screws securing the float tube to the tub bottom.
8. Remove the float tube and switch assembly from the service compartment.
9. Remove the two (2) screws securing the switch to the support.
10. Remove the electrical leads from the float switch.
11. Install the new float switch and check it for operation.
12. Reverse the above procedure to reassemble the unit.

Inner Door Panel Reinforcement

The inner door panel reinforcement support provides inner door panel structural support and mounting for the electrical components. See Figure 14–29.

The metal inner door reinforcement and the polypropylene inner door panel are serviced as an assembly. (To remove or replace the inner door panel and reinforcement, see the Inner Door Panel section presented later in this chapter.)

Tub and Door Seal

The tub and door seal is a one-piece molded vinyl rubber seal held in place by stainless-steel clips (clips hook over an internal support wire of the seal) and snapped into the tub top and sides. The lower section is secured to the front of the tub liner by a wire rod and screw mounted clip.

To remove or replace the tub and door seal, use the following procedure:

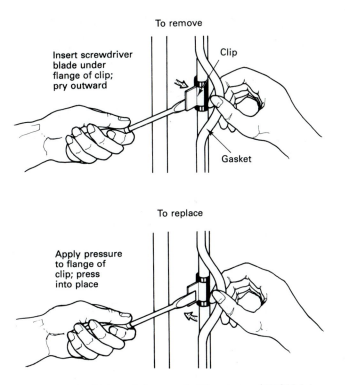

To remove

Insert screwdriver
blade under
flange of clip;
pry outward

Clip

Gasket

To replace

Apply pressure
to flange of
clip; press
into place

Figure 14–30 Tub and door seal. (Courtesy of WCI Major
Appliance Group.)

1. Disconnect the dishwasher from the electrical supply.

2. Open the dishwasher door and use a putty knife or a thin bladed screwdriver. Carefully apply pressure to remove the clips from the tub liner slots. See Figure 14–30.

3. Remove the two (2) ¼-inch hex-head screws securing the bottom of the service panel.

4. Open the dishwasher door and remove the two (2) ¼-inch hex-head screws (black) securing the top of the service panel, one (1) in each rear corner.

5. Remove the one (1) screw at the center of the bottom front, securing the seal wire retainer.

6. Remove the tub and door seal assembly.

7. Install a new tub and door seal assembly. *Note:* The service replacement tub and door seal is preformed to the tub shape with the wires in place.

8. Reverse the above procedure to reassemble the unit.

Fill Funnel and Air Gap Assembly

The funnel and air gap assembly is mounted onto the right side of the tub. See Figure 14–31. A threaded nipple extends through the tub wall. An O-ring seal and threaded nut holds the funnel securely to the tub. Its purpose is to provide a method of supplying water for the wash and rinse cycles through an air gap, as required by plumbing codes. The air gap prevents the siphoning of the wash water back into the water supply system should the water pressure drop to less than atmospheric. The dishwasher must be removed from under the counter to service the funnel.

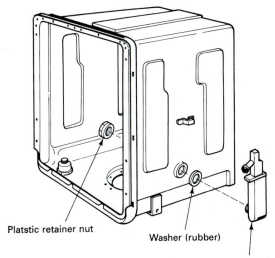

Platstic retainer nut

Washer (rubber)

Water fill funnel and air gap

Figure 14–31 Water inlet fill funnel and air gap assembly. (Courtesy of WCI Major Appliance Group.)

To remove or replace the fill funnel and air gap assembly, use the following procedure:

1. Disconnect the dishwasher from the electrical supply.

2. Remove the two (2) ¼-inch hex-head (black) screws securing the bottom of the service panel.

3. Open the dishwasher door and remove the two (2) ¼-inch hex-head screws (black), one (1) in each rear corner, securing the top of the service panel.

4. Remove the lower service panel by pulling it straight off.

5. Disconnect the water line union to the inlet fill valve.

6. Remove the two (2) screws securing the dishwasher to the countertop.

7. Position the washer to gain access to the right side.

8. Open the dishwasher door and remove the large plastic nut securing the water inlet funnel to the tub.

9. Remove the clamp securing the water line to the water inlet funnel and air gap assembly.

10. Remove the water inlet funnel and air gap assembly (snap-together design).

11. Install the new water inlet funnel and air gap assembly.

12. Reverse the above procedure to reassemble the unit.

Heating Element

The 800-watt heating element is used to heat the water during the wash and rinse cycles and to heat the air during the dry cycle. See Figure 14–32.

To check for operation, advance the timer to the dry cycle, set the selector switch for heated dry, close and latch the dishwasher door. Allow 1 or 2 minutes, open the dishwasher door and note if heat is present.

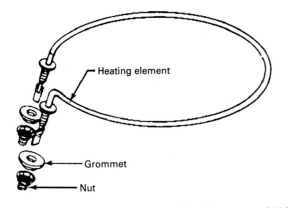

Figure 14–32 Heater element assembly. (Courtesy of WCI Major Appliance Group.)

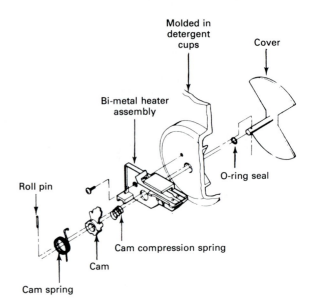

Figure 14–33 Detergent dispensing assembly. (Courtesy of WCI Major Appliance Group.)

To test the continuity of the heating element, use the following procedure:

1. Disconnect the dishwasher from the electrical supply.
2. Remove the two (2) hex-head (black) screws securing the bottom of the service panel.
3. Open the dishwasher door and remove the two (2) ¼-inch hex-head (black) screws, one in each rear corner, securing the top of the service panel.
4. Remove the lower service panel by pulling it straight off.
5. Remove the electrical lead from one side of the heating element. *Note:* The connectors have a lock tab to prevent the connectors from coming off. Depress the tab to remove it from the terminal.
6. Connect an ohmmeter across the element terminals.

To remove or replace the heater element, use the following procedure:

1. Disconnect the dishwasher from the electrical supply.
2. Remove the two (2) hex-head (black) screws securing the bottom of the service panel.
3. Open the dishwasher door and remove the two (2) hex-head (black) screws, one (1) in each rear corner, securing the top of the service panel.
4. Remove the lower service panel by pulling it straight off.
5. Remove the lower dishrack.
6. Remove the electrical leads from the heater.
7. Remove the two (2) ½-inch hex nuts securing the heater element to the bottom of the tub.
8. Remove the heater element.
9. Install the new heater element.
10. Reverse the above procedure to reassemble the unit.

Detergent Dispenser

The detergent dispenser is mounted on the inner door panel. See Figure 14–33. It consists of a circular molded housing with four (4) detergent cups. The detergent dispenser covers two (2) of the four (4) cups and dumps its detergent from the open cups into the tub as soon as the dishwasher door is closed. The cover is rotated by the user to close and cover the two (2) opposite cups. At the center of the dispenser cover is a shaft and bearing that extends down through the door panel. The bearing is sealed and anchored to the panel with an O-ring and retainer pin on the inner door panel side.

In the back side of the inner door panel, a cam and spring assembly fits over the detergent drawer shaft and is held in place by a retainer pin. See Figures 14–34 and 14–35. The cam has a quarter round boss on top, to which a coil drive spring attaches. The coiled spring tightens when the detergent cover is rotated and latched into the bimetal arm.

The bimetal arm is electrically connected in series with the heater element to limit the current flow through the bimetal. When this series circuit is energized by the timer during the wash cycle, the bimetal warps, moving the arm away from the cam. The drive spring then rotates the cam and cover to expose the detergent cups and releases the detergent into the dishwasher. When the timer contact opens, the bimetal cools, and returns to its original shape.

To test the bimetal arm, use the following procedure: The bimetal must be tested using an ohmmeter only. Never apply 120 volts directly across the bimetal terminals. A failed bimetal can often be traced to a bad connection or a broken lead.

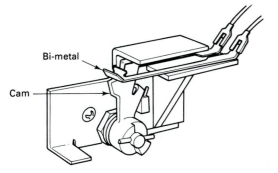

Figure 14–34 Detergent dispenser cam latched. (Courtesy of WCI Major Appliance Group.)

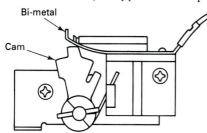

Figure 14–35 Bimetal arm activated, releasing the detergent dispenser cam. (Courtesy of WCI Major Appliance Group.)

1. Disconnect the dishwasher from the electrical supply.
2. Open the dishwasher door and remove the four (4) screws securing the outer door panel assembly.
3. Remove the ground wire from the outer door panel (male) terminal.
4. Remove the electrical leads to the bimetal and place the ohmmeter leads across the terminals. Continuity should exist. If it shows infinite resistance the bimetal must be replaced.

To remove or replace the bimetal heater, use the following procedure:

1. Disconnect the dishwasher from the electrical supply.
2. Open the dishwasher door and remove the four (4) screws securing the outer door panel assembly.
3. Remove the ground wire from the outer door panel (male) terminal.
4. Remove the electrical leads from the bimetal.
5. Remove the roll pin retainer securing the cam and spring to the shaft.
6. Remove the cam with the spring.
7. Remove the spring clip from the shaft.
8. Remove the Phillips-head screw securing the bimetal assembly.
9. Remove the bimetal assembly.
10. Install a new bimetal assembly.
11. Reverse the above procedure to reassemble the unit.

To remove or replace the detergent dispenser cover assembly, use the following procedure:

1. Disconnect the electrical supply from the dishwasher.
2. Open the dishwasher door and remove the four (4) screws securing the outer door panel assembly.
3. Remove the ground wire from the outer door panel (male) terminal.
4. Remove the wiring leads to the bimetal.
5. Remove the roll pin retainer securing the cam and spring to the shaft.
6. Remove the cam with the spring.
7. Remove the spring from the cup shaft.
8. Remove the dispenser cover, O-ring, and shaft assembly.
9. Install the new dispenser cover and O-ring.
10. Reverse the above procedure to reassemble the unit.

Terminal Board

Two (2) wiring terminal boards are utilized on this dishwasher; one is located in the service compartment behind the service panel, and one is located inside the control panel assembly. The connection point of these wiring terminals may be utilized in servicing the electrical circuitry. Complete schematic wiring diagrams and cycle charts are presented at the end of this chapter.

To remove or replace the control panel terminal board, use the following procedure:

1. Disconnect the dishwasher from the electrical supply.
2. Remove the $\frac{1}{4}$-inch hex-head screws securing the door latch knob to the door latch arm.
3. Open the door and remove the two (2) screws across the top of the door and the two (2) screws along both sides of the control panel. This will release the control panel from the door.
4. Carefully pull the control panel forward to gain access to the controls.
5. Remove the electrical wiring from the terminal board.
6. Remove the two (2) screws securing the terminal board.
7. Install the new terminal board.
8. Reverse the above procedure to reassemble the unit.

To remove or replace the main terminal board, use the following procedure:

1. Disconnect the dishwasher from the electrical supply.
2. Remove the two (2) $\frac{1}{4}$-inch hex-head (black) screws securing the bottom of the service panel.
3. Open the dishwasher door and remove the two (2) $\frac{1}{4}$-inch hex-head (black) screws, one in each rear corner, securing the top of the service panel.

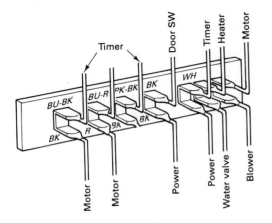

Figure 14–36 Main terminal board. (Courtesy of WCI Major Appliance Group.)

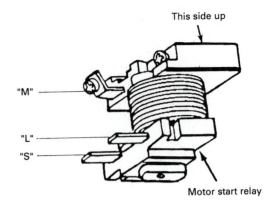

Figure 14–37 Motor start relay. (Courtesy of WCI Major Appliance Group.)

4. Remove the lower service panel by pulling it straight off.

5. Remove the electrical wiring from the terminal board. See Figure 14–36.

6. Remove the two (2) screws securing the terminal board.

7. Install the new terminal board.

8. Reverse the above procedure to reassemble the unit.

Motor Start Relay

The motor start relay handles the high starting current required by the start windings. See Figure 14–37. At the proper motor speed the relay drops out, allowing the motor to continue to operate on its run windings. The motor start relay is mounted on the dishwasher door hinge weld assembly, behind the service panel.

A magnetic-type motor start relay is used. The motor run winding is connected in series with the relay coil. When current is applied to the motor, it passes through the motor run winding and the start relay coil. The high current energizes the start relay coil to magnetically lift the relay armature and close the relay contacts L and S. This applies current across one of the start windings to start the motor.

The timer determines which start winding circuit is closed, to give either counterclockwise or clockwise rotation. As the motor comes up to speed, the run winding current decreases. When the current drops low enough, the motor start relay armature drops to open contacts L and S. This opens the start winding circuit. The motor continues to operate on the run winding.

The timer controls the electrical circuit to the run winding through contacts L and M (terminals) of the relay. The motor will come to a complete stop before reversing.

To test the relay, use the following procedure:

1. Disconnect the dishwasher from the electrical supply.

2. Remove the two (2) hex-head (black) screws securing the bottom of the service panel.

3. Open the dishwasher door and remove the two (2) ¼-inch hex-head (black) screws, one (1) in each rear corner, securing the top of the service panel.

4. Remove the lower service panel by pulling it straight off.

5. Remove the electrical wiring from the motor start relay.

6. Check the relay coil for continuity across terminals L and M. Replace the relay if the coil is open.

7. To check contacts L and S, remove the one (1) screw securing the motor start relay. *Note:* This relay is position sensitive, which means that when energized, the contacts are pulled up to close them. They are opened by gravity.

8. Turn the relay upside down and check the contacts for continuity. They should be closed indicating continuity in this position and infinite continuity when right-side up.

To remove or replace the relay, use the following procedure:

1. Disconnect the dishwasher from the electrical supply.

2. Remove the two (2) ¼-inch hex-head (black) screws securing the bottom of the service panel.

3. Open the dishwasher door and remove the two (2) hex-head (black) screws, one in each rear corner, securing the top of the service panel.

4. Remove the lower service panel by pulling it straight off.

5. Remove the electrical wiring from the motor start relay.

6. Remove the one (1) screw securing the motor start relay.

7. Install the new motor start relay. *Caution:* Be sure to mount the relay with the word "top" up.

8. Reverse the above procedure to reassemble the unit.

Motor and Pump Assembly

The drive motor is a ⅓-hp, 120-volt, 60-hertz, single-phase, 3450-rpm, internal thermal (automatic reset) over-

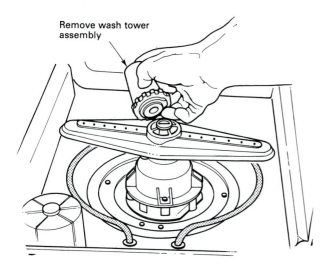

Remove wash tower
assembly

Figure 14–38 Wash tower removal. (Courtesy of WCI Major Appliance Group.)

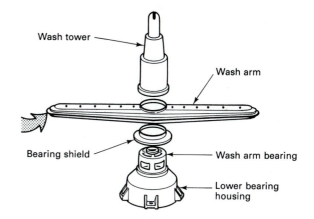

Wash tower

Wash arm

Bearing shield

Wash arm bearing

Lower bearing housing

Figure 14–39 Wash arm and impeller removal. (Courtesy of WCI Major Appliance Group.)

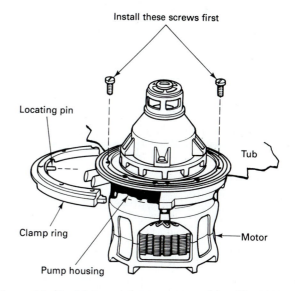

Install these screws first

Locating pin

Tub

Clamp ring

Motor

Pump housing

Figure 14–40 Motor and pump assembly. (Courtesy of WCI Major Appliance Group.)

load protected motor. It drives the impellers in the pump system, rotating counterclockwise for the wash and rinse cycles, and rotating clockwise for pumping water out of the dishwasher (as viewed from the top).

The drive motor has three (3) windings, one run winding and two (2) start windings. One start winding is used for counterclockwise rotation and the other start winding is used for clockwise rotation. A magnetic start relay is used to handle the high starting current (refer to the Motor Start Relay section presented earlier in this chapter).

To test the drive motor windings, use the following procedure: The drive motor can be tested at the lower main terminal board by using an ohmmeter and performing the following steps:

1. Disconnect the dishwasher from the electrical supply.
2. Remove the two (2) hex-head (black) screws securing the bottom of the service panel.
3. Open the dishwasher door and remove the two (2) ¼-inch hex-head (black) screws, one (1) in each rear corner, securing the top of the service panel.
4. Remove the lower service panel by pulling it straight off.
5. Remove each motor lead and check the resistance of the windings. The resistance should be within ±10 percent of the values listed in the table.
6. If any of the winding circuits are open, the motor must be replaced.

To remove the complete motor and pump assembly, use the following procedure:

1. Disconnect the dishwasher from the electrical supply.
2. Remove the two (2) ¼-inch hex-head (black) screws securing the bottom of the service panel.

3. Open the dishwasher door and remove the two (2) ¼-inch hex-head (black) screws, one (1) in each rear corner, securing the top of the service panel.
4. Remove the lower service panel by pulling it straight off.
5. Open the dishwasher door and remove the lower dishrack.
6. Remove the wash arm by turning the wash tower counterclockwise while holding the arm. See Figures 14–38 and 14–39.
7. Remove the wash arm and bearing shield.
8. Remove the six (6) $\frac{5}{16}$-inch hex-head screws around the outside edge of the pump housing that secure the motor assembly to the split ring clamps.
9. Remove the one (1) screw and clamp securing the wiring harness to the left split ring clamp.
10. Disconnect the pump motor electrical leads.
11. Disconnect the ground wire.
12. Disconnect the drain clamp and hose.
13. Remove the split ring clamps. See Figure 14–40.

Functional Components and Parts 165

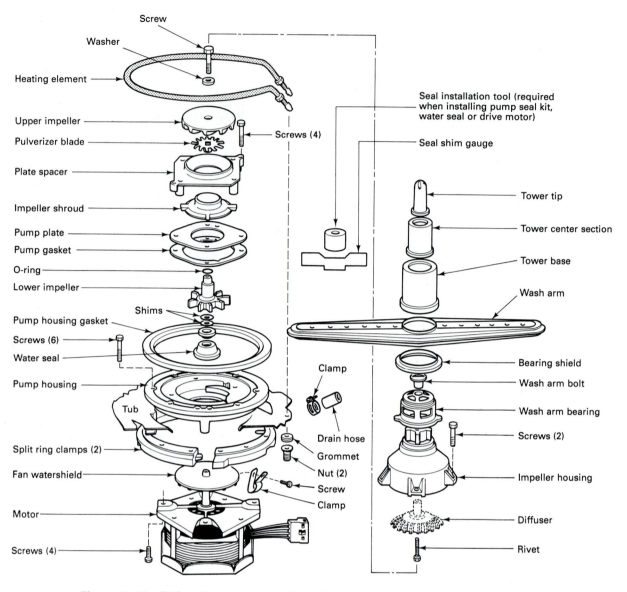

Figure 14–41 Dishwasher pump assembly and heating element. (Courtesy of WCI Major Appliance Group.)

14. Remove the motor and pump assembly from inside the dishwasher tub. *Caution:* Be careful not to hit the tub liner or drop the motor assembly when removing it.

15. Figure 14–41 provides an exploded view of the pump assembly.

16. Reverse the above procedure to reassemble the unit.

To replace the lower impeller and water seal, use the following procedure. This procedure can be performed without removing the motor and pump assembly. See Figure 14–41 for an exploded view of the pump.

1. Disconnect the dishwasher from the electrical supply.

2. Remove the lower dishrack.

3. Hold the wash arm and turn the center wash tower counterclockwise. Remove the wash arm and bearing shield.

4. Remove the front and rear $\frac{5}{16}$-inch hex-head screws securing the impeller housing to the pump housing. Lift off the impeller housing.

5. Remove the $\frac{5}{16}$-inch hex screw and washer securing the upper impeller to the motor shaft. Remove the upper impeller and pulverizer blade. (*Note:* An O-ring is installed between the upper and lower impellers. It usually remains attached to the impeller screw when the upper impeller is removed. A new O-ring should be installed during reassembly.)

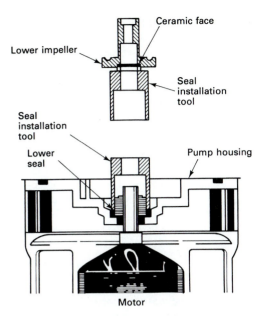

Figure 14–42 Seal installation details. (Courtesy of WCI Major Appliance Group.)

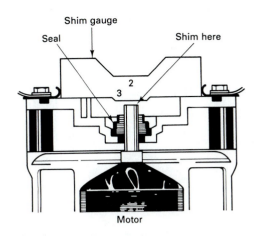

Figure 14–43 Using the shim gauge. (Courtesy of WCI Major Appliance Group.)

6. Remove the left- and right-side $\frac{5}{16}$-inch hex-head screws securing the spacer plate to the pump housing. Remove the spacer plate, impeller shroud, pump plate and gasket, and the lower impeller. (Note the number of shim washers used between the lower impeller and the motor shaft, if any. Check on top of the motor shaft and up inside the recess of the pump housing for shims.)

7. Carefully pry the water seal from the pump housing using a flat-blade screwdriver.

8. Install the new water seal using the seal installation tool. See Figure 14–42. Place the tool over the carbon-faced portion of the seal. Lubricate the outer edge of the seal with water. Position the assembly over the motor shaft and press the seal into the recess of the pump housing, ensuring that the seal is fully seated.

9. If the replacement ceramic-faced seal is not preinstalled into the recess of the replacement impeller as shown in Figure 14–42, lightly lubricate the outer edge of the seal with water and press it into position, ensuring that the seal is fully seated into the impeller recess with the ceramic surface facing out.

10. Install the new shims (use the same number of shims that were removed in step 6 above) on top of the motor shaft, ensuring that the flat side of the shims align with the slot in the motor shaft. Position the impeller over the motor shaft and align the key in the impeller with the slot in the motor shaft.

11. Install the pump plate and gasket (the pump housing has a raised tab to align the plate). Install the impeller shroud (the slotted projections on the shroud are placed front to back). Install the spacer plate (the inside stanchions of the spacer plate fit into the impeller shroud slots). Secure these components to the pump housing using the two (2) inside holes on the left and right side of the spacer plate.

12. Install the pulverizer blade and the upper impeller onto the lower impeller shaft, ensuring that the new O-ring is installed between the impellers. Tighten the impeller assembly to the motor shaft.

Note: At this time, manually rotate the impeller to check for any binding or rubbing condition. If the impellers are binding, disassemble them and use a shim gauge tool to determine if the correct number of shims was installed. See Figure 14–43.

13. Install the impeller housing with the two (2) remaining $\frac{5}{16}$-inch hex-head screws, front and back. Install the wash arm and the center wash tower.

14. Install the lower dishrack and pour approximately 1 quart of water into the pump area to prevent overheating of the water seal during the initial test.

15. Test the dishwasher for proper operation.

To replace the motor or pump housing, use the following procedure:

1. Remove the complete motor and pump assembly (refer to steps 1 through 15 of the To Remove Complete Motor and Pump Assembly procedure).

2. Follow steps 4 through 7 of the previous procedure, To Replace the Lower Impeller and Water Seal.

3. From the motor side, remove the four (4) $\frac{5}{16}$-inch hex-head screws securing the motor to the pump housing. Then remove the ground wire from the motor frame.

4. To reassemble, align the motor mounting holes with the holes in the pump housing. (*Note:* Wires from the motor should be adjacent to the drain port of the pump housing.)

5. Install, but only hand tighten, the four (4) mounting screws.

6. Insert the seal installation tool over the motor shaft and into the seal access of the pump housing as shown in Figure 14–44. This will center the motor to the pump housing. Tighten the motor mounting screws.

7. Follow steps 8 and 9 of the previous procedure, To Replace the Lower Impeller and Water Seal.

Important: Whenever the pump housing to motor is replaced, it is necessary to determine the number of shaft

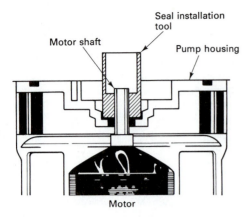

Figure 14–44 Centering the motor shaft. (Courtesy of WCI Major Appliance Group.)

shim washers needed to maintain the proper water seal spring pressure. Figure 14–43 illustrates the use of the shim gauge tool. The shim gauge must make contact with the pump housing (not the gasket).

8. Place the shim(s) on the top of the motor shaft until contact is made with the shim gauge. If the last (or only one) shim raises the gauge, do not use it. A gap less than the thickness of a shim is satisfactory.

9. Complete the reassembly following steps 10 through 13 of the previous procedure, To Replace the Lower Impeller and Water Seal.

10. Install the pump assembly into the dishwasher reversing steps 1 through 15 of the heading To Remove the Complete Motor and Pump Assembly procedure.

11. Pour approximately 1 quart of water into the pump area to prevent overheating of the water seal during the initial test.

12. Test the dishwasher for proper operation.

Wash Arm and Tower Assembly

The wash arm and tower assembly provide a rotating water spray for the washing and rinsing action.

To remove or replace the wash arm and tower, use the following procedure:

1. Disconnect the dishwasher from the electrical supply.

2. Open the dishwasher door and remove the lower dishrack.

3. Remove the wash arm assembly by turning the tower counterclockwise while holding the wash arm. The tower consists of three (3) pieces. See Figures 14–38 and 14–39.

4. Remove the wash arm and bearing shield.

5. Install the new wash arm or wash tower parts as required.

6. Reverse the above procedure to reassemble the unit.

Dishwasher Door Assembly

The dishwasher door assembly consists of the control panel, timer, selector switch, door latch, outer door panel, trims, decorator color panel insert, inner door panel and parts.

Outer Door Panel

The outer dishwasher door panel is white enameled steel. The decorator trim frames are mounted to the outer door panel sides and bottom. The outer door panel sides and bottom secure the colored aluminum decorator door insert and/or wood panel to match the kitchen cabinets.

To remove or replace the outer door panel and trims, use the following procedure:

1. Disconnect the dishwasher from the electrical supply.

2. Open the dishwasher door and remove the four (4) screws securing the outer door panel assembly.

3. Remove the ground wire from the outer door panel. *Note:* Place the outer door panel assembly on a protected surface.

4. Remove the screws securing the side and bottom door trims. Remove the trims.

5. Install the new outer door panel, insert, or trims.

6. Reverse the above procedure to reassemble the unit.

Inner Door Panel

The inner door panel assembly is a polypropylene panel attached to a metal panel for reinforcement. The control housing assembly, detergent dispenser, and the door latch are mounted to the inner door panel.

To remove or replace the inner door panel, use the following procedure:

1. Disconnect the dishwasher from the electrical supply.

2. Open the dishwasher door and remove the four (4) screws securing the outer door panel assembly.

3. Remove the ground wire from the outer door panel assembly.

4. Remove the $\frac{1}{4}$-inch hex-head screw securing the door latch knob to the door latch arm.

5. Open the door and remove the two (2) screws across the top of the door and the two (2) screws along both sides of the control panel. This will release the control panel from the door.

6. Remove the electrical leads from the door latch switch, detergent dispenser, and the ground leads.

7. Unhook the door counterbalance springs.

8. Remove the four (4) screws securing the inner door to the hinge assembly.

9. Lift the inner door panel from the hinge assembly.

10. Lay a new inner door panel on a protected surface.

11. Transfer the door latch assembly, detergent dis-

penser, electrical controls, control panel and parts.

12. Reverse the above procedure to reassemble the unit.

Control Panel

The timer, selector switch, upper terminal board, pilot lamp, vent slot, end cap trims, and escutcheon are mounted to the control panel.

To remove or disassemble the control panel, use the following procedure:

1. Disconnect the dishwasher from the electrical supply.

2. Remove the ¼-inch hex-head screw securing the door latch knob to the door latch arm.

3. Remove the timer control knob by pulling it straight off.

4. Open the door and remove the two (2) screws across the top of the door and the two (2) screws along both sides of the control panel. This will release the control panel from the door.

5. Carefully pull the control panel forward to gain access to the controls.

6. Remove the two screws securing the timer.

7. Remove the one (1) screw and the clamp securing the timer wiring.

8. Remove the two (2) screws securing the plastic wiring shield.

9. Remove the two (2) screws securing the upper terminal block.

10. Remove the one (1) screw securing the vent shield.

11. Remove the two (2) screws securing the selector switch.

12. Remove the two (2) screws (each end) securing the control panel end caps.

13. Remove the pilot lamps, lens, vent grill, and nameplate.

14. Install the new control panel.

15. Reverse the above procedure to reassemble the unit.

Dishwasher Tub

The dishwasher tub is constructed of porcelain on steel, with a built-in sump area. The drive motor and pump assembly mounts in the center bottom of the sump area.

To remove or replace the dishwasher tub liner, use the following procedure:

1. Disconnect the dishwasher from the electrical supply.

2. Remove the service panel.

3. Disconnect the dishwasher from the water supply.

4. Remove the two (2) screws securing the dishwasher to the countertop.

5. Position the dishwasher for removal of all components and parts.

6. Remove the outer door panel.

7. Remove the upper and the lower dishracks.

8. Remove the wash arm and lower bearing.

9. Remove the drive motor and pump assembly.

10. Disconnect all electrical leads and ground wires.

11. Disconnect the drain clamp and hose.

12. Remove the water float valve assembly.

13. Remove the door seal assembly.

14. Remove the upper dishrack support slide rollers.

15. Remove the door strike assembly.

16. Remove the water fill funnel and air gap.

17. Remove the heating element.

18. Remove the dishwasher door assembly.

19. Remove the structural support and leg assembly.

20. Remove any other miscellaneous parts.

21. Position the new tub on a padded surface to protect the tub liner while reassembling.

22. Reverse the above procedure to reassemble the unit.

HOME ECONOMIST DISHWASHING REVIEW

The following are some suggestions that will help the home economist to produce clean, spot-free dishes.

Factors in Good Dishwashing

Many of the factors that contribute to good automatic dishwashing are controlled by the user, such as water conditions, detergent usage, loading, and makeup of the load. They all affect the results.

Water Conditions

The following discussion of water conditions will aid the user in producing good dishwashing results.

Water temperature The water temperature should be between 140 and 160°F for good dishwashing results. Running hot water at the sink before operating the dishwasher will help to clear the pipes of cold water. The water temperature can drop several degrees as it circulates in a cold dishwasher. To check the temperature, wait until the wash fill is completed and the wash has started. Open the door and remove a glass of water from the tub. Any standard immersion-type thermometer can be used; if necessary, a good cooking thermometer would be best, if one is available. Many meat, candy, and deep-fat thermometers register temperatures in the 140 to 160°F range. Water hotter than 160°F in the wash cycle may cause certain food soils to bake on the dishes.

Water hardness The ideal water for automatic dishwashing would be from 4 to 8 grains of hardness. Most dishwasher detergents are designed to handle normal soil in water ranging from 4 to 8 grains per gallon. Extreme hardness or softness can cause spotting and filming of glasses and silverware. In some cases, this can be controlled by the use of detergent. If the water is 10 or more grains per gallon in hardness, the installation of a water-softening system is recommended.

Experiment with different detergents to see which one works best for the installation. The amount of detergent depends on the size of the load, the amount of soil, the water hardness, and whether the soil has been allowed to dry or harden on the dishes. Under average conditions—full mixed load, medium water—it is suggested that 2 or 3 tablespoons of detergent per cup be used. The cups filled will depend on the cycle selected. If the water is extremely hard (or the dishes are heavily soiled), more detergent may be required as shown later in this chapter.

Soft water (0 to 3.5 grains per gallon*) Use 2 tablespoons of detergent per cup ($\frac{1}{2}$ full).

Moderately hard water (3.6 to 7.0 grains per gallon) Use 2 to 2½ tablespoons of detergent per cup ($\frac{1}{2}$ to $\frac{3}{4}$ full).

Hard water (7.1 to 10.5 grains per gallon) Use 1 full cup of detergent.

Extremely hard water (10.6 grains per gallon and above) In this condition detergent alone is not sufficient. A water softener should be installed in the home plumbing system.

Preparation of the Dishes

Most normally soiled dishes do not require rinsing if washed promptly.

1. Rinse off the soluble food particles such as coffee grounds, salad greens, and citrus pulp.
2. Rinse heavy starch soils, such as mashed potatoes or cooked cereals.
3. Remove large and hard food scraps, bones, or pits.
4. Rinse egg, mayonnaise, and salad dressing off the dishes and silverware that will not be washed until later.
5. Presoak heavy baked-on soil.

Tips on Loading

The following tips will aid in producing cleaner dishes.

1. Place the soiled side of the dishes facing the spray column in the center of the spray arm at the bottom of the tub.
2. Tilt the cups, glasses, and utensils so that the water can drain out of the inside and off the bottom of them.

---

* Grains per gallon is the common unit used for expressing the concentration of materials dissolved in water. Water hardness is caused by high concentrations of calcium and magnesium that react with heat to form lime scale. If the water hardness is not known, have it tested or contact the local water company for an analysis.

3. Avoid the nesting of bowls, plates, or silverware.
4. Do not place a small item behind a large item that will block the water.
5. Do not place a large item against the spray column in the lower rack or water distribution could be affected.
6. Put heat-sensitive plastic items in the upper rack away from the heater.
7. Put large platters and utensils in the far right or left side of the lower rack.
8. Face heavily soiled utensils toward the spray arm below the rack.
9. Silverware can be loaded with the handles up or down. Handles up protects the hands and arms from sharp knives and forks and allows the most sanitary unloading procedure. Mix spoons, forks, and knives to prevent "nesting." Nesting causes improper circulation of wash water and prevents air circulation during the drying portion of the cycle. Also, make sure that the handles do not protrude through the bottom rack, blocking the wash arm action. Place silver and stainless-steel flatware in different compartments.

Dishwasher Detergents

Caution: Soap, laundry, or hand-washing detergents must not be used in any automatic dishwasher.

Dishwasher detergents contain ingredients to clean dishes, to protect metal and china, to help soften hard water, and to prevent foaming and oversudsing. Other types of detergent and/or soaps will block the washing action and cause the water to overflow onto the kitchen floor. *Note:* Dishwasher detergents and rinse agents should be stored out of the reach of children and away from moist or damp areas.

Use only the recommended dishwasher detergents. These include Calgonite, Cascade, Dishwasher All, Electra Sol, and Finish. Dishwasher detergents have a limited shelf life and should not be purchased in large quantities. Store them in a cool dry place (not under the kitchen sink) with the pouring spout closed.

Detergent performance varies in different kinds of water. Suggest that users try various brands to determine which one gives the best results. At least a full box of detergent should be used to evaluate the performance.

Liquid detergents are also available. Follow the directions on the package when using them.

Detergent dispensers should be dry when filled or caking can occur. Close the cover before loading the dishes to prevent liquids collecting in the dispenser. The detergent dispenser should be filled just before the dishwasher is operated. Avoid overfilling the cup. Otherwise the cover may not close securely and moisture will seep into the dispenser.

The recommended detergent usage listed in the Water Hardness section is based on the use of high-phosphate detergents and is only a suggestion. Since water conditions and soil loads vary, the best recommendation to give a customer is to increase the detergent until the results are satisfactory. In medium hard water, using additional detergent or adding a nonprecipitating water conditioner (Calgon or Spring Rain) will improve the washing results. The phosphate content of the detergent is listed on the package label. Some states will only allow a certain percentage of phosphate in detergents; therefore, use the highest-phosphate content possible for the area.

In nonrestricted areas, the phosphate content of recommended dishwasher detergents is

Calgonite	12.0%
Cascade	12.9%
Dishwasher All	12.9%
Electra Sol	8.7%
Finish	10.9%

Electra Sol and Finish are produced by the same manufacturer, but the formulas may differ in various parts of the country; they try to tailor the detergent to the local conditions. Electra Sol is designed for soft water and Finish for hard water.

Just about all dishes, flatware, pots and pans, and other cooking utensils can be washed in the dishwasher. Some items, however, are not designed for automatic dishwashing. Many items are labeled or stamped to indicate if they are dishwasher safe. If in doubt, check the manufacturer's instructions or test-wash just one piece in the dishwasher. Continue the test for about one month; compare the test piece with the rest of the set.

Removing Mineral Spots and Film

If dishes or glasses develop a white or hazy film, they may be cleaned by one of the following methods:

1. *Commercial product method* Use Glass Magic or some other commercial film and spot remover. Glass Magic can be used safely any number of times and is made to be used in a normal dishwashing cycle. Follow the directions on the package.

2. *Chlorine bleach and vinegar method*
   a. Set the dishwasher for heavy soil and allow the dishwasher to fill, drain, and refill.
   b. After the second fill, stop the dishwasher, open the door, and add one (1) cup of a liquid chlorine bleach to the wash water. Close the door and allow the dishwasher to complete the wash cycle.
   c. Allow the dishwasher to fill for the rinse cycle, then stop the dishwasher. Open the door, add two (2) cups of vinegar to the rinse water.
   d. Close the door and allow the dishwasher to complete the remaining rinse and dry cycles.

Caution: Do not place aluminum pots or pans, plastic ware, metal items, or any detergent in the dishwasher when performing this cleaning process.

To remove any heavy film on items or a buildup of film on the dishwasher interior, an acid treatment may be required.

Follow the steps below using citric-acid crystals which may be purchased at most drugstores.

3. *Citric-acid method*
   a. Run a soiled dishload through its normal cycle using detergent. Remove the metal items from the dishwasher.
   b. Set the dishwasher for heavy soil. Do not add detergent.
   c. Allow the dishwasher to fill, then stop the dishwasher, and open the door. Add $\frac{1}{4}$ to $\frac{1}{2}$ cup of citric-acid crystals to the wash water.
   d. Close the door and allow the dishwasher to complete the cycle.

4. *Lime-A-Way method* Lime-A-Way is a product designed to remove a lime scale buildup or iron from inside the dishwasher. It is safe to use periodically according to the directions on the package.

Note: Never mix vinegar and bleach; this mixture creates toxic fumes.

Glass Magic, an Economics Laboratory, Inc., product, is available in most supermarkets for use in removing spots and film from utensils in the dishwasher. Follow the directions on the package. It is not necessary to remove the silverware and other metal items from the load. To prevent recurrence, Glass Magic can be used occasionally with regular detergent. When used this way, use $\frac{1}{3}$ Glass Magic and $\frac{2}{3}$ detergent.

Lime-A-Way, another product of Economics Laboratory, Inc., can be used to remove rust stains or a heavy lime buildup from the dishwasher tub. Follow the directions on the package.

Rinse Aids

Rinse aids help promote faster, spot-free drying. Rinse aids cause the wash water to "sheet" or run off the dishes. "Sheeting" prevents water droplets from forming on the drying surface. Rinse aids are available in both liquid and solid forms.

The rinse aid dispenser is located inside the dishwasher door. The dispenser automatically dispenses a measured amount of the rinse aid into the dishwasher during the last rinse.

Before using the dishwasher for the first time, unscrew the dispenser cap and add the rinse aid agent to the dispenser.

For the best results, refill the rinse aid reservoir each month.

TROUBLESHOOTING

The troubleshooting checklist is common for all dishwasher models. They use different parts to accomplish the same task and the diagnosis will be similar.

When a problem arises and a possible cause is listed, follow the test, remove or replace procedures as outlined in the service section. The wiring diagram, schematic, and timer cycle chart are a necessity when making electrical checks. In most cases an ohmmeter will handle all the necessary tests.

For checking any particular cycle of operation, it is absolutely necessary that the cycle be set up as outlined in the product instruction manual.

15

Electric and Gas Dryers

We will combine electric and gas dryers in one chapter because there is so much material that is common to both types. Therefore, if we were to have a separate chapter on each type, much material would be duplicated.

INSTALLATION INSTRUCTIONS

The following are the installation instructions for the 240-volt electric clothes dryer.

Important Safeguards

Warning: Improper installation can cause electrical shock and/or ignition sources for fire, resulting in serious injuries or fire damage.

1. *Do not use flexible duct to exhaust the dryer!* Excessive lint can build up inside the ductwork and create a fire hazard and restrict the air flow. Restricted air flow will increase the amount of drying time. If the present system is made up of plastic duct, replace it with a rigid or flexible metal duct. Make sure that the present duct is free of any lint prior to installing the dryer.

2. *If the dryer is not exhausted outdoors, some fine lint will be expelled into the laundry area.* An accumulation of lint in any area of the home can create both a health and a fire hazard. Most dryers must be exhausted outside the dwelling.

3. *The dryer must be properly grounded.* Electrical shock can result if the dryer is not properly grounded. Follow the instructions in this chapter for proper grounding.

4. *Do not use an extension cord with the dryer.* Some extension cords are not designed to withstand the amounts of voltage and amperage that the dryers use. The extension cord can melt, creating an electrical shock and/or fire hazard. Make certain that the dryer is located within reach of the receptacle for the length of power cord purchased. Refer to the preinstallation requirements in this chapter for the proper power cord to be purchased.

5. *The strain relief must be installed onto the power cord.* If the strain relief is not attached to the power cord, the cord can be pulled out of the dryer. The power cord can be cut by any movement of the cord, resulting in an electrical shock. Install the strain relief as shown in the instructions. See Figure 15–3.

6. *Do not screen the exhaust end of the vent system or use screws or rivets to assemble the ductwork.* Lint can become caught in the screen and on the screws or rivets, clogging the ductwork and creating a fire hazard as well as increasing the drying time. Use an approved wall cap on the end of the ductwork. Seal the joints with duct tape. All female joint ends must be installed in the direction of the air flow.

7. *Do not exceed the length of duct or the number of elbows allowed.* Lint can accumulate in the system, plugging the ductwork and creating a fire hazard. Plugged ductwork will also increase the drying time.

8. *Do not use an aluminum-wired receptacle with a copper-wired power cord and plug (or vice versa).* A chemical reaction occurs between copper and aluminum which can cause electrical shorts. The proper wiring and receptacle is copper power cord and copper receptacle *or* aluminum power cord and aluminum receptacle.

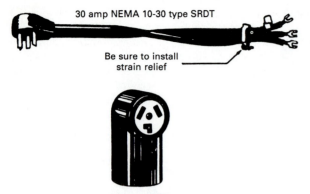

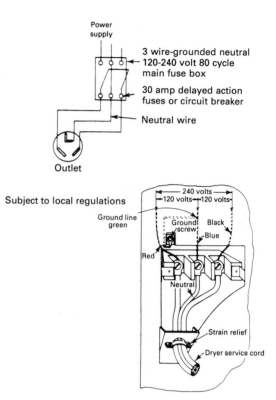

Figure 15–1 Thirty-amp power cord and receptacle. (Courtesy of WCI Major Appliance Group.)

9. *Destroy the carton, plastic bag, and metal band after the dryer is unpacked.* Children might use them for play. Cartons covered with rugs, bedspreads, or plastic sheets can become air-tight chambers causing suffocation. Place all materials in a garbage container or make the materials inaccessible to children.

Preinstallation Requirements

The following tools and equipment are required to complete the installation:

1. Phillips screwdriver.
2. Flat- or straight-blade screwdriver.
3. Adjustable pliers.
4. Channel-lock adjustable pliers.
5. Duct tape.
6. 8-inch or 10-inch adjustable wrench.
7. Carpenters level.
8. Putty knife.
9. 240-volt ac power cord kit. Three wire 30-amp. NEMA 10-30 type SRDT (if applicable).*
10. 240-volt ac power cord kit. Four wire 30 amp. NEMA 14-30 type SRDT or ST (if applicable).*
11. $\frac{5}{16}$-inch socket with ratchet or a $\frac{5}{16}$-inch nutdriver.
12. $\frac{1}{4}$-inch socket with ratchet or a $\frac{1}{4}$-inch nutdriver.
13. Rigid or flexible metal ductwork (if not presently installed).
14. Wall cap (if not presently installed).
15. UL approved ground bonding clamp.

Electrical Requirements

The following are the electrical requirements for an electric dryer:

1. Three (3) wire, 240-volt ac, single-phase, 60-hertz electrical power supply (the power supply should be checked and corrected if necessary by a qualified electrician).

---

* Some models will have factory-installed power cords.

Figure 15–2 Fuse box. (Courtesy of WCI Major Appliance Group.)

2. An individual 30-amp branch circuit from the main fuse box with a 30-amp maximum delayed-action fuse or circuit breaker. See Figures 15–1 and 15–2.

3. Power outlet with accessibility to the service cord plug when the dryer is installed.

4. Power supply cord kit rated at 240 volts ac minimum, 30 amps, with three open-end spade lug connectors *marked for use with clothes dryers.*

5. Manufactured (mobile) homes are constructed only for a four-wire electrical cord.

Exhaust System Requirements

Warning: If the dryer is not exhausted outdoors, some fine lint will be expelled into the laundry area. An accumulation of lint in any area of the home can create both a health and a fire hazard. The dryer must be exhausted outside the dwelling.

Warning: Do not use plastic flexible duct to exhaust the dryer! Excessive lint can build up inside the ductwork and create a fire hazard and restrict the air flow. Restricted air flow will increase the drying time. If the present system is made up of plastic duct, replace it with rigid or flexible metal duct. Make certain that the present duct is free of any lint prior to installing this dryer.

Use a 4-inch diameter (minimum) rigid or flexible metal duct only. A 4-inch wall cap is recommended.

Warning: Do not install the ductwork near combustible materials. A clothes dryer must not be exhausted into a

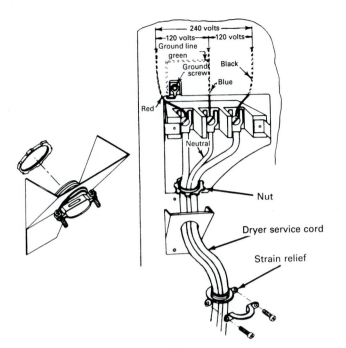

Figure 15–3 Service cord strain relief installation. (Courtesy of WCI Major Appliance Group.)

chimney, a wall, a ceiling, or any concealed space of a building where lint can accumulate, resulting in a fire hazard.

Warning: Do not exceed the length of duct pipe or the number of elbows allowed. Lint can accumulate in the system, plugging the ductwork and creating a fire hazard. Restricted air flow will increase the drying time. Install the exhaust duct and elbows as shown in the Exhausting Installation section presented later in this chapter. See Figure 15–4.

Exhaust duct locating dimensions *Note:* All dimensions are taken from the bottom of the dryer base. Add the length of the leveling legs, after leveling, to these dimensions and prior to cutting the hole.

Electrical Installation

Note: The electrical installation must conform to the local electrical code. When installing the dryer, call a licensed electrician and refer to Figure 15–2 along with the following information.

Warning: The dryer must be properly grounded. Electrical shock can result if the dryer is not properly grounded. Follow the instructions in this chapter for proper grounding.

240-volt ac three-wire connection power cord *Note:* If the dryer is connected to a 208-volt ac power circuit instead of a 240-volt ac power circuit, the drying time will be increased.

Warning: Disconnect the power to the dryer before making any adjustments. Electrical shock could result if the dryer is not unplugged. Do not use an extension cord with the dryer. Some extension cords are not designed to withstand the amounts of voltage and amperage that the dryer uses. The extension cord can melt and create an electrical shock and/or fire hazard. Make certain that the dryer is located within reach of the receptacle for the length of the power cord purchased. Refer to the preinstallation requirements in this chapter for the proper power cord to be purchased.

1. Remove the screws securing the terminal block access cover and the strain relief mounting bracket located on the back of the dryer.
2. Install a UL approved strain relief into the power cord entry hole of the mounting bracket. Finger tighten the nut only. See Figure 15–3.
3. Install a UL approved 30-amp power cord, NEMA 10-30 type SRDT, through the strain relief.
4. Attach the power cord middle open-end spade connector to the silver-colored center terminal on the terminal block. See Figure 18–11.
5. Attach the power cord outer open-end spade lug connectors to the outer brass-colored terminals on the terminal block. See Figure 15–3.
6. Reattach the strain relief mounting bracket to the back of the dryer with two (2) screws. Tighten the screws securely.
7. Tighten the screws that secure the power cord to the strain relief.
8. Tighten the strain relief nut securely to the mounting bracket to prevent turning of the strain relief.

Grounding instructions Use the following instructions for proper grounding of the dryer.

Grounded, cord-connected dryer This type of dryer must be grounded. In the event of a malfunction or breakdown, grounding will reduce the risk of electrical shock by providing a path of least resistance for the electrical current.

If the dryer is equipped with a service cord having an equipment-grounding conductor and a grounding plug, the plug must be plugged into an appropriate outlet that is properly installed and grounded in accordance with all local codes and ordinances.

If the dryer is not equipped with a service cord, see the instructions for proper installation.

Danger: Improper connection of the equipment-grounding conductor can result in an electrical shock. Check with a qualified electrician if in doubt as to whether the appliance is properly grounded.

Permanently connected appliance This type of dryer must be connected to a grounded metal, permanent wiring system, or an equipment-grounding conductor must be run with the circuit conductors and connected to the equipment-grounding terminal or lead on the appliance.

Note: The instructions appearing in this chapter and any literature which may be included with the dryer are not meant to cover every possible condition and situation that

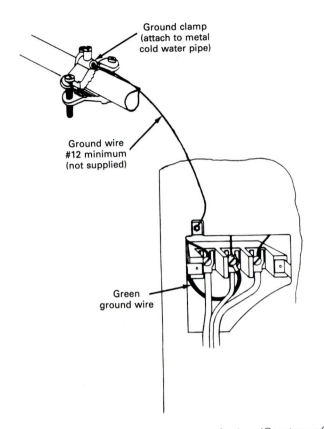

Ground clamp
(attach to metal
cold water pipe)

Ground wire
#12 minimum
(not supplied)

Green
ground wire

Figure 15–4 Using a separate ground wire. (Courtesy of WCI Major Appliance Group.)

may occur. Common sense and caution must be practiced when installing, operating, and maintaining any appliance.

Dryer is connected to a neutral conductor If the local codes do not permit neutral grounding, follow these steps:

1. Remove the green-colored ground wire from the green screw and fasten the wire under the center silver-colored screw on the terminal block.

2. Connect a separate grounding wire (No. 12 or heavier) between the green ground screw and a cold-water pipe. See Figure 15–4. Do not ground this appliance to a gas supply pipe or a hot-water pipe.

3. The grounded cold-water pipe must have metal continuity to an electrical ground and must register a maximum of 25 ohms of resistance. It must not be interrupted by plastic, rubber, or other electrical insulating connectors such as hoses, fittings, washers, or gaskets (including water meter or pump). Any electrically insulated connector should be jumped with a length of No.4 wire and securely clamped to bare metal at both ends. See Figure 15–5.

4. If a grounded water pipe is not available, a ground rod must be used. The ground rod must register a maximum of 25 ohms of resistance when installed in the ground. Drive the rod into the ground outside the dwelling and connect a separate grounding wire (12 AWG or heavier) between the green grounding screw and the grounding rod. It may take more than one ground rod to obtain the 25 ohms of resistance.

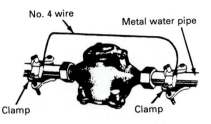

No. 4 wire

Metal water pipe

Clamp   Clamp

Figure 15–5 Providing a ground continuity on a cold water pipe. (Courtesy of WCI Major Appliance Group.)

Replace the terminal block access cover with the screws.

240-volt ac four-wire connection power cord *Note:* If the dryer is connected to a 208-volt ac power circuit instead of a 240-volt ac power circuit, the drying time will be increased.

Warning: Disconnect the power to the dryer before making any adjustments. Electrical shock could result if the dryer is not unplugged.

Warning: Do not use an extension cord with the dryer. Some extension cords are not designed to withstand the amounts of voltage and amperage that a dryer uses. The extension cord can melt, creating an electrical shock and/or fire hazard. Make certain that the dryer is located within reach of the receptacle for the length of power cord to be purchased.

1. Remove the terminal block access cover and the strain relief bracket located on the back of the dryer.

2. Install a UL approved strain relief into the power cord entry hole of the mounting bracket. Finger tighten the nut only. See Figure 15–6.

3. Remove the green grounding wire from the green ground screw.

4. Attach the green power cord connector to the cabinet using the green ground screw. Make certain that the grounding screw is tight.

5. Attach the white (neutral) power cord open-end spade lug connector. Connect the green dryer ground wire to the silver-colored center terminal on the terminal block.

6. Attach the red and black power cord open-end spade lug connectors to the outer brass-colored terminals on the terminal block.

7. Reattach the strain relief mounting bracket to the back of the dryer with the two screws. Tighten the screws securely.

8. Tighten the screws that hold the power cord to the strain relief.

9. Tighten the strain relief nut securely to the mounting bracket to prevent turning of the strain relief.

10. Replace the terminal block access cover with the screws.

Final Leveling and Testing

Use the following procedure to level and make the final test of the dryer installation:

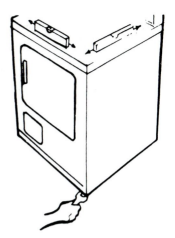

Figure 15-6 Leveling the dryer. (Courtesy of WCI Major Appliance Group.)

1. When the electrical and exhausting installations are complete, move the dryer into its final position and adjust one or more of the legs until the dryer is resting on all four (4) legs. Check to make certain that the dryer is level. See Figure 15-6. The dryer must be level and sitting on all four (4) legs.

Caution: Check to make certain that the electrical power is shut off at the circuit breaker before plugging the power cord into the receptacle.

2. Plug the power cord into the receptacle.

3. Turn on the electrical power at the circuit breaker or fuse.

4. Run the dryer through a function test to make certain that it is performing properly.

Installation Checklist

Check the following for proper operation of the dryer:

1. The proper electrical connections have been made.
2. The proper exhaust connections have been made.
3. The dryer is level.
4. The dryer has been put through a cycle check.
5. The wall cap operates properly.
6. The terminal block cover, terminal block mounting bracket, and the rear service panel are in place.

MECHANICAL AND ELECTRICAL TEST AND REPAIR

The following information describes the procedures used to test and repair the mechanical and electrical components of a clothes dryer.

Electrical Checks

Warning: To avoid electrical shock, disconnect the power supply before testing any component. When testing the drive motor, be sure that the motor and test cord are grounded.

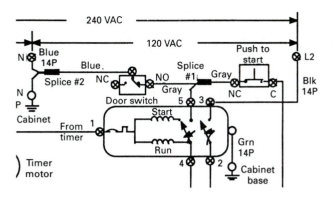

Figure 15-7 Motor circuit (electric dryer). (Courtesy of WCI Major Appliance Group.)

General The recommended method to check electrical components is with the aid of a voltmeter. When testing a component, remove the wires that are connected to it.

Heat-producing electrical components such as motors, solenoid coils, and heaters will show the resistance of the circuit in ohms. Contacts that are closed in timers or switches should not show any appreciable resistance between the terminals.

By following a timer chart and positioning the timer in a given cycle, the timer contacts can be checked. Switch charts show opened and closed contacts in the various selector switch positions. These charts should be followed when testing a switch.

Motor circuit Electric and gas dryers When diagnosing an electrical problem, always keep in mind the current flow needed to operate a dryer motor (120 volts). The electrical current must go through the power supply circuit protector (fuse or circuit breaker), door switch, the momentary switch in conjuction with the internal motor switch, the timer, selector switch, and the motor thermal overload protector. See Figure 15-7.

To test the drive motor, use the following procedure:

1. Mark the location or color code of the lead wires. Remove any external power source leads from the motor. (Do not remove the internal motor leads to the motor switch.)

2. With an external 120-volt ac grounded power supply, connect the wires to terminals 1 and 4. The motor should run. If it does not run, remove the motor from the unit. See Figure 15-8.

3. On open-framed motors, make a continuity test on the motor leads to terminals 1 and 4 of the switch, and between terminals 1 and 5 of the switch. If an open circuit is indicated on either test, replace the switch and repeat the above tests.

4. To test terminal 5, connect the 120-volt power leads to terminals 1 and 5. Apply power, and at the same time, jumper terminals 4 and 5 just long enough to bring the motor up to speed. Remove the jumper. The motor should continue to operate with the leads on terminals 1 and 5. If the motor stops, replace the switch. See Figure 15-9.

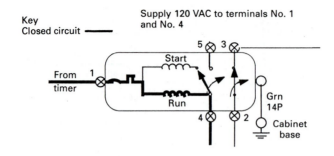

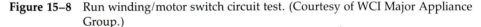

Figure 15–8 Run winding/motor switch circuit test. (Courtesy of WCI Major Appliance Group.)

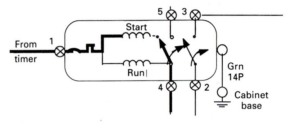

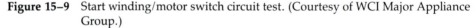

Figure 15–9 Start winding/motor switch circuit test. (Courtesy of WCI Major Appliance Group.)

Auxiliary contacts The auxiliary contacts are part of the safety features built into this circuit. These contacts are normally open (NO) and will close the circuit to the heating element or the gas burner only after the motor is up to speed.

To test the auxiliary contacts, use the following procedure:

1. Connect a continuity circuit to terminals 2 and 3.
2. Connect an external power supply to contacts 1 and 4. The motor should start and come up to speed. When this happens, the auxiliary contacts should close. The continuity test should then verify a circuit. If continuity is not complete, replace the switch. See Figure 15–10.

Temperature control thermostat The temperature control thermostat is located on the air discharge duct. See Figure 15–11. It is a biasing-heater-type thermostat, whose purpose is to fool the bimetal sensing disc inside the thermostat by supplying additional heat to it. By controlling this heat with the selector switch, it is possible to obtain three (3) levels of heat: regular, medium, and low.

To test the temperature control thermostat, use the following procedure:

1. Disconnect the dryer from the power supply.
2. Remove the rear access service panel.
3. Connect the leads of a voltmeter on the two (2) large terminals of the control thermostat.

Connect ohmmeter to terminals No. 2 and No. 3.
Supply 120 VAC to terminals No. 1 and No. 4.
Continuity should be present when motor is up
to speed

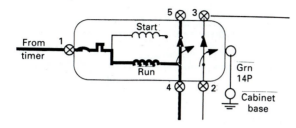

Figure 15–10 Auxiliary contacts circuit. (Courtesy of WCI Major Appliance Group.)

4. Install an accurate thermocouple thermometer in the center of the lint tray. Make certain that the thermometer sensing element is not touching any part of the lint housing or screen.

5. Reconnect the dryer to the power supply.

6. Turn the selector switch to the desired setting.

7. Set the timer for 30 minutes on the Timed Dry cycle.

8. Close the door carefully to prevent disturbing the sensing element. Start the dryer.

9. Allow the dryer to run through at least three (3) cycles before checking the temperature.

10. Note the temperature at which the thermostat opens (line voltage can be seen on the voltmeter) and when the thermostat closes (zero voltage is seen on the voltmeter).

Setting	Cut-in	Cutoff
Regular	120 ± 15°F	180 ± 15°F
Medium	107 ± 7°F	157 ± 7°F
Low	105 ± 5°F	147 ± 5°F

To replace the control thermostat, use the following procedure:

1. Disconnect the dryer from the power supply.

2. Remove the rear access panel.

3. Locate the control thermostat on the exhaust duct housing.

4. Remove the wiring and the two (2) mounting screws.

5. Remove the thermostat.

6. Reverse the above procedure to reassemble the unit.

Temperature control thermostat

Exhaust duct

Blower housing

Motor

Figure 15–11 Temperature control thermostat. (Courtesy of WCI Major Appliance Group.)

Electric models The following is an operating description of the control thermostat used on electric model clothes dryers.

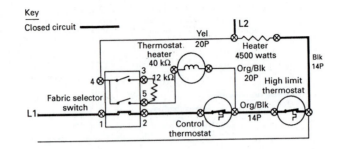

Figure 15–12 Regular heat circuit. (Courtesy of WCI Major Appliance Group.)

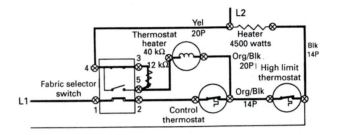

Figure 15–13 Medium-heat circuit. (Courtesy of WCI Major Appliance Group.)

Regular heat (120 to 180°F ± 15°F) When Regular Heat is selected, electrical current flows through contacts 1 and 2 of the selector switch, through the cycling thermostat, the high limit thermostat, and the heating element. The biasing heater is not in this circuit and, therefore, has no influence on the cutoff temperature. See Figure 15–12.

Medium heat (107 to 157°F ± 7°F) When Medium Heat is selected, a 240-volt circuit that is wired in parallel with the high limit thermostat is completed through contacts 3 and 4 in the selector switch, through the 12-kiloohm dropping resistor, the biasing heater, the control thermostat, the high limit thermostat, and the heating element. The heat created by the biasing heater lowers the cutoff temperature of the control thermostat. See Figure 15–13.

Low heat (105 to 147°F ± 5°F) When Low Heat is selected, electrical current flows through contacts 4 and 5 in the selector switch, through the biasing heater, the control thermostat, the high limit thermostat, and the heating element. Note that the dropping resistor is not included in this circuit, thus supplying full current to the biasing heater, which further lowers the control thermostat cutoff temperature. See Figure 15–14.

Air fluff (no heat) When Air Fluff is selected, no contacts are made in the selector switch. Therefore, no electrical current flows to the heating element. See Figure 15–15.

Gas models The following is an operating description of the control thermostat used on gas clothes dryers.

Regular heat (120 to 180°F ± 15°F) When Regular Heat is selected, electrical current flows through contacts 1 and 2 of the selector switch, through the control thermostat, the high limit thermostat, and the burner assembly. The biasing heater is not in this circuit. Therefore, it has no influence on the cutoff temperature of the control thermostat. See Figure 15–16.

Medium heat (107 to 157°F ± 7°F) When Medium Heat is selected, a 120-volt circuit that is wired in parallel with the heater and the high limit thermostat is completed through contacts 3 and 4 in the selector switch, through the 3.9-kiloohm dropping resistor, the biasing heater, the control thermostat, and the burner assembly. The heat created by the biasing heater lowers the cutoff temperature of the control thermostat. See Figure 15–17.

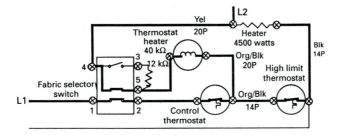

Figure 15–14 Low-heat circuit. (Courtesy of WCI Major Appliance Group.)

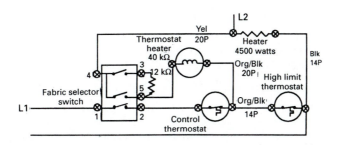

Figure 15–15 Air fluff (no heat) circuit. (Courtesy of WCI Major Appliance Group.)

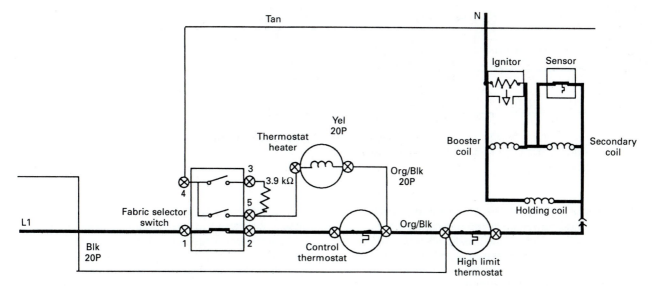

Figure 15–16 Regular heat circuit (gas models). (Courtesy of WCI Major Appliance Group.)

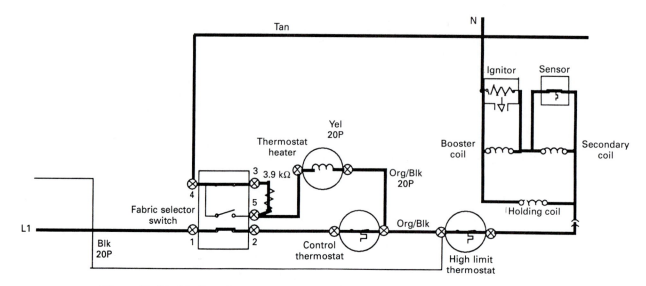

Figure 15–17 Medium-heat circuit (gas models). (Courtesy of WCI Major Appliance Group.)

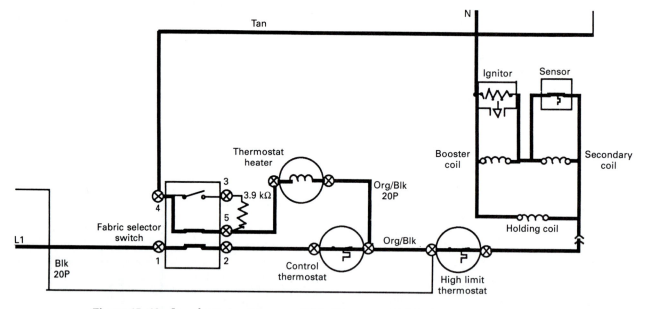

Figure 15–18 Low-heat circuit (gas models). (Courtesy of WCI Major Appliance Group.)

Low heat (105 to 147°F + 5°F) When Low Heat is selected, electrical current flows through contacts 4 and 5 in the selector switch, through the biasing heater, the control thermostat, the high limit thermostat, and the burner assembly. Note that the dropping resistor is not included in this circuit, thus supplying full current to the biasing heater and further lowering the cutoff temperature of the control thermostat. See Figure 15–18.

Air fluff (no heat) When Air Fluff is selected, no contacts are closed in the selector switch. Therefore, no electrical current flows to the burner assembly. See Figure 15–19.

High limit thermostat The high limit thermostat is designed to protect the dryer from overheating. This thermostat is located on the heater housing next to the terminal connections. See Figure15–20.

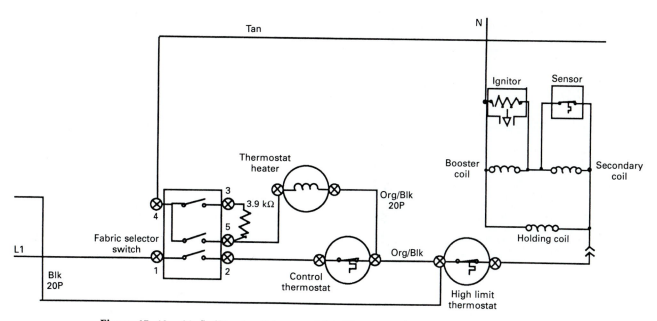

Figure 15–19 Air fluff (no heat) (gas models). (Courtesy of WCI Major Appliance Group.)

To test the high limit thermostat, use the following procedure:

1. Disconnect the dryer from the electrical supply.
2. Raise the top panel.
3. Use an ohmmeter to check the continuity of the thermostat contacts. The thermostat contacts must be closed at room temperature.
4. Replace the high limit thermostat if there is no continuity found through the contacts at room temperature.

Timer The timer controls the length of time that the dryer operates. The timer is mounted to the control panel cross brace. It consists of a series of cams and switches, driven by a synchronous motor. The timer is set by rotating the knob and dial clockwise.

To test the timer, use the following procedure:

1. Disconnect the dryer from the electrical supply.
2. Remove the rear service panel of the control panel.
3. Determine, by use of the timer sequence chart (located on the unit), when the particular circuit in question is to be completed through the timer.
4. Rotate the timer knob to that position and check the continuity at those contacts.
5. If the contacts are bad, replace the timer. (It may be necessary to check the timer motor. If so, refer to the electrical diagram for that particular model.) (See the Electrical Diagrams presented at the end of this chapter.)

To replace the timer, use the following procedure:

1. Disconnect the dryer from the electrical supply.

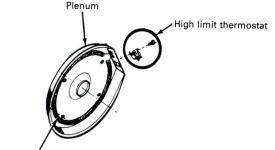

Figure 15–20 High limit thermostat. (Courtesy of WCI Major Appliance Group.)

2. Remove the service panel from the rear of the control panel.
3. Remove the timer knob and dial by turning the knob counterclockwise.
4. Reach through the dial hub opening in the escutcheon and remove the two (2) screws.
5. Install the new timer and transfer the wiring from the old to the new.
6. Replace the dial, knob, and the service panel.
7. Check the dryer for proper operation.

Push-to-start and door switch assemblies The push-to-start and the door switch are normally open (NO) switches that are wired in series with each other. They are connected to the hot side of the line between the timer and the drive motor. They are also wired in series with the timer motor. Neither motor will operate until both the dryer door is closed and the push-to-start switch is depressed. The

Plenum   Mounting screws (4)

Figure 15–21   Heating elements removal. (Courtesy of WCI Major Appliance Group.)

push-to-start switch is also in series with the centrifugal start switch in the motor. When the motor is up to speed, the centrifugal switch maintains the circuit that is momentarily established by the push-to-start switch.

To test the push-to-start switch, use the following procedure:

1. Disconnect the dryer from the electrical power supply.
2. Remove the rear service panel from the control panel.
3. Remove the wires from the push-to-start switch terminals and connect the ohmmeter leads across the switch terminals.
4. Press the button. There should be continuity as long as the button is pressed. If no continuity is indicated, replace the switch.

To test the door switch, use the following procedure:

1. Disconnect the dryer from the electrical power supply.
2. Raise the top panel to gain access to the door switch terminals.
3. Remove the wires from the terminals.
4. Connect the ohmmeter across the terminals.
5. Press the door switch actuator. The ohmmeter should show continuity. If no continuity is indicated, replace the switch.

Heating element (electric models only)   The 4500-watt heating element is located in the rear of the cabinet inside the plenum that surrounds the rear of the drum.

To test the heating element, use the following procedure:

1. Disconnect the dryer from the electrical power supply.
2. Remove the wires from the heating element terminals.
3. Connect an ohmmeter across the terminals and test for continuity. Continuity should be indicated. Also check for a grounded element by attaching one lead of the ohmmeter to one terminal and the other lead to the plenum. Repeat this procedure with the other terminal. Continuity should not be indicated.

To replace the heating element, use the following procedure:

1. Disconnect the dryer from the electrical power supply.
2. Raise the top panel and remove the front panel.
3. Remove the belt from the idler assembly and the motor pulley.
4. Remove the drum from the cabinet.
5. Disconnect the wires from the heating element terminal leads and the high limit thermostat.
6. Remove the heating element and the plenum by removing the four (4) screws that mount the plenum to the rear panel. See Figure 15–21.
7. Install the new heating element.
8. Reverse the above procedure to reassemble the unit.

Door switch   The door switch is wired into the electrical power circuit to the dryer. It interrupts the electricity to the dryer any time that the door is opened. It is a safety switch designed to protect the user against accidentally opening the door and putting a hand into the rotating dryer drum, causing an injury.

To replace the door switch, use the following procedure:

1. Disconnect the electrical power from the dryer.
2. Raise the top panel.
3. Remove the blue and gray wires from the door switch terminal leads.
4. Reach inside the cabinet and squeeze the mounting tabs on either side of the door switch. Push the switch toward the front of the cabinet.
5. Reverse the above procedure to replace the switch.

Cabinet light   The cabinet light receptacle is mounted to the front panel by one (1) screw. The bulb is a bayonet-type and must be pushed in slightly then turned to remove. See Figure 15–22.

Front panel   The front panel is a one-piece assembly that is mounted from the inside to the side panel flanges by four (4) screws. It is held in at the bottom by two (2) bracket hinges. *Note:* The front panel assembly must be removed to service the following components:

- Front drum seal
- Drum guides and rear drum support

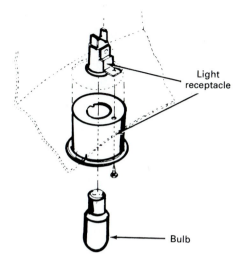

Figure 15–22 Cabinet light. (Courtesy of WCI Major Appliance Group.)

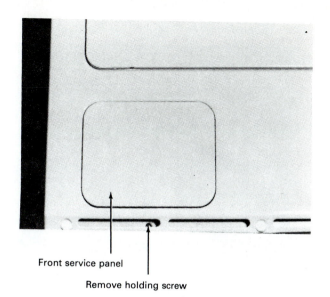

Front service panel

Remove holding screw

Figure 15–23 Front service panel (gas models). (Courtesy of WCI Major Appliance Group.)

- Air-duct-to-blower housing seal
- Drum assembly
- Rear drum seal (gas models)
- Heating element (electric models)
- Blower duct
- Drive belt

To remove the front panel, use the following procedure:

1. Disconnect the dryer from the electrical power supply.
2. Raise the top panel.
3. Disconnect the wires to the door switch and the drum light (if applicable).
4. Remove the four (4) $\frac{5}{16}$-inch hex-head screws securing the front panel to the side panel flanges.
Note: On gas dryers, remove the screw securing the front service panel to the base. See Figure 15–23.
5. Pull the top of the front panel forward to release the holding clips while supporting the dryer drum.
6. Lift the front panel up and off the hinge brackets.
7. Reverse the above procedure to reassemble the unit.

Front glides There are three (3) front glides that fit into the mounting bracket on the front panel. They are glued to the upper portion of the front seal. These glides are made of Teflon. See Figure 15–24.

To remove the front glides, use the following procedure:

1. Disconnect the dryer from the electrical power supply.
2. Remove the front panel.

3. Peel the front glides off the felt seal and remove them from the mounting bracket. Make certain that all the old glue is removed.

To replace the front guides, use the following procedure:

1. Insert the guides into the mounting bracket and apply the brown glue to the underside of the glide.
2. Press firmly and hold the seal until the glue is dried.
3. Reverse the above procedure to reassemble the front panel.

Air-duct-to-blower housing seal Should an air flow problem exist, or any time the front panel is removed, the air-duct-to-blower housing seal should be checked. See Figure 15–25.

To check the air-duct-to-blower seal, use the following procedure:

1. Disconnect the dryer from the electrical power supply.
2. Remove the front panel.
3. Remove the old seal and clean the mounting surface.
4. Use a heat-resistant adhesive to attach the new seal. *Note:* Be sure that full contact is made between the seal and the blower housing.
5. Replace the front panel and check the dryer for proper operation.

Drive belt The drive belt is routed around the drum, motor pulley, and the idler pulley. Tension is maintained

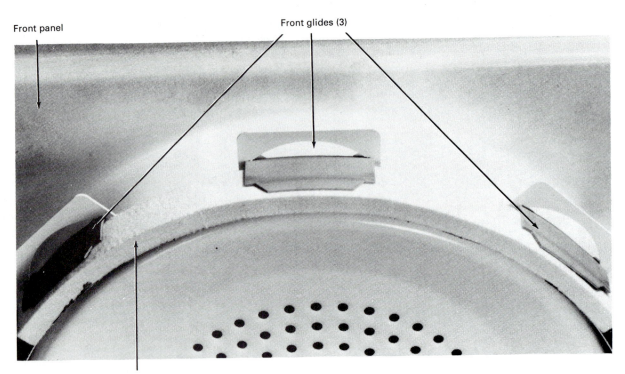

Front panel

Front glides (3)

Upper front seal

Figure 15–24 Front glides. (Courtesy of WCI Major Appliance Group.)

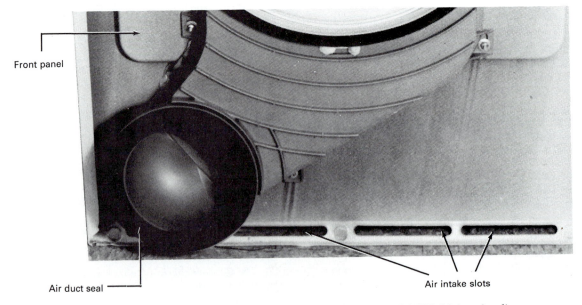

Front panel

Air duct seal

Air intake slots

Figure 15–25 Air-duct-to-blower housing seal. (Courtesy of WCI Major Appliance Group.)

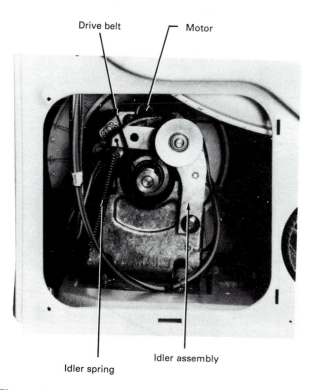

Drive belt — Motor

Idler spring — Idler assembly

Figure 15–26 Drive belt routing. (Courtesy of WCI Major Appliance Group.)

on the belt by the spring-loaded idler pulley. See Figure 15–26.

To remove the drive belt, use the following procedure:

1. Disconnect the dryer from the electrical supply.
2. Remove the rear service panel to gain access to the motor, idler assembly, and belt.
3. Release the tension on the belt by moving the idler assembly to the right while removing the belt from the idler pulley.
4. Remove the belt from the motor pulley.
5. Raise the top panel and remove the front panel.
6. Remove the belt from the drum.
7. Reverse the above procedure to reassemble the unit.

Drum assembly The drum assembly consists of the drum, three (3) sorting vanes, a heat deflector, and the rear drum support.

To replace the sorting vanes, use the following procedure:

1. Disconnect the dryer from the electrical supply.
2. Remove the rear service panel and free the drive belt from the idler and motor pulleys.
3. Raise the top panel.
4. Remove the front panel assembly.
5. Grasp the front of the drum flange and use the belt to lift the drum from the rear drum bearing, then guide the drum through the front opening. Be sure to keep track of

the small ball bearing located between the rear drum and the grounding clip.

6. Check the parts for wear. Replace any worn parts as needed.

Rear drum support The rear drum support, which resembles a trailer hitch, is mounted to the rear of the drum and rides in the rear drum bearing. See Figures 15–27 and 15–28.

To replace the rear drum support, use the following procedure:

1. Disconnect the dryer from the electrical supply.
2. Remove the drum using the procedure above.
3. Remove the three (3) screws securing the rear drum support to the drum.

Rear drum bearing The rear drum bearing is a U-shaped nylatron block that acts as the bearing surface for the rear drum support. This bearing assembly, which includes two (2) mounting screws, a mounting bracket, a grounding clip, and a small ball bearing, is mounted to the rear panel, and lubricated with Lubri-Plate. See Figures 15–29 through 15–31.

To replace the rear drum bearing, use the following procedure:

1. Disconnect the dryer from the electrical supply.
2. Remove the rear service panel. Remove the belt from the idler and motor pulleys.
3. Remove the front panel and drum, as described above.
4. Remove the two (2) screws securing the rear drum bearing bracket to the rear panel.
5. Remove the rear drum bearing. Be sure to keep track of the small ball bearing that is located between the rear drum support and the grounding clip. (This ball bearing completes a grounding path from the drum to the cabinet. It also prevents axial movement of the drum.)

Motor, blower, and idler assembly The drive motor is mounted on a support bracket with a ring-type resilient mount at either end. The motor pulley, which drives the belt, is mounted on the rear shaft. The blower wheel, which has a left-hand threaded hub, is screwed onto the front of the motor shaft. A circular cover plate is mounted over the motor shaft between the blower wheel and the motor end frame. This cover plate fits into the blower housing and is held in place by a large-diameter coil spring, positioned between the motor and the cover plate.

The assembly is mounted to the cabinet base with the blower wheel encased in the blower housing. The spring-loaded cover closes the rear opening of the housing when the assembly is installed. The motor and its support bracket are held in place by a tab at the front of the bracket that engages a slot in the cabinet base. Two (2) screws at the rear of the bracket secure the assembly to the cabinet base. See Figures 15–32 through 15–35.

To replace the motor assembly, use the following procedure:

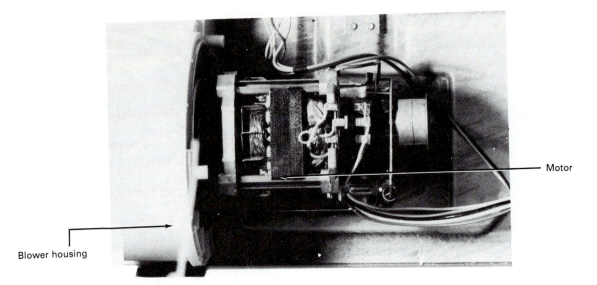

Figure 15–32   Motor and blower housing. (Courtesy of WCI Major Appliance Group.)

Figure 15–33   Blower housing and blower wheel. (Courtesy of WCI Major Appliance Group.)

Figure 15–34   Motor and blower wheel. (Courtesy of WCI Major Appliance Group.)

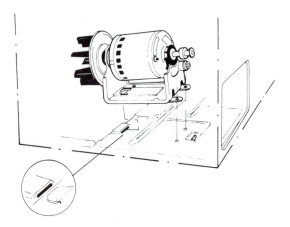

Figure 15–35   Motor mounting. (Courtesy of WCI Major Appliance Group.)

Figure 15–36 Idler pulley assembly. (Courtesy of WCI Major Appliance Group.)

1. Disconnect the dryer from the electrical power supply.
2. Remove the rear service panel.
3. Remove the belt from the idler pulleys.
4. Disconnect the wires from the motor switch. Note the wiring configuration.
5. Remove the two (2) screws mounting the motor support bracket to the cabinet base.
6. Remove the motor.
7. Reverse the above procedure to reassemble the unit.

Idler arm and pulley assembly   The idler arm can be removed without removing the motor from the dryer. The idler arm pivots on a stud mounted on the motor support bracket. See Figure 15–36.

To remove the idler assembly, use the following procedure:

1. Disconnect the dryer from the electrical power supply.
2. Remove the rear service panel.
3. Remove the belt from the motor and idler pulleys.
4. Lift the idler arm off the stud.
5. Install the new idler arm.
6. Reverse the above procedure to reassemble the unit.

INSTALLATION INSTRUCTIONS FOR THE GAS MODEL DRYERS

The following are some suggested installation procedures for the gas clothes dryer. See the installation procedures for electric dryers at the beginning of this chapter for a more complete listing of procedures. *Danger:* An improper installation can create ignition sources for fire, resulting in serious personal injuries or fire damages.

1. *Do not attempt to alter a gas orifice or adjust the burner air shutter.* Natural gas input may vary in some areas from 700 to 1200 Btu per cubic foot. If the gas orifice or the burner air shutter is incorrectly adjusted, serious personal injury and/or fire hazard can occur. Your local gas company will know the qualities of the gas in your area.
2. *Do not use an open flame when leak-testing for gas leaks.* Serious personal injury and/or fire hazard can result if an open flame is used to test for gas leaks. Use a soap and water solution to test all gas line fittings.
3. *Do not install a dryer to an LP gas supply without installing a conversion kit.* All dryers shipped out of the factory are equipped with a pressure regulator and a natural gas orifice. Using a natural gas orifice with an LP gas supply can result in personal injury, clothes damage, and/or fire hazard. If LP gas is to be used, be sure that a conversion kit is installed before completing the job.

Warning: An improper installation can cause electrical shock and/or ignition sources for fire, resulting in serious personal injuries or fire damages.

Electrical Requirements

The following are the electrical requirements for a gas dryer:

1. Separate, properly polarized, grounded three-wire circuit. See Figure 15–37.

Note: Do not under any circumstances remove grounding prong from plug

Grounding prong

Figure 15–37 Three-wire circuit plug and receptacle. (Courtesy of WCI Major Appliance Group.)

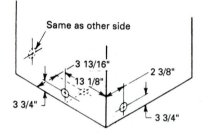

Same as other side

3 13/16"

13 1/8"

2 3/8"

3 3/4"

3 3/4"

Figure 15–38 Locating the exhaust hole. (Courtesy of WCI Major Appliance Group.)

2. 120-volt, 60-hertz, ac power supply.

3. 15-amp maximum, delayed-action fuse or circuit breaker.

4. Outlet with accessibility to the service cord plug when the dryer is installed.

Grounding instructions *The dryer must be grounded.* In the event of a malfunction or breakdown, grounding will reduce the risk of electrical shock by providing a path of least resistance for electrical current.

If the dryer is equipped with a service cord that has an equipment-grounding conductor and a grounding plug, the plug must be plugged into an appropriate outlet that is properly installed and grounded and in accordance with all local codes and ordinances. See Figure 15–40.

Exhaust System Requirements

Warning: If the dryer is not exhausted outdoors, some fine lint will be expelled into the laundry area. Excessive lint can build up inside the ductwork, creating a fire hazard and restricting the air flow. A restricted air flow will increase the drying time. If the present system is made up of plastic duct, replace it with a rigid or flexible metal duct. Make sure that the present duct is free of any lint prior to installing the dryer.

Use a 4-inch-diameter (minimum) rigid or flexible metal duct only. A 4-inch wall cap is recommended.

Warning: Do not install the ductwork near combustible materials. The dryer must not be exhausted into a chimney, a wall, a ceiling, or any concealed space of the building where lint can accumulate, resulting in a fire hazard.

Warning: Do not exceed the length of duct pipe or

the number of elbows allowed. Lint can accumulate in the system and plug the ductwork, creating a fire hazard. Clogged ductwork will also increase the drying time. Install the exhaust duct and elbows as instructed in the Exhausting Installation section of this chapter.

Exhaust duct locating dimensions *Note:* All dimensions are taken from the bottom of the dryer base. Add the length of the leveling legs, after leveling, to these dimensions and prior to cutting the hole. See Figure 15–38.

Gas Supply Line Installation

Do not attempt to alter the gas orifice or adjust the burner air shutter. Natural gas input may vary in some areas from 700 to 1200 Btu per cubic foot. If the gas orifice or the burner air shutter is incorrectly adjusted, serious personal injury and/or a fire hazard can occur. The local gas company will know the qualities of the gas in your area. Perform any adjustments with caution, making certain that they are correct.

Danger: Do not use an open flame for leak-testing gas piping. Serious personal injury and/or a fire hazard can result if an open flame is used to test for gas leaks. Use a soap and water solution to test all gas line fittings.

Danger: Do not install a dryer to an LP gas supply without installing a conversion kit. All dryers are shipped out of the factory with a pressure regulator and a natural gas orifice. Using a natural gas orifice with an LP gas supply can result in personal injury, clothes damage, and/or fire hazard. Be sure to properly install a gas conversion kit when the dryer is to be used with LP gas.

1. A $\frac{1}{8}$-inch NPT plugged tapping, accessible for test gauge connections, must be installed immediately upstream of the gas supply connection to the dryer.

2. Connect a $\frac{1}{2}$-inch ID semirigid tubing or approved pipe from the gas supply line to the $\frac{3}{8}$-inch pipe located on the back of the dryer. Use a $\frac{1}{2}$- to $\frac{3}{8}$-inch reducer for making this connection. Apply an approved thread sealer that is resistant to the corrosive action of LP gases on all pipe connections.

3. Leak-test all gas line connections that were made during the installation of the dryer, including those inside the dryer.

Note: The dryer and its individual shutoff valve must be disconnected from the gas supply piping system during any pressure testing. This is to protect the dryer valve and other components from damage.

Exhausting Installation

Refer to the ''EXHAUST SYSTEM REQUIREMENTS'' for electric dryers.

Exhaust Installation in Manufactured (Mobile) Homes

The following are specific dryer installation instructions for manufactured homes:

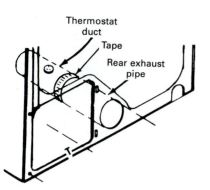

Figure 15–39 Exhausting a dryer from the side or bottom. (Courtesy of WCI Major Appliance Group.)

1. The dryer must be exhausted to the outside and secured to the floor.

2. The exhaust must not be terminated beneath a manufactured home.

3. The exhaust material must not support combustion.

4. The exhaust duct must not be connected to any other duct, vent, chimney, ceiling, or any concealed space of a manufactured (mobile) home.

5. The installation must conform to the current Manufactured Home Construction and Safety Standard (which is a Federal Regulation Title 24 CFR-Part 32-80), or when such a standard is not applicable, with the American National Standard for Mobile Homes.

Exhausting from the rear Most dryers are shipped from the factory for rear exhausting; however, they can also be exhausted from the right or the left side of the cabinet, or the bottom of the dryer. For rear exhausting, connect a 4-inch pipe to the duct fitting extending from the rear of the dryer.

Exhausting from the side or bottom The following instructions are to be used when a dryer is to be exhausted from either the side or the bottom:

1. Remove the rear service cover and the screw holding the exhaust pipe to the dryer.

2. Remove the tape holding the exhaust halves together and pull the rear half of the exhaust out of the dryer. See Figure 15–39.

3. The elbow and pipe can now be installed to the exhaust duct by retaping the joints. See Figure 15–40.

4. Replace the rear service cover.

5. Level the dryer to keep it from moving while completing the remainder of the installation. See Figure 15–41.

Dryer Function Test and Lighting Instructions:

Caution: Check to make certain that the electrical power is off at the circuit breaker or fuse before plugging the power cord into the receptacle.

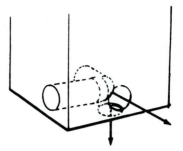

Figure 15–40 Connecting a vent to the exhaust duct. (Courtesy of WCI Major Appliance Group.)

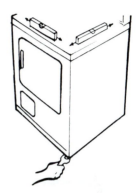

Figure 15–41 Leveling the dryer. (Courtesy of WCI Major Appliance Group.)

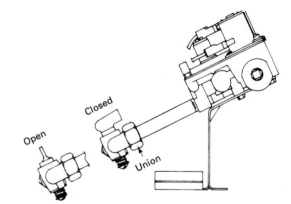

Figure 15–42 Operation of the gas shutoff valve. (Courtesy of WCI Major Appliance Group.)

1. Turn the timer knob on the console to the OFF position.

2. Remove the screw located under the front access panel (inside the dryer louvers) and tilt the panel up to gain access to the gas burner.

3. Make certain that the shutoff valve handle is in line with the gas supply pipe. If not, turn the handle. See Figure 15–42.

Installation Instructions for the Gas Model Dryers   193

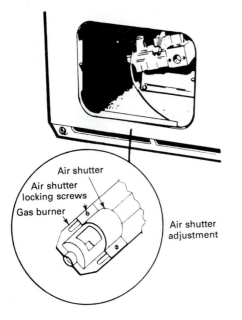

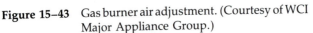

Air shutter

Air shutter
locking screws

Gas burner

Air shutter
adjustment

Figure 15–43   Gas burner air adjustment. (Courtesy of WCI Major Appliance Group.)

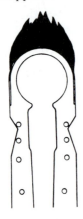

Figure 15–45   Correct adjustment of a gas burner air shutter. (Courtesy of WCI Major Appliance Group.)

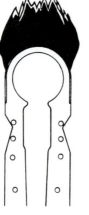

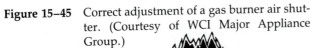

Figure 15–46   Incorrect adjustment of a gas burner air shutter. (Courtesy of WCI Major Appliance Group.)

Figure 15–44   Incorrect adjustment of a gas burner air shutter. (Courtesy of WCI Major Appliance Group.)

4. Secure the front access panel with the screw previously removed.

5. Plug the power cord into the receptacle.

6. Turn on the electrical power at the circuit breaker or the fuse.

7. Turn the timer knob on the console to a heat position and depress the push-to-start button.

8. Open the dryer door after 2 minutes of running time and note the heat inside the dryer. Make certain that the burner was lit before making this test.

Note: Before the burner will light, it is necessary for the gas line to be bled of air. If the burner does not light within 45 seconds, the first time the dryer is turned on, the safety switch will shut the burner off. If this happens, turn the timer to OFF and wait 5 minutes before making another attempt to light the burner.

Note: The burner adjustment must be checked for satisfactory operation. The air shutter is set at the factory and does not ordinarily require adjustment. If necessary, the air intake should be adjusted by a qualified technician.

To adjust the burner air shutter, use the following procedure:

1. Remove the screw located under the front access panel (inside the dryer louvers) and tilt the panel up to gain access to the gas burner.

2. Loosen the screws that secure the air shutter. See Figure 15–43.

3. Gradually slide the shutter forward or backward until the yellow tips of the burner flame just disappear. See Figures 15–44 through 15–46.

4. Secure the air shutter by tightening the screws previously loosened. See Figure 15–43.

5. Secure the front access panel with the screw previously removed.

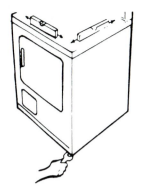

Figure 15–47 Leveling the dryer. (Courtesy of WCI Major Appliance Group.)

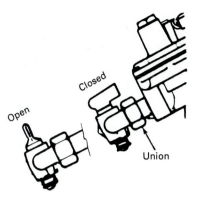

Figure 15–48 Operation of the gas shutoff valve. (Courtesy of WCI Major Appliance Group.)

Final Leveling and Testing

The following steps are used for the final leveling and testing procedure:

1. When the gas and exhausting installations are completed, move the dryer to its final position and adjust one or more of the legs until the dryer is resting on all four (4) legs. Make certain that the dryer is level. See Figure 15–47. The dryer must be level and solid on all four (4) legs.

2. Run the dryer through a function test to make certain of its performance.

Installation Checklist

Check the following for proper operation of the dryer:

1. The proper gas connections have been made.
2. The proper electrical connections have been made.
3. The proper exhaust connections have been made.
4. The dryer is level.
5. The dryer has been put through a cycle check.
6. The wall cap operates properly.

Service Conversion Kit

The following are instructions for conversion of the dryer to liquefied petroleum (bottled) gases.

Parts and instructions The following parts and instructions are for converting the White–Rodgers combination gas valve assembly to LP gases:

The kit includes the following parts:

1. 0600772 orifice (stamped No. 55).
2. 0622603 regulator (LP) blocking pin.
3. 0125518 stainless-steel burner.
4. Red tag.
5. Instruction sheet.

To remove the main burner assembly, use the following procedure:

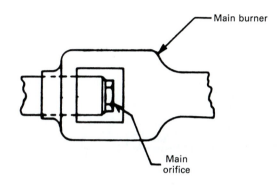

Figure 15–49 Location of the main orifice. (Courtesy of WCI Major Appliance Group.)

1. Disconnect the electrical power supply to the dryer before servicing or removing any panels.

2. Shut off the gas at the manual shutoff valve. Loosen the shutoff valve union. See Figure 15–48.

3. Remove the screw that secures the burner assembly to the base of the dryer. Disconnect the two (2) wires that connect to the burner assembly. Disconnect the orange and white wires that connect to the radiant sensor. Remove the burner and valve assembly from the dryer.

To install the parts in the kit, use the following procedure:

1. Remove the two (2) screws that fasten the burner to the main mounting bracket.

2. Remove the two (2) screws and speed nuts that fasten the burner air shutters. See Figure 15–51. Assemble the air shutters to the stainless-steel burner furnished in the kit.

3. Remove the two (2) screws that fasten the igniter bracket to the burner. Transfer the igniter bracket to the burner furnished in the kit.

4. Remove the original main orifice in the dryer and install the new orifice from this kit. *Note:* Save the original orifice in case conversion back to natural gas is desired later. See Figure 15–49.

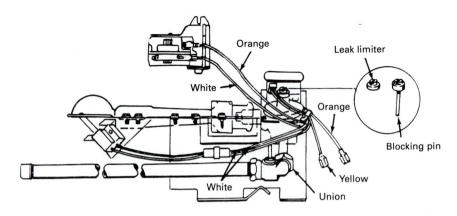

Figure 15–50   LP gas locking pin installation. (Courtesy of WCI Major Appliance Group.)

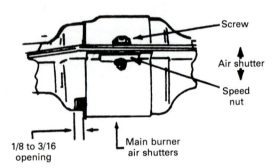

Figure 15–51   Main burner adjustment. (Courtesy of WCI Major Appliance Group.)

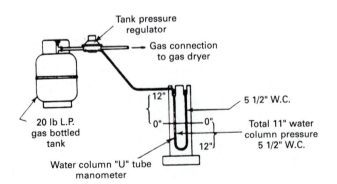

Figure 15–52   LP gas regulator location. (Courtesy of WCI Major Appliance Group.)

5. Reassemble the burner with the two (2) screws removed earlier. *Note:* Save the natural gas burner if conversion back to natural gas is desired later.

6. Remove the leak limiter and install the LP locking pin. See Figure 15–50.

7. Affix the red tag to the main valve for LP application. *Note:* Save the natural gas limiter if conversion back to natural gas is desired later.

8. Install the burner assembly into the dryer and fasten it with the screw. Tighten the union fitting and check the connections with a soap suds solution for leaks. Connect the two (2) wires from the dryer to the assembly. Connect the orange and white wires to the radiant sensor. Refer to the unit wiring diagram for the proper wiring connections.

Burner adjustment and operation   Operate the dryer with the air shutters closed on the main burner. Adjust the air shutters approximately $\frac{1}{8}$ to $\frac{3}{16}$ inch or until the yellow flame just disappears. See Figure 15–51. An orange-colored flame is acceptable. This is due to the burning of air impurities (dust).

The following is helpful information to use when installing the bottled gas and checking the tank's regulator water column pressures.

Observe all gas utility codes and contact the bottled gas dealer for any questions and installations.

Refer to the owner's manual for gas supply line connections.

Bottled gas (tanks) must be equipped with a gas pressure regulator before connecting to a gas-fired appliance as shown in Figure 15–52.

The bottled gas tank regulator outlet pressure should be a total of 11 inches of water column maximum or 8 inches of water column minimum.

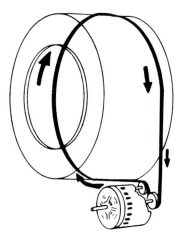

Figure 15–53 Drum rotation. (Courtesy of WCI Major Appliance Group.)

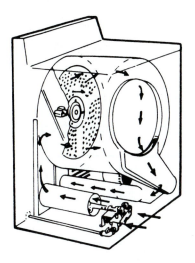

Figure 15–54 Air flow. (Courtesy of WCI Major Appliance Group.)

A water column, U-tube manometer should be used to check the tank regulator pressure as shown in Figure 15–52. The dryer should be operating and the gas should be burning in the dryer when checking the regulator pressure.

PRODUCT OPERATION

The mechanical operation, air flow, and timer cycles on both gas and electric dryers are basically the same; however, it is important that the technician become familiar with the operation of each type of unit.

Operation

The following is a description of the operation of a clothes dryer.

General Both gas and electric dryers are designed to dry clothes, sheets, blankets, towels, and small throw rugs. Users should follow the manufacturer's drying instructions found on each garment and the suggestions outlined in the owner's manual provided with each dryer.

Caution: Combustible materials should not be dried in these dryers. Items such as oil- or gas-soaked rags will ignite if dried in these dryers.

Safety All dryers meet UL safety requirements and are equipped with two (2) safety features built into their operating sequence. These safety features are: (1) the door switch, which is normally open and which must be closed before the dryer can be started; and (2) the high-temperature limiter thermostat designed to shut off the dryer should the internal temperatures exceed a predetermined safety limit. Technicians, as well as users, should heed all warnings related to the operation of these dryers.

Mechanical operation The parts of the dryer involved in the mechanical operation are the rotating drum or basket which is supported in the cabinet by means of a bearing in the rear of the cabinet and Teflon glides on the front panel, a motor, a belt, and a belt tightener or idler/spring assembly. The motor drives the belt that is mounted to the outside of the basket, causing it to rotate clockwise. The idler/spring assembly provides proper tension on the belt to ensure a drum rotation speed of 50 to 52 revolutions per minute. See Figure 15–53.

Air flow system In order to obtain the best drying action, air passing over the heated coils, or burner flame, is drawn through the basket and forced out the exhaust. Air drawn through the basket provides a more even drying action because it tends to travel around the garments, picking up moisture and carrying it out through the exhaust rather than being blocked by the garments and not drying at all. See Figure 15–54.

Heated air is the method by which the dryer operates to dry clothing. The air travels through the dryer in the following manner. When the dryer is started, the motor operates a blower wheel mounted on one end of the motor shaft. This blower wheel is located in the blower housing. The air system is closed in order to get the best air flow pattern and to ensure that a consistent volume of air is circulating at all times.

Air is drawn in the front bottom of the dryer. It passes around the outside of the dryer basket and enters the system at the rear of the basket. It then passes either over the burner flame or over the hot coils of the heater. It is then heated and drawn through holes at the rear of the basket. Inside the basket, the heated air passes around the clothes (picking up moisture and lint). It is then drawn through holes in the inner door panel, and down through the lint trap. Here, the lint is trapped so that it is forced out the exhaust system. The air enters the air duct as it is being pulled by the blower, and is forced out the exhaust duct where a biased thermostat measures the air temperature. The air then passes out of the dryer and into the exhaust system to the outside.

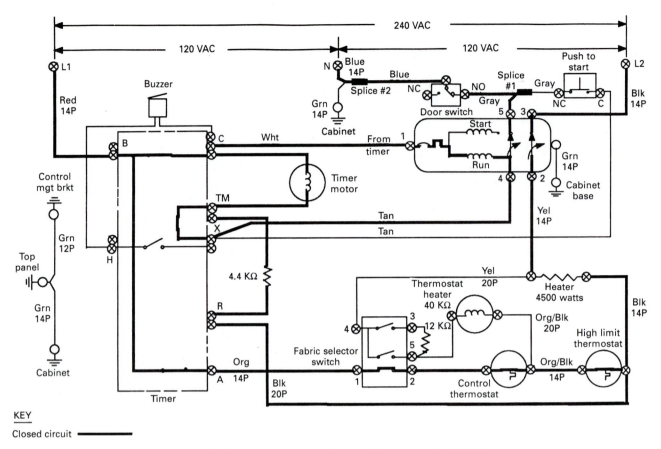

KEY

Closed circuit ━━━━━━━

Figure 15–55 Timed dry circuit (regular heat). (Courtesy of WCI Major Appliance Group.)

Cycles of Operation

The following is a description of the cycles of operation of a clothes dryer.

Timed Dry Refer to Figure 15–55. This cycle selection allows the user to vary the drying time up to 70 minutes and to determine the temperature setting by adjusting a separate temperature control switch. The user can choose from four (4) different temperatures: Regular (high), Medium, Low, and Air Fluff (no heat). Once the timer and temperature selection have been made, the start button must be depressed to start the dryer. One hundred and twenty volts is applied to the drive motor which, in turn, rotates the drum and the blower. As the motor comes up to speed: (1) the centrifugal switch in the motor opens the circuit to the motor run windings; (2) closes a holding circuit which parallels the circuit through the start switch that is closed only while the start button is pressed; and (3) closes

a set of contacts to complete a 240-volt circuit through the heating element, to complete a 120-volt circuit to the gas burner on gas models. (This switch arrangement in the motor is designed to open the heat circuit should the motor fail.) The temperature of the air in the drum will normally be kept between 100 and 180°F when the Regular heat is selected. A high limit thermostat is in series with the control thermostat to prevent overheating. Approximately 6 minutes before the timed cycle is complete, the circuit to the heating element or burner will open, permitting the clothes to cool down before the drum stops rotating. This feature helps to reduce wrinkling of the clothes.

Auto-Dry The temperature control thermostat and high limit thermostat are connected in parallel with the timer motor. The timer motor is, therefore, shunted out of the circuit by the thermostat as long as it is closed. This permits the dryer to heat air for an extended period of time until it reaches the selected cutout temperature. Once the cutout

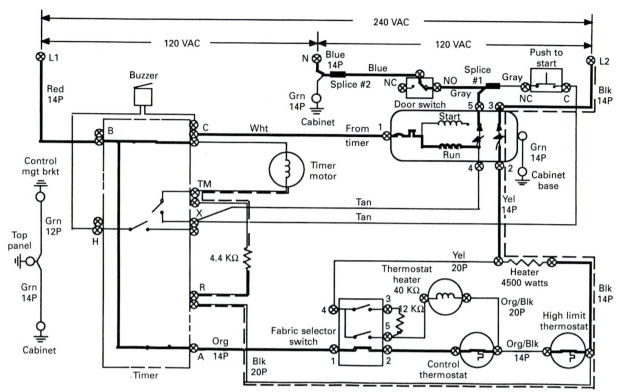

Figure 15–56 Auto-dry circuit (regular heat). (Courtesy of WCI Major Appliance Group.)

temperature is reached, the thermostat opens and the timer motor begins to advance. When the thermostat recloses, the timer motor will not operate until the air temperature again reaches the selected temperature setting. This permits the user to dry for a considerable length of time without having to reset the timer, as one might have to do with the Timed Dry cycle.

The user does not have to worry about mixed clothes being overdried during the Auto-Dry cycle because certain fabrics retain moisture longer and will be continually giving up moisture to those fabrics that dry more readily. Generally, the clothes are dry before the cycling thermostat is satisfied. Once it is satisfied, the timer will advance, and, as with the Timed Dry cycle, the last 6 minutes will be without heat. See Figure 15–56.

Permanent-Press/Knits Auto-Cycle The Permanent-Press/Knits Auto-Cycle is similar to the Auto-Dry cycle yet it has a much shorter period of time for which the heating element or the gas burner is energized (see timer sequence chart for details). Medium Heat must be selected for this operation as opposed to Regular Heat for Timed Dry. The

Auto-Cycle has a 10-minute cool-down period as opposed to a 6-minute cool-down period for Timed Dry. It will also continue to tumble with no heat for 30 minutes after the clothes are dry, alerting the user every 5 minutes with the end-of-cycle buzzer.

Timer motor circuit The circuit to the timer motor is quite different between the electric and gas models. Electric models have the timer motor connected to the 140-volt heater circuit. The 11.4-ohm heating element and the 4400-ohm dropping resistor are in series with the timer motor. The resistor drops the voltage by 119 volts. The remaining 119 volts operates the timer motor whenever it is energized.

Gas models have the timer motor connected across the 120-volt supply. No dropping resistor is required.

GAS COMPONENTS OPERATION AND REPAIR

The following is a description of the operation of the gas components of a clothes dryer and their repair procedures.

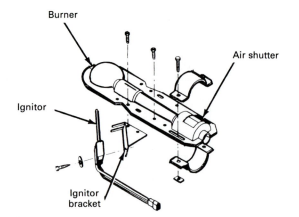

Figure 15-57 Burner assembly. (Courtesy of WCI Major Appliance Group.)

Burner Assembly

The heat source for the gas model dryer is the burner assembly shown in Figure 15-57. The burner assembly consists of a combination gas valve assembly, a burner, and a burner tube. The burner comes from the factory sized and adjusted to deliver heat at the rate of 20,000 Btu per hour, using natural gas.

Warning: No other fuel should be used with the burner unless the burner is properly modified. [See the Installation Instructions covering liquefied petroleum (LP) gas.] The burner assembly is mounted to the dryer base pan to the left front of the drum. It is accessible through the small, lower left front access panel.

Combination Gas Valve

The combination gas valve assembly includes two (2) solenoid shutoff valves, a gas pressure regulator, and a gas orifice assembled into a single-cast body. See Figure 15-58. The pressure regulator is factory set to maintain $3\frac{1}{2}$ inches of water column gas pressure at the orifice. The two solenoid valves are in series, so that the gas flowing through the pressure regulator must pass through both valves before getting to the orifice and burner. The first solenoid valve is referred to as the split-coil valve. The electrical coil, which operates the valve, is made up of two coils stacked in one housing. The upper, holding coil is capable of holding the valve open, but it is not capable of opening the valve from its normally closed position. To open the valve, the lower booster coil is energized.

The second solenoid valve is called the secondary valve. This valve is operated by a single electrical coil which opens the valve and holds it open when the gas burner is in operation. The orifice is a precisely drilled brass plug screwed into the outer port of the combination gas valve. The orifice extends into the burner. The combination of regulated gas pressure and orifice size provides the proper volume of gas for the heat rating of the burner.

Burner

The burner is a short metal tube, in which air is mixed with the gas fuel to provide a combustible mixture that will burn clean, without smoke or noxious fumes. An adjustable air shutter at the inlet controls the amount of mixing air. The shutter is preset and does not ordinarily require adjustment. A scoop-shaped flame spreader extends out over the open end of the burner. This is the area where combustion of the fuel/air mixture takes place. As installed in the dryer, the burner extends into the burner tube about 5 inches. All combustion takes place within the burner tube.

Ignitor

Ignition of the fuel/air mixture is accomplished through the use of a silicon-carbide ignitor (glow-bar). The ignitor is mounted to the burner at an angle, with the glow-bar extended into the flame area. Silicon carbide is a material capable of withstanding very high temperatures, but is fragile and susceptible to contamination from skin oils and other foreign material. It must be handled with care, using the insulated support. See Figure 15-57. Electrically, silicon carbide is a thermistor, that is, its electrical resistance changes (decreases) as its temperature increases. When energized, the ignitor glow-bar begins to heat, with this rise in temperature its resistance steadily decreases, allowing it to reach a temperature of 1800°F.

Flame Switch (Radiant Sensor)

This device is located on the side of the burner tube. It consists of a thermostatically operated single-pole, single-throw (SPST), normally closed (NC) switch, mounted over a window cut into the burner tube. The flame switch senses the intense heat, radiating initially from the ignitor and then from the flame. The flame switch opens within 15 to 90 seconds after the ignitor reaches the ignition temperature. The reaction time is longest during a cold start-up. See Figures 15-58 and 15-59.

Gas Valve Operating Sequence

At the time of dryer start-up and upon a call for heat from the control thermostat, the ignitor is energized. The split-coil valve opens as the following three (3) circuits are simultaneously completed. See Figure 15-60.

1. From the "hot" side of the line (OR) through the holding coil to "neutral" (T).
2. From the "hot" side of the line (OR) through the flame switch and the booster coil to "neutral" (Y).
3. From the "hot" side of the line (OR) through the flame switch and the ignitor to "neutral" (Y).

Within 15 to 90 seconds after the ignitor reaches the ignition temperature (1800°F), the flame switch opens. At this point, current for the ignitor passes through the secondary coil. Since the resistance of the ignitor is low when fully

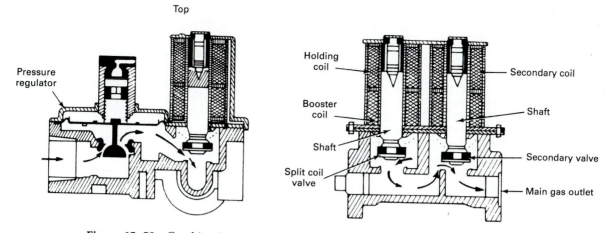

Top

Pressure regulator

Holding coil

Secondary coil

Booster coil

Shaft

Shaft

Secondary valve

Split coil valve

Main gas outlet

Figure 15–58 Combination gas valve. (Courtesy of WCI Major Appliance Group.)

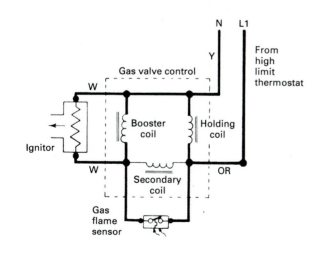

Figure 15–59 Flame switch (radiant sensor). (Courtesy of WCI Major Appliance Group.)

Flame switch (radiant sensor)

Figure 15–60 Gas valve operating sequence. (Courtesy of WCI Major Appliance Group.)

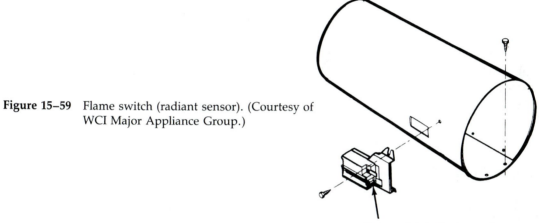

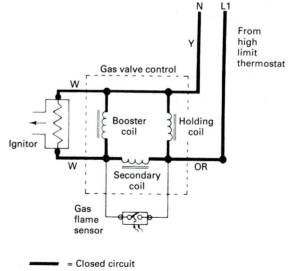

N L1

Y

From high limit thermostat

Gas valve control

W

Booster coil

Holding coil

Ignitor

W

OR

Secondary coil

Gas flame sensor

Call for heat

N L1

Y

From high limit thermostat

Gas valve control

W

Booster coil

Holding coil

Ignitor

W

OR

Secondary coil

Gas flame sensor

──── = Closed circuit

Flame switch open - flame established

heated, the secondary valve opens and gas flows through the orifice and into the burner. Ignition takes place. The impedance of the secondary coil reduces the current through the ignitor from approximately 4 amps to 1 amp. The current flow through the booster coil drops to almost zero. The split-coil valve remains open because the holding coil remains fully energized. The flame switch contacts are held open by the heat radiated by the flame. Burner operation will continue until the electrical power to the burner assembly is interrupted by either the timer, the control thermostat, the high limit thermostat, the door switch, the burner contacts on the motor switch (motor failure), blown fuse or circuit breaker tripped, or an interruption in the electrical service to the home. At such time, both gas valves will close and the flame switch will close. When the power is restored, the ignition cycle is repeated.

Fail-Safe Features

The following is a description of the fail-safe features used on clothes dryers.

Momentary power interruption Upon resumption of power, the flame switch contacts will open, permitting the secondary valve to remain open. The split-coil valve will remain closed because the pull-in circuit for the booster coil cannot be made through the open flame switch contacts. When the flame switch contacts do reclose, the secondary valve will close and the split-coil valve will open. The dryer will then go through the normal cycle for ignition.

Ignition failure If there is a momentary interruption in the gas supply or a flame is not established when the flame switch contacts open, the secondary valve will remain open until the flame switch contacts reclose. At that time, the secondary valve will reclose. The flame switch will continue to recycle the ignitor and the secondary valve (about once per minute) until normal operation is restored, the thermostat is satisfied, or the timer terminates the cycle.

Flame failure In the event of flame failure, the flame switch contacts will reclose in about 45 seconds, at which time the secondary valve will close (the split-coil valve will stay open) and a try for reignition will be made, as in a normal cycle.

Burner Assembly

The gas dryer is heated by a means of a gas burner assembly rated at 20,000 Btu per hour capacity. It is located behind the small, lower front access panel.

To remove the burner assembly, use the following procedure:

1. Disconnect the dryer from the electrical power supply.

2. Remove the small, front access panel at the lower left corner of the front panel by removing the mounting screw and lifting it out. The screw is located about 6 inches from the left side of the panel.

3. Shut off the gas supply at the manual shutoff valve, located adjacent to the gas valve assembly. Loosen the shut-off valve union. See Figure 15–57.

4. Disconnect the two (2) wire leads from the radiant sensor terminals.

5. Disconnect the plug connector from the color-coded cabinet harness.

6. Remove the single hex-head screw securing the burner assembly to the base of the dryer.

7. Slide the burner out the front access panel opening. Note the mounting bracket's rear tab projection and corresponding base slot.

8. Reverse the above procedure to reassemble the unit. Make sure that all gas connections are tight.

Note: After reinstalling the burner assembly, open the shutoff valve and check the union fitting for leaks with a soap suds solution. Never use an open flame to check for gas leaks.

Flame switch (radiant sensor) This device is designed to sense the radiant heat from the glowing ignitor and the gas flame. It is a thermostatically operated, single-pole, single-throw (SPST) switch with contacts that are normally closed (NC). The flame switch is mounted on the side of the burner tube. See Figure 15–58.

To remove the flame switch, use the following procedure:

1. Disconnect the dryer from the electrical supply.

2. Remove the small, lower left front access panel.

3. Shut off the gas at the manual shutoff valve, located adjacent to the gas valve assembly.

4. Disconnect the two (2) wire leads from the flame switch terminals.

5. Remove the $\frac{1}{4}$-inch hex-head screw securing the sensor to the burner tube. Rotate the sensor down and lift it out of its lower mounting slot.

6. Reverse the above procedure to reassemble the unit.

Ignitor The ignitor is a silicon-carbide thermistor. When it attains approximately 1800°F, the flame switch (mounted on the side of the burner tube) detects the high radiant heat and its contacts open. This energizes the secondary valve coil, allowing gas to flow through the gas valve orifice and impinge upon the hot glowing ignitor. The total sequence occurs within 15 to 90 seconds.

The ignitor is mounted to the burner at an angle with the silicon-carbide stem extended into the flame area. The stem is very fragile and susceptible to contamination from skin oils. Handle it with care by using the ignitor's insulated support. See Figure 15–57.

To test the ignitor, use the following procedure:

1. Disconnect the dryer from the electrical supply.
2. Remove the small, lower left front access panel.
3. Disconnect the plug connector from the ignitor-to-coil harness.
4. Check the resistance value of the ignitor. It should be approximately 50 to 400 ohms, depending on the temperature of the ignitor.

To replace the ignitor, use the following procedure:

1. Remove the burner assembly.
2. Remove the $\frac{1}{4}$-inch hex-head screw and washer securing the ignitor to its mounting bracket.
3. Install the new ignitor.
4. Reverse the above procedure to reassemble the unit.

Pressure regulator valve The purpose of the pressure regulator valve is to regulate the gas pressure to the dryer and to maintain this pressure within the specified limits of the dryer manufacturer.

To test a pressure regulator, use the following procedure:

1. Fully close the shutoff valve just ahead of the pressure regulator. Remove the small plug from the end of the combination valve body and securely install a water manometer adaptor. See Figure 15–60. Connect a water manometer to the adapter.
2. Fully open the gas shutoff valve. Set the timer to operate for more than 20 minutes. Start the dryer by pressing the start button. With the burner on, check the flowing gas pressure at the water manometer. If the flowing gas pressure is above or below $3\frac{1}{2}$ inches of water column, remove the screw on top of the regulator and gradually turn the adjustment screw beneath the cap while observing the change in the pressure on the manometer scale. Adjust the flowing gas pressure for $3\frac{1}{2}$ inches of water column.

Note: If the pressure is low and the regulator does not respond to adjustment, check to make certain that the pressure of the gas supply line is above 5 inches of water column. If the supply pressure is adequate, replace the combination gas valve assembly.

Caution: After completing the flowing gas pressure test, close the gas shutoff valve, remove the manometer tube, replace and tighten the plug in the combination valve body. Reopen the shutoff valve and leak-test the assembly with a soap and water solution.

Note: The screw cap on the regulator also serves as a vent to relieve air above the regulator diaphragm as the diaphragm moves up and/or down to regulate the gas pressure. Make certain that the vent opening in the cap is clear.

Valve coils There are two (2) solenoid coils mounted on the gas valve body, as shown in Figure 15–59. The secondary coil (the smaller one) has two (2) terminals. The split coil (the larger valve), consisting of the holding and booster coils molded into a single coil, has three (3) terminals.

Note: The resistance values stated in the following tests will vary with the temperature of the component being tested. All values are approximate.

To test the secondary coil, use the following procedure:

1. Disconnect the dryer from the electrical supply.
2. Remove the burner assembly.
3. Remove the connector plug from the coil.
4. Check the resistance value of the coil. It should be approximately 1200 ohms.

To test the split coil (holding and booster coils), use the following procedure:

1. Disconnect the dryer from the electrical supply.
2. Remove the burner assembly.
3. Remove the plug connector from the coil.
4. Note the numbers 1 and 3 marked near the terminals. Terminal 1 is the common terminal.
5. Check the resistance value between terminals 1 and 3. This is the booster coil and should be approximately 530 ohms.
6. Check the resistance value between terminal 1 and the center unmarked terminal. This is the holding coil and the resistance should be approximately 1320 ohms.

To replace the valve coils, use the following procedure:

1. Disconnect the dryer from the electrical supply.
2. Remove the burner assembly.
3. Disconnect the plug connections from the coils.
4. Remove the two (2) Phillips-head screws securing the coil hold-down bracket to the valve body.
5. Remove the coils, noting their positions. The secondary coil is mounted with its terminals toward the top and has a small embossment on the top that mates with a small hole in the hold-down bracket. The split coil is mounted with its terminals toward the bottom. It also has an embossment. In addition, the gas valve stem for this coil has a sleeve on it to ensure that the correct coil will be mounted in that position.
6. Reverse the above procedure to reassemble the unit.

Burner The burner acts as a mixing chamber by allowing a certain amount of air to mix with the gas exiting the orifice. At the scooped end of the burner is the flame spreader. The other end has an adjustable air shutter that is preset at the factory and does not ordinarily require adjustment.

To adjust the air shutter, use the following procedure:

1. Loosen the two (2) $\frac{1}{4}$-inch hex-head screws on both sides of the air shutter.
2. Gradually slide the shutter forward or backward until the yellow tips of the burner flame just disappear.
3. Retighten both shutter screws.

PRODUCT CUT-A-WAY VIEW
27" LAUNDRY CENTER

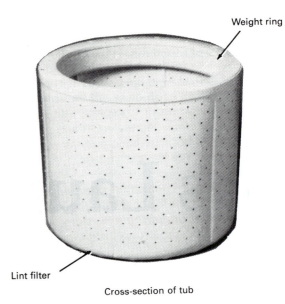

Weight ring

Lint filter

Cross-section of tub

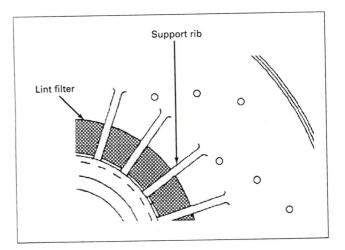

Support rib

Lint filter

Figure 16–2 View looking down into the tub. (Courtesy of WCI Major Appliance Group.)

Figure 16–1 Laundry center cut-away view. (Courtesy of WCI Major Appliance Group.)

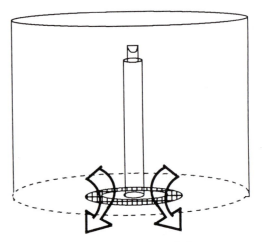

Water direction during agitation

Figure 16–3 Water flow direction during agitation. (Courtesy of WCI Major Appliance Group.)

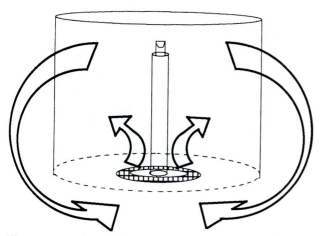

View looking down into tub

Figure 16–4 Water flow direction during spin. (Courtesy of WCI Major Appliance Group.)

2. Set the water level control according to the size of the wash load.

3. Load the clothes into the wash basket, measure and add the proper amount of detergent, and close the washer lid.

4. To set the timer, push in on the knob and turn it in a clockwise direction to the desired cycle and wash time.

5. Pull the timer knob out.

6. The washer will start operating.

Caution: Never turn the timer knob when it is in the out position. To do so could cause damage to the timer contacts and other electrical components.

Dryer Operation

Air is drawn into the heater housing (by a blower wheel mounted to one end of the dryer drive motor shaft) and across the heating section of the gas heater. It is then drawn through the tumbling clothes, picking up moisture and lint. The lint is filtered out as the air passes from the drum into the blower where it is discharged out the vent opening. The air temperature is controlled by the biased thermostat according to the setting of the fabric selector switch or the timer.

The length of the drying cycle is controlled by the number of minutes selected on the timer, or automatically controlled by the timer, in conjunction with the thermostat.

The clothes drying cycle of operation is controlled by the timer, fabric temperature selector switch, push-to-start switch, and thermostats. There are two thermostats in the dryer. The temperature control thermostat, mounted to the lower duct housing, senses the air temperature during operation. A biasing heater surrounds the thermostat and functions to "fool" the thermostat so that three different levels of heat can be obtained. A safety thermostat is mounted to the heater plenum. Its purpose is to open the heat circuit if excessive drum temperatures (260 ± 8°F) are reached.

Four safety devices control the dryer under certain conditions. These devices are:

1. *Door switch* The door switch stops all operation when the door is opened.

2. *Safety thermostat* The safety thermostat opens the heat circuit if the drum temperature exceeds a high limit.

3. *Motor centrifugal switch* The motor centrifugal switch opens the heater circuit when the motor is not up to normal operating speed.

4. *Push-to-start switch* The push-to-start switch must be momentarily closed by the user to start the dryer.

To operate the dryer, first check the lint screen and be certain that the screen is completely free of lint. Place the clothes in the dryer and close the door. (The dryer will not operate unless the door is closed.) Make the following settings:

1. Select the desired drying time, or automatic drying cycle, by turning the timer knob to the right.

2. Set the fabric selector for the type of fabric to be dried.

3. To start the dryer, push the start button and hold it in for 2 seconds.

4. At the end of the drying cycle, a signal will sound.

Safety Precautions

Always turn off the electrical power supply before servicing any electrical component, making ohmmeter checks, or replacing any parts. Refer to "Safe-Servicing Procedures" in the back of this chapter before servicing the washer or dryer.

All voltage checks should be made with a voltmeter having a full-scale range of 130 volts or higher.

After the service operations are completed, be sure that all safety grounding circuits are complete, all electrical connections are secure, and all panels are in place.

Washer Top Panel and Lid Assembly

Lid lock switch shield The lid lock shield is galvanized steel and surrounds the lid lock switch (a UL requirement for safety).

To remove or replace the lid lock switch shield, use the following procedure:

1. Disconnect the laundry center from the electrical supply.

2. Remove the washer front service panel.

3. Remove the one (1) screw securing the lid lock shield.

4. Install the new lid lock switch shield.

5. Reverse the above procedure to reassemble the unit.

Lid lock switch The lid lock switch prevents operation of the washing machine while the lid is open to provide safety to the user.

To remove or replace the lid lock switch, use the following procedure. (See the Washer Electrical Components Test and Repair section presented later in this chapter.)

Self leveling rear legs

Snubber

Counterbalance springs

Figure 16–5   Washer base assembly. (Courtesy of WCI Major Appliance Group.)

Install with arrow pointing up

Figure 16–6   Washer snubber ring. (Courtesy of WCI Major Appliance Group.)

1. Disconnect the laundry center from the electrical supply.
2. Remove the washer front service panel.
3. Disconnect the washer lid lock switch terminal connector (and light if applicable).
4. Remove the one (1) screw securing the lid lock shield.
5. Remove the two (2) screws securing the lid lock switch.
6. Install the new lid lock switch.
7. Reverse the above procedure to reassemble the unit.

To remove or replace the lid lock switch mounting bracket, use the following procedure:

1. Disconnect the laundry center from the electrical supply.
2. Remove the washer front service panel.
3. Disconnect the washer lid lock switch terminal connector (and light switch if applicable).
4. Remove the washer top panel and lid assembly.
5. Remove the one (1) screw securing the lid lock shield.
6. Remove the two (2) screws securing the lid lock switch.
7. Drill out the pop rivets securing the lid lock switch mounting bracket to the washer top panel.
8. Install the new lid lock switch mounting bracket.
9. Reverse the above procedure to reassemble the unit.

Washer Cabinet Base Pan Assembly

The base pan assembly serves as the support for the entire washer/dryer. It is constructed of heavy-gauge metal, embossed for strength.

The base assembly is supported by leveling legs, one on each front corner with rubber inserts. The rear is supported by a self-leveling leg assembly that is welded to the base pan. The raised portion at the center of the base provides a support surface for the snubber ring and stabilizes the tub and transmission during the spin cycle. See Figure 16–5.

The complete washer unit must be removed to service the washer base.

Snubber Ring

The snubber ring assembly is made of a polypropylene ring with Teflon strips secured to the upper and lower edges. The top edge of the ring is identified by an arrow molded into the outer surface. The snubber ring fits between the raised center portion of the washer base pan and the dome base of the tub and transmission assembly. See Figure 16–6.

To remove or replace the snubber ring, use the following procedure:

1. Disconnect the laundry center from the electrical supply.
2. Remove the washer front service panel.
3. Remove the washer top panel and lid assembly.
4. Disconnect the front vertical and horizontal springs from the base pan.
5. Place a 2-by-4-inch block under the drive motor section of the leg-and-dome assembly, carefully prying up the washer tub and transmission assembly off the base.
6. Using a wire with a hook on one end, reach under and pull out the snubber ring.

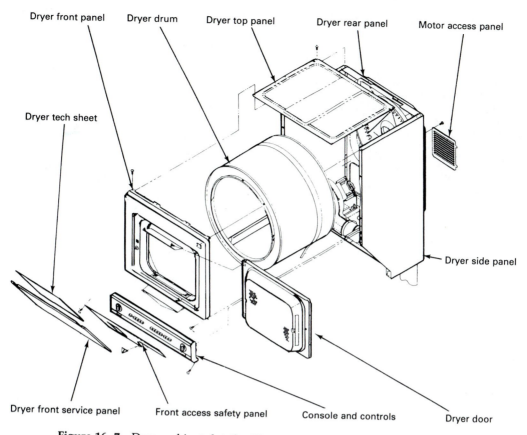

Dryer front panel   Dryer drum   Dryer top panel   Dryer rear panel   Motor access panel

Dryer tech sheet

Dryer side panel

Dryer front service panel   Front access safety panel   Console and controls   Dryer door

Figure 16–7 Dryer cabinet details. (Courtesy of WCI Major Appliance Group.)

7. Install the new snubber ring with the arrow pointing up.

8. Reverse the above procedure to reassemble the unit.

Note: The washer and base assembly can also be removed should the technician elect to do so.

Dryer Cabinet Assembly

The dryer cabinet consists of a base, front service access panel, dryer front panel, console electrical controls, dryer door, left and right cabinet side panels, top panel, rear panel, and a louvered ventilation panel. See Figure 16–7.

One or more of these parts must be removed to gain access to the other dryer components.

Dryer Front Access Safety Cover Panel

The front access safety cover panel allows access to the water pressure control switch and the dryer heat control thermostat.

To remove or replace the dryer front access panel, use the following procedure:

1. Disconnect the laundry center from the electrical supply.

2. Remove the two (2) screws at the upper corners securing the dryer front panel.

3. Tilt the front access panel out and lift it up to remove it from the unit.

4. Remove the six (6) screws securing the dryer front access safety cover panel.

5. Pull the panel out and away to remove it.

6. Reverse the above procedure to reassemble the unit.

Console and Controls

The console and controls assembly consists of the console panel, dryer timer, dryer temperature selector switch, push-to-start switch, washer timer, water temperature selector switch, water level control switch, control knobs, indicator light, and so on.

To remove or replace the console and controls, use the following procedure:

1. Disconnect the laundry center from the electrical supply.

2. Remove the dryer access panel.

3. Remove the dryer front access safety cover panel.

4. Remove the four (4) screws securing the console and controls assembly to the side panels.

5. Disconnect the electrical wiring connector blocks.

6. Remove the electrical leads for the console light (if applicable).

7. Lay the console and controls assembly on a protected surface and transfer all the remaining components and parts to the new panel.

8. Install the new console panel and/or controls.

9. Reverse the above procedure to reassemble the unit.

Note: See the Washer Electrical Components Test and Repair section presented later in this chapter.

Dryer Front Panel

The dryer front panel is made from steel and houses the dryer door assembly, door gasket, door hinges, lint filter, lint filter housing, door catch, door switch, interior drum light (if applicable), air duct, air grille, wiring clips, glide clips, and drum front felt seal.

To remove or replace the dryer front panel assembly, use the following procedure:

1. Disconnect the laundry center from the electrical supply.
2. Remove the dryer front access panel.
3. Remove the dryer front access safety cover panel.
4. Remove the console and controls.
5. Reach through and release the drum belt from the idler pulley to relieve the tension on the front panel.
6. Remove the two (2) screws (lower front) securing the dryer front panel assembly to the side panels.
7. Remove the four (4) screws (upper top front) securing the dryer front panel assembly to the top panel.
8. Carefully remove the dryer front panel assembly.
9. Lay the front panel on a protected surface and transfer all the remaining components and parts to the new panel.
10. Install the new dryer front panel.
11. Reverse the above procedure to reassemble the unit.

Dryer Door Seal

The extruded vinyl door seal is secured to the inner door panel lip and provides a seal between the dryer front panel and the dryer door.

To remove or replace the dryer door seal, use the following procedure:

1. Disconnect the laundry center from the electrical supply.
2. Open the dryer door and remove the four (4) screws securing the door assembly to the door hinges.
3. Remove the six (6) screws securing the inner door panel to the outer door panel.
4. Lift the inner door panel and seal out from the bottom to remove it from the lip of the outer door panel top.
5. Remove the door seal from the inner door panel.
6. Install the new dryer door seal.
7. Reverse the above procedure to reassemble the unit.

Dryer Door Lint Filter Housing

The polypropylene dryer door lint filter housing is mounted in the dryer door assembly and houses the lint filter screen.

To remove or replace the dryer door lint filter housing, use the following procedure:

1. Disconnect the laundry center from the electrical supply.
2. Open the dryer door and remove the four (4) screws securing the door assembly to the door hinges.
3. Remove the six (6) screws securing the inner door panel to the outer door panel.
4. Lift the inner door panel and seal out from the bottom to remove it from the lip of the outer door panel top.
5. Remove the lint filter housing and the lint filter screen.
6. Install the new lint filter housing.
7. Reverse the above procedure to reassemble the unit.

Dryer Lint Filter Screen

The lint filter screen is in the dryer door lint filter housing (exhaust) opening. It is made of 100 percent Dacron Polyester Leno Weave molded in a plastic frame. The lint filter screen should be cleaned after every load.

To remove or replace the dryer lint filter screen, use the following procedure:

1. Disconnect the laundry center from the electrical supply.
2. Open the dryer door and remove the lint filter screen.
3. Install the new lint filter screen.
4. Reverse the above procedure to reassemble the unit.

Dryer Door Switch

The dryer door switch assembly is mounted in the recessed door area of the dryer front panel. See Figure 16–8.

To remove or replace the dryer door switch, use the following procedure:

1. Disconnect the laundry center from the electrical supply.
2. Remove the dryer front access panel.
3. Remove the dryer front access safety cover panel.
4. Remove the console and controls.
5. Reach through and release the drum belt from the idler pulley to relieve the tension on the front panel.
6. Remove the two (2) screws (lower front) securing the dryer front panel assembly and sides.
7. Loosen the four (4) screws (upper top front) securing the dryer front panel assembly.
8. Carefully pull out the bottom of the dryer front

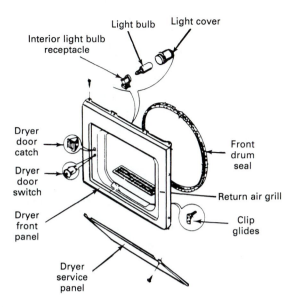

Figure 16–8 Dryer front panel details. (Courtesy of WCI Major Appliance Group.)

panel assembly to gain access to the dryer door switch.

9. Release the spring side clips securing the dryer door switch.

10. Remove the electrical leads from the switch.

11. Install the new dryer door switch.

12. Reverse the above procedure to reassemble the unit.

Dryer Interior Drum Light Bulb and Cover

The dryer interior light bulb and cover are mounted to the interior of the dryer front panel and can be removed through the dryer drum opening.

To remove or replace the dryer drum light bulb and cover, use the following procedure:

1. Disconnect the laundry center from the electrical supply.

2. Open the dryer door.

3. Remove the interior drum light cover by turning it counterclockwise.

4. Remove the interior dryer drum light bulb by pushing it in and turning it counterclockwise.

5. Install the new interior drum light bulb and/or cover.

6. Reverse the above procedure to reassemble the unit.

Dryer Interior Drum Light Bulb and Receptacle

The dryer interior drum light bulb receptacle is mounted to a bracket on the interior of the dryer front panel and can be accessed by removing the dryer top panel.

To remove or replace the dryer drum light bulb receptacle, use the following procedure:

1. Disconnect the laundry center from the electrical supply.

2. Remove the dryer front access panel.

3. Remove the dryer front access safety cover panel.

4. Remove the console and controls.

5. Reach through and release the drum belt from the idler pulley to relieve the tension on the front panel.

6. Remove the two (2) screws (lower front) securing the dryer front panel assembly to the side panels.

7. Remove the four (4) screws (upper top front) securing the dryer fornt panel assembly to the top panel.

8. Carefully remove the dryer front panel assembly.

9. Open the dryer door.

10. Remove the interior drum light cover by turning it counterclockwise.

11. Remove the interior drum light bulb by pushing it in and turning it counterclockwise.

12. Remove the electrical leads from the interior of the dryer drum light bulb receptacle.

13. *Caution:* Use care when removing the dryer drum light bulb receptacle.

14. Install the new dryer drum light bulb receptacle.

15. Reverse the above procedure to reassemble the unit.

Dryer Front Drum Seal

The felt seal is secured (glued) to the inner drum flange of the drum opening in the dryer front panel.

To remove or replace the dryer front drum seal, use the following procedure:

1. Disconnect the laundry center from the electrical supply.

2. Remove the dryer front access panel.

3. Remove the dryer front access safety cover panel.

4. Remove the console and controls.

5. Reach through and release the drum belt from the idler pulley to relieve the tension on the front panel.

6. Remove the two (2) screws (lower front) securing the dryer front panel assembly to the side panels.

7. Remove the four (4) screws (upper top front) securing the dryer front panel assembly to the top panel.

8. Carefully remove the dryer front panel assembly.

9. Remove the felt drum seal.

10. Install a new dryer front drum seal; the thicker portion of the felt seal is positioned at the top.

11. Reverse the above procedure to reassemble the unit.

Dryer Air Duct

The dryer air duct is secured to the inner drum flange of the dryer front panel. See Figure 16–9.

To remove or replace the dryer air duct, use the following procedure:

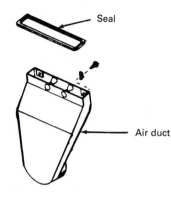

Figure 16–9 Dryer air duct. (Courtesy of WCI Major Appliance Group.)

1. Disconnect the laundry center from the electrical supply.
2. Remove the dryer front access panel.
3. Remove the dryer front access safety cover panel.
4. Remove the console and controls.
5. Reach through and release the drum belt from the idler pulley to relieve the tension on the front panel.
6. Remove the two (2) screws (lower front) securing the dryer front panel assembly to the side panels.
7. Remove the four (4) screws (upper top front) securing the dryer front panel assembly to the top panel.
8. Carefully remove the dryer front panel assembly.
9. Remove the two (2) screws securing the air duct to the inner dryer front panel.
10. Remove the air duct by sliding it to the side.
11. Remove the two Tinnerman-type clip nuts.
12. Install the new dryer air duct.

Note: Check the foam seal inside the dryer front panel; replace it if necessary.

13. Reverse the above procedure to reassemble the unit.

INSTALLATION INSTRUCTIONS

Use the installation instructions that are packaged with the unit that is being installed. The following instructions should be used as a minimum requirement for a satisfactory installation.

Preinstallation Requirements

The following is a listing of the necessary requirements for a satisfactory installation of this type of washer/dryer.

Tools and materials required for an installation   The following is a listing of the tools and materials generally required for a satisfactory installation of a washer/dryer combination:

1. Phillips screwdriver.
2. Channel-lock adjustable pliers.
3. Duct tape.

NOTE Do not under any circumstances remove grounding prong from plug.

Figure 16–10 Electrical outlet. (Courtesy of WCI Major Appliance Group.)

4. Approved pipe thread sealer (resistant to the corrosive action of liquefied petroleum gases).
5. 8- or 10-inch adjustable wrench.
6. Carpenters level.
7. $\frac{1}{2}$-inch ID semirigid tubing or pipe with $\frac{1}{2}$- to $\frac{3}{8}$-inch reducer.
8. Rigid or flexible metal ductwork (if not presently installed).
9. Wall cap (if not presently installed).
10. $\frac{3}{8}$-inch socket with a ratchet or a $\frac{3}{8}$-inch nutdriver.

Electrical requirements   The following is a listing of the electrical requirements. See Figure 16–10.

1. Separate, properly polarized, grounded three-wire circuit.
2. 115-volt, 60-hertz, ac power supply.
3. 10-amp minimum delayed-action fuse or circuit breaker.
4. Outlet with accessibility to the service cord when the laundry center is installed.

Water supply requirements   Hot- and cold-water faucets must be installed within 42 inches of the inlet of the washer. The faucets must be $\frac{3}{4}$-inch garden hose, so that the inlet hoses can be connected. The water pressure must be within 10 to 120 psi (maximum unbalanced pressure, hot vs. cold flowing, 10 psi). The local water department can provide the available water pressure. The hot-water temperature should be about 140°F.

Drain requirements   The drain or standpipe must be capable of accepting a discharge rate of 17 gallons per minute. The drain height must be between 33 and 58 inches maximum. For a drain facility less than 33 inches high, a siphon break kit *must* be installed so that the washer tub will retain the water.

The standpipe diameter must be 1 inch minimum. There must be an air gap around the drain hose in the standpipe. A snug fit can cause a siphoning action.

Locating the Laundry Center

If the laundry center is located in a garage or a garage-type building, where flammables are kept or stored (including automobiles), be sure that the installation conforms to ANSI Z223.1-1988 National Fuel Gas Code, local codes, and ordinances for Class 1, Division 2 hazardous locations.

A laundry center that is installed in a closet or recessed area must be exhausted to the outside and have enough air around it to operate properly. See Figures 16–15 through 16–17.

Do not install the laundry center in an area where the temperatures will fall below freezing. If the laundry center is to be stored or transported in freezing temperatures, be sure that all the water has been drained from the fill and drain systems.

Do not install the laundry center up against curtains or draperies.

Do not install the laundry center on carpet.

See Figures 16–11 and 16–12 for rough in dimensions.

Gas Supply Line Installation

The following precautions must be taken when installing the gas line to the laundry center:

1. *Do not attempt to alter the gas orifice or adjust the burner air shutter.* Natural gas input may vary in some areas from 700 to 1200 Btu per cubic foot. If the gas or the burner air shutter is incorrectly adjusted, serious personal injury and/ or fire hazard can occur. The local gas company will know the qualities of the gas in your area. Contact the local servicing dealer for burner adjustment or if orifice changes are necessary.

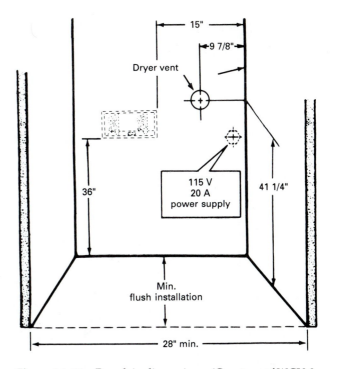

Figure 16–11   Rough in dimensions. (Courtesy of WCI Major Appliance Group.)

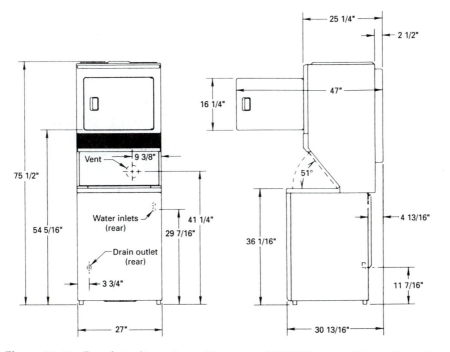

Figure 16–12   Rough in dimensions. (Courtesy of WCI Major Appliance Group.)

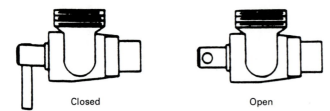

Figure 16–13 Gas valve. (Courtesy of WCI Major Appliance Group.)

2. *Do not use an open flame for leak-testing.* Serious personal injury and/or fire hazard can result if an open flame is used to test for gas leaks. Use a soap and water solution to test all gas line fittings.

3. *Do not install the laundry center to an LP gas supply without installing a gas conversion kit.* All laundry centers are shipped with a pressure regulator and a natural gas orifice. Using a natural gas orifice with LP gas can result in personal injury, clothes damage, and/or fire hazard. Have a qualified gas technician install a conversion kit in the laundry center before using it.

a. A $\frac{1}{8}$-inch natural gas thread plugged tapping, accessible for a test gauge connection, must be installed immediately upstream of the gas connection to the laundry center.

b. Connect a $\frac{1}{2}$-inch ID semirigid tubing or approved pipe from the gas supply line to the $\frac{3}{8}$-inch pipe located on the back of the laundry center. Use a $\frac{1}{2}$- to $\frac{3}{8}$-inch reducer for connecting the piping. Apply an approved thread sealer that is resistant to the corrosive effect of liquefied gases on all pipe connections.

c. Test all gas line connections that were made during the installation of the laundry center for leaks, including those inside the dryer.

Warning: The laundry center and its individual shutoff valve must be disconnected from the gas piping system during any pressure-testing procedure when the test pressures exceed $\frac{1}{2}$ psig. See Figure 16–13.

Warning: The laundry center must be isolated from the gas supply system by closing its individual manual shutoff during any testing of the gas supply piping system at a test pressure equal to or less than $\frac{1}{2}$ psig.

Installing the Exhaust

Warning: The following are specific requirements for proper and safe installation of the laundry center. Failure to follow these instructions can create excessive drying times and fire hazards.

1. *Do not use plastic flexible duct to exhaust the dryer.* Excessive lint can build up inside the ductwork, creating a fire hazard and restricting the air flow. Restricted air flow will increase the drying time. If the present system is made up of plastic duct, *replace it* with a rigid or flexible metal duct. Ensure that the present duct is free of any lint *prior to* installing the laundry center dryer duct.

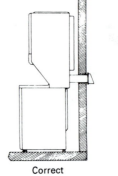

Figure 16–14 Exhaust installation. (Courtesy of WCI Major Appliance Group.)

2. *If the dryer is not exhausted outdoors, some fine lint will be expelled into the laundry area.* An accumulation of lint in any area of the home can create both a health and a fire hazard. The dryer ductwork must be exhausted outside the dwelling.

3. *Do not allow combustible materials (e.g., clothing, draperies/curtains, paper) to come in contact with the ductwork.* The dryer *must not* be exhausted into a chimney, a wall, a ceiling, or any concealed space of a building in which lint can accumulate, resulting in a fire hazard.

4. *Do not exceed the length of duct pipe or the number of elbows allowed.* Lint can accumulate in the system, plugging the ductwork and creating a fire hazard, as well as increasing the drying time. Install the duct and elbows as illustrated in Figure 16–14.

5. *Use extreme care when installing ductwork.* There could be sharp edges that could result in personal injuries.

6. *Do not screen the exhaust end of the vent system or use screws or rivets to assemble the ductwork.* Lint can become caught in the screen and on the screws or rivets, clogging the ductwork and creating a fire hazard as well as increasing the drying time. Use an approved vent hood (as described in the Exhaust System Requirements section) to terminate the duct outdoors, and seal the joints with duct tape. All male pipe fittings must be installed downstream with the flow of air. See Figure 16–14.

Exhaust system installation in manufactured (mobile) homes Use the following recommendations when installing a laundry center in a mobile home:

1. The dryer *must* be exhausted to the outside (outdoors).

2. The exhaust *must not* be terminated beneath the manufactured (mobile) home. See Figure 16–15.

3. The exhaust duct *must not* support combustion. Use only rigid or flexible metal duct. Do not use plastic duct.

4. The exhaust duct must not be connected to any other duct, vent, chimney, or any concealed space of a manufactured (mobile) home.

5. The installation *must* conform to current Manufactured Home Construction and Safety Standard (which is a Federal Regulation Title 24 CFR-Part 32-80), or when such standard is not applicable, with the American National Standard for Mobile Homes.

Installation in a recess or closet The laundry center needs space around it for ventilation. Figure 16–16 shows, in inches, the minimum clearance dimensions for proper operation of the laundry center in a recess or closet installation. *Do not install the laundry center in a closet with a solid door.*

A minimum of 120 square inches of opening, equally divided between the top and the bottom of the door is required. Air openings are required to be unobstructed when a door is installed. A louvered door with the equivalent air openings for the full length of the door is acceptable. See Figures 16–17 and 16–18.

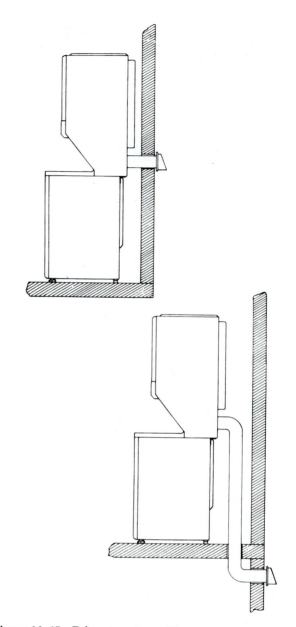

Figure 16–15 Exhaust system. (Courtesy of WCI Major Appliance Group.)

Other Installation Requirements

Additional installation requirements are listed below. These requirements should also be given top priorty when installing a laundry center.

1. The installation must conform with state and local laws and ordinances, or in the absence of laws and ordinances, with the National Fuel Gas Code ANSI Z223.1-1988.

Warning: Take note of the requirements when installing the unit in a residential garage which is classified as a Class 1, Division 2 hazardous location.

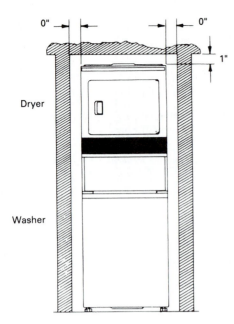

Figure 16–16 Dimensions required for recess or closet installation. (Courtesy of WCI Major Appliance Group.)

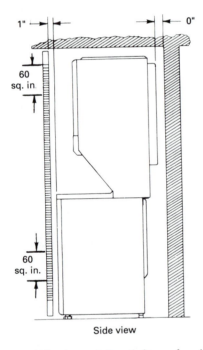

Side view

Figure 16–18 Side view of closed door, showing ventilation air openings location. (Courtesy of WCI Major Appliance Group.)

2. A laundry center installation in a bedroom, bathroom, or closet *must be* exhausted to the outdoors.

3. If the laundry center is to be installed in a closet, no other fuel-burning appliance shall be installed in the same closet.

4. Do not install or store the laundry center in an

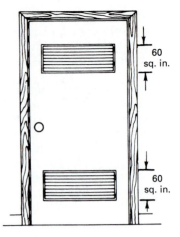

Figure 16–17 Ventilation air openings in door. (Courtesy of WCI Major Appliance Group.)

area that is exposed to dripping water or outside weather conditions.

5. *Warning:* The laundry center is designed under ANSI Z21.5.1, for home use only.

Final Installation Checklist

The following is a final checklist for the installation:

1. Complete the proper gas connections and test them for leaks.

2. Check to make sure that the maximum lengths of duct pipe and the number of elbows are not exceeded and the proper connections were made.

3. Adjust the laundry center legs so it sits solidly on the floor.

4. Ensure the proper operation of an approved vent hood.

5. Test the functional operation of the laundry center.

INSTALLATION INSTRUCTIONS (GAS CONVERSION KITS)

The following are the procedures for installing a gas conversion kit in the laundry center for different types of gas.

Gas Conversion Kit LP to Natural Gas Laundry Center Dryer Input 20,000 Btu

Warning: Failure to comply with the following proper instructions and use of the correct conversion kit can result in a fire, explosion, serious bodily injury, and/or property damage.

Conversion is at the risk of and the responsibility of the person making the conversion and it shall be made by a *qualified gas technician only.*

Use only the correct conversion kit specified in the parts catalog for the model number of the laundry center dryer being converted.

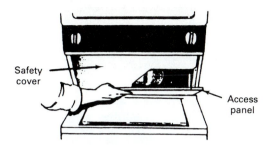

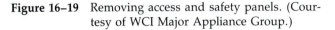

Safety cover

Access panel

Figure 16–19 Removing access and safety panels. (Courtesy of WCI Major Appliance Group.)

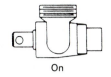

Off

On

Figure 16–20 Operation of gas shutoff valve. (Courtesy of WCI Major Appliance Group.)

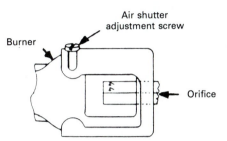

Air shutter adjustment screw

Burner

Orifice

Figure 16–21 Main burner parts location. (Courtesy of WCI Major Appliance Group.)

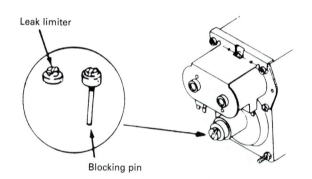

Leak limiter

Blocking pin

Figure 16–22 Installation of leak limiter. (Courtesy of WCI Major Appliance Group.)

Parts and instruction Check the kit to assure that it contains the following parts:

1. Orifice (stamped No. 44).
2. Regulator leak limiter.
3. Red warning tag.
4. Installation instructions.
5. Gas rating plate.

To remove the main burner assembly, use the following procedure:

1. *Warning:* Disconnect the laundry center from the electrical power supply.
2. Remove the dryer front access panel and the dryer safety cover. See Figure 16–19.
3. *Warning:* Turn the gas shutoff valve to the OFF position (90° to the gas supply pipe) and loosen the shutoff valve union. See Figure 16–20.
4. Disconnect the wires that connect to the flame radiant sensor and to the burner assembly.
5. Remove the four (4) screws that fasten the burner assembly to the access shield of the dryer.
6. Remove the burner and valve assembly from the unit.

To remove the parts supplied in the kit, use the following procedure:

1. Remove the two (2) screws that fasten the burner to the main burner mounting bracket and remove the burner.
2. Remove the LP orifice and install the new orifice (stamped No. 44). See Figure 16–21.
3. Reassemble the main burner to the main burner mounting bracket with the two (2) screws removed in the previous step.
4. Remove the LP blocking pin from the valve and install the leak limiter provided in the new kit. See Figure 16–22.
5. Remove the red warning tag from the shutoff valve and affix the self-adhesive rating plate (20,000 Btu) directly on top of the existing LP appliance serial plate located on the laundry center.

6. Reinstall the burner and valve assembly into the laundry center dryer. Tighten the union fitting and turn on the gas at the shutoff valve. Test the gas line connection by brushing on a soapy water solution.

Warning: Do not leak-test with an open flame—use the soap test method to leak-test. Fire, explosion, serious bodily injury, and/or property damage can result from leak-testing a gas line for leaks with an open flame.

7. Reconnect the wires to the flame radiant sensor and burner assembly. Refer to the appliance wiring diagram, located on the dryer access panel, for the proper wiring connections.

Burner adjustment and operation Use the following procedure when making adjustments and checking the operation of the main burner:

1. Reconnect the laundry center to the electrical supply.

Installation Instructions (Gas Conversion Kits) 217

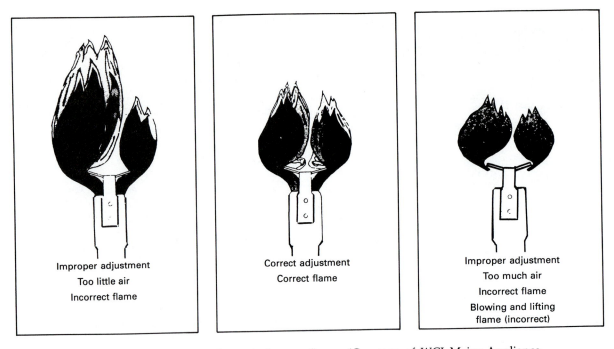

Improper adjustment
Too little air
Incorrect flame

Correct adjustment
Correct flame

Improper adjustment
Too much air
Incorrect flame
Blowing and lifting
flame (incorrect)

Figure 16–23 Adjusting the main burner flame. (Courtesy of WCI Major Appliance Group.)

2. Operate the dryer in the heat cycle and adjust the main burner air shutter (See Figure 16–21, if needed, to obtain the correct flame. See Figure 16–23.

3. Reinstall the dryer safety cover and the dryer front access panel.

4. Make sure that the appliance wiring diagram is returned to the envelope on the dryer access panel.

Gas Conversion Kit Natural to LP Gas Laundry Center Dryer Input 20,000 Btu

Warning: Failure to comply with the following proper instructions and the use of the correct conversion kit can result in fire, explosion, serious bodily injury, and/or property damage.

The conversion is at the risk of and the responsibility of the person making the conversion and should be made by a *qualified gas technician only.*

Use only *the current conversion kit* specified in the parts catalog for the model of laundry center dryer being converted.

Parts and instruction Check the kit to assure that it contains the following parts:

1. Orifice (stamped No. 55).
2. Regulator (LP) blocking pin.
3. Red warning tag.
4. Installation instructions.
5. Gas rating plate.

To remove the main burner assembly, use the following procedure:

1. *Warning:* Disconnect the laundry center from the electrical power supply.

2. Remove the dryer front access panel and the dryer safety cover. See Figure 16–24.

3. *Warning:* Turn the gas shutoff valve to the OFF position (90° to the gas supply pipe) and loosen the shutoff valve union. See Figure 16–25.

4. Disconnect the wires that are connected to the flame radiant sensor and to the burner assembly.

5. Remove the four (4) screws that fasten the burner assembly to the access shield of the dryer.

6. Remove the burner and the valve assembly from the unit.

Safety cover

Access panel

Figure 16–24 Access and safety panels removal. (Courtesy of WCI Major Appliance Group.)

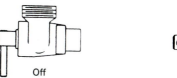

Off   On

Figure 16–25 Operation of gas shutoff valve. (Courtesy of WCI Major Appliance Group.)

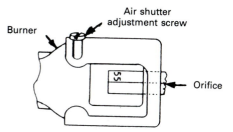
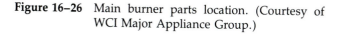

Burner

Air shutter adjustment screw

Orifice

Figure 16–26 Main burner parts location. (Courtesy of WCI Major Appliance Group.)

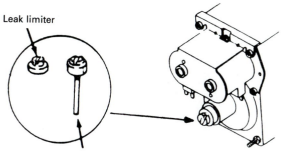

Leak limiter

Blocking pin

Figure 16–27 Installation of leak limiter. (Courtesy of WCI Major Appliance Group.)

To install the parts kit, use the following procedure:

1. Remove the two (2) screws that fasten the burner to the main burner mounting bracket and remove the burner.

2. Remove the original orifice and install the new orifice (stamped No. 55). See Figure 16–26.

3. Reassemble the main burner to the main burner mounting bracket with the two (2) screws removed.

4. Remove the leak limiter from the valve and install the LP blocking pin provided in the kit. See Figure 16–27.

5. Affix the red warning tag to the shutoff valve and affix the self-adhesive rating plate (20,000 Btu) to the side of the appliance serial plate which is located on the laundry center.

6. Reinstall the burner and the valve assembly into the laundry center dryer. Tighten the union fitting and turn on the gas at the shutoff valve. Leak-test the gas line connection by brushing on a soapy water solution.

Warning: Do not leak-test with an open flame—use the soap test method to leak-test. Fire, explosion, serious bodily injury, and/or property damage can result from testing a gas line for leaks with an open flame.

7. Reconnect the wires to the flame radiant sensor and burner assembly. Refer to the appliance wiring diagram, located on the dryer access panel, for proper wiring connections.

Burner adjustment and operation   Use the following steps to adjust the burner flame:

1. Reconnect the laundry center to the electrical supply.

2. Operate the dryer in the heat cycle and adjust the main burner air shutter (refer to Figure 16–26), if needed, to obtain the correct flame. See Figure 16–28.

Improper adjustment
Too little air
Incorrect flame

Improper adjustment
Too much air
Incorrect flame
Blowing and lifting flame (incorrect)

Correct adjustment
Correct flame

Figure 16–28 Adjusting the main burner flame. (Courtesy of WCI Major Appliance Group.)

Installation Instructions (Gas Conversion Kits)   219

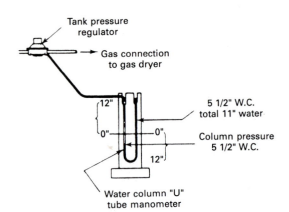

Figure 16–29 Checking gas pressures. (Courtesy of WCI Major Appliance Group.)

3. Reinstall the dryer safety cover and the dryer front access panel.

4. Make sure that the appliance wiring diagram is returned to the envelope on the dryer access panel.

The following is helpful information to use when installing the tank gas and checking the tank's regulator water column pressures.

Observe all local codes and the American National Standard, National Fuel Gas Code ANSI Z223.1-1988. Contact the tank gas dealer for any questions and installations.

Refer to the product installation instructions for gas supply line connections to the laundry center dryer.

Gas tanks must be equipped with a gas pressure regulator before connecting them to a gas-fired appliance. See Figure 16–29. The tank regulator outlet pressure should be a total of 11 inches of water column maximum (or) 8 inches of water column minimum.

A water column U-tube manometer should be used to check the tank regulator pressure. The gas burner should be ignited in the dryer and the dryer in operation when checking the tank regulator pressure.

WASHER COMPONENTS AND PARTS

The following is a description of the washer components and parts of the laundry center.

Washer Cabinet Unit and Base

The base pan assembly, which serves as the support for the entire washer/dryer, is constructed of heavy-gauge metal, embossed for strength.

The base assembly is supported by leveling with rubber inserts, one on each front corner. The rear is supported by a self-leveling assembly welded to the base pan. The raised portion at the center of the base provides a support surface for the snubber ring and stabilizes the tub and transmission during the spin cycle.

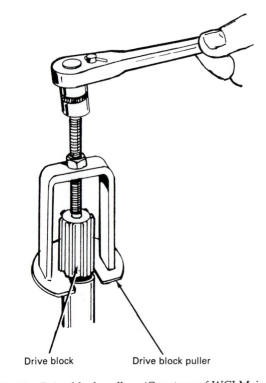

Drive block   Drive block puller

Figure 16–30 Drive block pulley. (Courtesy of WCI Major Appliance Group.)

The complete washer unit must be removed to service the washer base.

Agitator and Cap

The polypropylene agitator and cap are centered in the inner wash basket. The cap is threaded into the agitator drive block. It can be removed by turning counterclockwise. The agitator can be lifted off the drive block after the cap is removed. The drive block is splined and the agitator fits tight onto the block. Upward pressure will be needed to free the agitator from the drive block.

To remove or replace the agitator and cap, use the following procedure:

1. Disconnect the laundry center from the electrical supply.
2. Raise the washer lid.
3. Remove the agitator cap by turning it counterclockwise.
4. Lift out the agitator.
5. Install the new agitator and/or cap.
6. Reverse the above procedure to reassemble the unit.

Drive Block

The drive block is fastened to the agitator drive shaft with a stud bolt. It drives the agitator and provides a means of adjusting the clearance between the lower edge of the

Tub splash cover

Tub splash cover mounting tabs (12)

Figure 16–31 Tub splash cover. (Courtesy of WCI Major Appliance Group.)

agitator and the tub. The drive block is adjusted with shims ($\frac{3}{16}$-inch minimum, $\frac{5}{16}$-inch maximum). A drive block puller is required to remove the drive block. See Figure 16–30.

Tub Splash Cover

The polypropylene tub splash cover, which prevents water from splashing over the top of the tub, has a molded rear lip to prevent water coming from the fill hose and the nozzle from splashing out of the tub. The splash cover includes a bleach cup drain area. A sealing gasket in a groove around the circumference of the splash cover snaps over the outer tub assembly. See Figure 16–31.

To remove or replace the tub splash cover, use the following procedure:

1. Disconnect the laundry center from the electrical supply.
2. Remove the dryer front service panel.
3. Remove the washer front service panel.
4. Remove the washer top panel and lid assembly.
5. Release the water level pressure control tube from the retainer of the tub splash cover.
6. Use a flat-blade screwdriver to pry outward the 12 locking tabs securing the splash cover to the outer tub assembly.
7. Remove the tub splash cover and gasket seal assembly.
8. Install the new splash cover, making sure that the sealing gasket is in place before snapping the cover into place. *Note:* The tub splash cover can be installed incorrectly. Position the bleach cup drain area directly over the washer drive motor assembly and make sure that *all* mounting tabs are snapped securely. Failure to do this may result in a leak

during the spin cycle. Securing the three rear mounting tabs first will make the installation easier.
9. Reverse the above procedure to reassemble the unit.

Tub Splash Cover Gasket

A sealing gasket (plastic rubber), in a groove around the circumference of the splash cover, snaps over the outer tub assembly for a water-tight seal. See Figure 16–31.

To remove or replace the tub splash cover gasket, use the following procedure:

1. Disconnect the laundry center from the electrical supply.
2. Remove the dryer front service panel.
3. Remove the washer front service panel.
4. Remove the washer top panel and lid assembly.
5. Release the water level pressure control tube from the retainer of the tub splash cover.
6. Use a flat-blade screwdriver to pry outward the 12 locking tabs securing the splash cover to the outer tub assembly.
7. Remove the tub splash cover and gasket seal assembly.
8. Install a new tub splash cover, making sure that the sealing gasket is in place before snapping the cover into place. *Note:* The tub splash cover can be installed incorrectly. Position the bleach cup area directly over the washer drive motor assembly and make sure that *all* mounting tabs are snapped securely or a leak will occur during the spin cycle. Securing the three rear mounting tabs first will make the installation much easier.
9. Reverse the above procedure to reassemble the unit.

Inner Wash Basket

The inner wash basket is made of polypropylene, perforated at the bottom and sides to separate lint and soil from the clothes during the agitation cycle and to extract the water during the spin cycle. The inner wash basket has a built-in upper weight ring (liquid filled). The inner wash basket is supported by the trunnion and is held in place by five (5) bolts. The trunnion is fastened to the spin shaft with a lockplate.

The lockplate is located in a milled area on the spin shaft and is fastened to the trunnion with two hex-head bolts. See Figure 16–32.

To remove or replace the inner wash basket, use the following procedure:

1. Disconnect the laundry center from the electrical supply.
2. Remove the dryer front service panel.
3. Remove the washer front service panel.
4. Remove the washer top panel and lid assembly.
5. Release the water level pressure control tube from the splash cover.

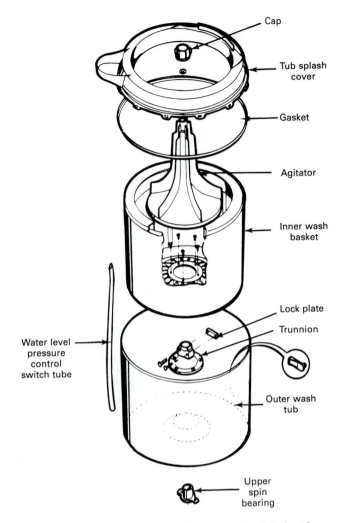

Figure 16–32 Inner wash basket mounting details. (Courtesy of WCI Major Appliance Group.)

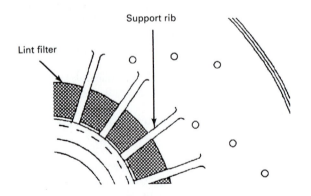

View looking down into tub

Figure 16–33 Inner wash basket lint filter cross section. (Courtesy of WCI Major Appliance Group.)

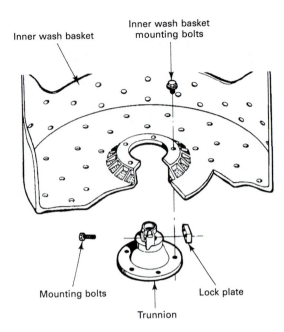

Figure 16–34 Inner wash basket and trunnion. (Courtesy of WCI Major Appliance Group.)

6. Remove the tub splash cover.

7. Remove the cap and agitator.

8. Remove the five (5) $\frac{3}{8}$-inch hex bolts securing the inner wash basket trunnion.

9. Slide the inner wash basket up and off the transmission shaft.

10. Install the new inner wash basket.

11. Reverse the above procedure to reassemble the unit.

Inner Wash Basket Lint Filter

The inner wash basket has a self-cleaning lint filter incorporated into the base of the inner wash basket. See Figure 16–33. It is not necessary to clean the filter, the filter is rinsed clean during the spin cycle. (See the Washer Operation section presented earlier in this chapter for more details.)

Trunnion

The trunnion supports the inner wash basket and is mounted to the spin shaft by a locking plate. See Figure 16–34.

To remove or replace the trunnion, use the following procedure. See Figure 16–35.

1. Disconnect the laundry center from the electrical supply.

2. Remove the dryer front service panel.

3. Remove the washer front service panel.

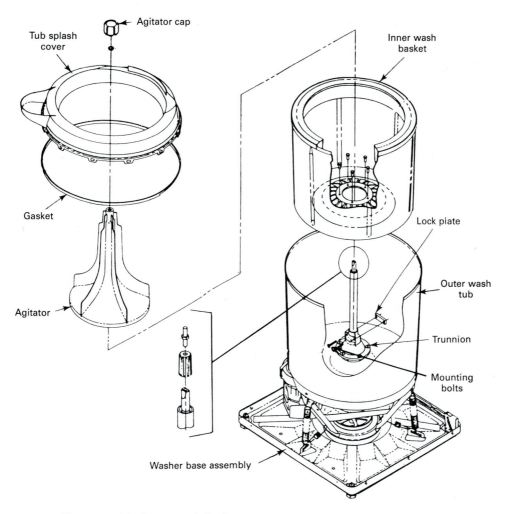

Figure 16–35 Inner wash basket mounting details. (Courtesy of WCI Major Appliance Group.)

4. Remove the washer top panel and lid assembly.

5. Release the water level pressure control tube from the splash cover.

6. Remove the tub splash cover.

7. Remove the cap and the agitator.

8. Remove the five (5) ⅜-inch hex bolts securing the inner wash basket to the trunnion.

9. Slide the inner wash basket up and off the transmission shaft.

10. Loosen the two (2) ½-inch hex bolts securing the trunnion lockplate to the milled flat side of the spin shaft.

11. Remove the trunnion.

12. Install the new trunnion and apply Loctite.

13. Reverse the above procedure to reassemble the unit.

Water Seal Assembly

The water seal assembly sits underneath the trunnion and inside the upper bearing housing to the transmission. It seals the water inside the outer tub. The assembly consists of a rubber and bronze seal and a carbon mating seal surface which is mounted on a spring-loaded rubber bellows. The bellows seal is pressed into the upper bearing housing that is mounted in the outer tub. A slinger washer completes the seal assembly. See Figure 16–36.

To remove or replace the water seal assembly, use the following procedure:

1. Disconnect the laundry center from the electrical supply.

2. Remove the dryer front service panel.

3. Remove the washer front service panel.

4. Remove the washer top panel and lid assembly.

5. Release the water level pressure control tube from the splash cover.

6. Remove the tub splash cover.

7. Remove the cap and the agitator.

8. Remove the five (5) ⅜-inch hex bolts securing the inner wash basket to the trunnion.

9. Slide the inner wash basket up and off the transmission shaft.

10. Loosen the two (2) ½-inch hex bolts securing the lockplate to the milled flat of the spin shaft.

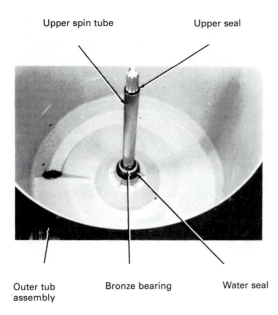

Upper spin tube   Upper seal

Outer tub assembly   Bronze bearing   Water seal

Figure 16–36   Water seal assembly. (Courtesy of WCI Major Appliance Group.)

Upper spin bearing   Tub to pump house

Figure 16–37   Outer wash tub (upper spin bearing and drain hose). (Courtesy of WCI Major Appliance Group.)

11. Remove the trunnion.

12. Remove the bronze washer from the spin shaft.

13. Remove the water seal slinger from the bearing housing and slide it up and off the spin shaft.

14. Install the new water seal assembly.

Note: Do not use any lubricant on this seal when installing.

15. Reverse the above procedure to reassemble the unit.

Outer Tub Assembly

The outer tub and air dome is made of polypropylene and is mounted to the leg-and-dome assembly. The drain hose (tub to pump) and the upper spin bearing are installed onto the outer tub. The air dome on the tub side provides the connection for the pressure switch hose. See Figure 16–37.

To remove or replace the outer tub assembly, use the following procedure:

1. Disconnect the laundry center from the electrical supply.

2. Remove the dryer front service panel.

3. Remove the washer front service panel.

4. Remove the washer top panel and lid assembly.

5. Release the water level pressure control tube from the splash cover.

6. Remove the washer level pressure control tube from the water level pressure control switch.

7. Remove the tub splash cover.

8. Remove the cap and the agitator.

9. Remove the five (5) ⅜-inch hex bolts securing the inner wash basket to the trunnion.

10. Slide the inner wash basket up and off the transmission shaft.

11. Loosen the two (2) ½-inch hex bolts securing the trunnion lockplate to the milled flat of the spin shaft.

12. Remove the trunnion.

13. Remove the washer drive belt.

14. Remove the outer tub-to-pump clamp and hose.

15. Loosen the Allen screw and remove the transmission drive pulley.

16. Remove the two (2) ⅜-inch hex bolts securing the transmission to the leg-and-dome assembly.

17. Remove the three (3) ½-inch hex bolts securing the outer wash tub to the leg-and-dome assembly.

18. Lift the washer transmission and outer wash tub assembly from the leg-and-dome assembly.

19. Remove the bronze washer from the spin shaft.

20. Remove the water seal slinger from the bearing housing and slide it up and off the shaft.

21. Lift the outer wash tub up and off the transmission shaft.

22. Remove the two (2) screws securing the upper spin bearing to the bottom of the outer wash tub.

23. Remove the outer tub-to-pump hose (seal with adhesive upon reassembly).

Note: Adhesive Primer and Red Adhesive are available from the parts supplier.

24. Install the new outer wash tub assembly.

25. Reverse the above procedure to reassemble the unit.

Leg-and-Dome Assembly

The leg-and-dome assembly is constructed of galvanized steel. The entire washer assembly is mounted to and supported by the leg-and-dome assembly. The leg-and-dome assembly pivots on a snubber ring that is supported and stabilized by three (3) sets of horizontal and vertical springs.

To remove or replace the leg-and-dome assembly, use the following procedure:

1. Disconnect the laundry center from the electrical supply.
2. Remove the washer front service panel.
3. Remove the washer top panel and lid assembly.
4. Release the water level pressure control tube from the splash cover.
5. Remove the water level pressure control tube from the water level pressure control switch.
6. Remove the tub splash cover.
7. Remove the cap and the agitator.
8. Remove the five (5) ⅜-inch hex bolts securing the inner wash basket to the trunnion.
9. Slide the inner wash basket up and off the transmission shaft.
10. Loosen the two (2) ½-inch hex bolts securing the trunnion lockplate to the milled flat on the spin shaft.
11. Remove the trunnion.
12. Remove the washer drive belt.
13. Remove the outer tub-to-pump clamp and hose.
14. Loosen the Allen set screw and remove the transmission drive pulley.
15. Remove the two (2) ⅜-inch hex bolts securing the transmission to the leg-and-dome assembly.
16. Remove the three (3) ½-inch hex bolts securing the outer wash tub to the leg-and-dome assembly.
17. Lift the washer transmission and the outer wash tub assembly up and off the leg-and-dome assembly.
18. Release the three (3) sets of horizontal and vertical springs securing the leg-and-dome assembly to the cabinet base pan.
19. Remove the four (4) hex nuts securing the washer drive motor.
20. Remove the washer drive motor belt idler arm.
21. Install the new leg-and-dome assembly.
22. Reverse the above procedure to reassemble the unit.

Stabilizing Springs

Each vertical stabilizing spring hooks into an adjustable tab formed in the washer base pan and to the upper end of the leg-and-dome assembly. These three (3) springs keep the washer tub centered during operation and hold the washer mechanism against the snubber ring. The red-stained spring must be installed opposite the washer drive motor and pump assembly.

Each horizontal stabilizer spring hooks into the leg-and-dome assembly and a raised tab on the washer base pan. These three (3) springs aid in keeping the mechanism stable and centered on the crown of the washer base pan.

To adjust the vertical stabilizer springs: When the washer is in the spin cycle, the tub splash cover (subtop)

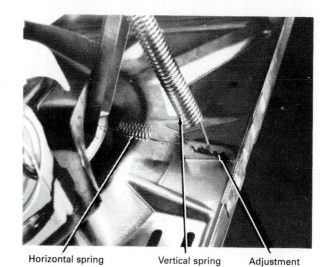

Horizontal spring   Vertical spring   Adjustment

Figure 16–38 Stabilizer springs. (Courtesy of WCI Major Appliance Group.)

should be centered in the cabinet. If it is off center, adjust each vertical spring as required. Extremely unbalanced loads may cause the washer splash cover (subtop) to strike the cabinet. Pressure of the leg-and-dome assembly against the snubber can be increased or decreased by equally adjusting the spring tension of the vertical springs. Use a pair of vise grip pliers to adjust the springs. See Figure 16–38.

Snubber Ring

The snubber ring assembly is made of a polypropylene ring with Teflon strips secured to the upper and the lower edges. The top edge of the ring is identified by an arrow molded into the outer surface. The snubber ring fits between the raised center portion of the washer base pan and the dome base of the tub and transmission assembly. See Figure 16–6.

To remove or replace the snubber ring, use the following procedure:

1. Disconnect the laundry center from the electrical supply.
2. Remove the washer front service panel.
3. Remove the washer top panel and lid assembly.
4. Disconnect the front vertical and horizontal springs from the base pan.
5. Place a 2-by-4-inch block under the drive motor section of the leg-and-dome assembly, carefully prying the washer tub and transmission assembly off the base.
6. Using a wire with a hook on the end, reach under and pull out the snubber ring.
7. Install the new snubber ring with the arrow pointing up.
8. Reverse the above procedure to reassemble the unit.

Note: The washer and base assembly can also be removed, should the service technician elect to do so.

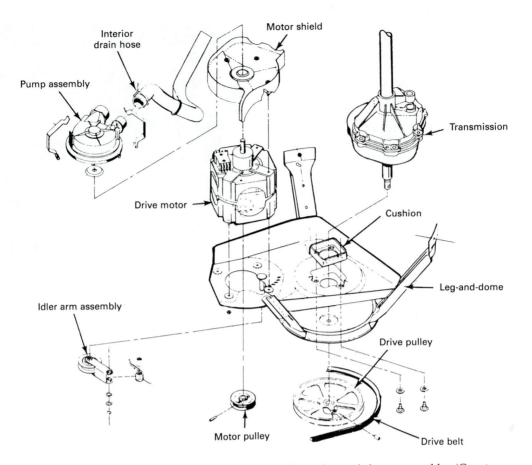

Figure 16–39 Washer motor, pump, drive pulleys, leg-and-dome assembly. (Courtesy of WCI Major Appliance Group.)

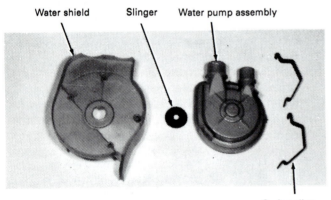

Figure 16–40 Water pump and water shield. (Courtesy of WCI Major Appliance Group.)

Water Pump Assembly

The water pump (a one-piece, sealed assembly) is mounted above a water shield on top of the washer drive motor by two spring clips. See Figures 16–39 and 16–40.

Note: Should the washer outer wash tub be full or partially full of water and the washer drive motor is inopera-

tive, remove the inlet hose clamp. Insert a small bladed screwdriver alongside the drain inlet hose, allowing a small amount of water into the water shield. Place a pan under the overflow spout to catch the water.

To remove or replace the water pump assembly, use the following procedure:

1. Disconnect the laundry center from the electrical supply.

2. Remove the washer front service panel.

3. Remove the drain hose, inlet and outlet hoses, and the clamps from the water pump assembly.

Note: To gain working space for removing the water pump, remove the front outer wash tub to leg-and-dome front mounting bolt, lift up on the outer wash tub assembly, and reinstall the bolt into the leg-and-dome assembly bolt hole.

4. Remove the two spring mounting clips securing the water pump assembly to the washer drive motor.

Note: Upon reassembly, be sure that the electrical leads do not become pinched under the spring clips.

5. Lift the pump up and off the washer drive motor shaft.

6. Install the new water pump assembly.

7. Reverse the above procedure to reassemble the unit.

Drive Motor Water Shield

The water pump (a one-piece, sealed assembly) is mounted above the polypropylene water shield on top of the washer drive motor by two spring clips. The drive motor water shield has molded legs that fit into holes in the drive motor upper end plate.

To remove or replace the drive motor water shield, use the following procedure:

1. Disconnect the laundry center from the electrical supply.
2. Remove the washer front service panel.
3. Remove the drain hose, inlet and outlet hoses, and the clamps from the water pump assembly.

Note: To gain working space for removing the water pump, remove the front outer wash tub to leg-and-dome front mounting bolt, lift up on the outer wash tub assembly, and reinstall the bolt into the leg-and-dome assembly bolt hole.

4. Remove the two spring mounting clips securing the water pump assembly to the washer drive motor.

Note: Upon reassembly, be sure that the electrical leads do not become pinched under the spring clips.

5. Remove the washer drive motor shaft slinger.
6. Lift the pump up and off the washer drive motor shaft.
7. Remove the drive motor water shield.
8. Install the new drive motor water shield.
9. Reverse the above procedure to reassemble the unit.

Washer Drive Motor

The washer drive motor is a two-speed, $\frac{1}{2}$-hp, capacitor start, reversible motor rated at 120 volts, 60 hertz. The electrical connections to the motor are quick connect-type terminal blocks. The washer drive motor is mounted to the leg-and-dome assembly and drives the transmission by a pulley mounted on its shaft, a drive belt, and a belt idler arm assembly.

To remove or replace the washer drive motor, use the following procedure:

1. Disconnect the laundry center from the electrical supply.
2. Remove the washer front service panel.
3. Remove the drain hose, inlet and outlet hoses, and clamps from the water pump assembly.

Note: To gain working space for removing the water pump, remove the front outer wash tub to leg-and-dome assembly and reinstall the bolt into the leg-and-dome assembly bolt hole.

4. Remove the two spring mounting clips securing the water pump assembly to the washer drive motor.

Note: Upon reassembly, be sure that the electrical leads do not become pinched under the spring clips.

5. Lift the pump up and off the washer drive motor shaft.

6. Remove the drive motor belt (roll off).
7. Disconnect the washer drive motor electrical terminal connector blocks.
8. Remove the four (4) hex nuts securing the washer drive motor.
9. Remove the washer drive motor.
10. Loosen the set screw and remove the drive pulley from the motor shaft.
11. Install the new washer drive motor.
12. Reverse the above procedure to reassemble the unit.

Washer Drive Motor Belt

The washer drive motor belt is used to transmit power from the washer drive motor to the wash and spin basket. An idler arm pulley maintains the proper belt tension.

To remove or replace the washer drive motor belt, use the following procedure:

1. Disconnect the laundry center from the electrical supply.
2. Remove the washer front service panel.
3. Remove the drive motor belt (roll off).
4. Install the new washer drive motor belt.
5. Reverse the above procedure to reassemble the unit.

Washer Drive Motor Pulley

The washer drive motor pulley is secured to the washer drive motor shaft with a roll pin.

To remove or replace the washer drive pulley, use the following procedure:

1. Disconnect the laundry center from the electrical supply.
2. Remove the washer front service panel.
3. Remove the drive motor belt (roll off).
4. Using an $\frac{1}{8}$-inch punch and a hammer, drive out the roll pin securing the pulley to the washer drive motor shaft.
5. Remove the washer drive motor pulley.
6. Install the new washer drive motor pulley.
7. Reverse the above procedure to reassemble the unit.

Washer Drive Motor Belt Idler Arm

The washer drive motor belt is used to transmit power from the washer drive motor to the washer and spin basket. The idler arm pulley maintains the proper drive belt tension.

The drive system provides positive drive in the agitation (clockwise rotation) and variable-speed drive in the spin (counterclockwise rotation). This dual-drive function is the result of the direction of drive. In agitation, the direction of pull is from pulley to pulley. This causes the belt to ride tight in the pulley and provide a positive drive. In spin, the direction of pull is across the spring-loaded idler pulley. The

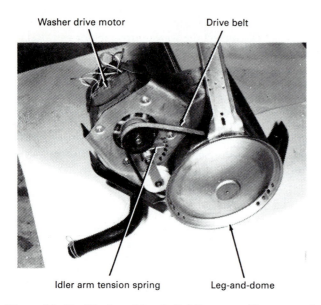

Washer drive motor   Drive belt

Idler arm tension spring   Leg-and-dome

Figure 16–41 Washer drive belt idler arm. (Courtesy of WCI Major Appliance Group.)

idler arm "senses" the load and controls the belt tension to provide a gradual increase in the speed as the water is extracted. With an extremely unbalanced load, the tub will spin at a reduced speed throughout the spin cycle. This permits the washer to complete the full cycle.

Proper operation of the drive system is dependent on correct adjustment of the idler arm tension spring. See Figure 16–41.

To remove or replace the washer drive motor belt idler arm, use the following procedure:

1. Disconnect the laundry center from the electrical power supply.
2. Remove the washer front service panel.
3. Remove the drive motor belt (roll off).
4. Remove the C clip retainer securing the idler arm assembly to the leg and dome.
5. Install the new washer drive motor belt idler arm.
6. Place the idler arm tension spring into the third hole from the motor.

Note: The idler arm adjustment is determined by the wattage draw of the motor. (See the procedure To Adjust the Drive Belt Idler Arm Tension.)

7. Reverse the above procedure to reassemble the unit.

To adjust the drive belt idler arm tension, use the following procedure:

1. Check the pump out wattage. The wattage must be within the specifications recommended with a full water load.
2. Tighten the adjustment spring to increase the wattage draw. Loosen the spring tension to decrease the wattage draw.

Note: If a wattage meter is not available, use the following procedure:

  a. Fill the inner wash basket with water to the high level (maximum) and advance the timer to the spin cycle. *Warning:* Do not restrict the lid lock arm when manually depressing or damage to the lid switch bimetal element will occur and affect the lock and the unlock times.

  b. Start the spin cycle. If the water begins pumping out before the spin basket starts to spin, the belt tension is too loose. Tighten the spring two (2) holes for a new belt and one (1) hole for an existing belt. Recheck by starting the spin cycle again.

  c. If, during the spin cycle, water moves up the outer tub and out between the tub cover and the inner wash basket, the belt tension is too tight. Adjust the spring back one (1) hole and recheck by starting the spin cycle again.

Proper adjustment of the idler arm tension is necessary for trouble-free operation of the drive system.

Water Inlet Hose and Nozzle

The water inlet hose and polypropylene nozzle provide an antisiphon device for the incoming water supply. Water flows from the water inlet valve through the inlet hose, inlet nozzle, and then into the inner wash basket. The water inlet nozzle is attached to the dryer bottom pan.

To remove or replace the water inlet and nozzle, use the following procedure:

1. Disconnect the laundry center from the electrical power supply.
2. Remove the dryer front service panel.
3. Remove the washer front service panel.
4. Remove the washer top panel and lid assembly.
5. Loosen the screw securing the water inlet valve shield.
6. Remove the water inlet valve shield by pulling it up.
7. Remove the water inlet fill valve clamp and hose.
8. Remove the screw securing the water inlet nozzle to the dryer base pan.
9. Install the new water inlet hose and nozzle.
10. Reverse the above procedure to reassemble the unit.

Drain Hose Check Valve

A check valve is installed in the drain system to reduce sudsing by preventing air intake by the pump during agitation. See Figure 16–42.

The check valve is installed in the plastic drain hose coupling located at the rear of the washer. It is positioned inside the coupling. The semicircular lip of the valve should be positioned as illustrated in Figure 16–42.

To remove or replace the drain hose check valve, use the following procedure:

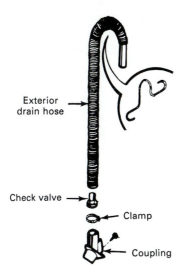

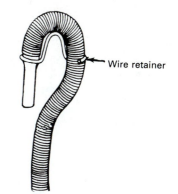

Figure 16–43 Exterior drain hose. (Courtesy of WCI Major Appliance Group.)

Figure 16–42 Drain hose check valve. (Courtesy of WCI Major Appliance Group.)

1. Disconnect the laundry center from the electrical power supply.

2. Loosen the clamp and remove the exterior drain hose from the cabinet coupling.

3. Remove the drain hose check valve; be sure of the correct position of the check valve.

4. Install the new drain hose check valve; be sure of the correct position of the check valve.

5. Reverse the above procedure to reassemble the unit.

Interior Drain Hose

The interior drain hose connects to the water pump and connects to the cabinet exterior connector.

To remove or replace the interior drain hose, use the following procedure:

1. Disconnect the laundry center from the electrical power supply.

2. Remove the washer front service panel.

3. Loosen the clamp and remove the interior drain hose from the cabinet coupling.

4. Loosen the clamp and remove the interior drain hose from the water pump.

5. Install the new interior drain hose.

6. Reverse the above procedure to reassemble the unit.

Drain Hose Connector

The interior drain hose attaches to the water pump and connects to the exterior cabinet connector.

To remove or replace the cabinet exterior drain hose connector, use the following procedure:

1. Disconnect the laundry center from the electrical power supply.

2. Remove the washer front service panel.

3. Loosen the clamp and remove the interior drain hose from the cabinet coupling.

4. Remove the exterior drain hose and clamp.

5. Remove the one (1) screw securing the drain hose connector to the cabinet.

6. Install the new exterior drain hose connector.

7. Reverse the above procedure to reassemble the unit.

Exterior Drain Hose

The exterior drain hose attaches to the cabinet exterior connector and completes the drain to the household drain system. See Figure 16–42.

To remove or replace the cabinet exterior drain hose, use the following procedure:

1. Disconnect the laundry center from the electrical power supply.

2. Loosen the clamp securing the exterior drain hose and bushing to the exterior cabinet drain connector.

3. Remove the exterior drain hose, wire retainer, and the clamp.

4. Install the new exterior drain hose and tighten the clamp.

5. Form a U shape in the end of the drain hose and install the drain hose wire retainer. See Figure 16–43.

6. Reverse the above procedure to reassemble the unit.

WASHER TRANSMISSION MECHANICAL OPERATION

Warning: The entire transmission body turns during the spin cycle. Keep hands and tools out of the lower cabinet area when the washer is operating.

Transmission Drive System

Before discussing the transmission tear down, it is best to know the operation of the drive system.

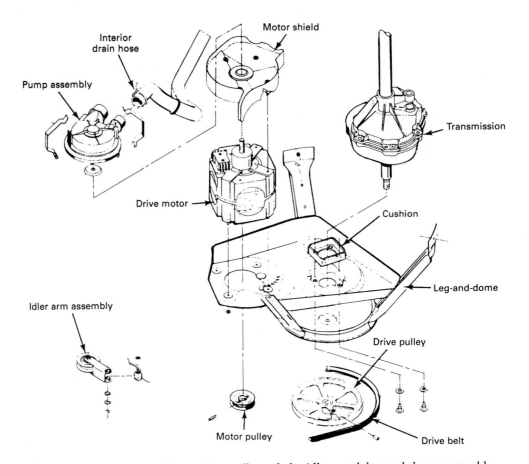

Figure 16–44 Washer drive motor pulleys, belt, idler, and leg-and-dome assembly. (Courtesy of WCI Major Appliance Group.)

The transmission drive system consists of the following major components. See Figure 16–44.

1. A ½-hp, 60-hertz, 120-volt, single- or two-speed, capacitor start, reversible motor.
2. A motor pulley, which is attached to the lower motor shaft and water pump assembly and to the upper motor shaft.
3. A ventilated transmission drive pulley, which is attached to the transmission shaft with an Allen set screw.
4. A spring-loaded belt tension idler arm, with a V groove pulley.
5. A V belt.

Note: All references to the direction of rotation are made as viewed from below.

The drive system provides a positive drive in the agitation (clockwise rotation) and a variable-speed drive in the spin (counterclockwise rotation). This dual-drive function is a result of the direction of drive. In agitation, the direction of pull is from pulley to pulley. This causes the belt to ride tight in the pulley and provide a positive drive. In spin, the direction of pull is across the spring-loaded idler pulley. The idler arm "senses" the load and controls the belt tension to provide a gradual increase in the speed as the water is

extracted. With an extremely unbalanced load, the tub will spin at a reduced speed throughout the spin cycle. This permits the washer to complete the full cycle.

Proper operation of the drive system is dependent on the correct adjustment of the idler arm tension spring. See Figure 16–45. Also, refer to the Washer Drive Motor Belt Idler Arm section presented earlier in this chapter.

To adjust the drive belt idler arm tension, use the following procedure:

1. Check the pump out wattage. The wattage must be within the specifications recommended with a full water load.
2. Tighten the adjustment spring to increase the wattage draw. Loosen the spring tension to decrease the wattage draw.

To remove or replace the drive belt idler arm, use the following procedure:

1. Disconnect the laundry center from the electrical power supply.
2. Remove the washer front service panel.
3. Remove the drive motor belt (roll off).
4. Remove the C clip retainer securing the idler arm assembly to the leg and dome.

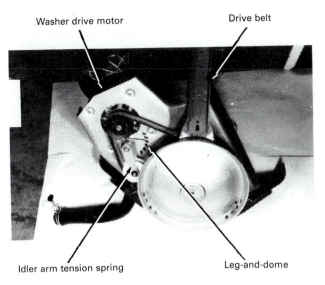

Figure 16–45 Idler arm adjustment. (Courtesy of WCI Major Appliance Group.)

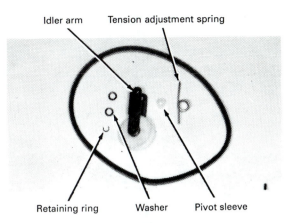

Figure 16–46 Idler arm details. (Courtesy of WCI Major Appliance Group.)

Note: Another method of removing and replacing the idler arm is to remove the pump and motor and then reach through the motor opening.

5. Install the new washer drive motor belt idler arm.

6. Place the idler arm tension spring into the third hole from the motor. (This is the factory-set position.)

Note: Idler arm adjustment, if needed, is determined by the wattage draw.

7. Reverse the above procedure to reassemble the unit.

Note: The idler arm has a pivot sleeve and bushings. Lightly lubricate the pivot sleeve to prevent vibration. The pivot sleeve and bushings provide the bearing surface for the idler arm. See Figure 16–46.

Transmission Operation

Warning: The entire transmission body turns during the spin cycle. Keep hands and tools out of the lower cabinet area when the washer is operating.

The transmission consists of an upper and lower housing which encases the gears, connecting rod, and a 32-ounce charge of special oil. The transmission is supported by a lower bearing housing assembly.

A 210° arc of oscillation is created by the gear train during the wash agitation. This mode of operation is dependent on the direction of motor rotation and the function of two (2) LGS clutch springs.

During the agitation period, the smaller LGS clutch spring inside the transmission opens slightly and slips on the counterclockwise turning input shaft. Simultaneously, the large LGS clutch spring, which is located outside the transmission housing, tightens over the base of the lower transmission housing hub to prevent turning of the transmission and the inner wash basket. See Figure 16–47.

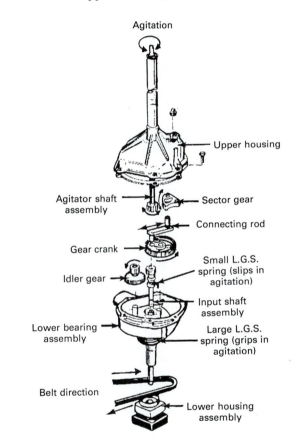

Figure 16–47 Transmission details (belt in clockwise direction). (Courtesy of WCI Major Appliance Group.)

The counterclockwise rotation of the input shaft pinion is transmitted through the idler gear to the crank gear. A connecting rod (or crank), operated by the crank gear, moves the sector gear back and forth against the agitator shaft drive gear. This action oscillates the agitator back and forth through a 210° arc.

During the spin cycle, the small spring inside the transmission tightens on the input shaft when the motor is

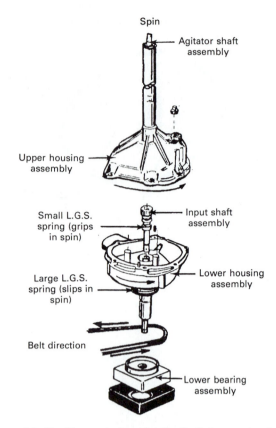

Spin

Agitator shaft assembly

Upper housing assembly

Small L.G.S. spring (grips in spin)

Input shaft assembly

Large L.G.S. spring (slips in spin)

Lower housing assembly

Belt direction

Lower bearing assembly

Figure 16–48 Transmission details (belt in counterclockwise direction). (Courtesy of WCI Major Appliance Group.)

turning clockwise. The large spring on the lower housing opens slightly, and slips, allowing the entire transmission and spin basket to turn. See Figure 16–48. *Note:* Do not condemn the gear train or other components prior to a careful check of all the drive components.

Drive Component Check

Note: All references to the direction of rotation of the drive system are made as viewed from below.

To check the drive components, use the following procedure:

1. Disconnect the laundry center from the electrical supply.
2. Remove the washer front service panel.
3. Remove the drive motor belt (roll off).
4. Turn the drive pulley counterclockwise by hand. The large spring must act as a brake and the gear case *must not* revolve. If the transmission rotates with the pulley, the lower transmission housing, the bearing housing, or the large LGS spring may be worn or damaged. Examine all these parts for wear and replace if necessary.
5. When the drive pulley is turned clockwise, the transmission should revolve without gear train movement. If the transmission rotates too *slowly* and the agitator oscillates, the small LGS spring is slipping or broken and is not locking on the input shaft.

Transmission Repair

Only after careful check of the components, proceed with the required transmission repair.

Transmission Removal

Removal of the transmission is required to replace the input shaft assembly. For a complete tear down of the transmission, it must be completely removed from the washer base.

To remove or replace the transmission assembly, use the following procedure:

1. Disconnect the laundry center from the electrical supply.
2. Remove the washer front service panel.
3. Remove the drive motor belt (roll off).
4. Remove the washer top panel and lid assembly.
5. Release the water level pressure control tube from the splash cover.
6. Remove the water level pressure control tube from the water level pressure control switch.
7. Remove the tub splash cover.
8. Remove the cap and the agitator.
9. Remove the five (5) hex bolts securing the inner wash basket to the trunnion.
10. Slide the inner wash basket up and off the transmission shaft.
11. Loosen the two (2) $\frac{1}{2}$-inch hex bolts securing the trunnion lockplate to the milled flat of the shaft.
12. Remove the trunnion.
13. Remove the washer drive belt.
14. Remove the outer tub-to-pump clamp and hose.
15. Loosen the Allen set screw and remove the transmission drive pulley.
16. Remove the two (2) $\frac{3}{8}$-inch hex bolts securing the transmission to the leg-and-dome assembly.
17. Remove the three (3) $\frac{1}{2}$-inch hex bolts securing the outer wash tub to the leg-and-dome assembly.
18. Lift the inner washer transmission and outer tub assembly from the leg-and-dome assembly.
19. Remove the bronze washer from the spin shaft.
20. Remove the water seal slinger from the bearing housing and slide it up and off the spin shaft.
21. Lift the outer wash tub up and off the transmission shaft.
22. Remove the oil plug from the top of the transmission. This allows the removal of the oil. The oil can be removed with a common roasting baster (at room temperature) and a small piece of plastic hose.
23. Remove the six (6) screws and take off the upper transmission housing, including the spin tub and the agitator shaft.
24. Remove the sector gear connecting rod, crank gear, and idler gear.
25. Reverse the above procedure to reassemble the unit.

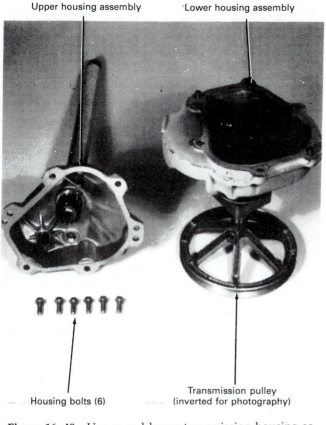

Figure 16–49 Upper and lower transmission housing assemblies. (Courtesy of WCI Major Appliance Group.)

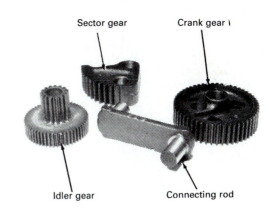

Figure 16–50 Transmission gears and connecting rod. (Courtesy of WCI Major Appliance Group.)

Figure 16–51 Transmission details. (Courtesy of WCI Major Appliance Group.)

Complete Transmission Repair

Whenever the transmission is repaired, carefully examine all components for wear and always replace all gaskets and seals. Use the following procedure:

1. Remove the plug from the top of the housing and drain the oil.

2. Remove the six (6) bolts and separate the upper and lower housings. See Figure 16–49.

3. Remove the sector gear, connecting rod, crank gear, and the idler gear from the lower housing. See Figure 16–50.

Note: The sector gear and crank gear are subject to improper positioning during reassembly. The housing cannot be bolted together if the crank gear is inverted.

4. Remove the snap ring, spacer washer, and thrust washer from the input shaft. See Figure 16–51. Pull the drive shaft up and out of the tube in the lower housing. (A bronze thrust washer is located between the pinion gear and the hub inside the lower assembly; be certain that this bronze thrust washer is in place during reassembly.) Remove the lower lip seal at the end of the spin tube.

5. Remove the upper lip seal from the upper end of the agitation shaft. See Figure 16–52.

6. Remove the snap ring located under the lip seal.

7. Pull the agitator shaft assembly down and out from the spin shaft. Steel spacer washers and a stainless-steel thrust washer are located in the upper cavity of the spin

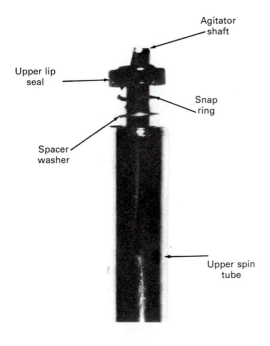

Figure 16–52 Upper spin tube and agitator shaft details. (Courtesy of WCI Major Appliance Group.)

Figure 16–53 Lower transmission. (Courtesy of WCI Major Appliance Group.)

shaft. The spacers are to take up end play of the agitator shaft which should be no more than 0.020 inch.

8. Reassembly is easily done by supporting the lower transmission housing in an upright position. Place the gears and the connecting rod in the lower housing. Holding the sector gear in place in the upper housing, line up the transmission halves by turning the agitator shaft and fit them together. See Figure 16–53.

Specifications for Reassembly

The following are the specifications for reassembling the transmission:

1. All parts are to be free of chips, dirt, and foreign material, including cleaning solvents.

2. *Lip seal installation*   The upper and lower lip seals and their respective shafts should be well lubricated with transmission oil to provide maximum protection for the seals during the reassembly process.

3. The torque required to turn the input shaft counter-clockwise after reassembly is to be a maximum of 6 inch-pounds steady and 16 inch-pounds for tight spots. In addition, the input shaft assembly must withstand a minimum of 150 inch-pounds in a clockwise direction without turning or slipping. (Hold the transmission case stationary for this test.)

4. Tighten the housing bolts to 120 to 180 inch-pounds of torque. A minimum of 80 inch-pounds must be maintained after the gasket has set. See Figure 16–54.

WASHER ELECTRICAL COMPONENTS TEST AND REPAIR

Caution: Always disconnect the unit from the electrical power source before making any electrical tests.

Washer Drive Motor

The washer drive motor is either a single-speed or a two-speed, $\frac{1}{2}$-horsepower, capacitor start, reversible motor, rated at 120 volts, 60 hertz. (See the specific features for the model being serviced.) The connections at the motor switch are quick disconnect-type terminal blocks. The terminal blocks are molded in such a way that they cannot be reversed. See Figure 16–55. An internal overload protector is built into the motor and will open due to excessively high temperatures and/or excessive current. The overload protector is wired electrically in series with the washer circuitry, and if it opens all washer operations will stop. The protector is self-resetting and in some cases takes as long as 30 seconds to reset.

Two-Speed Drive Motor

The high speed starts on the four-pole main winding in conjunction with the four-pole phase (start) winding. After the motor has accelerated sufficiently to activate the start switch, the start winding is deenergized. When the low speed is selected, the motor starts on the four-pole main

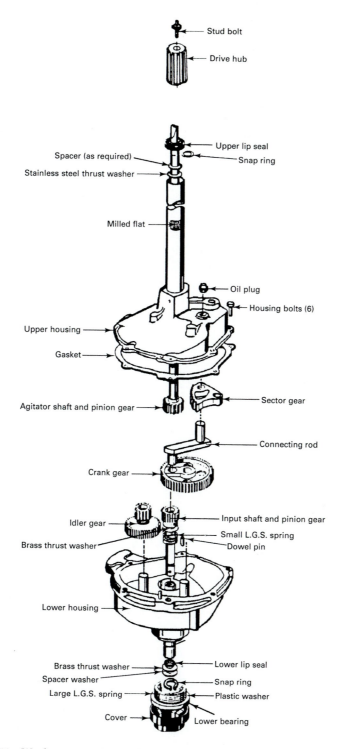

Stud bolt

Drive hub

Upper lip seal

Spacer (as required)

Snap ring

Stainless steel thrust washer

Milled flat

Oil plug

Housing bolts (6)

Upper housing

Gasket

Sector gear

Agitator shaft and pinion gear

Connecting rod

Crank gear

Input shaft and pinion gear

Idler gear

Small L.G.S. spring

Brass thrust washer

Dowel pin

Lower housing

Brass thrust washer

Lower lip seal

Spacer washer

Snap ring

Large L.G.S. spring

Plastic washer

Cover

Lower bearing

Figure 16–54 Washer transmission details. (Courtesy of WCI Major Appliance Group.)

winding and switches to the low-speed six-pole winding after the motor has accelerated to open the start switch.

If an open circuit exists in the low-speed six-pole winding, the motor will repeatedly start and stop on the low-speed selection.

Washer Motor Testing

Caution: Disconnect the unit from the electrical power supply before making any continuity test.

If the motor runs in one direction but will not reverse,

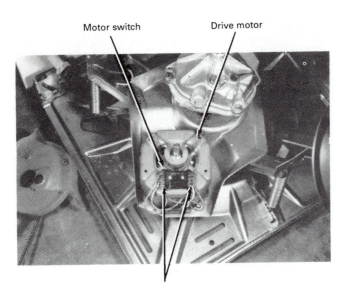

Figure 16–55 Washer drive motor. (Courtesy of WCI Major Appliance Group.)

Figure 16–55 Washer drive motor. (Courtesy of WCI Major Appliance Group.)

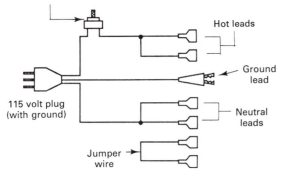

Figure 16–56 Test cord. (Courtesy of WCI Major Appliance Group.)

the timer or wiring harness is at fault because the same motor components are used when the motor runs in either direction.

Use a 120-volt test cord with four male terminals, two wires on the live side, two wires on the neutral side, and a fifth wire for the ground to test the motor. See Figure 16–56. Connections are made through the motor switch located at the top of the motor. See Figure 16–57.

Figure 16–58 illustrates the internal wiring diagram of the one- and two-speed capacitor start-type motors. If the motor fails to operate during any of the tests outlined above, replace the motor.

Washer Drive Motor Replacement

The washer drive motor is either a one- or a two-speed ½-hp, capacitor start, reversible motor rated at 120 volts, 60 hertz. The connections to the motor are quick disconnect-

Test cord	Motor leads shaft	Rotation & speed
Line Neutral	2 & 5 1 & 4	Counterclockwise - Fast
Line Neutral	7 & 5 1 & 4	Counterclockwise - Slow
Line Neutral	2 & 4 1 & 5	Clockwise - Fast
Line Neutral	7 & 4 1 & 5	Clockwise - Slow

Figure 16–57 Test cord lead connections. (Courtesy of WCI Major Appliance Group.)

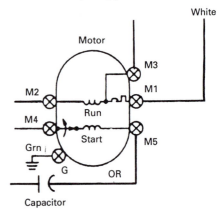

One speed motor

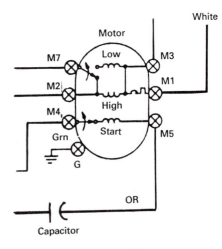

Two speed motor

Figure 16–58 Washer drive motor connections. (Courtesy of WCI Major Appliance Group.)

type terminal blocks. The washer drive motor is mounted to the leg-and-dome assembly and drives the transmission by a pulley mounted on its lower shaft, a drive belt, and an idler arm assembly. The water pump is operated by the upper washer drive motor shaft.

To remove or replace the washer drive motor, use the following procedure:

1. Disconnect the laundry center from the electrical supply.
2. Remove the washer front service panel.
3. Remove the drain hose, inlet and outlet hoses, and clamps from the water pump assembly.

Note: To gain working space for removing the water pump, remove the front outer wash tub leg-and-dome front mounting bolt, lift up on the outer wash tub assembly, and reinstall the bolt into the leg-and-dome assembly bolt hole.

4. Remove the two (2) spring mounting clips securing the water pump assembly to the washer drive motor.

Note: Upon reassembly, be sure that the electrical leads do not become pinched under the spring clips.

5. Lift the pump up and off the washer drive motor shaft.
6. Remove the drive motor water shield.
7. Remove the drive motor belt (roll it off).
8. Disconnect the washer drive motor electrical terminal connector blocks.
9. Remove the four (4) hex nuts securing the washer drive motor.
10. Remove the washer drive motor.
11. Loosen the set screw and remove the drive pulley from the motor shaft.
12. Install the new washer drive motor.
13. Reverse the above procedure to reassemble the unit.

Washer Drive Motor Start Switch

The washer drive motor start switch is mounted externally to the top of the washer drive motor and is not replaceable.

Capacitor

A 189- to 210-microfarad, 120-volt, 60-hertz-rated starting capacitor is used with either the one- or the two-speed washer drive motor, because torque assistance is required to get the motor started.

The start capacitor is mounted to a bracket that is mounted to the washer interior cabinet rear. See Figure 16–59. *Note:* If the motor will not run if it tries to start and shuts off, the problem may be a faulty capacitor.

Capacitor testing *Caution:* Disconnect the laundry center from the electrical supply before making any continuity tests. The capacitor can be tested by using an ohmmeter set on the highest scale. (Most digital ohmmeters do not have sufficient battery strength to test a capacitor; the test may be misleading.) With the power disconnected, disconnect one of the terminal wires and connect the ohmmeter across the terminals.

Caution: Before checking any capacitor, make certain that it is discharged. To discharge a capacitor, remove the leads and short the capacitor terminal with a jumper wire. Remove the jumper wire.

Figure 16–59 Motor start capacitor. (Courtesy of WCI Major Appliance Group.)

To test the capacitor, use the following procedure:

1. Discharge the capacitor.
2. Measure the resistance between the two capacitor terminals. If the capacitor is good, the meter should indicate continuity for a short period of time (while the capacitor is charging), and then it should indicate an open circuit once the capacitor is fully charged.
3. If the ohmmeter indicates a constant continuity between the terminals, the capacitor is shorted. If the ohmmeter has no initial reading, then the capacitor is open. In either case, the capacitor should be replaced.

To remove or replace the washer drive motor start capacitor, use the following procedure:

1. Disconnect the laundry center from the electrical supply.
2. Remove the washer front service panel.
3. Remove the capacitor from its mounting bracket; the capacitor mounting bracket clips to the washer cabinet rear.
4. Remove the wiring.
5. Install the new washer motor start capacitor.
6. Reverse the above procedure to reassemble the unit.

WASHER TIMER

The washer timer controls the sequence of operation of each of the electrical components of the washer used during a complete washing cycle. It is made up of three assemblies: the timer motor, a series of cams, and switches.

The motor is a synchronous-type motor, similar to those used in electric clocks, geared down to operate a pawl and ratchet mechanism that rotates the cams a few degrees every 90 seconds. The actual movement of the cam is quite rapid, to ensure a quick, positive opening and closing of the switch contacts and to reduce arcing. Certain switch contacts operate at a subinterval time (less than 90 seconds). Refer to the timer sequence chart found with the schematic wiring

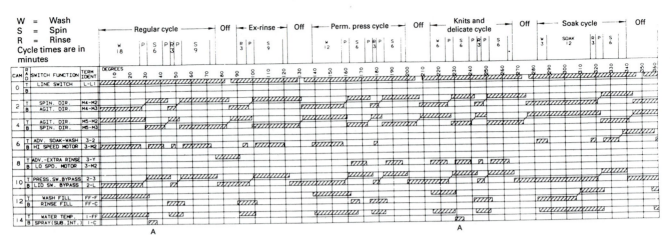

Figure 16-60 shows the washer timer sequence chart with the following legend:

W = Wash
S = Spin
R = Rinse
Cycle times are in minutes

Cycles across top: Regular cycle — Off — Ex-rinse — Off — Perm. press cycle — Off — Knits and delicate cycle — Off — Soak cycle — Off

Switch functions (CAM / Switch Function / Term Ident):
- 0 — LINE SWITCH — L-L1
- 2 — SPIN. DIR. — M4-M2 / AGIT. DIR. — M4-M3
- 4 — AGIT. DIR. — M5-M2 / SPIN. DIR. — M5-M3
- 6 — ADV. SOAK-WASH — 3-2 / HI SPEED MOTOR — 3-M2
- 8 — ADV.-EXTRA RINSE — 3-Y / LO SPD. MOTOR — 3-M2
- 10 — PRESS.SW.BYPASS — 2-3 / LID SW. BYPASS — 2-L
- 12 — WASH FILL — FF-F / RINSE FILL — FF-C
- 14 — WATER TEMP. — I-FF / SPRAY(SUB.INT.) — I-C

Degrees across top: 0 through 360

Figure 16-60 Washer timer sequence chart. (Courtesy of WCI Major Appliance Group.)

Washer notes:

1. Timer line switch shown in off position. Temperature selector switch at hot wash/cold rinse, and water level selector switch at minimum level.

diagram for full details of the timer switching sequence for all washing cycles.

The user must push in the timer knob and turn it clockwise to advance the timer. Pulling the timer knob out will start the washer; pushing the knob in will stop the washer. *Caution:* Advancing the timer with the knob in the out position (with 120 volts across it or while it is running) can destroy the switch contacts and possibly damage other washer components.

Washer Timer Testing

Caution: Disconnect the laundry center from the electrical supply before making any continuity tests.

To check the timer motor operation, disconnect the two leads from the timer motor and connect them to a properly fused 120-volt service cord. Plug the cord into the wall receptacle and check to see if the motor gear is turning.

Multiple-switch contacts activated by a series of cams control the water delivery, water temperature, motor, and the lid lock. Determine, if possible, which period of the cycle is inoperative.

When it has been determined what portion of the cycle is not functioning, check the circuit from the timer to the nonfunctioning component with a continuity tester. With the knob of the timer out (running position), turn the timer knob slowly through the portion of the cycle needing to be tested. If the continuity test shows positive, the circuit will be closed. It may be necessary to operate the timer with a test cord allowing the motor to rotate the cam normally.

The timer should be replaced as a complete assembly and *should not be repaired.*

Timer Sequence Chart

The information contained in the timer sequence chart is necessary to properly test the operation of the timer. See Figure 16-60.

The numbers 0 through 360 across the top of the chart are degrees of rotation at which the timer switches operate (except for subinterval contacts; see the subinterval chart next to the timer chart and Figure 16-61).

Note: The timer sequence chart should be used in conjunction with the schematic diagram to diagnose electrical problems. Using the chart to indicate the timer switch contacts that should be closed at any point in the cycle, the electrical circuits in use can be traced on the schematic diagram and the same circuit traced and checked on the washer.

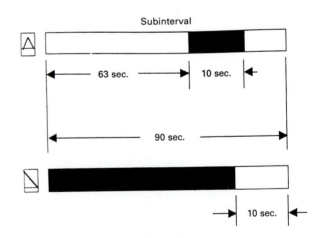

Figure 16-61 Subinterval timer chart. (Courtesy of WCI Major Appliance Group.)

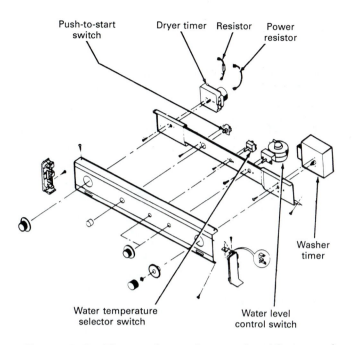

Push-to-start switch   **Dryer timer**   **Resistor**   **Power resistor**

Washer timer

Water temperature selector switch

Water level control switch

Figure 16–62   Timer and controls mounting. (Courtesy of WCI Major Appliance Group.)

Console and Controls

The console and controls assembly consists of the console panel, dryer timer, dryer temperature selector switch, push-to-start switch, washer timer, water temperature selector switch, water level control switch, control knobs, indicator lamps, and so on.

To remove or replace the console and controls assembly, use the following procedure. See Figure 16–62.

1. Disconnect the laundry center from the electrical supply.
2. Remove the dryer front access panel.
3. Remove the dryer front access safety cover panel.
4. Remove the four (4) screws securing the console and controls assembly to the side panels.
5. Disconnect the electrical wiring connector blocks.
6. Remove the electrical wiring leads for the console light (if applicable).
7. Lay the console and controls assembly on a protected surface and transfer all remaining components and parts.
8. Install the new console panel and/or controls.
9. Reverse the above procedure to reassemble the unit.

To remove or replace the washer timer, use the following procedure:

1. Disconnect the laundry center from the electrical supply.
2. Remove the dryer front access panel.

3. Remove the dryer front access safety cover panel.
4. Remove the four (4) screws securing the console and controls assembly to the side panels.
5. Disconnect the electrical wiring connector blocks.
6. Remove the electrical wiring leads for the console light (if applicable).
7. Lay the console and controls panel assembly on a protected surface.
8. Remove the timer knob by pushing the inner knob in and turning it counterclockwise until the knob threads off the shaft; pull the outer dial off the timer shaft.
9. Remove the two (2) screws securing the washer timer to the control bracket.
10. Remove and transfer all electrical wiring to the new timer.
11. Reverse the above procedure to reassemble the unit.

Water Level Pressure Control Switch

The water level control is a pressure-operated switch that controls the water level inside the inner wash basket. As the water level rises, air in the pressure tube is compressed (the pressure tube is located between the water level control and the air dome) and is forced against the diaphragm in the water level control. Do not attempt to adjust the range of the water level control switch.

Two different types of water level control switches are (1) rotary and (2) remote. These switches are used with the 27-inch laundry centers. See Figures 16–63 and 16–64.

The remote water level pressure control is a snap-action, triple-pole, double-throw switch rated at 12 amps, at 120 volts. It is mounted to the dryer base pan. See Figure

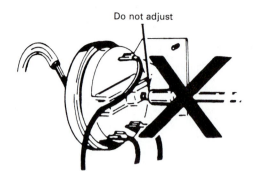

Do not adjust

Figure 16–63 Water level pressure control switch. (Courtesy of WCI Major Appliance Group.)

16–64. The remote water level pressure control switch is controlled by the push-button selector switch on the control console.

The rotary (dial operated) water level pressure control switch is a single-pole, double-throw, manual-reset, normally open, pressure-activated switch, rated at 12 amps, at 120 volts.

Do not condemn a water level pressure control switch until the entire system has been examined and tested. For example, if the air dome, located in the outer wash tub area, and the pressure tube have filled with water, the water level control will overfill the washer.

To test the water level pressure control switch, use the following procedure:

1. If any water is remaining is the inner wash basket, empty it by advancing the timer to a drain cycle.

2. After the water has drained out, turn off the washer and remove the front service panel.

3. Examine the pressure control switch tube. There should not be any water visible in the pressure tube. If water is present, go to step 4. If water is not present, go to step 7.

4. Disconnect the laundry center from the electrical supply and gain access to the water level pressure control switch. Remove the pressure tube at the water level control and blow into the tube until it is clear. Quickly crimp the tube with your fingers and reattach the tube to the water level pressure control tube. There should now be no water in it.

5. Reconnect the washer to the electrical supply, set the water level control to MIN, and turn the timer to a fill cycle. After the water level control has been satisfied, push and turn the timer to the OFF position, open the washer, and measure the depth of water. On the MIN setting, the depth of the water should be 5.2 to 6.4 inches.

6. Close the lid, advance the water level control to the LARGE setting, and turn the timer to a fill cycle. After the water level control has been satisfied, push and turn the timer to the OFF position, open the lid, and measure the depth of water. On the LARGE setting, the depth of water should be 11.9 to 13.5 inches. Set the timer to the drain cycle and drain the washer.

7. Disconnect the laundry center from the electrical supply.

Water level pressure control switch

Figure 16–64 Remote water level pressure control. (Courtesy of WCI Major Appliance Group.)

8. Remove the wire from terminal 2 on the water level control.

9. Check the continuity between terminals 2 and 3. Continuity should exist. If no continuity exists, replace the water level control.

10. Remove the pressure tube from the water level control and attach a short piece of scrap pressure tube to the water level control. Blow into the tube until the water level control "trips," then tightly clamp the end of the tubing shut.

11. Recheck the continuity between terminals 2 and 3. No continuity should exist. Leave the crimped scrap tubing attached to the water level control for a few minutes. If the water level control trips during this time frame, the internal diaphragm is leaking and the water level control must be replaced.

To remove or replace the water level control switch (rotary dial operated), use the following procedure:

1. Disconnect the laundry center from the electrical supply.

2. Remove the dryer front access panel.

3. Remove the dryer front access safety cover panel.

4. Remove the four (4) screws securing the console and controls assembly to the side panels.

5. Disconnect the electrical wiring connector blocks.

6. Remove the electrical wiring leads for the console light (if applicable).

7. Lay the console and controls assembly on a protected surface.

8. Remove the water level pressure control switch knob.

9. Remove the two (2) screws securing the water level pressure control switch to the control bracket.

10. Remove and transfer all electrical wiring and the pressure tube to the new water level pressure control switch.

11. Reverse the above procedure to reassemble the unit.

To remove or replace the water level pressure control switch (remote operated), use the following procedure:

1. Disconnect the laundry center from the electrical supply.

2. Remove the dryer front access panel.

3. Remove the dryer front access safety cover panel.

4. Remove the water level pressure control switch tube.

5. Remove the water level pressure control switch by pulling it out from its slide-in slot.

6. Remove and transfer all electrical wiring to the new water level pressure control switch.

7. Reverse the above procedure to reassemble the unit.

Water Level Pressure Control Switch Tube

In order to provide enough air pressure to activate the water level pressure control switch, a large volume of air must be compressed and transferred to the water level control switch. This is accomplished by utilizing an air dome on the side of the outer wash tub and clear plastic tubing.

The air stored in the air dome is compressed by the weight of the water in the wash/spin basket. The resulting pressure is then transferred to the water level pressure control switch through the clear plastic tubing.

To remove or replace the water level pressure control switch tube, use the following procedure:

1. Disconnect the laundry center from the electrical supply.

2. Remove the washer front service panel.

3. Remove the dryer front access panel.

4. Remove the dryer front access safety cover panel.

5. Remove the console and controls assembly.

6. Disconnect the electrical wiring connector blocks.

7. Remove the electrical wiring leads for the console light (if applicable).

8. Lay the console and controls assembly on a protected surface.

9. Remove the water level pressure control tube from the switch and air dome on the left rear side of the outer wash tub.

Note: Use caution when removing the tube from the outer wash tub so as not to break off the connector to the air dome.

10. Install the new water level pressure control switch tube.

11. Reverse the above procedure to reassemble the unit.

Water Temperature Selector Switch

The water temperature selector switch controls the temperature of the water entering the machine. It is a four-

Switch Position	TS1 - TS5	TS2 - TS5	TS2 - TS6
Warm wash - warm rinse	X	X	X
Cold wash - cold rinse	X	O	O
Warm wash - cold rinse	X	O	X
Hot wash - cold rinse	O	O	X

X = Closed
O = Open

Figure 16–65 Water temperature selector switch chart. (Courtesy of WCI Major Appliance Group.)

position switch that provides four cycle combinations of water temperature: hot wash and cold rinse, cold wash and cold rinse, warm wash and warm rinse, warm wash and cold rinse.

The electrical check for the water temperature switch is listed in the temperature switch chart on the electrical diagram. See Figure 16–65.

The water temperature selector switch may be either a rotary dial control or a push-button design.

To test the water temperature selector switch, check the continuity between the terminals as indicated on the temperature selector chart.

To remove or replace the water temperature selector switch, use the following procedure:

1. Disconnect the laundry center from the electrical supply.

2. Remove the dryer front access panel.

3. Remove the dryer front access safety cover panel.

4. Remove the four (4) screws securing the console and controls assembly to the side panels.

5. Disconnect the electrical wiring connector blocks.

6. Remove the electrical wiring leads for the console light (if applicable).

7. Lay the console and controls assembly on a protected surface.

8. Remove the water temperature control switch knob.

9. Remove the washer timer control knob and dial.

10. Remove the dryer timer control knob.

11. Remove the screws securing both end caps.

12. Remove the control panel escutcheon.

13. Remove the two (2) screws securing the water temperature control selector switch to the control bracket.

14. Remove and transfer all the electrical wiring to the new water temperature control selector switch.

15. Reverse the above procedure to reassemble the unit.

Water Inlet Mixing Valve

The water inlet mixing valve is actually two (2) solenoid valves in one body. A hot-water inlet and a cold-water inlet valve discharge into a common mixing chamber; the flow of water out of the chamber is controlled by a rubber

Water inlet
valve

Figure 16–66 Water inlet valve. (Courtesy of WCI Major
Appliance Group.)

flow washer out of the chamber capable of maintaining
a flow rate of 5 gallons per minute (±10 percent) with an
incoming water pressure of 30 to 120 psi. The inlet valves
are controlled by the timer and water temperature selector
switch, individually or together, to provide hot, cold, or
warm water for washing. The temperature of the warm
mixture is dependent on the temperature and pressure of
the hot- and cold-water supply lines. See Figure 16–66.

Valve operation Both inlet valves are identical in
construction and operation. The valve body provides an air
inlet connection and passage with a large orifice and seat
where the water flow can be stopped. The outlet of the valve
body empties into the mixing chamber. A movable rubber
diaphragm operates against the valve seat to start and stop
the flow of water. The diaphragm is operated by water
pressure. It has a small bleed orifice at its center. The armature of the solenoid serves to open and close the main orifice.
The armature operates within a closed metal tube (valve
guide) which is sealed by the outer edge of the diaphragm
to the valve body. A coil spring holds the armature down
against the diaphragm main orifice when the solenoid is not
energized.

The following line drawings and text explain the basic
valve operation:

Figure 16–67a shows a valve in the closed position and
not energized. The water has passed through the diaphragm
bleed orifice, placing the incoming line water pressure on
top of the diaphragm. The bottom of the diaphragm is essentially at atmospheric pressure (open to the outlet). This
pressure differential will hold the valve shut.

When the solenoid is energized, the resulting magnetic field pulls the armature up into the valve guide. See
Figure 16–67b.

The armature spring is compressed by this action.
When the armature moves up, it allows the water on the
top side of the diaphragm to flow through the main orifice.

The diaphragm orifice is much smaller than the main
orifice and will not admit enough water to maintain this
pressure on the top side of the diaphragm; thus the pressure
on the top side of the diaphragm is reduced to almost zero.
Therefore, the pressure under the bleed orifice lifts the diaphragm off of the valve seat, allowing a full flow of water.
See Figure 16–67c.

When the solenoid is deenergized, the armature drops
down closing the diaphragm main orifice. See Figure
16–67d.

Water continues to flow through the diaphragm bleed
orifice, building up pressure until it equalizes on both sides
of the diaphragm. The spring then forces the diaphragm
down against the valve seat. See Figure 16–67a.

To test the water valve, use the following procedure:

1. Disconnect the laundry center from the electrical
supply.

2. Use an ohmmeter to make a continuity check of
the electrical harness to determine whether or not a circuit
exists.

3. The resistance of the solenoid coil should be approximately 880 ohms ± 10 percent at 77°F.

4. If the harness and solenoid coil tests are OK, then
simulate a normal valve operation by directly testing the
solenoid coil using a separate 120-volt power supply, with
a properly fused and grounded service cord.

5. If the water valve operates on both solenoids, check
the timer, water level pressure control switch, and the water
temperature selector circuits. If the water valve fails to operate, check the valve inlet screens for debris and/or replace
the water inlet valve.

*To remove or replace the water valve, use the following
procedure.* The water inlet fill valve is mounted on the inner
rear panel of the washer cabinet.

1. Disconnect the laundry center from the electrical
supply.

2. Shut off the water faucets.

3. Remove the washer front service panel.

4. Remove the washer top panel and lid assembly.

5. Loosen the one (1) ⅜-inch hex bolt securing the
water inlet valve shield.

6. Remove the water inlet valve shield by lifting
upward.

7. Remove the ⅜-inch hex bolt that was loosened in
step 4.

8. Remove the electrical connector blocks from the
solenoids.

9. Pull the water inlet valve and fill hoses into the
cabinet interior.

10. Remove the hot- and cold-water hoses to the
valve.

11. Remove the clamp and the fill hose.

12. Install the new water inlet fill valve and check for
proper operation.

13. Reverse the above procedure to reassemble the
unit.

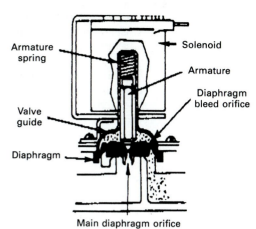

(a)

Figure 16–67a Water valve closed. (Courtesy of WCI Major Appliance Group.)

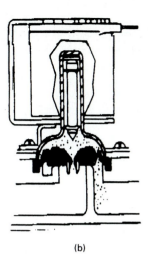

(b)

Figure 16–67b Water valve open. (Courtesy of WCI Major Appliance Group.)

(c)

Figure 16–67c Water valve (diaphragm up). (Courtesy of WCI Major Appliance Group.)

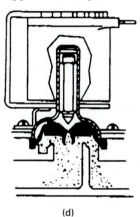

(d)

Figure 16–67d Water valve closing. (Courtesy of WCI Major Appliance Group.)

Lid Switch and Lock Assembly

Caution: Disconnect the unit from the electrical power source before making any continuity tests.

The lid switch and lock assembly is mounted to the top right corner of the washer cabinet top. It is wired in series with the drive motor and opens the circuit to the motor whenever the lid is opened. However, the lid switch is effective only during the spin cycle. During the wash and rinse cycles the lid switch is bypassed by a parallel circuit through a timer contact, and the motor continues to operate even when the lid and the lid switch are open.

This assembly incorporates the lid switch, bimetal, and a shunt switch in a phenolic case that is mounted to a bracket. The lid locks approximately 3 seconds after the machine starts to spin, when the bimetal strip energizes and pulls the locking arm into the lid strike. The shunt switch helps to protect the bimetal from excessive current draw. See Figure 16–68.

Lid switch and lid lock testing *Note:* Do not attempt to check the lid lock by placing line voltage on the device.

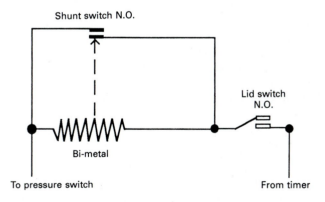

Figure 16–68 Lid switch and lid lock wiring details. (Courtesy of WCI Major Appliance Group.)

The resistance of this circuit is less than 1 ohm. A check can be made by using a jumper wire across the terminals of the switch when testing the washer under repeated spin starts. This will prevent excessive overheating of the bimetal.

Washer Timer 243

When the bimetal is deenergized, the unlock time may vary from 18 to 50 seconds, depending on the ambient conditions, length of time the unit has been running, water temperature, and the load size.

It is important to check the bimetal operation when servicing the washer. Always check for continuity of the bimetal when replacing a new lock assembly.

Lid lock switch shield The lid lock shield is galvanized steel and surrounds the lid lock switch, a UL requirement for safety.

To remove or replace the lid lock switch shield, use the following procedure:

1. Disconnect the laundry center from the electrical supply.
2. Remove the washer front service panel.
3. Remove the one (1) screw securing the lid lock shield.
4. Install the new lid lock switch shield.
5. Reverse the above procedure to reassemble the unit.

To remove or replace the lid switch and lid lock assembly, use the following procedure:

1. Disconnect the laundry center from the electrical supply.
2. Remove the washer front service panel.
3. Disconnect the washer lid lock switch terminal connector (and light switch if applicable).
4. Remove the one (1) screw securing the lid lock shield.
5. Remove the two (2) screws securing the lid lock switch.
6. Install the new lid lock switch.
7. Reverse the above procedure to reassemble the unit.

DRYER COMPONENTS AND PARTS

The following is a description of the dryer components and parts.

Dryer Structural Components

See the Structural Components section presented earlier in this chapter for the following dryer cabinet components: dryer front access panel, front access safety cover panel, top panel, side panels, console and controls, dryer front panel, door panels, door seal, strike, door hinges, catch, lint filter, drum light bulb and receptacle, drum seal, air duct grille, and dryer cabinet bottom pan.

One or more of these parts and/or panels must be removed to gain access to the other dryer components.

The dryer consists of a white, 5.5-cubic-foot, cold-rolled steel, rotating drum, a blower capable of moving a large volume of air, and an electrical heat source. A one-speed, dual-shaft, ¼-hp motor drives both the blower and the drum.

The drum is a metal cylinder driven by a flat belt that rides on its outer surface, and a series of pulleys that turn the drum in a counterclockwise direction (point of view: facing the front of the dryer). Three (3) vanes mounted to the drum interior cause the clothes to tumble. The open ends of the cylinder are closed by a perforated rear panel, front panel, and the door. The drum support system consists of a plastic bearing that snaps into the inside of the front drum flange and a felt seal that is glued to the front panel assembly. The rear of the drum is supported by a ball-and-socket bearing that resembles a trailer hitch. A heater element is mounted to the back side of the rear panel. The lint filter screen extends full width inside the dryer door. An air duct mounted to the rear of the front panel extends down to the blower intake. The blower discharges the moisture-laden air to the rear vent duct system. The dryer is mounted on top of the washer by means of marriage brackets. See Figure 16–69.

Dryer Drum

The dryer drum is a cylinder formed of heavy-gauge, cold-rolled steel. It rotates on a stationary drum support system consisting of a plastic (two piece) bearing that snaps into the inside of the front drum flange and a felt seal that is glued to the front panel assembly. The rear of the drum is supported by a ball-and-socket bearing that resembles a trailer hitch. A heater baffle and shaft support are mounted to the rear of the drum. See Figure 16–70.

To remove or replace a dryer drum, use the following procedure:

1. Disconnect the laundry center from the electrical supply.
2. Remove the dryer front access panel.
3. Remove the dryer access safety cover panel.
4. Remove the console and controls assembly.
5. Reach through and release the drum belt from the idler pulley to relieve the tension on the front panel.
6. Remove the two (2) lower front screws securing the dryer front panel assembly to the side panels.
7. Remove the four (4) upper top front screws securing the dryer front panel assembly.
8. Carefully remove the dryer front panel assembly.
9. Remove the dryer top panel.
10. Remove the dryer drum by lifting up to release it from the rear bearing support (trailer hitch).
11. Install the new dryer cabinet drum.
12. Reverse the above procedure to reassemble the unit.

Dryer Drum Sorting Vanes

The dryer drum has three (3) clothes-sorting vanes mounted 120° apart around the inside of the drum. The

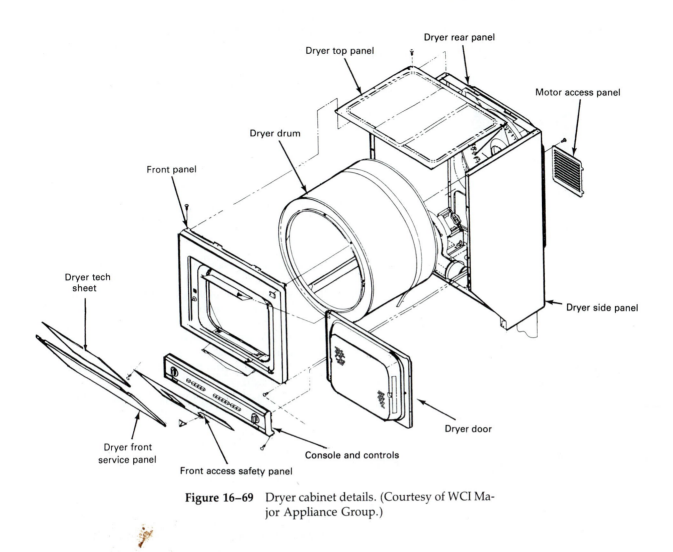

Figure 16–69 Dryer cabinet details. (Courtesy of WCI Major Appliance Group.)

vanes tumble the clothes during operation to allow thorough air circulation. See Figure 16–70.

To remove or replace the dryer drum sorting vane(s), use the following procedure:

1. Disconnect the laundry center from the electrical supply.
2. Remove the dryer top panel.
3. Open the dryer door and hold the vane while removing the two (2) screws from the dryer drum exterior.
4. Install the new dryer sorting vane(s).
5. Reverse the above procedure to reassemble the unit.

Dryer Drum Front Ring Bearings

The dryer drum has two (2) ring bearings made of Celcon M90, mounted 180° apart around the inside of the drum. The bearings snap into holes (five for each bearing) around the front drum opening and are supported by a felt

seal attached to the dryer front panel, inner door opening, and rim flange. See Figure 16–70.

To remove or replace the dryer drum front ring bearing(s), use the following procedure:

1. Disconnect the laundry center from the electrical supply.
2. Remove the dryer front access panel.
3. Remove the dryer front access safety cover panel.
4. Remove the console and controls.
5. Reach through and release the drum belt from the idler pulley to relieve the tension on the front panel.
6. Remove the two (2) lower front screws securing the dryer front panel assembly to the side panels.
7. Remove the four (4) upper top front screws securing the dryer front panel assembly.
8. Carefully remove the dryer front panel assembly.
9. Remove the dryer drum front ring bearing(s).
10. Install the new dryer drum front ring bearing(s).
11. Reverse the above procedure to reassemble the unit.

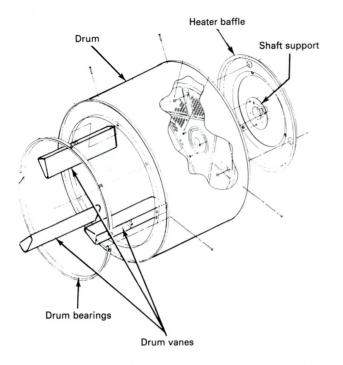

Heater baffle

Drum

Shaft support

Drum bearings

Drum vanes

Figure 16–70   Dryer drum assembly. (Courtesy of WCI Major Appliance Group.)

Dryer Drum Heater Baffle

The dryer drum has a heater baffle made of galvanized steel mounted to the rear of the drum. See Figure 16–70.

To remove or replace the dryer drum heater baffle, use the following procedure:

1. Disconnect the laundry center from the electrical supply.
2. Remove the dryer front access panel.
3. Remove the dryer front access safety cover panel.
4. Remove the console and the controls.
5. Reach through and release the drum belt from the idler pulley to relieve the tension on the front panel.
6. Remove the two (2) lower front screws securing the dryer front panel assembly.
7. Remove the four (4) upper top front screws securing the dryer front panel assembly.
8. Carefully remove the dryer front panel assembly.
9. Remove the dryer top panel.
10. Remove the dryer drum by lifting it up to release it from the rear bearing support (trailer hitch).
11. Remove the three (3) screws securing the dryer drum heater baffle.
12. Install the new dryer drum heater baffle.
13. Reverse the above procedure to reassemble the unit.

Dryer Drum Rear Support Shaft

The dryer rear drum support shaft (resembles a trailer hitch) is mounted to the dryer drum rear and rides in the rear support bearing. See Figure 16–71.

To remove or replace the dryer drum rear support shaft, use the following procedure:

1. Disconnect the laundry center from the electrical supply.
2. Remove the dryer front access panel.
3. Remove the dryer front access safety cover panel.
4. Remove the console and controls.
5. Reach through and release the drum belt from the idler pulley to relieve the tension on the front panel.
6. Remove the two (2) lower front screws securing the dryer front panel assembly to the side panels.
7. Remove the four (4) upper top front screws securing the dryer front panel assembly.
8. Carefully remove the dryer front panel assembly.
9. Remove the dryer top panel.
10. Remove the dryer drum by lifting it up to release it from the rear bearing support (trailer hitch).
11. Remove the three (3) screws securing the dryer drum rear support shaft.
12. Install the new dryer drum rear support shaft.
13. Reverse the above procedure to reassemble the unit.

Dryer Drum Rear Support Bearing

The dryer drum rear support bearing is a U-shaped nylatron block that acts as the bearing surface for the dryer drum rear support (trailer hitch). The bearing assembly, which includes the mounting screws, a mounting bracket, a grounding clip, and a small ball bearing, is mounted to the cabinet rear panel and is lubricated with Lubri-Plate. The ball bearing completes a grounding path from the drum to the cabinet and prevents axial movement of the dryer drum. See Figure 19–84.

To remove or replace the dryer drum rear support bearing, use the following procedure. Refer to Figure 16–72.

1. Disconnect the laundry center from the electrical supply.
2. Remove the dryer front access panel.
3. Remove the dryer front access safety cover panel.
4. Remove the console and controls.
5. Reach through and release the drum belt from the idler pulley to relieve the tension on the front panel.
6. Remove the two (2) lower front screws securing the dryer front panel assembly to the side panels.
7. Remove the four (4) upper top front screws securing the dryer front panel assembly.
8. Carefully remove the dryer front panel assembly.
9. Remove the dryer top panel.
10. Remove the dryer drum by lifting it up to release it from the rear bearing support (trailer hitch).
11. Remove the two (2) screws securing the dryer drum rear support shaft bearing assembly.

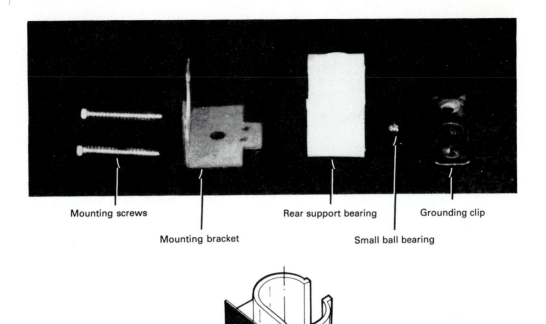

Figure 16–71 Dryer drum rear support bearing assembly. (Courtesy of WCI Major Appliance Group.)

12. Tape the grounding clip in the correct position on the back of the cabinet.

13. Apply a small amount of Lubri-Plate to the ball bearing surface and insert the ball bearing into the recess.

Note: The small ball bearing must be positioned between the support bearing and the grounding clip.

14. Install the new dryer drum support bearing assembly.

15. Apply a liberal amount of Lubri-Plate to the grooves of the support bearing.

16. Reverse the above procedure to reassemble the unit.

Dryer Drive Motor

The dryer drive motor is a single-speed, dual-shaft, 1725-rpm, ⅛-hp motor, rated at 120 volts, 60 hertz. An internal overload protector is built into the motor and will open due to an excessive temperature rise and/or excessive current draw. If the overload protector opens, all dryer operations will stop. The motor protector is self-resetting and may take as long a 30 seconds to reset. See Figure 16–73.

The dryer drive motor has a multiple-terminal, snap-on, electrical wiring connector block.

The centrifugal switch in the dryer motor performs three functions. When the motor is at rest, a single-pole, double-throw (SPDT) contact closes a circuit to the motor start windings. When the dryer is started, power is momentarily supplied to the motor through the push-to-start switch. As the motor comes up to speed, the SPDT contact disconnects the start winding and supplies power to the run winding. This provides a holding circuit to keep the motor operating after the momentary push-to-start switch is released. As the motor switch operates, a second set of contacts closes a circuit to the heating section. This portion of the motor switch prevents the heater from operating until the motor is operating at normal speed. The motor start switch *is not replaceable.*

To remove or replace the dryer drive motor, use the following procedure:

1. Disconnect the laundry center from the electrical power supply.

2. Remove the dryer front access panel.

3. Remove the dryer front access safety cover panel.

4. Remove the console and controls.

5. Reach through and release the drum belt from the idler pulley to relieve the tension on the front panel.

6. Remove the two (2) lower front screws securing the dryer front panel assembly to the side panels.

7. Remove the four (4) upper top front screws securing the dryer front panel assembly.

8. Carefully remove the dryer front panel assembly.

9. Remove the dryer top panel.

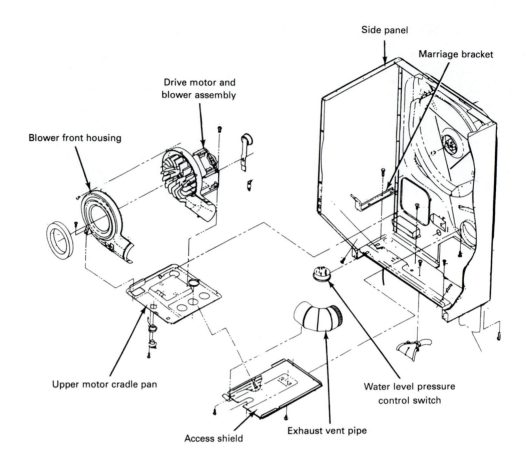

Figure 16–72 Dryer cabinet (drive motor and parts). (Courtesy of WCI Major Appliance Group.)

10. Remove the dryer drum by lifting it up to release it from the rear bearing support (trailer hitch).

11. Remove the two (2) screws securing the exhaust vent elbow to the blower housing.

12. Remove the six (6) screws securing the dryer front blower housing to the rear blower housing.

13. Loosen the hex bolt securing the blower wheel and slide the side blower wheel off the drive motor shaft.

14. Remove the three (3) screws securing the rear blower housing to the drive motor cradle.

15. Release the electrical wiring connector block to the drive motor.

16. Remove the two (2) clamps securing the motor to the mounting cradle.

17. Remove the dryer drive motor belt pulley (press-on fit, flush with the end of the motor shaft).

18. Install the new dryer drive motor.

19. Reverse the above procedure to reassemble the unit.

Dryer Drive Motor Pulley

The dryer drive motor pulley is a press-on type, mounted flush with the end of the motor shaft. The dryer drive motor pulley has a smooth belt surface with a small V groove that indicates its outer end position on the shaft.

To remove or replace the dryer drive motor pulley, use the following procedure:

1. Disconnect the laundry center from the electrical supply.

2. Remove the dryer front access panel.

3. Remove the dryer front access safety cover.

4. Remove the console and controls.

5. Reach through and release the drum belt from the idler pulley to relieve the tension on the front panel.

6. Remove the two (2) lower front screws securing the dryer front panel assembly to the side panels.

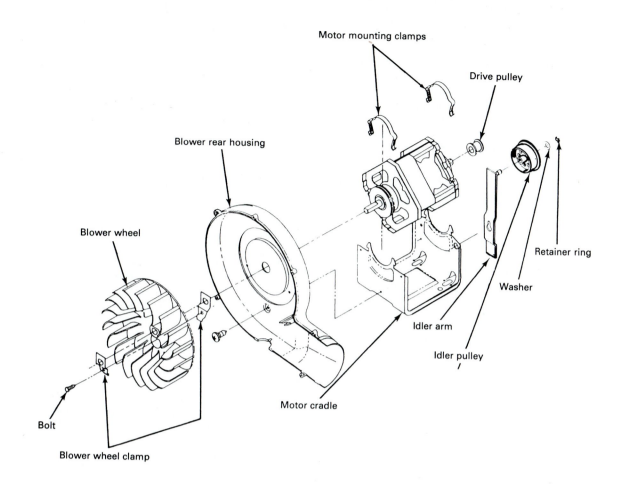

Figure 16–73 Dryer drive motor, blower wheel, blower housing, idler arm, and parts. (Courtesy of WCI Major Appliance Group.)

7. Remove the four (4) upper top front screws securing the front panel assembly.

8. Carefully remove the dryer front panel assembly.

9. Remove the dryer top panel.

10. Remove the dryer drum by lifting it up to release it from the rear bearing support (trailer hitch).

11. Remove the two (2) screws securing the exhaust vent elbow to the blower housing.

12. Remove the six (6) screws securing the dryer front blower housing to the rear blower housing.

13. Loosen the hex bolt securing the blower wheel and slide the blower wheel off the drive motor shaft.

14. Remove the three (3) screws securing the rear blower housing to the drive motor cradle.

15. Release the electrical wiring connector block to the drive motor.

16. Remove the two (2) clamps securing the drive motor to the mounting cradle.

17. Remove the dryer drive motor belt pulley (press-on fit, flush with the end of the motor shaft).

18. Install the new dryer drive motor.

19. Reverse the above procedure to reassemble the unit.

Dryer Drum Drive Belt

The dryer drum drive belt is a neoprene, fiberglass, and nylon tire cord reinforced flat drive belt. The dryer drum drive belt is driven by the dryer drive motor pulley located around the dryer drum. The proper tension is maintained by an idler pulley and arm assembly.

To remove or replace the dryer drum drive belt, use the following procedure:

1. Disconnect the laundry center from the electrical supply.

2. Remove the dryer front access panel.

3. Remove the dryer front access safety cover panel.

4. Remove the console and controls.

5. Reach through and release the drum belt from the idler pulley to relieve the tension on the front panel.

6. Remove the two (2) lower front screws securing the dryer front panel assembly to the side panels.

7. Remove the four (4) upper top front screws securing the dryer front panel assembly.

8. Carefully remove the dryer front panel assembly.

9. Remove the dryer top panel.

10. Remove the dryer drum by lifting it up to release it from the rear bearing support (trailer hitch).

11. Install the new dryer drive belt.

12. Reverse the above procedure to reassemble the unit.

Idler Assembly

The idler assembly consists of an idler arm, a pulley, and a spring to maintain a constant tension on the drive belt for proper drum speed. The idler arm bracket has a key hole slot and mounts over a stud on the rear of the drive motor cradle.

To remove or replace the idler assembly, use the following procedure:

1. Disconnect the laundry center from the electrical supply.

2. Remove the dryer front access panel.

3. Remove the dryer front access safety cover panel.

4. Remove the console and controls.

5. Reach through and release the drum belt from the idler pulley to relieve the tension on the front panel.

6. Remove the idler arm tension spring.

7. Remove the idler arm and roller assembly from the stud on the motor cradle.

8. Install the new idler arm assembly.

9. Reverse the above procedure to reassemble the unit.

Blower Housing

The blower housing is used to control the exhaust air leaving the dryer. It is located behind the lower part of the dryer front panel.

The blower wheel is contained in the blower housing. A foam seal is installed between the exhaust duct and the blower housing inlet. The control thermostat and exhaust tubing are secured to the dryer base pan and extend into the blower housing outlet.

To remove or replace the blower housing, use the following procedure:

1. Disconnect the laundry center from the electrical supply.

2. Remove the dryer front access panel.

3. Remove the dryer front access safety cover panel.

4. Remove the console and controls.

5. Reach through and release the drum belt from the idler pulley to relieve the tension on the front panel.

6. Remove the two (2) lower front screws securing the dryer front panel assembly to the side panels.

7. Remove the four (4) upper top front screws securing the dryer front panel assembly.

8. Carefully remove the dryer front panel assembly.

9. Remove the dryer top panel.

10. Remove the dryer drum by lifting it up to release it from the rear bearing support (trailer hitch).

11. Remove the two (2) screws securing the exhaust vent elbow to the blower housing.

12. Remove the six (6) screws securing the dryer front blower housing to the rear blower housing.

13. Loosen the hex bolt securing the blower wheel and slide the blower wheel off the drive motor shaft.

14. Remove the three (3) screws securing the rear blower housing to the drive motor cradle.

15. Remove the two (2) screws securing the exhaust vent elbows to the blower housing.

16. Install the new dryer blower housing(s).

17. Reverse the above procedure to reassemble the unit.

Blower Wheel

The dryer is equipped with a centrifugal-type blower wheel secured to the D-shaped motor shaft with a double spring clip and bolt. The blower wheel mounts over the end of the motor shaft. See Figure 16–72.

To remove or replace the dryer blower wheel, use the following procedure:

1. Disconnect the laundry center from the electrical supply.

2. Remove the dryer front access panel.

3. Remove the dryer front access safety cover panel.

4. Remove the console and controls.

5. Reach through and release the drum belt from the idler pulley to relieve the tension on the front panel.

6. Remove the two (2) lower front screws securing the dryer front panel assembly to the side panels.

7. Remove the four (4) upper top front screws securing the dryer front panel assembly.

8. Carefully remove the dryer front panel assembly.

9. Remove the dryer top panel.

10. Remove the dryer drum by lifting it up to release it from the rear bearing support (trailer hitch).

11. Remove the two (2) screws securing the exhaust vent elbow to the blower housing.

12. Remove the six (6) screws securing the dryer front blower housing to the rear blower housing.

13. Loosen the hex bolt securing the blower wheel and slide the blower wheel off the drive motor shaft.

14. Install the new blower wheel (blower wheel and rear spring clamp to be bottomed out on the motor shaft flat).

15. Reverse the above procedure to reassemble the unit.

Exhaust Vent Tube

The flexible, metal exhaust vent tube elbow connects to the blower housing and to the external exhaust vent system. The exhaust can be routed either to the rear or to the right side for venting after it exits the cabinet rear. (See the Exhaust-Venting Installation section).

DRYER ELECTRICAL COMPONENTS TEST AND REPAIR

Caution: Always disconnect the laundry center from the electrical supply before making any continuity tests.

Push-to-Start and Door Switch

The push-to-start switch is a single-pole, single-throw (SPST), normally open (NO) switch mounted in the control panel. The user momentarily presses the switch button to activate the dryer drive motor.

The push-to-start switch on some models is part of the fabric temperature selector switch.

The door switch is s single-pole, single-throw (SPST), normally open (NO) switch that is mounted in the dryer front panel. It is activated by the dryer door.

Both switches are normally open. They are wired electrically in series with each other and are connected to the hot side of the line between the timer and the drive motor. They are also in series with the timer motor. Neither motor will operate until the dryer door switch is closed and the start switch is momentarily depressed (closed) by the user.

The push-to-start switch is also wired in parallel with the centrifugal start switch inside the motor. When the motor is up to speed, the centrifugal switch maintains the circuit that was momentarily established by depressing the push-to-start switch.

To test the push-to-start switch, use the following procedure:

1. Disconnect the laundry center from the electrical supply.
2. Remove the dryer front access panel.
3. Remove the dryer front access safety cover panel.
4. Remove the four (4) screws securing the console and controls assembly to the side panels.
5. Disconnect the electrical wiring connector blocks.
6. Remove the electrical wiring leads for the console light (if applicable).
7. Lay the console and controls assembly on a protected surface.
8. Remove the wires from the push-to-start switch and connect the ohmmeter leads across the terminals.
9. Press the push-to-start button, the ohmmeter should show continuity.
10. If no continuity is present, replace the push-to-start switch.

To remove or replace the push-to-start switch, use the following procedure:

1. Disconnect the laundry center from the electrical supply.
2. Remove the dryer front access panel.
3. Remove the dryer front access safety cover panel.
4. Remove the four (4) screws securing the console and controls assembly to the side panels.
5. Disconnect the electrical wiring connector blocks.
6. Remove the electrical wiring leads for the console light (if applicable).
7. Lay the console and controls assembly on a protected surface.
8. Remove the water temperature control switch knob.
9. Remove the washer timer control knob and dial.
10. Remove the dryer timer control knob.
11. Remove the push-to-start switch push buttons by pulling them off the stem (if applicable).
12. Remove the screws securing both end caps.
13. Remove the control panel escutcheon.
14. Remove the two (2) screws securing the push-to-start switch to the control bracket.
15. Remove and transfer all electrical wiring to the new push-to-start switch.
16. Reverse the above procedure to reassemble the unit.

Door switch The door switch is a single-pole, single-throw (SPST), normally open (NO) switch that is mounted in the dryer front panel. It is activated by the dryer door. See Figure 16–8.

To test the door switch, use the following procedure:

1. Disconnect the laundry center from the electrical supply.
2. Remove the dryer front access panel.
3. Remove the dryer front access safety cover panel.
4. Remove the four (4) screws securing the console and controls assembly to the side panels.
5. Disconnect the electrical wiring connector blocks.
6. Remove the electrical wiring leads for the console light (if applicable).
7. Lay the console and controls assembly on a protected surface.
8. Remove the wires from the door switch and connect an ohmmeter across the terminals.
9. Press the door switch button, the ohmmeter should show continuity.
10. If no continuity is found, replace the door switch.

To remove or replace the dryer door switch, use the following procedure:

1. Disconnect the laundry center from the electrical supply.
2. Remove the dryer front access panel.
3. Remove the dryer front access safety cover panel.
4. Remove the console and controls.
5. Reach through and release the drum belt from the idler pulley to relieve the tension on the front panel.

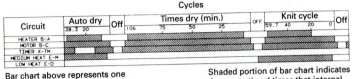

Bar chart above represents one complete revolution of timer shaft.

Shaded portion of bar chart indicates the proportional times that internal timer contacts are closed.

Figure 16–74 Timer bar chart. (Courtesy of WCI Major Appliance Group.)

6. Remove the two (2) screws (lower front) securing the dryer front panel assembly to the side panels.

7. Loosen the four (4) screws (upper top front) securing the dryer front panel assembly.

8. Carefully pull out on the bottom of the dryer front panel assembly to gain access to the dryer door switch.

9. Release the spring side clips securing the dryer door switch.

10. Remove the electrical leads from the switch.

11. Install a new dryer door switch.

12. Reverse the above procedure to reassemble the unit.

Dryer Timer

The dryer timer primarily controls the length of time that the dryer operates. The timer consists of a series of cams and switches driven be a synchronous motor. It is set by rotating the knob and dial clockwise. There is no push/pull switch in the dryer timer.

The timer motor operates on 120 volts, 60 hertz. In the timed cycle, the timer motor is controlled by the internal contacts of the timer. In the automatic cycle, the timer motor is controlled by the back-side contact 2 on the heat control thermostat.

Dryer timer testing When checking the timer, follow the information contained in the timer sequence chart. The chart is found on the wiring diagram for the unit being serviced.

Read from left to right, the chart shows the actions of the various contacts through 360 degrees of the timer cam rotation. At the left of the chart, each function and each set of switch contacts is identified by the terminals to which they are connected. The horizontal bars to the right indicate when the switch contacts are closed. The description of the cycles is listed across the bottom of the chart. The maximum cycle time in minutes is stated (except for the automatic cycle, which is variable in length depending on the load size, wetness, and the types of fabrics being dried).

Dryer timer motor test Caution: Disconnect the laundry center from the electrical supply before attaching test cord leads.

If the timer does not advance in the timed cycle(s), disconnect the timer motor leads from the timer and/or other terminals and connect them to a fused test cord. Plug the test cord into a 120-volt outlet. If the timer does not advance, replace it. If the timer does not advance through the automatic cycle, test the timer as described above. If it still does not advance, replace it. If the timer does advance when connected to a test cord, but does not advance through the automatic cycle, check and make sure that the control thermostat is cycling, because the timer motor operates only when the thermostat is open.

Testing timer switch contacts Caution: Disconnect the laundry center from the electrical supply before attaching the test cord leads.

Refer to the timer sequence chart on the unit wiring diagram to determine which set of contacts should be open or closed. See Figure 16–74. To check for continuity, remove the leads from the selected terminal and connect a continuity tester to the selected terminal. Slowly turn the timer knob one revolution to activate the contacts. While turning the timer knob, check to make sure that the contacts open and close according to the timer chart.

To remove or replace the dryer timer, use the following procedure:

1. Disconnect the laundry center from the electrical supply.

2. Remove the dryer front access panel.

3. Remove the dryer front access safety cover panel.

4. Remove the four (4) screws securing the console and controls assembly to the side panels.

5. Disconnect the electrical wiring connector blocks.

6. Remove the electrical wiring leads for the console light (if applicable).

7. Lay the console and controls assembly on a protected surface.

8. Remove the dryer timer control knob.

9. Remove the two (2) screws securing the dryer timer to the control bracket.

10. Remove and transfer all electrical wiring to the new dryer timer.

11. Reverse the above procedure to reassemble the unit.

Fabric Temperature Selector Switch

The fabric temperature selector switch is either a rotary or a push-button switch that allows the user to select the proper temperature for each clothes load. This is accomplished in conjunction with the heat control thermostat. All fabric temperature selector switches have a set of contacts that directly control the electrical power to the dryer gas burner. These contacts are open for no heat drying. Some fabric selector switches have an additional contact that supplies electric current to the control thermostat biasing heater as follows:

Low heat	Direct connection to the bias heater
Medium heat	Through a 3.0-kiloohm resistor
High heat	No current to bias heater

The proper temperature is controlled through the timer and is selected when the timer drying cycle is set by the user.

Note: On the push-button-operated fabric temperature selector switch, the push-to-start switch is also part of this switch. The push-to-start switch is momentarily closed by the user to complete the circuit to the dryer drive motor.

Fabric temperature selector switch testing The continuity of the fabric temperature selector switch can be checked by first removing all electrical wires and the resistor from the switch terminals. Then refer to the switch circuitry chart on the unit wiring diagram to determine which set of contacts should be open or closed at each position. See Figure 16–75 for an example.

To remove or replace the fabric temperature selector switch, use the following procedure:

1. Disconnect the laundry center from the electrical supply.
2. Remove the dryer front access panel.
3. Remove the dryer front access safety cover panel.
4. Remove the four (4) screws securing the console and controls assembly to the side panels.
5. Disconnect the electrical wiring connector blocks.
6. Remove the electrical wiring leads for the console light (if applicable).
7. Lay the console and controls assembly on a protected surface.
8. Remove the water temperature control switch knob.
9. Remove the washer timer control knob and dial.
10. Remove the dryer timer control knob.
11. Remove the screws securing both end caps.
12. Remove the control panel escutcheon.
13. Remove the two (2) screws securing the water temperature control selector switch to the control bracket.
14. Remove and transfer all electrical wiring to the new water temperature control selector switch.
15. Reverse the above procedure to reassemble the unit.

Fabric Temperature Selector Switch 3.0-kiloohm Resistor

On some models, a 3.0-kiloohm resistor is placed in series with the bias heater to reduce the voltage to the bias heater and to provide another level of drum air temperature. Connection of the bias heater and resistor to the circuit is controlled by the fabric selector switch and/or timer. Refer to the product wiring diagram for each model unit.

To remove or replace the fabric temperature selector switch 3.0-kiloohm resistor, use the following procedure:

1. Disconnect the laundry center from the electrical supply.
2. Remove the dryer front access panel.
3. Remove the dryer front access safety cover panel.
4. Remove the four (4) screws securing the console and controls assembly to the side panels.
5. Disconnect the electrical wiring connector blocks.
6. Remove the electrical wiring leads for the console light (if applicable).

Switch position	SOB-Y	SWB-R	SWB-SR
High heat	X	O	O
Medium heat	X	X	O
Low heat	X	O	X
No heat	O	O	O

X = Closed
O = Open

Figure 16–75 Fabric temperature selector switch chart. (Courtesy of WCI Major Appliance Group.)

7. Lay the console and controls assembly on a protected surface.
8. Remove the fabric temperature selector switch 3.0-kiloohm resistor.
9. Install the new fabric temperature selector switch 3.0-kiloohm resistor.
10. Reverse the above procedure to reassemble the unit.

Safety Thermostat

The safety thermostat is a snap-action, automatic-reset, enclosed-disc type. It provides protection for the dryer in the event of a blocked exhaust, clogged lint screen, or an overloaded drum. It is mounted on the top side of the heater housing, in the air stream, and senses a heat buildup.

To test the safety thermostat, use the following procedure. Caution: Disconnect the laundry center from the electrical supply before making any continuity tests.

1. Disconnect the laundry center from the electrical supply.
2. Remove the dryer top panel.
3. Remove the leads to the safety thermostat.
4. Connect the ohmmeter across the safety thermostat terminals.
5. The ohmmeter should show continuity between the two (2) thermostat terminals at room temperature.
6. To check for stuck contacts in the thermostat, run the dryer on high heat with the exhaust duct completely blocked. The contacts must open within 3 minutes.

Like the control thermostat, the safety thermostat is very reliable and seldom fails or drifts from its temperature setting.

To remove or replace the safety thermostat, use the following procedure:

1. Disconnect the laundry center from the electrical supply.
2. Remove the dryer front access panel.

Thermostat Temperature (°F)

Control thermostat	Open	Closed
High	170° ± 5°	145° ± 5°
Medium	160° ± 5°	135° ± 5°
Low	150° ± 5°	125° ± 5°

Figure 16–76 Heat control thermostat cut-in, cut-out chart. (Courtesy of WCI Major Appliance Group.)

3. Remove the dryer front access safety cover panel.

4. Remove the console and controls.

5. Reach through and release the drum belt from the idler pulley to relieve the tension on the front panel.

6. Remove the two (2) lower front screws securing the dryer front panel assembly to the side panels.

7. Remove the four (4) upper top front screws securing the dryer front panel assembly.

8. Carefully remove the dryer front panel assembly.

9. Remove the dryer top panel.

10. Remove the dryer drum by lifting it up to release it from the rear bearing support (trailer hitch).

11. Remove the electrical wiring to the safety thermostat switch.

12. Install the new safety thermostat switch.

13. Reverse the above procedure to reassemble the unit.

Heat Control Thermostat

The heat control thermostat is a single-pole, single-throw, normally closed (but opens on temperature rise), bimetal, snap-action switch. It opens and closes internal contacts to control the temperature of the air flowing through the dryer. The control thermostat is located in the blower housing and is in series with the gas burner. On some laundry centers, the control thermostat has a low-wattage biasing heater mounted close to the bimetal. When the bias heater is energized, the bimetal heats up more quickly and opens the circuit to the gas burner at a lower drum air temperature. On some models, a 3.0-kiloohm resistor is placed in series with the bias heater, reducing the voltage to the bias heater and providing another level of drum air temperature. Connection of the bias heater and resistor to the circuit is controlled by the fabric selector switch and/or timer. Refer to the wiring diagram for each model.

Heat control thermostat testing When testing the heat control thermostat, use the following procedure. *Caution:* Disconnect the laundry center from the electrical supply before making any continuity tests.

1. Disconnect the laundry center from the electrical supply.

2. Remove the dryer front access panel.

3. Remove the dryer front access safety cover panel.

4. Remove the wires from the heat control thermostat terminals. Check the continuity between the terminals. The ohmmeter should show continuity between them at room temperature.

5. To check the cutout temperature, insert a thermocouple sensing lead into the lint trap. Allow the dryer to run long enough to permit the thermostat to cycle three or four times. The temperature after the third or fourth cycle should comply with the temperature chart. See Figure 16–76.

To remove or replace the heat control thermostat, use the following procedure:

1. Disconnect the laundry center from the electrical supply.

2. Remove the dryer front access panel.

3. Remove the front access safety cover panel.

4. Remove the electrical wiring to the heat control thermostat.

5. Remove the two (2) screws securing the heat control thermostat and the biasing heater to the mounting bracket.

6. Install the new heat control thermostat.

7. Reverse the above procedure to reassure the unit.

Heat Control Thermostat Biasing Heater

Use the following procedures when testing or replacing the heat control thermostat biasing heater.

To test the heat control thermostat biasing heater, use the following procedure:

1. Disconnect the laundry center from the electrical supply.

2. Remove the dryer front access panel.

3. Remove the dryer front access safety cover panel.

4. Remove the electrical wiring to the heat control thermostat biasing heater. Connect an ohmmeter across the terminals.

5. The ohmmeter should show 10,000 ohms ± 10 percent.

To remove or replace the heat control thermostat biasing heater, use the following procedure:

1. Disconnect the laundry center from the electrical supply.

2. Remove the dryer front access panel.

3. Remove the dryer front access safety cover panel.

4. Remove the electrical wiring to the heat control thermostat biasing heater.

5. Remove the two (2) screws securing the heat control thermostat and the biasing heater to the mounting bracket.

6. Install the new heat control thermostat biasing heater.

7. Reverse the above procedure to reassemble the unit.

End-of-Cycle Buzzer

The end-of-cycle buzzer sounds for approximately 5 seconds at the end of any drying cycle, thus indicating to the user that the clothes should be removed from the dryer at that time to avoid wrinkling.

The Auto-Dry Cycle has a 10-minute cool-down period as opposed to the 6-minute cool-down period for the Timed Dry. Contact W to BL will close, allowing the current to flow through the buzzer coil. The pulser will allow the buzzer to sound for 5 seconds out of each 50 seconds for a 5-minute interval, thus alerting the user every 5 minutes to remove the clothes.

The end-of-cycle buzzer is part of the dryer timer.

Dryer Drive Motor

The dryer drive motor is a single-speed, dual-shaft, 1725-rpm, $\frac{1}{8}$-horsepower motor, rated at 120 volts, 60 hertz. An internal overload protector is built into the motor and will open due to excessive temperature rise and/or excessive current. If the overload protector opens, all dryer operations will stop. The motor protector is self-resetting. It may take as long as 30 minutes for it to reset.

The dryer drive motor has a multiple-terminal, snap-on, electrical wiring connector block.

The centrifugal switch in the dryer motor performs three functions. When the motor is at rest, a single-pole, double-throw (SPDT) contact closes a circuit to the motor start windings. When the dryer is started, power is momentarily supplied to the motor through the push-to-start switch. As the motor comes up to speed, the SPDT contact disconnects the start winding and supplies power to the run winding. This provides a holding circuit to keep the motor operating after the momentary push-to-start switch is released. As the motor switch operates, a second set of contacts closes a circuit to the gas burner. This portion of the motor switch prevents the gas burner from operating until the motor is operating at full speed. The motor start switch *is not replaceable.*

Dryer drive motor testing The motor, motor protector, and centrifugal motor switch are the three (3) components that must function correctly in order for the motor to run. To determine the cause of a motor that is not operating properly, it is necessary to isolate the related components from the rest of the circuitry. The following service test procedure is suggested to facilitate locating a defective component in the motor circuit.

1. Disconnect the laundry center from the electrical supply.
2. Remove the dryer front access panel.
3. Remove the dryer front access safety cover panel.
4. Remove the console and controls.
5. Reach through and release the drum belt from the idler pulley to relieve the tension on the front panel.

6. Remove the two (2) lower front screws securing the dryer front panel assembly to the side panels.
7. Remove the four (4) upper top front screws securing the dryer front panel assembly.
8. Carefully remove the dryer front panel assembly.
9. Remove the dryer top panel.
10. Remove the dryer drum by lifting it up to release it from the rear bearing support (trailer hitch).
11. Disconnect all wires coming from the wiring harness to the motor switch. Do not disconnect the wires that come from inside the motor itself.
12. Connect a properly fused test cord to terminals 1 and 4 of the motor start switch. If the motor runs, the problem is in the dryer wiring or another control. If it does not run, continue the test procedure.

The motor overload protector is not available as a replacement part. Should the protector fail, the motor must be replaced. The contacts of the protector are normally closed. They open in response to a temperature rise, an excessive current draw, or a combination of both. The contacts reset automatically when the protector cools down. Do not condemn a motor for having a defective motor protector while the motor is hot. Allow it to cool down and then check it again.

To test the motor protector continuity, use the following procedure:

1. Disconnect the laundry center from the electrical supply.
2. Remove the dryer front access panel.
3. Remove the dryer front access safety cover panel.
4. Remove the console and controls.
5. Reach through and release the drum belt from the idler pulley to relieve the tension on the front panel.
6. Remove the two (2) lower front screws securing the dryer front panel assembly to the side panels.
7. Remove the four (4) upper top front screws securing the dryer front panel assembly.
8. Carefully remove the dryer front panel assembly.
9. Remove the dryer top panel.
10. Remove the dryer drum by lifting it up to release it from the rear bearing support (trailer hitch).
11. Remove the white wire from terminal 1 of the motor start switch.
12. Connect an ohmmeter across terminals 1 and 4 of the centrifugal switch (2 to 4.5 ohms ± 10 percent indicates a closed motor protector). The test is made through the start and run windings of the motor.

If the motor protector opens repeatedly during the dryer operation, check the following:

1. A binding condition in the motor, blower wheel, or drum suspension system.
2. Poor electrical connections or a low voltage condition.
3. Dust or lint in the ventilating openings of the motor, causing the motor to overheat.

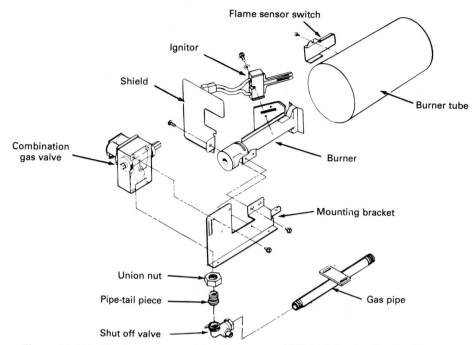

Figure 16–77 Gas burner assembly. (Courtesy of WCI Major Appliance Group.)

If the motor protector is found open and the motor is cool, the motor must be replaced.

To remove or replace the dryer drive motor, use the following procedure:

1. Disconnect the laundry center from the electrical supply.
2. Remove the dryer front access panel.
3. Remove the dryer front access safety cover panel.
4. Remove the console and controls.
5. Reach through and release the drum belt from the idler pulley to relieve the tension on the front panel.
6. Remove the two (2) lower front screws securing the dryer front panel assembly to the side panels.
7. Remove the four (4) upper top front screws securing the dryer front panel assembly.
8. Carefully remove the dryer front panel assembly.
9. Remove the dryer top panel.
10. Remove the dryer drum by lifting it up to release it from the rear bearing support (trailer hitch).
11. Remove the two (2) screws securing the exhaust vent elbow to the blower housing.
12. Remove the six (6) screws securing the dryer front blower housing to the rear blower housing.
13. Loosen the hex bolt securing the blower wheel and slide the blower wheel off the drive motor shaft.
14. Remove the three (3) screws securing the rear blower housing to the drive motor cradle.
15. Release the electrical wiring connector block to the drive motor.
16. Remove the two (2) clamps securing the drive motor to the mounting cradle.

17. Remove the dryer drive motor belt pulley (press-on fit, flush with the end of the motor shaft).
18. Install the new dryer drive motor.
19. Reverse the above procedure to reassemble the unit.

GAS COMPONENT OPERATION AND REPAIR

The following is a description of the operation and repair procedures for the gas components of the laundry center.

Operation

The heat source for the gas dryer section of the laundry center model is the burner assembly. It consists of a combination gas valve, ignitor, flame sensor switch, burner, and burner tube. The burner comes from the factory sized and adjusted to deliver heat at the rate of 20,000 Btus per hour, using natural gas. *Warning:* No other fuel should be used with the burner assembly unless the burner assembly is properly modified. [See the Installation Instructions covering liquefied petroleum LP gas.]

The burner assembly is mounted to the dryer base access shield and to the right of the drum. It is accessible through the dryer lower front access panel.

Combination Gas Valve

The combination gas valve assembly includes two (2) solenoid shutoff valves, a pressure regulator, and a gas orifice assembled into a single-cast body. See Figures 16–77 and 16–78. The pressure regulator is factory set to maintain $3\frac{1}{2}$ inches of water column gas pressure at the orifice. The regulator is service adjustable.

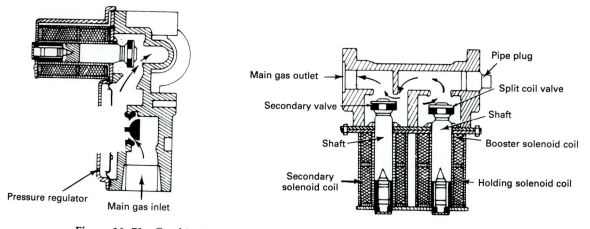

Figure 16–78 Combination gas valve assembly. (Courtesy of WCI Major Appliance Group.)

The two (2) solenoid valves are in series, so that the gas flowing through the pressure regulator must pass through both valves to get to the orifice and burner. The first solenoid valve is referred to as the split-coil valve [two (2) coils stacked in one (1) housing]. The upper holding coil is capable of holding the valve open, but it is not able to open the valve from its normally closed position. To open the valve, the lower booster coil must be energized. The second solenoid valve is called the secondary valve. This valve is operated by a single electrical coil that opens the valve and holds it open when the gas burner is in operation.

The orifice is a precisely drilled brass plug screwed into the outlet port of the combination valve. The orifice extends into the burner. The combination of regulated pressure and orifice size provides the proper volume of gas for the heat rating of the burner.

Burner

The burner is a short metal tube in which air is mixed with the gas fuel to provide a combustible mixture that will burn clean, without smoke or noxious fumes. An adjustable air shutter at the inlet controls the amount of air to be mixed. See Figure 16–79. The shutter is preset and does not ordinarily require adjustment. A flame spreader extends out over the open end of the burner. This is the area where combustion of the fuel/air mixture takes place. As installed in the dryer, the burner extends into the burner tube about 5 inches. All combustion takes place within the burner tube.

Ignitor

Ignition of the fuel/air mixture is accomplished through the use of a silicon-carbide ignitor (glow-bar). The ignitor is mounted to the burner at an angle, with the glow-bar extended into the flame area. Silicon carbide is a material capable of withstanding very high temperatures. However, it is fragile and susceptible to contamination from skin oils and other foreign material. It must be handled with care, using the insulated support. See Figure 16–79.

Electrically, silicon carbide is a thermistor. Its electrical

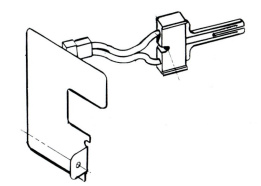

Figure 16–79 Ignitor. (Courtesy of WCI Major Appliance Group.)

resistance changes (decreases) as its temperature increases. When energized, the ignitor begins to heat (steadily decreasing its resistance) and it eventually reaches a temperature of 1800°F.

Flame Sensor Switch

The flame sensor switch is located on the left side of the burner tube. See Figure 16–80. It consists of a thermostatically operated single-pole, single-throw (SPST), normally closed (NC) switch, mounted over a window cut into the burner tube. The flame sensor switch senses the intense heat, radiating initially from the ignitor, and then from the burner flame. The sensor switch opens within 15 to 90 seconds after the ignitor reaches the 1800°F ignition temperature. The reaction time is longest during a cold start-up.

Gas Valve Operating Sequence

At the time of dryer start-up and upon a call for heat from the heat control thermostat, the ignitor is energized and the split-coil valve opens as the following circuits are simultaneously completed. See Figure 16–81.

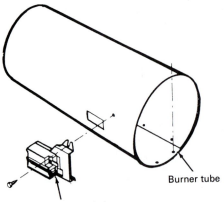

Figure 16–80 Flame sensor switch. (Courtesy of WCI Major Appliance Group.)

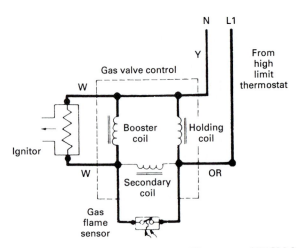

Figure 16–81 Gas valve operation. (Courtesy of WCI Major Appliance Group.)

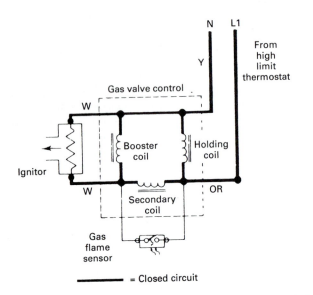

Figure 16–82 Gas valve operation (flame sensor switch open and flame established). (Courtesy of WCI Major Appliance Group.)

1. From the L1 (Hot) side of the line (Yellow) through the holding coil to N (Neutral) (Orange).
2. From the L1 (Hot) side of the line (Yellow) through the flame sensor switch and the booster coil to N (Neutral) (Orange).
3. From the L1 (Hot) side of the line (Yellow) through the flame sensor switch and the ignitor to N (Neutral) (Orange).

Within 15 to 90 seconds after the ignitor reaches the ignition temperature (1800°F), the flame sensor switch opens. At this point, current for the ignitor passes through the secondary coil. Because the resistance of the ignitor is low when fully heated, the secondary valve opens. The gas flows through the orifice into the burner, and ignition takes

place. See Figure 16–82. The impedance of the secondary coil reduces the current through the ignitor from approximately 4 amps to 1 amp. The current flow through the booster coil drops to almost zero. The split-coil valve remains open because the holding coil remains fully energized. The flame sensor switch contacts are held open by the heat radiated from the flame.

Burner operation will continue until the electrical power to the burner assembly is interrupted by either the timer, the heat control thermostat, the high limit safety thermostat, the door switch, the auxiliary contacts on the motor switch (motor failure), a blown fuse or tripped circuit breaker, or an interruption in the electrical service to the home. At such time, both gas valves will close and the flame sensor switch will close. When power is restored, the ignition cycle is repeated.

Fail-Safe Features

The following is a listing of the fail-safe features on most laundry centers.

Momentary power interruption Upon restoration of the power, the flame sensor switch contacts will open, permitting the secondary valve to remain open. The split-coil valve will remain closed, because the pull-in circuit for the booster coil cannot be made through the open flame switch contacts. When the flame sensor switch contacts do reclose, the secondary valve will close and the split-coil valve will open. The dryer will then go through the normal cycle for ignition.

Ignition failure If there is a momentary interruption in the gas supply or if a flame is not established as the flame sensor switch contacts open, the secondary valve will remain open until the flame switch contacts reclose. At that time, the secondary valve will reclose. The flame sensor switch will continue to recycle the ignitor and the secondary valve (once per minute) until normal operation is restored, the thermostat is satisfied, or the timer terminates the cycle.

Flame failure In the event of flame failure, the flame sensor switch contacts will reclose in about 45 seconds. At that time, the secondary valve will close (the split-coil valve will stay open) and a try for reignition will be made as in a normal cycle.

Component Replacement

The following is a description of the replacement of various components on a laundry center.

Gas valve and burner assembly The dryer gas valve and burner assembly is rated at 20,000 Btu per hour capacity. It is located behind the dryer lower front access panel.

To remove or replace the burner assembly, use the following procedure:

1. Disconnect the laundry center from the electrical supply.
2. Remove the two (2) screws at the upper corners securing the dryer front access panel.
3. Tilt the front access panel out and lift it up to remove it.
4. Remove the six (6) screws securing the dryer front access safety cover panel.
5. Remove the four (4) screws securing the console and controls assembly.
6. Disconnect the electrical wiring connector blocks to the console and controls assembly.
7. Remove the electrical wiring leads for the console light (if applicable).
8. Lay the console and controls assembly on a protected surface.
9. Shut off the gas supply at the manual shutoff valve, located adjacent to the gas valve assembly.
10. Remove the four (4) screws securing the gas valve and burner assembly to the dryer base access shield.
11. Loosen the shutoff valve union completely. (Refer to Figure 16–78.)
12. Disconnect the two (2) wire leads from the flame sensor terminals.
13. Disconnect the electrical connector blocks to the ignitor and gas valve assembly solenoids.
14. Remove the single hex-head screw securing the burner assembly to the base of the dryer.
15. Slide the burner assembly out the front access opening.
16. Reverse the above procedure to reassemble the unit. Make sure that all the connections are tight.

Note: After reinstalling the burner assembly, open the shutoff valve. Check the union fitting for leaks with a soap suds solution. *Caution:* Never use a lighted match or lighter to check for gas leaks.

Flame sensor switch The flame sensor switch senses the intense heat radiating from the ignitor, and then from the burner flame. The flame sensor switch consists of a thermostatically operated single-pole, single-throw (SPST), normally closed (NC) switch. It is mounted over a window cut into the left side of the burner tube. See Figure 16–79.

To remove or replace the flame sensor switch, use the following procedure:

1. Disconnect the laundry center from the electrical supply.
2. Remove the two (2) screws at the upper corners securing the dryer front access panel.
3. Tilt the front access panel out and lift it up to remove it from the dryer.
4. Remove the six (6) screws securing the dryer front access safety cover panel.
5. Shut off the gas supply at the manual shutoff valve, located adjacent to the gas valve assembly.
6. Disconnect the two (2) wire leads from the flame sensor switch terminals.
7. Remove the $\frac{1}{4}$-inch hex-head screw securing the flame sensor switch to the burner tube, rotate the sensor down, and lift it out of its lower mounting slot.
8. Install the new flame sensor switch.
9. Reverse the above procedure to reassemble the unit.

Ignitor The ignitor is a silicon-carbide thermistor. The sensor is mounted on the left side of the burner at an angle so that the silicon stem will extend into the flame area and sense the high temperature radiating from the flame. When it reaches approximately 1800°F, its contacts open. This energizes the secondary solenoid valve coil. The valve then opens, allowing gas to flow through the gas valve orifice and impinge upon the hot, glowing ignitor. The total sequence occurs within 15 to 90 seconds.

The ignitor stem is very fragile and susceptible to contamination from skin oils. *Handle with care* by using the insulated support of the ignitor. See Figure 16–79.

To test the ignitor, use the following procedure:

1. Disconnect the laundry center from the electrical supply.
2. Remove the two (2) screws at the upper corners securing the dryer front access panel.
3. Tilt the front access panel out and lift it up to remove it from the dryer.
4. Remove the six (6) screws securing the dryer front access safety cover panel.
5. Shut off the gas supply at the manual shutoff valve, located adjacent to the gas valve assembly.
6. Disconnect the plug from the ignitor-to-coil harness.
7. Check the resistance value of the ignitor. It should be approximately 50 to 400 ohms at 75°F room temperature.

To remove or replace the ignitor, use the following procedure:

1. Disconnect the laundry center from the electrical supply.
2. Remove the two (2) screws at the upper corners securing the dryer front access panel.

3. Tilt the front access panel out and lift it up to remove it from the dryer.

4. Remove the six (6) screws securing the dryer front access safety cover panel.

5. Shut off the gas supply at the manual shutoff valve, located adjacent to the gas valve assembly.

6. Disconnect the wiring connector block to the ignitor.

7. Loosen the $\frac{1}{4}$-inch hex-head screws with the washer securing the ignitor to its mounting bracket.

8. Install the new ignitor.

9. Reverse the above procedure to reassemble the unit.

Pressure regulator valve The pressure regulator is part of the combination gas valve assembly, which includes two-solenoid shutoff valves and a gas orifice assembled into a single-cast body. See Figures 16–77 and 16–78. The pressure regulator is factory set to maintain $3\frac{1}{2}$ inches of water column gas pressure at the orifice. The regulator is service adjustable.

To test the pressure regulator valve, use the following procedure:

1. Fully close the gas shutoff valve just ahead of the pressure regulator. Remove the small plug from the end of the combination valve body (see Figure 16–78) and securely install a water manometer adaptor. Connect a water manometer to the adaptor.

2. Fully open the gas shutoff valve. Set the timer to operate for more than 20 minutes. Start the dryer by pressing the START button. With the burner on, check the flowing gas pressure at the water manometer connection. If the flowing gas pressure is above or below $3\frac{1}{2}$ inches of water column, remove the screw cap on top of the regulator. Gradually turn the adjustment screw beneath the cap, observing the change in pressure on the manometer scale. Adjust the flowing gas pressure for $3\frac{1}{2}$ inches of water column.

Note: If the pressure is low and the regulator does not respond to the adjustment, check to make certain that the pressure of the gas supply line is above 5 inches of water column. If the supply pressure is adequate, replace the combination gas valve assembly. (See the procedure To Remove or Replace the Gas Valve Assembly.)

After completing the flowing gas pressure test, close the gas shutoff valve and remove the manometer tube. Replace and tighten the plug in the combination gas valve body. Reopen the shutoff valve and leak-test the assembly with a soap solution.

Note: The screw cap on the regulator also serves as a vent to release air above the regulator diaphragm as the diaphragm moves up or down to regulate the gas pressure. Make certain that the vent opening in the cap is clear.

Gas valve solenoid coils There are two solenoid coils mounted on the gas valve body. See Figure 16–77. The secondary coil (the smaller one) has two terminals. The split coil (the larger one), consisting of the holding and booster coils molded into a single coil, has three terminals.

Note: The resistance values stated in the following component tests will vary with the room temperature. All values are approximate.

To make a quick test of all valve coils, use the following procedure:

1. Disconnect the laundry center from the electrical supply.

2. Remove the two (2) screws at the upper corners securing the front access panel.

3. Tilt the front access panel out and lift it up to remove it from the dryer.

4. Remove the six (6) screws securing the dryer front access safety cover panel.

5. Shut off the gas supply at the manual shutoff valve, located adjacent to the gas valve assembly.

6. Disconnect the ignitor/burner assembly wiring harness and one lead of the flame switch.

7. Measure the resistance at the burner assembly terminal connection. The resistance should be 800 ohms ± 10 percent at 77°F. If the resistance falls outside of this range, check each individual coil as described below.

To test the secondary coil, use the following procedure:

1. Disconnect the laundry center from the electrical supply.

2. Remove the two (2) screws at the upper corners securing the dryer front access panel.

3. Tilt the front access panel out and lift it up to remove it from the dryer.

4. Remove the six (6) screws securing the dryer front access safety cover panel.

5. Shut off the gas supply at the manual shutoff valve, located adjacent to the gas valve assembly.

6. Remove the plug connector from the coil.

7. Check the resistance value of the coil. It should be approximately 1200 ohms ± 10 percent at 77°F.

To test the split coil (holding and booster coils), use the following procedure:

1. Disconnect the laundry center from the electrical supply.

2. Remove the two (2) screws at the upper corners securing the dryer front access panel.

3. Tilt the front access panel out and lift it up to remove it from the dryer.

4. Remove the six (6) screws securing the dryer front access safety cover panel.

5. Shut off the gas supply at the manual shutoff valve, located adjacent to the gas valve assembly.

6. Remove the plug connector from the coil.

7. Note the numbers 1 and 3 marked near the terminals. Number 1 is the common terminal.

8. Check the resistance value between terminals 1 and 3. This is the booster coil and it should have a resistance of approximately 530 ohms ± 10 percent at 77°F.

To remove or replace the gas valve coils, use the following procedure:

1. Disconnect the laundry center from the electrical supply.
2. Remove the two (2) screws at the upper corners securing the dryer front access panel.
3. Tilt the access panel out and lift it up to remove it from the dryer.
4. Remove the six (6) screws securing the dryer front access safety cover panel.
5. Shut off the gas supply at the manual shutoff valve, located adjacent to the gas valve assembly.
6. Disconnect the plug connectors from the coils.
7. Remove the two (2) Phillips-head screws securing the coil hold-down bracket to the valve body.
8. Remove the coils, noting their positions. The secondary coil has two (2) terminals and a small embossment that mates with a small hole in the hold-down bracket.

The split coil has three (3) terminals, as well as an embossment. In addition, the gas valve stem for this coil has a sleeve on it to ensure that the correct coil will be mounted in that position.

9. Reverse the above procedure to reinstall the coils on the valve body.

To remove or replace the gas valve assembly, use the following procedure:

1. Disconnect the laundry center from the electrical supply.
2. Remove the two (2) screws at the upper corners securing the dryer front access panel.
3. Tilt the front access panel out and lift it up to remove it from the dryer.
4. Remove the six (6) screws securing the dryer front access safety cover panel.
5. Remove the four (4) screws securing the console and controls assembly to the side panels.
6. Disconnect the electrical wiring connector blocks to the console and controls assembly.
7. Remove the electrical wiring leads for the console light (if applicable).
8. Lay the console and controls assembly on a protected surface.
9. Shut off the gas supply at the manual shutoff valve, located adjacent to the gas valve assembly.
10. Remove the four (4) screws securing the gas valve and burner assembly to the dryer base access shield.
11. Loosen the shutoff valve union completely. (Refer to Figure 16–76.)
12. Disconnect the two (2) wire leads from the flame sensor switch terminals.
13. Disconnect the electrical connector blocks to the ignitor and gas valve assembly solenoids.
14. Slide the burner assembly out the front access opening.
15. Remove the one (1) screw securing the burner tube to the mounting bracket.

16. Remove the one (1) $\frac{1}{4}$-inch hex screw securing the flame shield to the mounting bracket.
17. Loosen the $\frac{1}{4}$-inch hex-head screw with the washer securing the ignitor to its mounting bracket.
18. Remove the two (2) hex screws securing the burner and air shutter to the mounting bracket.
19. Remove the three (3) $\frac{1}{4}$-inch hex screws securing the gas valve to the mounting bracket.
20. Transfer the union nut and tail piece.
Note: Be sure to use an approved pipe and thread sealer that is resistant to the corrosive action of liquefied petroleum gases.
21. Install the new gas valve assembly.
22. Reverse the above procedure to reassemble the unit. Make sure that all gas connections are tight.

Note: After reinstalling the burner assembly, open the shutoff valve and check the union fitting for leaks with a soap suds solution. *Caution:* Never use a lighted match or lighter to check for gas leaks.

Burner and air shutter assembly The burner acts as a mixing chamber by allowing a certain amont of air to mix with the gas leaving the orifice. At the end of the burner is a flame spreader. The other end has an adjustable air shutter, preset at the factory. It does not ordinarily require adjustment.
To adjust the air shutter, use the following procedure:

1. Loosen the one (1) $\frac{1}{4}$-inch hex-head screw on top of the air shutter.
2. Gradually slide the shutter forward or backward until the yellow tips of the burner flame disappear.
3. Retighten the shutter screw.

To remove or replace the burner, use the following procedure:

1. Disconnect the laundry center from the electrical supply.
2. Remove the two (2) screws at the upper corners securing the dryer front access panel.
3. Tilt the front access panel out and lift it up to remove it from the dryer.
4. Remove the six (6) screws securing the dryer front access safety cover panel.
5. Remove the four (4) screws securing the console and controls assembly to the side panels.
6. Disconnect the electrical wiring connector blocks to the console and controls assembly.
7. Remove the electrical wiring leads for the console light (if applicable).
8. Lay the console and controls assembly on a protected surface.
9. Shut off the gas supply at the manual shutoff valve, located adjacent to the gas valve assembly.
10. Remove the four (4) screws securing the gas valve and burner assembly to the dryer base access shield.
11. Loosen the shutoff valve union completely. (Refer to Figure 16–77.)

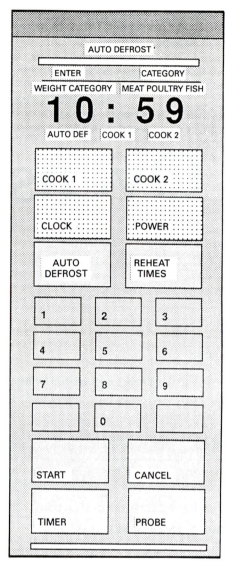

Figure 17–1 Typical touch control. (Courtesy of WCI Major Appliance Group.)

Table 17–1 Conversions of ounces to tenths of a pound.

Weight in Ounces	Tenths of a Pound
1-2 ounces	enter as .1
3 ounces	enter as .2
4-5 ounces	enter as .3
6-7 ounces	enter as .4
8 ounces	enter as .5
9-10 ounces	enter as .6
11 ounces	enter as .7
12-13 ounces	enter as .8
14-15 ounces	enter as .9

(Courtesy of WCI Major Appliance Group.)

Clock The Clock pad is used to set the time of day or to check the time of day during the cooking operation.

To set the clock, use the following procedure:

1. Touch the Clock pad.
2. Touch the number pads corresponding to the time of day.
3. Touch the Start pad.

Auto Defrost The automatic defrost function automatically calculates the exact amount of time and the correct power level needed to defrost the food. Select one of the three food categories (Meat, Poultry, or Fish) and enter the weight of the food being placed in the microwave. A preprogrammed standing time of 5 minutes is included in the calculated time to allow complete defrosting before the cooking process begins.

To use automatic defrost, use the following procedure:

1. Touch the Auto Defrost pad. The Auto Defrost indicator light will appear and flash in the display window. The Enter Category indicator light will also flash.

2. Touch the Auto Defrost pad again to select the food category. The indicator light will move across the display window under three choices (Meat, Poultry, or Fish) as the Auto Defrost pad is touched.

3. Touch the appropriate number pads to enter the weight of the food being placed in the oven. The food weight must be entered in pounds and tenths of pounds. Refer to Table 17–1 for the conversion of the package weights in ounces into decimal numbers. If the weight is not entered within 3 seconds after choosing the food category, the Enter Weight indicator light will appear and flash in the display window as a signal to enter the weight of the food.

If an entered weight is not within the range of the minimum and maximum weights for each category, the control will beep, signaling that the food is too large or too small to be defrosted using the automatic defrost function of the microwave. The maximum and the minimum weight for each food category is listed in Table 17–2.

4. Touch the Start pad. The Auto Defrost indicator light will continue to flash in the display window. When the preprogrammed defrosting time has run out, the oven will beep and the word End will appear in the display window.

Defrost (models with five power levels) For defrosting foods the control board automatically operates the oven at 30 percent power.

Table 17–2 Minimum and maximum food weights allowed.

Food	Minimum Weight	Maximum Weight
Meat	.1 lbs.	8.0 lbs.
Poultry	.1 lbs.	13.0 lbs.
Fish	.1 lbs.	8.0 lbs.

(Courtesy of WCI Major Appliance Group.)

Table 17–3 Setting reheat times.

10 seconds	touch the pad once
30 seconds	touch the pad twice
1 minute	touch the pad three times
1 minute, 30 seconds	touch the pad four times
2 minutes	touch the pad five times
3 minutes	touch the pad six times

(Courtesy of WCI Major Appliance Group.)

To select an automatic defrosting operation, use the following procedure:

1. Touch the Defrost pad.
2. Set the defrost time by touching the appropriate number pads.
3. Touch the Start pad.

Power This oven function may be used in conjunction with Cook 1 and Cook 2. The Power pad allows the user to select a percentage of operating time for the magnetron tube. The percentage of power that may be selected ranges from 10 to 90 percent. To select a percentage of power, select the amount of operating time in either Cook 1 or Cook 2. Touch the Power pad and any number from 1 through 9 (1 being 10 percent and 9 being 90 percent). The magnetron tube will cycle on and off for the selected percentage of time based on a 20-second time base.

Reheat Times The reheat times function is a time-saving convenience when heating small amounts of food or beverages. The reheat times function is preprogrammed for six different reheating times. Touch the pad from one to six times to select one of the preprogrammed settings. See Table 17–3.

Cancel If at any time during operation of the oven it is desired to terminate an operation, touch the Cancel pad and the unit will stop its count and cancel all programs except the time of day.

Probe The automatic temperature probe feature measures the internal temperature of the food, then either turns the oven off automatically when the food is done, or holds the food at that temperature for delayed serving.

To use the automatic temperature probe, use the following procedure:

1. Insert the probe into the food as close to the center as possible.
2. Plug the probe into the microwave oven jack.
3. Touch the probe pad on the oven control panel.
4. Touch the last two numbers of the desired temperature for the food being cooked.
Note: The number 1 is automatically placed in by the control panel for the first number.
5. Touch the Start pad on the control panel.

Timer The timer can be used regardless of whether or not the oven is off or on. The timer can be set during either the cooking or the defrosting process to be used as a reminder to turn or stir foods in the middle of the cooking process. The timer does not control the cooking function. If using the timer during the cooking function, enter the function and start the oven before setting the timer.

To set the timer, use the following procedure:

1. Touch the timer pad. "Time?" will appear in the upper display window.
2. Touch the number pads to set the amount of time desired. The timer can be set for any time from 1 second to 99 minutes and 99 seconds.
3. Touch the Start pad.

Turntable (if applicable) Models that are equipped with a turntable feature have a round glass tray in the bottom of the oven cavity that can be rotated by a motor during the oven cooking operation. This feature is used in conjunction with the stirrer to give the oven an even cooking pattern.

To operate the turntable, use the following procedure:

1. Install the round glass tray into the oven cavity.
2. Set a cooking time on the control panel.
3. Press in on the Turn switch (if applicable).
4. Touch the Start pad.

Note: Turntable models come with a square glass tray and a round glass tray.

INSTALLATION INFORMATION

Microwave ovens must be installed in the proper manner for safe and efficient operation. The following are some guidelines that may be used during the installation process.

ELECTRICAL GROUNDING

Microwave ovens are equipped with a three-prong grounding plug for protection against a shock hazard. These ovens must be plugged directly into a properly grounded receptacle. See Figure 17–2. Do not cut or remove the grounding prong from this plug. Use a properly polarized and grounded three-hold receptacle as required by the National Electrical Code on all new construction.

The Appliance Must Be Grounded

For the safety of the user, it is required that all major appliances requiring electricity for operation be electrically grounded in accordance with the National Electric Code ANSI-1981.

Check all code rules and regulations for connecting the appliance to be certain that the installation conforms with all local, municipal, and state codes as well as local utility regulations.

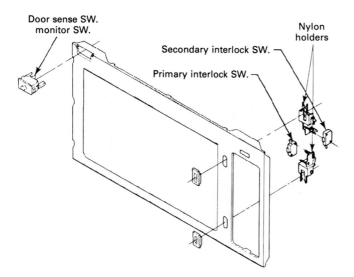

Door sense SW.
monitor SW.

Secondary interlock SW.

Primary interlock SW.

Nylon holders

Figure 17–13 Primary interlock switch replacement. (Courtesy of WCI Major Appliance Group.)

2. Lift off the air duct.
3. The bulb can now be removed.
4. Replace the bulb with a 40-watt bulb and reinstall the air duct.

Note: The cavity light can be replaced by the customer.

Light Socket

To replace the light socket, use the following procedure:

1. Disconnect the electrical power from the unit, remove the wrapper, and discharge the capacitor.
2. Remove the light bulb.
3. Depress the prong and pull the mounting bracket out.
4. Replace the socket wire for wire to prevent any mistake in rewiring the unit. Replace the socket into the cabinet by pushing the socket into the hole provided for it.

Line Fuse

Important: If the 20-amp fuse blows, refer to the troubleshooting section presented later in this chapter. Do not replace the fuse until the problem is corrected. Also, if for any reason a fuse blows, internal or external, *all interlock switches must be replaced.* (See the interlock switches section presented later in this chapter.)

To replace the line fuse, use the following procedure:

1. Disconnect the electrical power from the unit, remove the wrapper, and discharge the capacitor.
2. Remove the fuse from the holder.
3. A blown fuse can be diagnosed visually. The fuse will be discolored.
4. Determine what caused the fuse to blow before replacing it.
5. Replace the fuse with a 20-amp fuse.

Interlock Switches

Use the following procedures to check the primary interlock switch, the safety monitor and sense switch, and the secondary interlock switches:

To check the primary interlock switch, use the following procedure:

1. Disconnect the electrical power from the unit, remove the wrapper, and discharge the capacitor.
2. Remove the control panel to gain access to the interlock switch.
3. Spread the mounting bracket tabs and lift the mounting bracket from the bezel.
4. To remove the switch, spread the switch holding tabs and lift out the switch.
5. Disconnect the electrical leads to the switch.
6. Reverse the above procedure to reassemble the unit. See Figure 17–13.

Important: Be sure that the switch holding tabs of nylon holders are not bent, broken, or otherwise deficient in the power to hold the switches. If they are deficient, replace the holder as well as the switches.

To replace the safety monitor and sense switch, use the following procedure:

1. Disconnect the electrical power from the unit, remove the wrapper, and discharge the capacitor.
2. Remove the wires from the switch.
3. Squeeze the tabs on the sides of the switch and push the switch through the front frame. See Figure 17–13.

To replace the secondary interlock switch, use the following procedure:

1. Disconnect the electrical power from the unit, remove the wrapper, and discharge the capacitor.
2. Remove the control panel to gain access to the switch.
3. Spread the mounting bracket tabs and lift the mounting bracket from the bezel.
4. To remove the switch, spread the switch holding tabs and lift out the switch.
5. Disconnect the electrical leads from the switch.
6. Reverse the above procedure to reassemble the unit.

Important: Be sure that the switch holding tabs of nylon holders are not bent, broken, or otherwise deficient in the power to hold the switches. If they are deficient, replace the holder as well as the switches.

Important: The interlock switches must meet strict federal requirements for reliable operation. It is extremely important that any questionable switches be replaced with the exact same type switch. Defective switches must be rendered inoperative by twisting off one of the terminals.

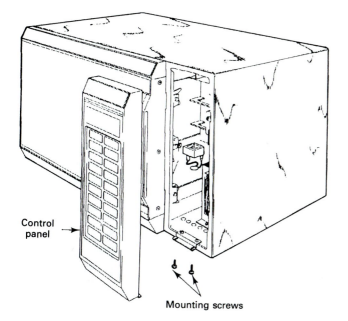

Figure 17–14 Control panel removal. (Courtesy of WCI Major Appliance Group.)

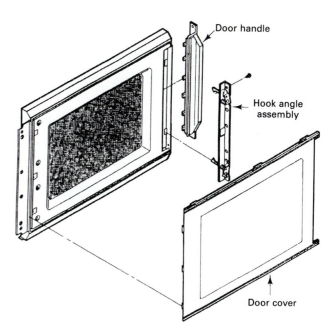

Figure 17–15 Oven door removal. (Courtesy of WCI Major Appliance Group.)

Control Panel Assembly

To remove the control panel, use the following procedure:

1. Disconnect the electrical power from the unit, remove the wrapper, and discharge the capacitor.
2. Tag and disconnect the wiring from the control board.
3. Remove the two (2) screws from the control panel assembly. See Figure 17–14.
4. Open the oven door and swing the bottom of the control panel out about 2 inches, then pull gently downward.
5. To separate the touch panel/circuit board from the control panel, remove the two (2) screws at the top and two (2) screws at the bottom of the control panel.

Door Assembly

To remove the door assembly, use the following procedure:

1. Disconnect the electrical power from the unit, remove the wrapper, and discharge the capacitor.
2. Remove the six (6) screws that secure the hinge of the chassis and remove the door. See Figure 17–15.
3. Remove the three (3) screws that secure the door handle to the door and remove the handle.
4. Lift the door cover from the assembly.
5. Remove the two (2) screws that secure the hook angle assembly to the door and remove the hook angle assembly.

Door Adjustment

The door adjustment on all microwave ovens is very important to prevent any excessive leakage. The RF leakage levels should be checked after all door, interlock switches, and associated repairs are made. The following door adjustment must be performed for proper door alignment.

Note: The primary and secondary interlock switches must open before the door gap exceeds 0.10 inch.

To adjust the hinge side of the door, use the following procedure:

1. Disconnect the electrical power from the unit, remove the wrapper, and discharge the capacitor.
2. Loosen the six (6) screws on the hinge side until the door can be easily moved back and forth.
3. Swing the door 90°. Lift up on the end of the door until it is level. Then hold the door in that position and tighten the screws.

Important: After the repairs have been made, it is very important to make a microwave leakage check of the door.

Magnetron Assembly

To remove the magnetron assembly from the unit, use the following procedure:

1. Disconnect the electrical power from the unit, remove the wrapper, and discharge the capacitor.

Disassembly, Adjustments, and Replacement of Parts 271

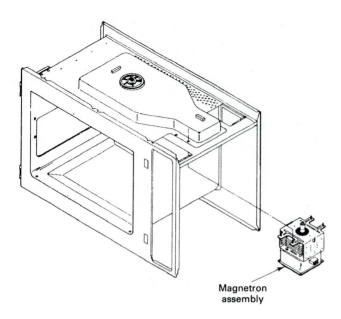

Figure 17–16 Magnetron assembly removal. (Courtesy of WCI Major Appliance Group.)

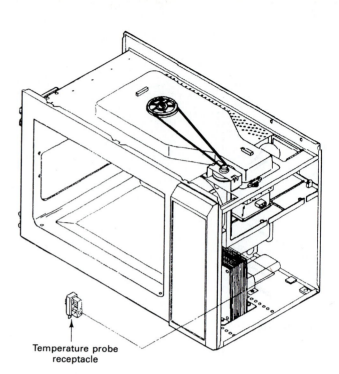

Figure 17–17 Probe receptacle removal. (Courtesy of WCI Major Appliance Group.)

2. Disconnect the electrical wires from the magnetron filament terminals.

3. Remove the two (2) screws that hold the rear of the air duct.

4. While supporting the magnetron, remove the four $\frac{5}{16}$-inch mounting nuts. The magnetron can now be pulled down and out of the unit. See Figure 17–16.

Note: The magnetron studs have metric threads, do not loosen the nuts.

Important: Do not operate the oven with the magnetron disassembled from the waveguide.

Important: Care must be taken to make sure that the RF gasket remains between the magnetron and the waveguide. Replacement magnetrons should be equipped with a new RF gasket to replace the old one.

Probe Receptacle

To remove the probe receptacle, use the following procedure:

1. Disconnect the electrical power from the unit, remove the wrapper, and discharge the capacitor.

2. Remove the electrical wires from the receptacle.

3. While holding the body of the receptacle, loosen the nut on the inside of the oven cavity using a wrench. Once the nut is loose, unscrew it from the receptacle.

4. Reverse the above procedure to install a new receptacle.

5. Insert the probe and check to see that the temperature probe function is working. See Figure 17–17.

Bottom Plate (Turntable Models)

To remove the bottom plate, use the following procedure:

1. Disconnect the electrical power from the unit. Remove any loose items from inside the oven cavity and set the oven on the hinge side.

2. Remove the two (2) screws from the rear of the bottom plate and slide the plate toward the rear and out of the unit. See Figure 17–18.

Turntable Motor (Turntable Models)

To remove the turntable motor, use the following procedure:

1. Disconnect the electrical power from the unit, remove the wrapper, and discharge the capacitor.

2. Remove the bottom plate from the unit.

3. Disconnect the two (2) electrical leads from the turntable motor.

4. Remove the two (2) screws holding the turntable motor from the inside of the oven cavity.

5. Lift the square drive from the motor shaft and pull the motor out of the unit. See Figure 17–18.

TROUBLESHOOTING

Checking Power Wattage

There are two power wattages indicated on the model/serial plate of each microwave oven. For example, the fol-

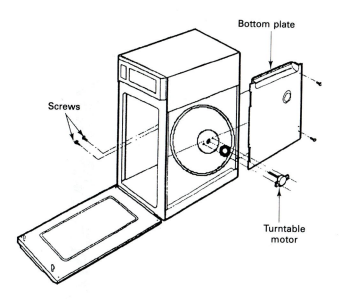

Figure 17–18 Bottom plate removal. (Courtesy of WCI Major Appliance Group.)

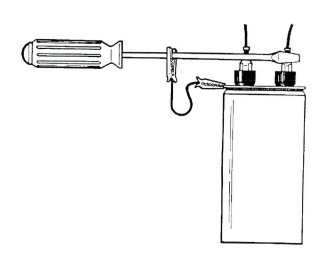

Figure 17–19 Discharging the capacitor. (Courtesy of WCI Major Appliance Group.)

lowing information may be printed below the model/serial number:

> 120 V, 60 HZ, OUTPUT 700 W
> FREQUENCY 2450 MHZ
> 1500 W, 3 WIRE AC ONLY

From the above example, "1500 W" is the amount of power wattage needed to operate the oven during a full power cooking cycle (input power wattage). "OUTPUT 700 W," on the other hand, is the amount of power wattage produced by the magnetron assembly during a full-power cooking cycle (output power wattage).

When testing a microwave oven for power wattage, the output power wattage must be checked. To perform this test and obtain more accurate results, two 1-liter (microwave safe) containers and a glass Centigrade scale thermometer should be used.

To check the output power wattage, use the following procedure:

1. Fill the two 1-liter containers with tap water.
2. Stir the water in each container with the thermometer and record the temperature of each container.
3. Find the average temperature of the two containers by adding the two temperatures together and dividing the total by 2.
4. Place both containers in the center of the microwave oven cavity.
5. Set the oven power level to the highest setting and operate the oven for exactly 2 minutes.

Note: When testing an oven with a mechanical timer, use an accurate watch for better timing accuracy.

6. When 2 minutes have elapsed, stir each container again with the thermometers and record the temperatures.
7. Find the average temperature of the two containers by adding the two temperatures together and dividing the total by 2.
8. Subtract the average temperature determined in step 3 from the average temperature determined in step 7 to determine the temperature rise.
9. Multiply the temperature rise by a factor of 70 to determine the approximate output power wattage.
10. Compare the result in step 9 with the output power wattage printed on the model/serial plate of the unit tested. If the result is within 10 percent of the stated output, the oven has sufficient cooking capabilities.*

Discharging the Capacitor

To discharge the capacitor, use an insulated handle screwdriver and a 29-inch jumper wire with alligator clips.

1. Connect one end of the wire to the screwdriver blade and the other end to the case of the capacitor, or chassis of the oven.
2. Place the screwdriver blade across the terminals of the capacitor. See Figure 17–19.

---

* The 10 percent allowance in output power wattage is due to the voltage operating range of the microwave oven (110 to 120 volts). This is noted in the owner's manual by referring to a "range of cooking time" (i.e., "2 to 3 minutes") when heating food.

18

Electric Wall Ovens

There are several differences between gas and electric ovens and other heating surfaces. Most people prefer one over the other. Both of them have their advantages and disadvantages, depending on how they are viewed by the persons using them.

Electric ovens are a bit slower in both heating up and cooling off than gas ovens. However, they both seem to cook well and the preference is strictly with the user.

In reality, the electric oven is merely a sophisticated hot plate, with the exception that the cooking utensil never touches the element proper. The basic operation of electric cooking units is very simple; turn on the heating elements when heat is wanted and turn them off when heat is not wanted.

The heating element used in electric ovens is of the tubular-cased type. These elements have a nichrome wire encased in a stainless-steel tube and are insulated from it by an insulator. The insulator is usually magnesium-oxide powder. See Figure 18–1. No external insulation is required to prevent electrical shock. Food has very little effect on the stainless-steel tubing that encases the electric heating element.

The amount of heat produced by the nichrome wire is directly proportional to the amount of current that the element is drawing. If we remember Ohm's law, the formula for wattage is

$$W = E^2/R$$

From this we can determine that the wattage is dependent on the square of the voltage. Thus, if the heating element is energized with 240 volts rather than 120 volts, the amount of heat produced by the same heating element is therefore increased four times rather than just being doubled. Also, if the amount of resistance that is in the electrical circuit is doubled, the amount of heat produced is reduced. Thus, we can use this information to our advantage. Having two voltages and two heating elements, we can have as many as eight different amounts of heat produced by the same unit, depending on how the elements are wired to the different voltages. These different amounts of heat are possible by switching in and out the different voltages and the different resistances with the desirable controllers. See Figure 18–2.

A thermostat is used for controlling the temperature in an electric oven. These thermostats are generally of the fluid-filled type rather than the bimetal type. In this type of thermostat the fluid expands with an increase in temperature and causes a set of contacts to open or close, depending on the temperature inside the oven. A selector switch offers a choice between upper and lower heating elements (either the bake position or the broil position) or the use of both heating elements (preheat), but the same thermostat controls the temperature whether the switch is in the bake position or the broil position. The wattage rating of these heating elements is 2000 and 3000 watts when energized with a continuous supply of 240 volts.

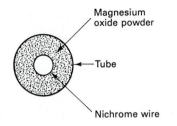

Figure 18–1 Tubular-cased heating element.

Another type of oven control is the timer. Timers are quite popular and most ovens are equipped with some type of device to turn the oven on and off automatically. There are two different types of oven timers: the clock-operated switch and an electric alarm.

The switch timer actually turns the heating element on and off at preset times. The alarm may be either a bell or a buzzer. The alarm has no control over the operation of the heating element; it just warns the user that the amount of preset time has elapsed and that some attention is required.

Most of the problems that are encountered in an electric oven can be related to the heating element, the controller, and wiring faults. There are other problems that can occur with the operation of the accessories, such as the lights and timers which tend to malfunction after some period of usage; however, they have no bearing on the basic operation of the oven. It is problems with the basic operation of the oven, such as the heating elements and the thermostat, that cause the user to call for a service technician.

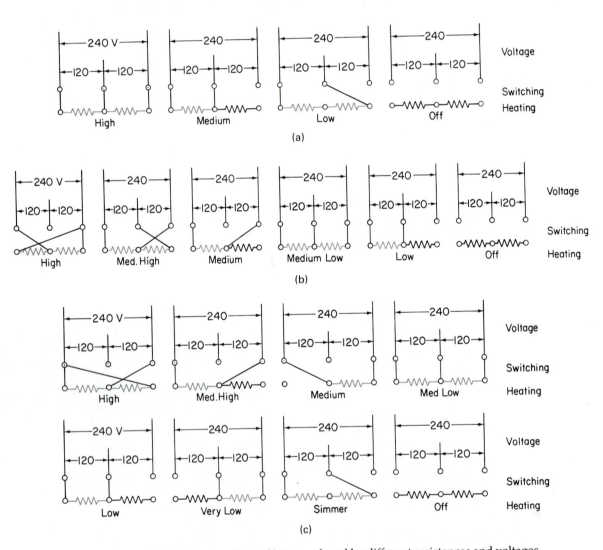

Figure 18–2 The different values of heat produced by different resistances and voltages. (Courtesy of George Meyerink, *Appliance Service Handbook*, 2nd ed., 1988, pp. 40–67, 171–179. Reprinted by permission of Prentice Hall, Englewood Cliffs, NJ.)

OVEN OPERATION

We will use the Frigidaire "F" line electric wall oven for our discussion of this appliance.

Control Panel

The control panel provides a visual indication of all operations that are taking place within the oven. Control panels vary in their style and in their operation.

One type of control panel has an electronic oven control (EOC). The EOC is a solid-state time and temperature control. The oven functions of Bake, Timed Bake, Broil, and Self-Clean are controlled for both time and temperature by a solid-state microprocessor on the circuit boards. There are two types of controllers used on EOCs: an eight-button switch with a knob or a touch pad with a knob system. The knob on both controllers adjusts only the digital display. EOCs are not field repairable.

The microprocessor constantly monitors its internal circuitry, as well as several critical oven circuits, to ensure that they are operating properly. If any of these systems fail, the controller will immediately shut down oven operation, a beeping sound will occur, and an error code will start flashing in the display window. If this condition should occur, simply push the Cancel button.

The oven temperature input to the EOC microprocessor is provided by a sensor located in the oven. The room temperature device (RTD) sensor is a stainless-steel tube with a thin film of platinum at the end. The RTD is a nonadjustable assembly.

Another type of control has an electronic timer control (ETC). The ETC microprocessor controls the timing operation for a selected oven function (top oven only on double-oven models). The selections of Bake, Timed Bake, Broil, and Self-Clean are accomplished by a selector switch. The oven temperature is controlled by a thermostat (hydraulic type) with a thermal bulb. The thermostat temperature range for baking is 150 to 550°F with a broil and self-clean position. The circuit board is a nonfield-repairable assembly.

The selector switch functions of Timed Bake and Clean are available only on selected models. Check the owner's manual or the features listing for that model.

The oven indicators vary for each type of oven controller. The EOC indicates the operator-selected function in the display. The ETC uses a remote light in the selector thermostat circuitry and cycles on and off with the heating element showing that the desired oven temperature is being maintained. The EOC displays the oven temperature as it warms initially, then "locks" onto the selected temperature. The EOC display will not change while the microprocessor circuitry is cycling the oven element.

The lock light operates during self-cleaning and indicates that the oven door cannot be opened.

Broil

The selection of the broil operation allows the heating of the broil element. The food is cooked by direct exposure to the radiated heat from the element. It is not necessary to preheat the broiler. If the recipe calls for preheating, preheat for only 5 minutes.

In selecting the broil operation, the operation is not limited to only high-temperature broiling. To broil slowly, adjust the temperature control to a lower temperature. Also, the height of the broiling rack can be adjusted to vary the heat level of the food. It is suggested that the oven door be left ajar during the broiling process.

On some models the selector switch also connects to the oven indicator. The indicator light will cycle with the operation of the thermostat if the oven door is closed while broiling. It will also cycle if the thermostat is on a temperature other than broil.

Bake

In selecting the bake operation, the circuitry provides full power to the baking element and some power to the broil element.

When preheating the oven, the indicator light turns on and remains on until the oven is preheated to the selected temperature. Preheating the oven will take about 12 to 15 minutes. When the light goes out, the oven is ready for use. Preheating is recommended when baking, but is not necessary for roasting or cooking casseroles.

The oven indicator cycles on and off with the bake heating element showing that the desired oven temperature is being maintained. Some models are not equipped with an oven indicator.

Timed Bake

The timed bake selection allows the user to start the baking operation immediately (or delayed) and to stop after a selected length of cooking time.

Self-Cleaning

The pyrolytic method of cleaning raises the oven temperature to approximately 900°F, carbonizing the food particles resulting in a small amount of white-colored ash on the oven liner, which can easily be wiped out with a wet cloth when the oven is cool.

In setting the self-cleaning operation, remove the broil pan and cooking utensils, remove the oven racks, close the oven door, and set the clock, thermostat, and selector switch, or program the oven controller.

The oven door will automatically lock and begin heating. The oven requires about 60 minutes to reach 900°F. The cycle may be interrupted at any time before the oven lock indicator light turns on. Once the lock light is on, the oven requires a cool-down period before the oven door can be opened.

Oven Vent

The oven is vented on top of the control panel trim. When the oven is on, warm air is released from the oven

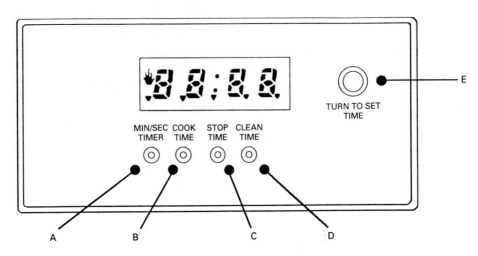

Figure 18–3 Electronic oven control. (Courtesy of WCI Major Appliance Group.)

through the unit. This venting is necessary for proper air circulation in the oven and good cooking results. Do not block this vent opening.

Electronic Timer Control

An electronic timer control (EOC) controls the timing functions of Time of Day, Min/Sec Timer, Timed Bake, and Clean Time. See Figure 18–3. Some products are not equipped with all of the above features. Consult the owner's manual or the feature listing for the model being serviced.

The Time of Day and Min/Sec Timer are set with the same control.

The Min/Sec Timer does not start or stop the cooking process. It simply serves as a timer that buzzes when the set time has run out.

Note: A glowing arrow indicates which time/function is being shown in the display window. A flashing arrow indicates that a particular time/function has been set, but is not being shown in the display window. To check a set time/function, touch the button below the flashing arrow. The set time will appear in the display window.

How to check the clock When the product is first plugged in or when the power supply to the product is disconnected, "88:88" will flash in the display window. Flashing arrows will appear above the Min/Sec Timer and the Cook Time button as a signal to set the clock.

To set the clock, use the following procedure:

1. Push in and hold buttons A and B. See Figure 18–3. Turn knob E in a clockwise direction until the correct time of day appears in the display window.

2. When the correct time is reached, release buttons A and B.

To set the Min/Sec Timer, use the following procedure:

1. Push in and hold button A. Turn knob E in a clockwise direction until the desired time appears in the display window. The timer can be set for any amount of time from 1 second to 99 minutes.
   - When setting the timer for 1 to 59 seconds, each second will show in the display.
   - When setting the timer for 1 to 10 minutes, the display will increase by 10-second increments.
   - When setting the timer for 10 to 99 minutes, the display will increase by 1-minute increments.

2. When the desired time is reached, release button A.

The arrow above the Min/Sec Timer will glow as the set time counts down in the display. When the set time has run out, the timer will beep continuously for about 2 minutes. Touch button A to turn off the control.
   - To check the time of day while the timer is in use, touch and hold buttons A and B. The arrow above the Min/Sec Timer will flash, indicating that the timer is in use. Touch button A to return the minute timer display.
   - To check the minute timer when it is in use, turn E clockwise to increase the time or counterclockwise to decrease the time or turn the timer off.

To set the timed bake feature, use the following procedure:

1. Be sure that the clock shows the correct time of day.

2. Place the food in the oven.

3. Set the temperature control to the desired temperature.

4. Set the selector to Time Bake.

5. Touch and hold the Cook Time button B. Turn knob E until the desired amount of cooking time (in hours and minutes) appears in the display window. The arrow above the Cook Time button will glow and a steaming pot symbol will appear in the upper left-hand corner of the display.

The oven will now start and shut off automatically. In order for the oven to be started at a delayed time and shut off automatically, the stop time must be changed by the control.

6. Touch and hold the Stop Time button C. The display will show the earliest stop time that is possible. Turn knob E until the desired stop time appears in the display window. The stop time can only be increased from the earliest stop time.

The arrow above the Stop Time button C will glow. The arrow above the Cook Time button B will flash, indicating that the cook time has been set, but is not visible in the display window. The steaming pot symbol will go out.

When a delayed start has been set by changing the stop time, the automatic control will calculate backward from the set stop time to determine when cooking should begin. The oven will come on automatically at the calculated time. The steaming pot symbol will glow in the display window.

When the stop time is reached, the oven will turn off and the control will beep continuously for about 2 minutes. Touch any control button A, B, or C to turn the control off.

- To cancel a set start or stop time, touch the Cook Time button B and turn knob E until "0:00" appears in the display window.
- To check the time of day when the automatic timer is in use, touch and hold buttons A and B. The time of day will appear in the display window. The arrows above the Cook Time button B and the Stop Time button C will flash, indicating that the automatic timer is in use.

Touch button B or C to return the Cook Time or the Stop Time to the display window.

To set the self-clean feature, use the following procedure:

1. Close the oven door.
2. Be sure that the clock shows the correct time of day.
3. Set the selector and temperature control to the clean position.

The motor-driven lock begins to close as soon as the controls are set. It takes about 15 seconds for the lock to close. The door can be opened if the self-clean program is canceled before the oven reaches 560°F. After the oven reaches this temperature and the lock light comes on, the door cannot be opened until the oven cools below 560°F.

4. Touch the Clean Time button D. "3:00" will appear in the display window. Three hours has been preprogrammed into the cleaning cycle and is the recommended amount of cleaning time for a moderately soiled oven.

The length of the cleaning time can be changed to a longer or shorter period. To adjust the amount of cleaning time, touch and hold the Clean Time button D. Turn knob B to increase or decrease the cleaning time. The maximum amount of time that the cleaning cycle can be set is 4 hours.

The arrow above the Stop Time button C will flash, indicating that the control has automatically set a stop time based on the length of the cleaning time and the time of day.

The clean cycle will start now and shut off automatically. If it is desired to start the clean cycle at a delayed time and shut off automatically, the stop time which is set by the control must be changed.

5. Touch and hold the Stop Time button C. The display window will show the stop time determined by the control. Turn knob E until the desired stop time appears in the display window. The stop time can be increased from the stop time set by the control.

The arrow above the Stop Time button C will glow. The arrow above the Stop Time button D will flash, indicating that the cleaning time has been set, but is not visible in the display window.

When a delayed start has been set by changing the stop time, the automatic control will calculate backward from the set stop time to determine when the cleaning cycle should begin.

The steaming pot symbol will glow in the upper left corner of the display. The oven signal will come on and remain on until the oven reaches the cleaning temperature. It will then cycle on and off to maintain the proper cleaning temperature.

The oven will heat until it reaches a very high temperature. Soil and spillovers will burn off of the oven surfaces. When the oven reaches the cleaning temperature, the lock light will come on, indicating that the oven door is now locked and cannot be opened until the oven temperature cools below 560°F.

Do not try to open the door when the lock light is on.

When the oven cleaning time has elapsed, the oven must cool for about 1 hour or until the interior temperature has dropped below 560°F. Only then can the oven door be opened.

To interrupt or cancel a clean cycle, use the following procedure:

1. Touch and hold the Clean Time button D. Turn knob E until "0:00" appears in the display window.
2. If the lock light stays on, the oven is too hot to open the door. Wait until the oven has cooled and the lock light goes out. *Do not try to force the oven door.* This can damage the automatic door-locking system.

If the door will not open, use the following procedure:

1. Turn the temperature control and selector to Clean.
2. Wait about 15 seconds for the door to completely lock.
3. Turn the selector to OFF. Wait for the lock light to go off. The door will not unlock until the oven temperature drops below 560°F.
4. After the door is unlocked, turn the temperature control to OFF.

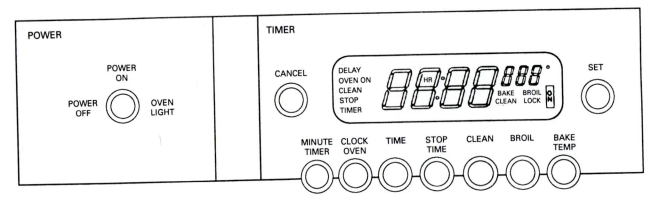

Figure 18–4 Twenty-four- and twenty-seven-inch single ovens. (Courtesy of WCI Major Appliance Group.)

If the door will not close, use the following procedure:

1. Turn the temperature control and the selector to Clean.

2. Open the oven door and push in the oven light switch for about 15 seconds. The door locking latch will move to the closed (locked) position. The oven light is located at the top left of the oven door opening.

3. Release the oven light switch. Turn the selector to OFF.

4. Push in and hold the oven light switch for another 15 seconds. The door locking latch will return to the open position. Wait until the latch is completely opened.

5. Close the oven door. Turn the temperature control to the off position.

When the clean cycle is completed, the following will occur:

1. "0:000" will appear in the display window.

2. When the oven cools below approximately 560°F, the lock light goes out. After the lock light goes out, the door can be opened.

3. Turn both the Selector and the Temperature Control to the OFF position. After about 15 seconds, the oven signal light will go out.

4. Use caution and open the oven door. The oven may still be very hot. A light or dark powdery ash will be found wherever there was heavy soil in the oven. This ash may be wiped up with a damp cloth when the oven has cooled. Fine hair lines may appear on the oven interior, but they will not affect the operation of the oven.

It is normal for some soil to remain at the bottom front and corners of the oven. This remaining soil is difficult to remove because it was exposed to high heat. To remove, scrub the areas with a new steel wool pad containing soap.

Electronic Oven Control

Some ovens are equipped with a state-of-the-art electronic oven controller. The oven can be programmed to function in Bake, Timed Bake with automatic on and off times, Broil, and Self-Clean. See Figure 18–4.

Among its features is a full-time oven diagnosis system. The controller constantly monitors its internal circuitry as well as several critical oven circuits, to ensure that they are all operating correctly. If at any time one of these systems fails, the controller will immediately shut down operation of the oven, beeping and flashing an error code in the display window. If this condition should arise, simply push the Cancel button.

Silent control panel When choosing a function, a beep will be heard each time a button is pushed. If desired, the controls can be programmed for silent operation. Push and hold the Cancel button. The control will beep. This will block the controls from sounding when a button is pushed. To return the sound, push and hold the Cancel button again until one beep is heard.

Power switch The power switch controls two operations of the control panel. It is used to turn the power on and off to the electronic oven control and the oven light.

The power switch must be in the power-on position in order to program the oven. The switch should only be turned to the power-off position when the oven will not be operated for an extended period of time. The power switch can be used as a power safety switch to shut off power to the oven when leaving the home unattended for extended periods of time.

Set knob The Set knob is used along with a function button to select the oven temperature, cooking time, stop time (when programming and automatic stop time), time of day, clean time, and the minute timer. After pushing one of the function buttons, turn the knob until the desired time or temperature appears in the display window.

If the wrong button is selected, push the Cancel button. This will erase all information previously entered except the time of day.

At the end of any cooking function, push Cancel to clear the display window.

Table 18–1 Minute timer.

Settings	Display
One second to two minutes.	In five second increments.
Two minutes to one hour.	In ten second increments.
One to two hours.	In one minute increments.
Two hours to nine hours and 50 minutes.	In five minute increments.

(Courtesy of WCI Major Appliance Group.)

Selecting oven control functions Note: Before using the oven the power switch must be in the power-on position in order to program the oven. The switch should be turned to the power-off position when the oven will not be operated for an extended period of time.

Setting the clock The time of day operated even when the oven is not in use. The function of a timed oven operation requires the use of stop and start times. Therefore, the clock must always be set to the correct time.

To set the clock when first connected or power supply is interrupted, use the following procedure:

1. Be sure that the power switch is in the power-on position.
2. The display window will flash. The word LOCK will glow in the lower right corner of the display. The locking and unlocking of the oven door takes about 1 minute. This is a normal part of the initial start-up and does not mean that the oven will begin a self-clean cycle.
3. Push the Clock button. The display will stop flashing. The word TIME will glow in the lower left corner of display window.
4. Turn the Set knob in a clockwise direction until the correct time of day appears in the display window.
5. Push the Cancel button. The word TIME will disappear. The clock will automatically start 30 seconds after the Set knob is rotated.

To change the set time, use the following procedure:

1. Push the Clock button. The word TIME will glow in the lower left corner of the display window.
2. Turn the Set knob in a clockwise direction until correct time of day appears in the display window.
3. Push the Cancel button. The word TIME will disappear.

Minute timer The minute timer does not start or stop a function. It serves as an extra timer that will beep when the set time has run out. The minute timer can be used independently during any of the oven functions. The timer can be set for any amount of time from 5 seconds to 9 hours and 50 minutes. See Table 18–1.

To set the minute timer, use the following procedure:

1. Push the Minute Timer button. ":00" will glow in the display window. The word TIMER will glow in the lower left corner of the display.
2. Turn the Set knob in a clockwise direction until the desired time appears in the display window. As soon as the Set knob is released, the time will begin to count down in the display window. The word TIMER will continue to glow.
3. When the set time has run out, the timer will beep three times. It will then continue to beep once every 10 seconds for 10 minutes (or until the Cancel button is pushed).

To change the minute timer while the oven is in use, use the following procedure:

1. Turn the Set knob clockwise to increase the time or counterclockwise to decrease the time.

To cancel the minute timer, use the following procedure:

1. To cancel the set time before the time has run out, push the Minute Timer button, then turn the Set knob counterclockwise until ":00" appears in the display window. Or the minute timer may be canceled by pushing the Minute Timer button and holding it for 2 seconds.

Bake The oven can be programmed to bake at any temperature from 170 to 550°F. Preheating the oven is recommended when baking, but not always necessary.

To preheat the oven for regular baking, use the following procedure:

1. Push the Bake Temp button. "_____°" will glow in the upper right corner of the display window.
2. Turn the Set knob in a clockwise direction until the desired baking temperature appears in the display window. The display can be changed in 5°F increments.

As soon as the Set knob is released, the oven will begin heating to the selected temperature. The word BAKE and ON will appear in the display window. When the display temperature reaches the desired baking temperature, the oven is ready for baking.

To change the baking temperature, use the following procedure. If the baking operation has begun, repeat steps 1 and 2.

To cancel the baking function, use the following procedure. Push the Cancel Button.

Timed bake The timed bake is controlled by the Oven Time and Stop Time buttons. The automatic timer will turn the oven on and off at the times that are selected in advance. Before setting the oven for the timed bake operation, be sure that the clock shows the correct time of day.

The oven can be programmed to start immediately and shut off automatically, or to start at a delayed time and shut off automatically.

To start the oven now and stop automatically, use one of the following methods:
Method 1

1. The clock should show the correct time of day.
2. Push the Oven Time button. "OHR:00" will glow in the display window. The words OVEN TIME will glow in the left side of the display window.
3. Turn the Set knob in a clockwise direction until the desired baking time appears in the display window. The baking time can be set for any amount of time from 10 minutes to 11 hours and 55 minutes. When setting the baking time, the time will appear in the display window and increase in 5-minute increments.
4. Push the Bake Temp button. "_____°" will flash in the upper right corner of the display window. The word BAKE will glow in the lower right corner of the display window.
5. Turn the Set knob in a clockwise direction until the desired baking temperature appears in the display window. When setting the baking temperature, the temperature will appear in the display window and increase in 5°F increments.

If the baking temperature is not set, the control will beep. The word BAKE and "_____°" will appear and flash in the display window.

As soon as the controls are set, the oven will come on and begin heating to the selected baking temperature. The words BAKE and ON will appear in the lower right corner of the display window. The main display will show the oven baking time and will count down by minutes until it reaches "OHR:00." When the oven Set Time runs out, the control will beep three times. "0:00" will appear in the display window. The control will then continue to beep once every 10 seconds for 10 minutes, or until the Cancel button is touched.

Method 2

1. The clock should show the correct time of day.
2. Push the Stop Time button. The current time of day will show in the display window. The words STOP TIME will glow in the left side of the display window.
3. Turn the set knob in a clockwise direction until the desired stop cooking time appears in the display window.
4. Push the Bake Temp button. "_____°" will glow in the upper right corner of the display window. The word BAKE will glow in the lower right corner of the display window.
5. Turn the Set knob in a clockwise direction until the desired baking temperature appears in the display window. When setting the baking temperature, the temperature will appear in the display window and will increase in 5°F increments.

If the baking temperature is not set, the control will beep. The word BAKE and "_____°" will appear and flash in the display window.

As soon as the controls are set, the oven will come on and begin heating to the selected baking temperature. The words BAKE and ON will appear in the lower right corner of the display window. The main display will show the time of day that the cooking is programmed to stop. When the set Stop Time is reached, the control will beep three times. "0:00" will appear in the display window. The control will continue to beep once every 10 seconds for 10 minutes or until the Cancel button is touched.

Timed bake: delayed start Using the Oven Time and Stop Time buttons, the oven may be programmed for a delayed start and shut off automatically.

To set the oven for a delayed starting time and automatic shutoff, use the following procedure:

1. Be sure that the clock shows the correct time of day.
2. Push the Oven Time button. "OHR:00" will glow in the display window. The words OVEN TIME will glow in the left side of the display window.
3. Turn the Set knob in a clockwise direction until the desired baking time appears in the display window.
Note: The baking time can be set for any amount of time from 10 minutes to 11 hours and 55 minutes. When setting the baking time, the time will appear in the display window and increase in 5-minute increments.
4. Push the Stop Time button. The time of day and the set baking time will appear in the display window. The words STOP TIME will glow in the left side of the display window.
5. Turn the Set knob in a clockwise direction until the desired cooking time appears in the display window.
6. Push the Bake Time button. "_____°" will flash in the upper right corner of the display window. The word BAKE will glow in the lower right corner of the display window.
7. Turn the Set knob in a clockwise direction until the desired baking temperature appears in the display window. When setting the baking temperature, the temperature will appear in the display window and increase in 5°F increments.

If the baking temperature is not set, the control will beep. The Word BAKE and "_____°" will appear and flash in the display window. The control will calculate backward from the set stop time to determine when the cooking should begin. The words DELAY OVEN, STOP TIME, and BAKE will glow in the display. To check the time that the oven will come on, push and hold the Oven Time button until the words OVEN ON appear in the display window. The calculated start time will appear in the display window for a few seconds.

The oven will come on automatically at the calculated time. At that time, the words DELAY OVEN will go out and BAKE ON will appear. When the set Oven Time runs out, the control will beep three times. "0:00" will appear in the display window. The control will then continue to beep once every 10 seconds for 10 minutes or until the Cancel button is touched.

Broiling Broiling is a quick method of cooking by direct radiant heat. Foods are placed under the upper heating element of the oven. Broiling results and timing depend on the distance between the food and the broiling element. Preheating the broil element is an option.

To set the broil feature, use the following procedure:

1. Place the oven rack in the desired position under the broil element.

2. Push the Broil button. "_____°" will appear in the upper right corner of the display window. The word BROIL will glow in the lower right corner of the display window.

3. Turn the Set knob in a clockwise direction until the desired broiling level appears in the display window. Set the broiling temperature for any of six levels: The display indicates "1," "2," "3," "4," "5," and HI for selected heating levels.

4. Leave the door open during broiling at the broil stop position.

5. When the broiling is completed, push the Cancel button.

Self-cleaning A self-cleaning oven cleans itself electrically. It heats to a very high temperature to burn off baked-on splatters. When the cleaning cycle is in operation, the oven heats to temperatures higher than those used for normal cooking. Sounds of metal expansion and odor are also normal because the food spatters are being carbonized. Smoke may appear from the oven vent, located on the top of the oven control panel.

A smoke eliminator in the oven converts most of the smoke into colorless vapor. If heavy spillovers are present and are not wiped up before cleaning, they may flame and cause more smoke and odor than usual. This is normal and safe and should not cause alarm. If available, use an exhaust fan during the self-cleaning cycle.

To set the controls for self-cleaning, use the following procedure:

1. Close the oven door. Be sure that the clock shows the correct time of day.

2. Push the Clean button. "_____:_____" will appear in place of the time of day in the display window. The words CLEAN TIME will appear in the lower left corner of the display window.

3. Turn the set knob in a clockwise direction until "3:00" appears in the display window. Three hours of cleaning is the recommended amount of cleaning time for a moderately soiled oven. You may wish to set the control for a longer or a shorter cleaning period, depending on the amount of soil in the oven. The cleaning time can be set for any amount of time from 2 to 4 hours. When turning the Set knob to select the cleaning time, the time will increase or decrease in 5-minute increments.

As soon as the controls are set, the motor-driven door lock begins to close and the cleaning cycle begins. It takes about 15 seconds for the door to completely lock, and the words CLEAN LOCK will appear in the display window.

To set the controls for a delayed cleaning cycle, follow steps 1 through 3 above, then proceed with the following for a delayed cleaning cycle:

4. Push the Stop Time button. The time of day and the cleaning time will appear in the display window. The words CLEAN STOP TIME will glow in the left side of the display window.

5. Turn the Set knob in a clockwise direction until the desired cleaning time appears in the display window. The control will calculate backward from the set stop time to determine when the cleaning cycle should begin. The words DELAY and CLEAN STOP TIME will glow in the display window. The calculated start time will appear in the display window for a few seconds.

Note: The clean cycle will come on automatically at the calculated time. At that time, the words DELAY TIME will go out and the word ON will appear. The oven will continue to heat until it reaches a very high temperature. Soil and spillovers will burn off of the oven surfaces. Do not try to open the oven door without canceling the cleaning cycle. When the set cleaning time has elapsed, the oven must cool for about 1 hour or until the interior temperature has dropped below 560°F. The word LOCK will remain in the display window until the oven cools below 560°F. Only then can the door be opened.

To interrupt or cancel a cleaning cycle, use the following procedure:

1. Push the Cancel button.

2. If the word LOCK remains in the display window, the oven is too hot to open the door. Wait until the word disappears from the display window before trying to open the door.

Do not try to force the oven door open. This can damage the automatic door locking system.

Oven temperature adjustment To adjust the oven temperature, use the following procedure:

1. Push the Bake button.

2. Set the temperature to 550°F by rotating the Set knob.

3. Quickly (within 2 seconds) press and hold the Bake button until the special digit display appears. Release the Bake button.

The display now indicates the amount of degrees offset between the original factory setting and the current temperature setting. If the oven control has the original factory calibration, the display will read "00."

4. The temperature can now be adjusted up or down 35°F, in 5°F increments, by rotating the Set knob. Adjust the Set knob until the desired amount of offset appears in the display window. A minus sign (−) will appear before the number to indicate that the oven will be cooler by the displayed amount of degrees.

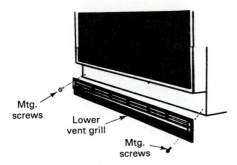

Figure 18–5 Lining up the mounting holes and the grille. (Courtesy of WCI Major Appliance Group.)

5. When the desired amount of adjustment is entered, push the Clock button to return to the time of day display.

Note: The self-cleaning temperature will not be changed by adjustments made for the baking temperature.

INSTALLATION INSTRUCTIONS

When installing single, double, and combination microwave oven models be sure to follow the manufacturer's recommendations and the local and national codes.

Electrical Connection

Unless expressed otherwise, it is the personal responsibility and obligation of the customer to contact a qualified installer to assure that the electrical connection is adequate and is in conformance with the National Electrical Code and the local code ordinances.

An electrical ground is required on electric ovens Most electric ovens are equipped with copper lead wires. If connection is made to aluminum house wiring, use only special connectors which are approved for joining copper and aluminum wires in accordance with the National Electrical Code and local codes and ordinances.

Generally, electric ovens are manufactured with a white neutral power supply wire and a frame-connected bare ground wire.

1. If local codes permit connection of the frame grounding conductor to the neutral (white wire), connect the bare wire and the white wire from the supply cable of the appliance together and to the neutral (white) wire in the junction box. Connect the remaining wires from the supply cable, matching colors to the wires in the junction box.

2. If the electric oven is to be used in a mobile home or if local codes do not permit frame grounding to the neutral wire, separate the white and bare ground wires that extend out of the end of the supply cable of the appliance. Connect the white wire from the supply cable to the neutral white wire in the junction box. Connect the black wire from the supply cable to the black wire in the junction box. Connect the red wire from the supply cable to the red wire in the

junction box. The bare wire must now be used to ground the appliance in accordance with local electrical codes. Connect the bare copper ground wire to a grounded cold water pipe* or to the grounded lead in the service panel. Do not ground this wire to a gas supply pipe. Do not connect the oven to the electrical power supply until the appliance is permanently grounded. Connect the ground wire before turning on the electrical power. See Figure 18–5.

Caution: If connecting the oven to a four-wire electrical system (mobile home), the appliance frame must not be connected to the neutral wire of the four-wire electrical system. Separate the white and the bare wires that extend out of the end of the supply cable of the appliance. Connect the white, red, and black wires from the supply cable, matching the colors to the corresponding wires in the junction box. Connect the bare ground wire from the supply cable to the ground wire in the junction box.

Checking Oven Operation

To check the different model ovens and their functions, use the following procedures.

Single-oven model To check the bake and broil operations, use the following procedures.

Bake Turn the oven selector knob to bake and set the Oven Temperature knob to 350°F. The lower oven element should become red and the Heating Signal Light should glow. When the oven reaches the desired temperature, the light will go out. The Heating Signal Light is designed to turn on and off during the baking process as the Bake Element cycles on and off.

Broil Turn the Oven Selector knob to Broil and set the Oven Temperature knob to Broil. The top element should become red and the Signal Light should glow.

Double-oven model To check the operation of both the upper and lower ovens, use the following procedures.

Upper oven Use the instructions for checking the bake and broil operations covered under Single-Oven Model above.

Lower oven To check the operation of both the bake and broil operation on these models, use the following procedures.

Bake Turn the lower thermostat knob to 350°F. The lower oven element should become red and the Signal Light should glow. When the oven reaches the desired tempera-

---

* The cold water pipe must have metal continuity to the electrical ground and must not be interrupted by plastic, rubber, or other electrically insulating connectors (including water meter or pump) without adding a jumper wire at these connections.

ture, the Signal Light will go out. The oven Signal Light is designed to turn on and off during the baking procedure as the bake element cycles on and off.

Broil   Turn the lower Oven Thermostat knob to Broil. The top element should become red and the Signal Light should glow.

OPERATION AND REPAIR

The following is a discussion of the operation and repair of the various oven components.

Thermostats (non-EOC models)   Wall ovens incorporate two types of thermostats: a high-temperature thermostat for the self-cleaning operation and a low-temperature thermostat for regular oven operation. Common to each thermostat is an internal circuit for providing reduced power to the broil element during the baking procedure. The internal circuit connects the broil element to half the normal operating voltage, thus reducing the radiated heat to one-fourth of the rated value.

Thermostat operation   The thermostat controls the oven temperature by opening and closing the electrical circuit to the oven heating elements. The spring-loaded contacts of the thermostat are controlled by the pressure of the thermostat bellows. The bellows are connected to, and share fluid with, the thermostat sensing bulb that is located inside the oven. When the stem of the thermostat is rotated from the off position, pressure that holds the contacts open is removed and the contacts snap closed. The bellows now must expand to force the contacts open. The higher the temperature setting on the dial, the more the bellows must expand to open the contacts.

With the contacts of the thermostat closed, electrical current flows through the oven bake or broil circuit. As the temperature inside the oven increases, the fluid in the thermostat sensing bulb expands. The expanding fluid causes the bellows inside the thermostat to expand. When the temperature in the oven reaches approximately 20°F above the temperature set on the thermostat dial, the expanding bellows opens the thermostat contacts and causes the current flow to the elements to stop. When the temperature inside the oven drops approximately 20°F below the dial setting, the fluid in the sensing bulb has contracted enough to remove sufficient pressure from the contacts to allow them to close and start the current flow in the oven bake or broil circuit.

Thermostat testing   Use the following procedure for testing the temperature of the thermostat.

Temperature testing   The following procedure is recommended for obtaining the oven thermostat average temperature for calibration purposes:

1. Use a thermocouple-type temperature tester.
2. Place the thermometer sensor as close as possible to the center of the oven cavity using an empty oven rack to support the thermometer sensor.
   3. Set the thermostat to 350°F.
   4. Set the oven selector control to Bake.
   5. After the initial off cycle, allow a minimum of three cut-in and three cut-out cycles to permit all oven cavity surfaces to stabilize at 350°F.
   6. Record the last cycle and calculate the average oven temperature using the following formula:

Average Temperature =
$$\text{Cut-in Temperature} + \text{Cutout Temperature}/2$$

For example, the last cycle recorded was 335°F cut-in and 385°F cutout.

Average Temperature = Cut-in Temperature + Cutout
$$\begin{aligned} &\text{Temperature}/2 \\ &= 335°F + 385°F/2 \\ &= 720°F/2 \\ &= 360°F \end{aligned}$$

Note: A 120°F differential between the cut-in and the cutout temperature is not excessive in an empty oven. The mass of the heating elements will cause a slight lowering of the oven temperature after the thermostat cuts in and a slight rise after it cuts out. This adds up to approximately 10°F to the differential and is unavoidable.

7. If the average oven temperature is above 50°F from the selected setting, the thermostat should be replaced.

8. If the average oven temperature is more than 25°F, but less than 50°F, from the selected setting, the thermostat should be calibrated. See the specific thermostat instructions for proper calibration.

Self-clean thermostat   The self-clean thermostat controls baking temperatures from 200 to 500°F, the broil temperature, and the clean temperature.

Contact operation   Refer to Figures 18–6 and 18–7 and Table 18–2 while following the explanation. During the baking operation, contacts 1 and 6 cycle the bake element and contacts 2 and 12 cycle the broil element. The broil element is supplied with half voltage, thus reducing the radiated power output to one-fourth of its normal value. The door lock contacts 3 and 4 are closed during the baking operation.

In the broiling operation, contacts 2 and 12 cycle the broil element at full power. The bake contacts 1 and 6 also cycle in the thermostat but the bake element is not connected to a circuit. The door lock contacts 3 and 4 are closed during the broiling operation.

The self-clean operation uses contacts 1 and 7 to cycle the bake element. The contacts cycle between 850 and 900°F while the oven is in the cleaning operation. The door lock contacts 3 and 4 open when the oven temperature has reached 550°F and close at 500°F.

Caution: Do not break the bulb or the capillary tube.

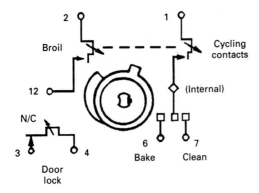

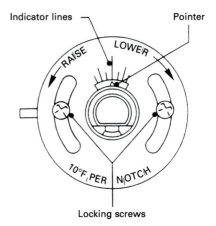

Figure 18–6 Self-clean thermostat wiring diagram. (Courtesy of WCI Major Appliance Group.)

Regular oven thermostat
calibration bezel

Locking screws

Figure 18–8 Temperature control calibration. (Courtesy of WCI Major Appliance Group.)

The self-clean thermostat bulb and capillary contents are dangerous to the eyes, skin, clothing, and combustibles. They contain sodium and potassium which produces lye (sodium and potassium hydroxide). If the contents should come in contact with the skin, remove all material with a dry cloth or towel, then wash with milk, soap, and water.

Regular oven thermostat The regular oven thermostat is a two-section assembly: a thermostat and a selector switch. The thermostat section is controlled by a sensing bulb and bellows in the housing and operates two cycling contacts. The selector switch section connects the element to the proper circuit supply voltage. Both controls are operated by a single shaft.

In the baking operation, selecting a temperature closes the internal contacts in the circuit to heat both the bake element and the broil element. The broil element is heated to one-fourth its normal rated wattage value. This is accomplished by providing half the normal voltage to the internal contacts. *Note:* The increased heat saturation of the oven improves the browning and the baking performance. The bake operation controls oven temperatures ranging from 150 to 500°F.

The broil position provides full power to the broil element. During the broil operation, the element will glow red.

Contact operation Refer to Figure 18–8 and Table 18–3 while following the explanation. During the baking operation, thermostat section contact 1 is cycling to contacts 2 and 3. The selector switch section contacts 4 to 6 and 5 to 7 are closed. Thermostat contacts 1, 2, and 3 will cycle together as the selected temperature is reached and maintained in the oven. Selector switch contacts 4 to 6 connect the bake element, and contacts 5 to 7 connect the broil element. The bake element circuit is provided with full voltage (240 volts ac), while the broil element circuit is connected to one-half full voltage (120 volts ac).

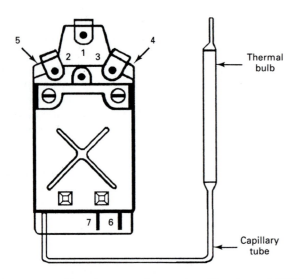

Figure 18–7 Regular oven thermostat. (Courtesy of WCI Major Appliance Group.)

Table 18–2 Self-clean thermostat internal switching.

Dial Setting	Function & Contacts			
	Bake 1-6	Clean 1-7	Broil 2-12	Door Lock 3-4
Off	O	O	O	X
Bake	Cycle	O	Cycle	X
Broil	Cycle	O	Cycle	X
Clean	O	Cycle	Cycle	O
X = Contacts closed. O = Contacts open.				

(Courtesy of WCI Major Appliance Group.)

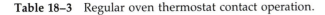

Table 18–3 Regular oven thermostat contact operation.

Dial Setting	Thermostat Section Contacts	Selector Switch Section Contacts
Off	Open	Open
Bake	1 to 2 1 to 3	4 to 6 5 to 7
Broil	1 to 2 1 to 3	4 to 7

(Courtesy of WCI Major Appliance Group.)

The broil operation closes thermostat section contacts 1 to 2 and 1 to 3, and the selector section contacts 4 to 7 connect the broil element to full voltage (240 volts ac).

Calibration The self-clean and regular oven thermostats are externally calibrated by an adjustable knob on the control panel. If calibration should be needed, set the thermostat to 350°F and obtain an average oven temperature (see the Temperature Testing section presented earlier in this chapter).

If the maximum or minimum allowable range of calibration is exceeded, replace the thermostat. The thermostat has no external adjustments.

To adjust the temperature control, use the following procedure:

1. After obtaining an average oven temperature, turn the temperature control to the off position and pull the knob off the shaft.
2. Examine the back of the knob. See Figure 18–8. There are a series of small indicator lines and a tiny pointer near the center stem. Note the position of the pointer and the words RAISE or LOWER near the outside edge of the knob.
3. Using a screwdriver, loosen the two (2) locking screws located on the back of the knob.
4. To increase or decrease the temperature, hold the knob in one hand and use the other hand to turn the colored part of the knob. Gently turn the knob until the pointer moves from the center position. Move the pointer to one of the indicator lines, toward the direction of the word RAISE to increase the temperature or LOWER to decrease the temperature. Each line increases the temperature by approximately 10°F. The pointer will click into place on each of the lines as the knob is turned.
5. When the desired adjustment is reached, tighten the screws.
6. Replace the temperature control knob onto the shaft on the control panel. Push the knob securely into place.

Note: Once an adjustment is made, the off position will vary from the reference mark on the control panel, depending on the amount of adjustment.

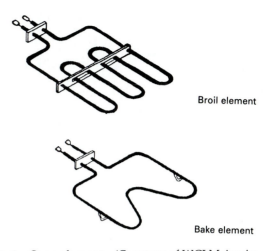

Figure 18–9 Oven elements. (Courtesy of WCI Major Appliance Group.)

Broil element The broil element is a 240-volt, 3400-watt heating unit. See Figure 18–9. The element is accessible for replacement from inside the oven cavity. The oven door may be removed during removal for testing or replacement of the element.

To remove the oven heating elements, use the following procedure:

1. Disconnect the electrical power from the oven.
2. Remove the mounting screws from the bracket holding the element to the back wall of the oven cavity. Pull the element forward into the oven.
3. Disconnect the push-on wire terminals.
4. Install the replacement element by reversing the procedure.

Adjustment There is no adjustment that can be made on the heating elements.

To test the heating element, use the following procedure:

1. Disconnect the electrical power from the oven.
2. Perform the above removal procedure.
3. Disconnect the push-on wire terminals.
4. Check the element resistance and compare the results with the resistance recommended by the manufacturer.
5. For a ground check, set an ohmmeter on the highest resistance scale. Check from one terminal to the case of the element. A reading of infinite resistance is a good reading.
6. If the element is found to be open, shorted, or grounded, replace it.

Bake element The bake element is a 240-volt, 2100-watt heating element. See Figure 18–9. The element is removed from the inside of the oven cavity. The oven door may be removed during the removal for testing or replacement.

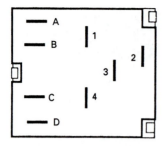
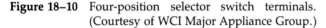
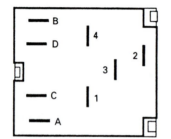

Figure 18–10 Four-position selector switch terminals. (Courtesy of WCI Major Appliance Group.)

Figure 18–11 Five-position selector switch terminals. (Courtesy of WCI Major Appliance Group.)

To remove the bake element, use the following procedure:

1. Disconnect the electrical power from the oven.
2. Remove the mounting screws from the bracket holding the element to the back wall of the oven cavity. Pull the element forward into the oven.
3. Disconnect the push-on wire terminals.
4. Install the replacement by reversing the above procedure.

Adjustment   There is no adjustment possible on heating elements.

To test the heating element, use the following procedure:

1. Disconnect the electrical power from the oven.
2. Perform the above removal procedure.
3. Disconnect the push-on wire terminals.
4. Check the element resistance and compare the results with the resistance recommended by the manufacturer.
5. For a ground check, set an ohmmeter on the highest resistance scale. Check from one terminal to the case of the element. A reading of infinite resistance is a good reading.
6. If the element is open, shorted, or grounded, replace it.

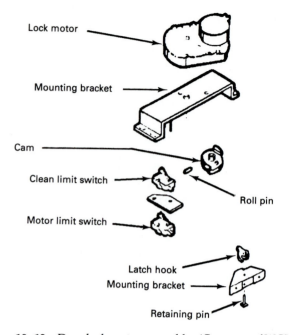

Figure 18–12 Door lock motor assembly. (Courtesy of WCI Major Appliance Group.)

Selector switch   The selector switch is located on the control panel. The switch controls the L1 side of the line voltage to the heating elements.

The functions available on a selector switch will depend on the model features. The selector switch tables will provide the closed contacts during selected functions.

Adjustment   There is no adjustment possible on this selector switch.

Testing   If a mechanical failure has occurred, replace the switch.

To test the selector switch, disconnect the electrical power from the oven and select the type of switch from Figures 18–10 and 18–11. Using Tables 18–4 and 18–5 and an ohmmeter set on the Rx1 scale, check for continuity across each pair of terminals as indicated by the selected function.

Door lock motor   All self-clean ovens are equipped with a door lock motor that operates on 120 volts ac. The motor has a cam mounted to its output shaft that operates two limit switches and a linkage system for the door latch. The motor is attached to its mounting bracket with four (4) screws, and is located in the back rear upper corner. See Figure 18–12.

Motor limit switch   The motor limit switch is a snap-action, single-pole, double-throw switch and is mounted to the latch motor mounting bracket. This switch controls the operation of the latch motor. The switch is removed first by removing the latch motor assembly.

Table 18–4 Four-position switch.

Contacts	4 Position Switch			
	Off	T-Bake	Bake	Broil
D to 4		X	X	X
C to 2				X
C to 3			X	
B to 2			X	
B to 1		X		
A to 2			X	X
A to 1		X		

(Courtesy of WCI Major Appliance Group.)

Table 18–5 Five-position switch.

Contacts	5 Position Switch				
	Off	T-Bake	Bake	Broil	Clean
D to 2			X		
D to 4					X
C to 2				X	
C to 3		X	X		
B to 2	X	X	X	X	
A to 1					X

(Courtesy of WCI Major Appliance Group.)

Clean limit switch The clean limit switch is a snap-action, single-pole, single-throw switch and its mounted to the latch motor mounting bracket. This switch controls the operation of the clean cycle. The switch is removed by first removing the latch motor assembly.

Door lock latch The door lock latch is operated by a cam mounted on the motor shaft. The latch assembly is mounted to the front frame of the oven wall. It is held in place by two (2) screws and a link arm from the lock motor cam.

The door latch switch has a roller catch that is mounted to the inner door panel.

Door switch Two types of door switches are installed on wall ovens. The first type is a plunger-operated switch. Plunger-type switches are located at the top left corner of the oven front frame. They are door operated and may

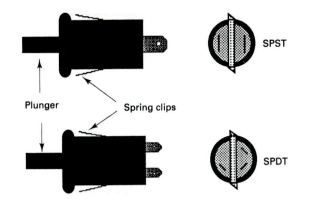

Figure 18–13 Plunger door switches. (Courtesy of WCI Major Appliance Group.)

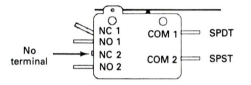

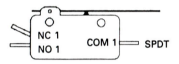

Figure 18–14 Lever door switches. (Courtesy of WCI Major Appliance Group.)

control two circuits. In double ovens both lights turn on when either oven door is opened.

Plunger-type switches are mounted by spring clips, and may be removed by squeezing the spring clips and pulling forward and disconnecting the wires.

Refer to Figure 18–13 for plunger door switch identification.

The second type is the lever-operated switch. The lever switch is attached to a bracket mounted on the back rear panel, upper right corner (as viewed from the back). It is activated by a rod when the oven door is opened or closed. The rod is mounted to the front frame upper left corner and extends to the back rear panel.

Refer to Figure 18–14 for lever door switch identification and terminal arrangement.

Electronic oven control (EOC) The electronic oven control (EOC) is a solid-state time and temperature control. See Figure 18–15. The oven functions of Bake, Timed Bake, and Self-Clean are controlled for time and temperature by a solid-state microprocessor on the circuit boards. EOCs are not field repairable.

The microprocessor constantly monitors its internal circuitry and the oven temperature to ensure that they are

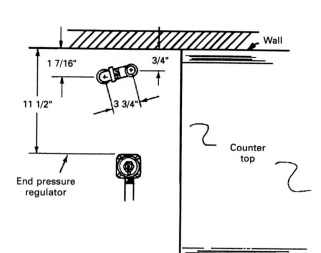

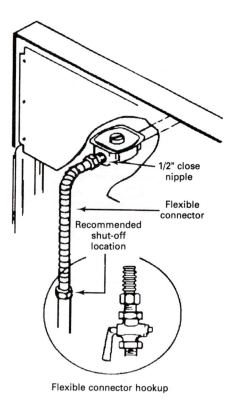

1/2" close
nipple

Flexible
connector

Recommended
shut-off
location

Flexible connector hookup

Figure 19–5 Flexible connector hookup. (Courtesy of WCI Major Appliance Group.)

Top View

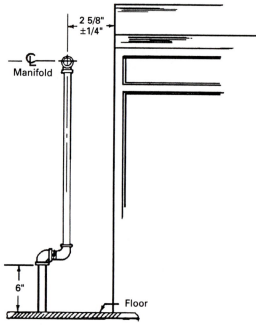

Front View

Figure 19–7 Connecting the range with rigid pipe. (Courtesy of WCI Major Appliance Group.)

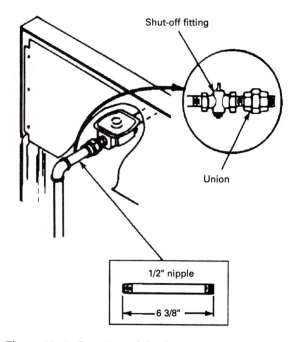

Shut-off fitting

Union

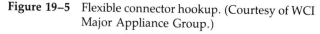

1/2" nipple

6 3/8"

Figure 19–6 Location of flexible connector. (Courtesy of WCI Major Appliance Group.)

location to install a shutoff valve is between the incoming steel gas supply pipe and the flexible connector. See Figure 19–5. On the rigid pipe hookup, the recommended location of the shutoff valve is between the pipe elbow and the ground joint union. See Figure 19–6.

2. When natural gas is to be used, a ¾-inch flexible connector or rigid pipe should be installed to connect the range to the gas supply. Either a ¾- or a ½-inch flexible connector or pipe may be used when the range is to be operated with LP gas.

3. Remove the main range top to gain access to the inlet of the gas pressure regulator.

4. It is important to connect the piping or the flexible connector to the range without placing any strain on any part of the assembly.

5. Figure 19–5 illustrates the proper procedure for using a flexible connector to connect the range to the gas supply.

6. The normal procedure for hooking up the range with rigid pipe is to build the hookup from the stubout in the floor up to a ground joint union. See Figure 19–7. Using the nipple lengths as suggested in the figure, build the hookup to and include half of the ground joint union.

7. Install a close nipple and the other half of the ground joint union to the inlet of the gas pressure regulator on the range.

8. Place the range in the proper position and complete the hookup by joining the two halves of the ground joint union. Be sure that the two halves of the union are aligned with each other. If the union is tightened without proper alignment, the union may be ruined. Turn the gas supply on and check the complete installation for gas leaks, using only an approved bubble leak test solution. Never use an open flame for leak testing gas piping.

Note: Any opening in the wall behind the range or in the floor under the range must be sealed before the range is placed in position against the wall. Unsealed holes in floors or walls are considered to be a fire hazard.

LEAKS

Suitable compounds which are resistant to the action of LP gas should be used on all threads to prevent leaks, except brass-to-brass fittings which require no sealing compounds. After the range has been connected to the gas supply, make certain that all appliances and gas valves are shut off before turning on the gas at the main gas inlet valve at the gas meter. To test for leaks, apply only an approved bubble solution to all connections in the supply line and the range piping.

Warning: Never attempt to test for leaks with a lighted match or any other type of flame.

If any leaks are present, bubbles will form at the point of leakage. If any leaks are found, turn off the main gas supply before attempting to stop them. After the leaks have been stopped, turn on the gas and recheck all joints for leaks before adjusting the range as described in the Adjusting Instructions. Any other gas appliance that was shut off may be relighted.

GAS ADJUSTMENTS

When making adjustments to any of the gas components on the range, use the following steps for each component.

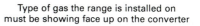

Type of gas the range is installed on must be showing face up on the converter

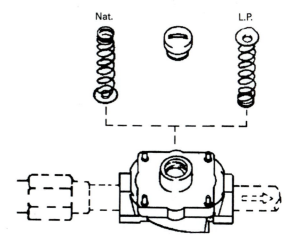

Figure 19–8 Gas pressure regulator. (Courtesy of WCI Major Appliance Group.)

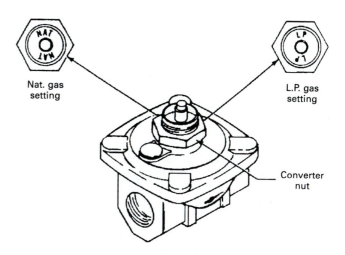

Figure 19–9 Gas pressure regulator. (Courtesy of WCI Major Appliance Group.)

Gas Pressure Regulator

On most ranges, the gas pressure regulator is equipped with a convertible feature and can be used with either natural or LP gas, it may be one of the three shown in Figures 19–8 through 19–10. The gas pressure regulator is preadjusted for the type of gas specified on the range number plate and will need no readjustment if the range is installed on the specified gas supply.

To convert the range to another type of gas, refer to Figures 19–8 through 19–10. If the type of regulator shown in Figure 19–8 is used, take the cap screw off and remove and invert the spring with the type of gas (natural or LP) face up.

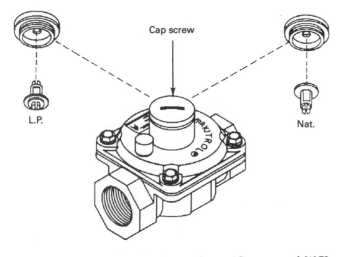

Figure 19–10 Gas pressure regulator. (Courtesy of WCI Major Appliance Group.)

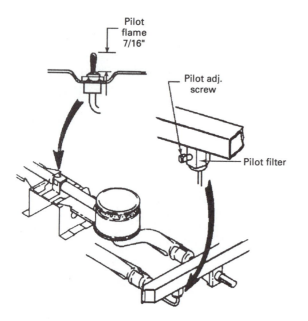

Figure 19–11 Adjusting the pilot flame. (Courtesy of WCI Major Appliance Group.)

If the type shown in Figure 19–9 is used, remove and invert the converter nut with the type of gas used (natural or LP) face up.

When the type shown in Figure 19–10 is used, remove the cap screw and snap out the red plastic gas indicator by bushing it sideways. Turn the red plastic gas indicator for the type of gas used (natural or LP) and snap it back into the cap screw. Put the cap screw back on the gas pressure regulator.

The maximum allowable gas supply pressure to the regulator is 14 inches of water column. The minimum supply pressure required for checking the pressure regulator setting is 5 inches of water column for natural gas and 11 inches of water column for LP gas.

Top and Griddle Burner Pilots

Note: There is no top pilot on electronic ignition systems.

Use the following steps when adjusting the top and griddle burner pilots:

Raise the main range top. Adjust the pilot flame by turning the adjustment screw on each pilot filter until the flame is about $\frac{7}{16}$-inch high and has a slightly yellow tip. See Figure 19–11. The pilot filters are located between the top burner halves.

If there is some delay or failure in the lighting of the top burner, it cannot be corrected by increasing the height of the pilot flame to more than $\frac{1}{2}$ inch. Spilled food or other foreign matter may be clogging the gas ports of the burner. This will cause a lighting failure, especially if the lighter ports are clogged. Clean these ports with a needle using care not to damage the burner port. Also recheck the gas and air shutter adjustment for the top burner.

Note: Be sure that each burner is level and is attached to the valve at the front.

Top Burner Adjustment

Use the following steps when adjusting standard valves.

When the range is being installed on a natural gas system:

If the inner cone of the flame is set correctly, do not adjust the flame length. If some adjustment is required, place a $\frac{1}{2}$-inch wrench on the orifice hood and turn it until the proper flame is obtained for the high setting. See Figure 19–12. (Facing the range, turn the hood to the right to increase the flame and to the left to decrease the flame.) The complete flame should be blue in color. The small cone in the center of the flame should be sharp in shape, greenish-blue in color, and about $\frac{5}{8}$-inch long.

When the range is being installed on LP gas using HI-MED-LOW valves, use the following steps:

The HI-MED-LO valve is a three-position valve with a definite flame setting for each position. If adjustment is required for the three-position valve, proceed according to the following instructions. See Figure 19–13.

1. *HI position* Turn the valve to the HI position and adjust as described above for the standard valve.

2. *MED position* No adjustment is necessary for the MED position. The flame from this position should be about $\frac{1}{16}$ inch above the burner cap.

3. *LO position* Turn the valve counterclockwise to the LO setting. Pull the knob off the valve stem and adjust the flame by turning the adjustment screw marked "A" in Figure 19–13. Turn the screw clockwise to decrease and counterclockwise to increase the flame size. The LO setting should barely be visible in a well-lighted room.

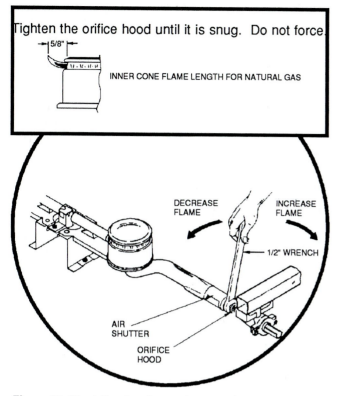

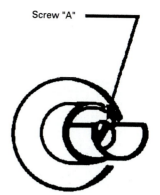

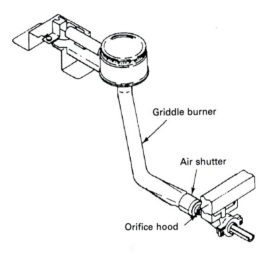

Figure 19–13   Flame adjustments for the HI-MED-LO valve. (Courtesy of WCI Major Appliance Group.)

Figure 19–12   Adjusting the top burner flame. (Courtesy of WCI Major Appliance Group.)

Figure 19–14   Griddle burner adjustment. (Courtesy of WCI Major Appliance Group.)

Air Shutter Adjustment

Turn on each burner and examine the flames. Each flame should be blue in color and not blow away from the burner. If not, an air shutter adjustment will be necessary. To make the adjustments, lift or remove the main range top. Turn on the burners and adjust the air shutter to obtain a flame that is blue in color and does not blow away from the burner.

Follow the manufacturer's recommendations when making these adjustments.

Griddle Burner

Use the following steps when making natural gas orifice hood adjustments. See Figure 19–14.

Place a ½-inch wrench on the orifice hood and turn it counterclockwise to increase the size of the burner flame, or clockwise to decrease it. Proceed to adjust the air shutter, then readjust the orifice hood, if necessary, to obtain the proper flame length. The inner cone flame length should measure about ½ inch long.

The proper air adjustment must be made before the flame length can be determined. Loosen the air shutter screw and adjust the air shutter until the flame has distinct inner cones that do not blow away from the burner ports.

Oven

Use the following procedures to light a standby pilot. See Figure 19–15.

The pilot must be lighted with the control dial in the off position.

1. Turn the main gas supply on. Bleed all air from the gas lines by turning the top valves on for a few seconds.
2. Remove the oven bottom and hold a lighted match to the oven pilot burner.

If the pilot is extinguished accidentally by a strong draft, the burner cannot be turned back on until the pilot burner is lighted again. Even when the control dial is turned on, no gas can flow to the burner unless a pilot flame is there to ignite it. This is a safety device for the protection of the user.

Note: The standby pilot flame burns continuously once it is lit, unless it is accidentally extinguished.

Note: There is no pilot on electronic ignition systems.

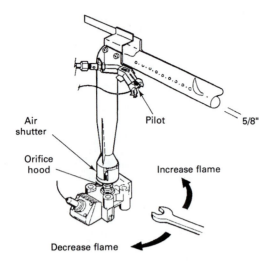

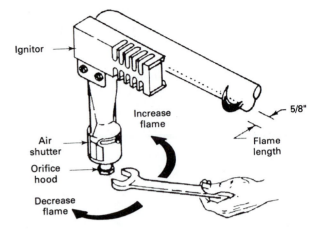

Figure 19–17 Natural gas orifice. (Courtesy of WCI Major Appliance Group.)

Figure 19–15 Lighting standby oven pilot. (Courtesy of WCI Major Appliance Group.)

If the Range Is Being Installed on LP Gas

Use the following steps to accomplish this procedure.

Tighten the orifice hood until it is snug. Do not force it. To observe the oven burner flame for adjustment, it is necessary to remove the oven bottom and the oven burner baffle on some models.

Electronic Ignition System

Use the following steps when adjusting the oven burner flame on these units.

The proper air adjustment must be made before the flame length can be determined. See Figure 19–17. Loosen the air shutter screw and adjust the air shutter until the flame has distinct inner cones that do not blow away from the burner ports. The length of these cones is the flame length.

Important Safety Warning: The range must be securely fastened to the floor by brackets that are supplied with the range. Failure to secure the range could allow the range to accidentally tip over if excessive weight is placed on an open door or if a child climbs upon it. Serious injury might result from spilled hot liquids or from the range itself.

Note: If the range is ever moved to a different location, the anchor brackets must also be moved and installed with the range.

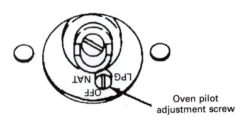

Figure 19–16 Oven pilot adjustment. (Courtesy of WCI Major Appliance Group.)

Oven Pilot Adjustment (Pilot Models Only)

To adjust the oven pilot, use the following procedure.

Pull the oven temperature knob straight off the control shaft. Use a ⅛-inch flat-blade screwdriver to turn the oven pilot adjustment screw. To change the settings, turn the adjusting screw clockwise to the stop position for the natural gas position and counterclockwise to the stop position for LP gas. The Off position is midway between the natural and LP gas positions. The NAT and LPG appear upside down. See Figure 19–16.

A range equipped with an electronic ignition system must be connected to a source of electricity before the oven burner will ignite.

Natural Gas: Orifice Adjustment

Use the following steps if any adjustment is necessary.

Place a ½-inch wrench on the orifice hood and turn it counterclockwise to increase the size of the burner flame, or clockwise to decrease the pilot flame. Proceed to adjust the air shutter, then readjust the orifice hood, if necessary, to obtain the proper flame length. See Figure 19–17. The inner cone length should measure about ⅝ inch long.

GAS CONTROLS

The following is a description of most of the various controls used on gas ranges.

Gas Pressure Regulator

Most gas ranges are equipped with convertible gas pressure regulators. The regulator is preset for 4 inches of water column for use on natural gas, and may be converted to 10 inches of water column for use on LP gas. The regulator is not otherwise adjustable.

Type of gas the range is installed
on must be showing face up on
the converter

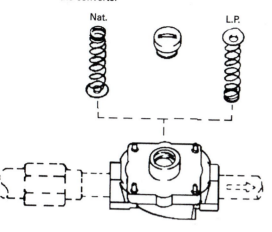

Figure 19–18   Gas pressure regulator conversion. (Courtesy of WCI Major Appliance Group.)

Type of gas the range is installed on
must be showing face up on the converter nut

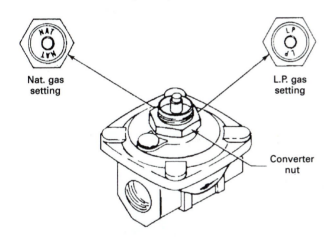

Figure 19–19   Gas pressure regulator conversion. (Courtesy of WCI Major Appliance Group.)

Caution: Never attempt to use these regulators as the sole pressure regulator for LP gas. They are secondary-type regulators and must be used in conjunction with a high-pressure regulator at the main LP gas supply tank.

The maximum allowable supply pressure to the regulator is 14 inches of water column. The minimum supply pressure required for checking the regulator setting is 5 inches of water column for natural gas and 11 inches of water column for LP gas.

Three types of regulators are used. To convert a range to a different gas, find the regulator type used (refer to Figures 19–18 through 19–20) and use that illustration to make the conversion.

Type 1: Remove the cap screw and remove and invert the spring with the type of gas (natural or LP) facing up. See Figure 19–18.

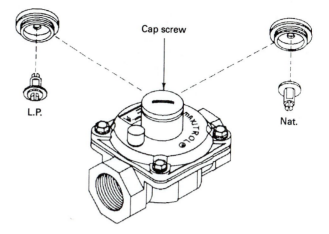

Figure 19–20   Gas pressure regulator conversion. (Courtesy of WCI Major Appliance Group.)

Type 2: Remove and invert the converter nut with the type of gas (natural or LP) facing up. See Figure 19–19.

Type 3: Remove the cap screw and snap out the red plastic gas indicator by pushing it sideways. Turn the red plastic gas indicator for the type of gas (natural or LP) and snap it back into the screw cap. Put the screw cap back on the gas pressure regulator. See Figure 19–20.

Surface Burners

The four surface burners are made of plated steel and aluminum. The ignitor ports on the side of the burner head must be cleaned and in alignment with the flash tube for proper burner ignition. The burner air shutter is a spring steel self-locking type. Care should be taken during cleaning and handling to prevent changing the shutter setting.

Air shutter adjustment   Turn on each burner and examine its flame. It should be blue in color and not blow away from the burner. If not, an air shutter adjustment will be necessary. Lift or remove the main top. Turn on the burner and adjust the air shutter to obtain a flame that is blue in color and does not blow away from the burner head.

Surface Burner Valves

Most ranges use either a standard or HI-MED-LO surface burner valves. Both types are the push-to-turn type and use a universal gas orifice.

Universal gas orifices are designed to operate with either natural or LP gas.

The universal orifice is a combination of a fixed orifice and an adjustable orifice. The needle is fixed in a position, while the orifice hood is adjustable. The needle has an orifice drilled through it that is sized for LP gas operation.

In order to operate the range on LP gas, the orifice hood is screwed all the way down until the tapered needle point completely blocks off the orifice in the hood. The LP gas rate is now set by the size of the fixed orifice through the needle and the gas pressure. See Figure 19–21.

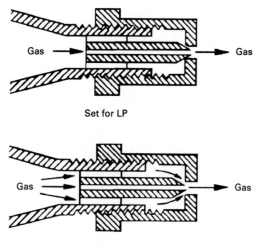

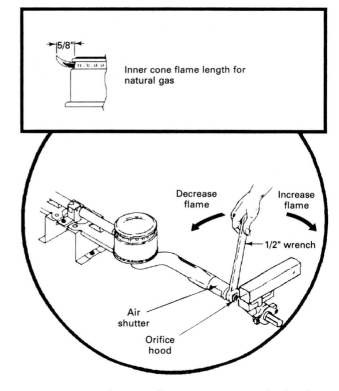

Figure 19–21 Universal orifices. (Courtesy of WCI Major Appliance Group.)

To operate the burners on natural gas, the orifice hood is screwed away from the needle. Gas then flows through the fixed orifice in the needle and through the opening between the shoulder of the needle and the edge of the orifice in the hood.

With this arrangement, the natural gas rate is adjustable, while the LP gas is not. To change the LP gas flow rate remove the needle and replace the orifice hood with one of the desired orificing.

Standard valves   For ranges being installed on natural gas use the following procedure.

If the inner cone flame is set correctly, do not adjust the flame length. If adjustment is required, place a $\frac{1}{2}$-inch wrench on the orifice hood and turn it until the desired flame is obtained for the high setting. See Figure 19–22. (Facing the range turn the orifice hood clockwise to increase the flame size and counterclockwise to decrease the flame size.) The complete flame should be blue in color. The small cone in the center of the flame should be sharp in shape, greenish-blue in color, and about $\frac{5}{8}$-inch long.

For ranges being used on LP gas use the following procedure.

Using a $\frac{1}{2}$-inch wrench on the orifice hood as illustrated in Figure 19–22, tighten the orifice hood until it is snug. Do not force the orifice hood.

HI-MED-LO valves   The HI-MED-LO valve is a three-position valve with a definite setting for each position.

If adjustment is required for the three-position valve, proceed according to the following instructions.

HI position   Turn the valve knob to HI and adjust as described for the standard valve.

MED position   No adjustment is necessary for the MED position. The flame from this setting should be about $\frac{1}{16}$ inch above the burner cap.

Figure 19–22 Adjusting flame using a standard valve. (Courtesy of WCI Major Appliance Group.)

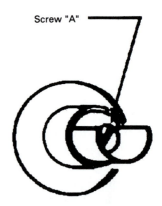

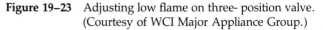

Figure 19–23 Adjusting low flame on three- position valve. (Courtesy of WCI Major Appliance Group.)

LO position   Turn the valve counterclockwise to the LP setting.

Pull the valve knob from the control shaft and adjust the flame by turning the adjustment screw marked "A." See Figure 19–23. Turn the screw clockwise to decrease and counterclockwise to increase the flame size. The LO setting should be barely visible in a well-lighted room.

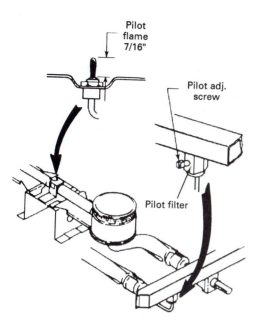

Figure 19–24 Surface burner pilot adjustment. (Courtesy of WCI Major Appliance Group.)

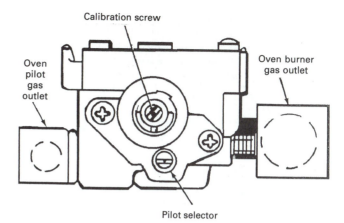

Figure 19–25 Selecting type of gas for an oven thermostat. (Courtesy of WCI Major Appliance Group.)

Surface Burner Pilots

On ranges using surface burner pilots, the orifice is fixed. The flame size is adjusted by turning the adjustment screw on each pilot filter until the flame is about $\frac{7}{16}$ inch high and has a yellow tip. Pilot filters are located between the top burner valves.

If there is some delay or failure in the lighting of the top burners, this condition cannot be corrected by increasing the height of the pilot flame to more than $\frac{1}{2}$ inch. Spilled food or other foreign matter may be clogging the gas ports of the burner. This will cause a lighting failure, especially if the lighter ports are clogged. Clean these ports with a needle, using care to prevent damage to the port. Also recheck the gas and air shutter adjustments for the top burner. See Figure 19–24.

Note: Be sure that each burner is level and is attached to the valve at the front.

On ranges with electronic ignition, a high-voltage spark electrode assembly is used in place of the gas pilot. The spark ignites the gas at the flash tube just as the gas pilots do.

Surface Burner Spark Module

A solid-state module having two ignition points are used. This module produces a high voltage (about 15 kilovolts) by a capacitor discharging into a pulse transformer which sends impulses to the electrodes at the rate of 2 sparks per second. It is rated for 120 volts, 60 hertz, but will operate with 85 to 110 percent of the rated voltage.

The module can be tested by directly connecting 120 volts to the L1 and N terminals.

Modules with a griddle burner have three electrode assemblies, and are equipped with a spark module having four ignition points. The fourth unused spark output must be grounded to the range body for proper operation of the module.

Oven Thermostats

The following is a description of the gas control system thermostat.

The gas thermostat is used on all gas systems. It is a hydraulic control which responds to any expansion and contraction of a liquid in the oven bulb. It operates a valve within the thermostat which opens to increase the gas supply to the pilot unit when heat is required, and closes to shut off this increase rate to the pilot when the oven reaches the set temperature.

In addition to controlling the heater pilot gas, the thermostat also provides a positive shutoff of the main burner gas. The gas supply to the oven burner is routed through the shutoff valve portion of the thermostat to ensure that no gas can flow to the oven burner when the oven control knob is in the off position.

The thermostat adjustment screw is located in the center of the thermostat shaft. To increase the temperature, turn the adjustment screw counterclockwise. To lower the temperature, turn the screw clockwise.

To select the proper gas operation, rotate the screw clockwise to a full stop for natural gas, and counterclockwise to a full stop for LP gas. See Figure 19–25.

ELECTRIC CONTROL SYSTEM

The following is a description of the thermostats used on non-self clean self-clean oven models.

Thermostat used on non-self-clean oven models An electric thermostat is used on the all electrical control systems. See Figure 19–26.

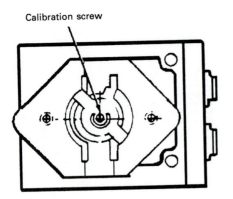

Figure 19–26 Thermostat for non-self-cleaning oven models. (Courtesy of WCI Major Appliance Group.)

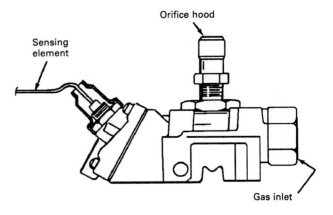

Figure 19–28 Oven safety valve. (Courtesy of WCI Major Appliance Group.)

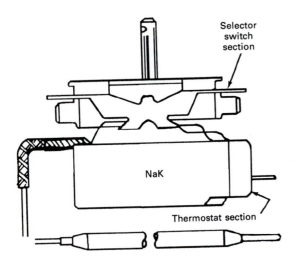

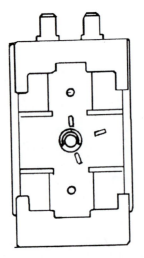

Figure 19–27 Thermostat for self-clean oven models. (Courtesy of WCI Major Appliance Group.)

This thermostat is similar to those used in many electric ranges. A set of contacts open and close depending on the internal pressure of the capillary tube (sensing bulb). As the oven heats up, the internal pressure within the sensing bulb increases until the thermostat contacts open. As the oven cools, the internal pressure decreases and the thermostat contacts close.

The thermostat contacts open and close the circuit to the safety valve and the Norton glow-bar ignitor.

The thermostat adjustment screw is located in the center of the thermostat shaft. To increase the temperature, turn the adjustment screw counterclockwise. To lower the temperature, turn the adjustment screw clockwise.

Thermostat used on self-clean oven models The Robertshaw NaK electric thermostat is used on all electric self-clean oven control systems. See Figure 19–27.

This thermostat is operated by a single sodium-potassium (NaK) charged sensing element.

Caution: If the bulb or capillary is broken, the contents are dangerous to the eyes, skin, clothing, and combustibles. The bulb contains sodium potassium which produces lye (sodium and potassium hydroxide). If it comes into contact with the skin, remove all material with a dry cloth or towel, then wash with a mild soap and water.

The thermostat assembly has a temperature control section consisting of a single-pole, snap-action switch. This switch provides temperature control for all cooking and cleaning operations.

The selector switch section selects the bake or broil operations. In the Clean setting, all selector switch contacts are open.

The original factory calibration settings for all cooking and cleaning temperatures are "locked" in place. The thermostat cannot be field calibrated.

Oven Safety Valves

The following is a description of the gas controlled safety valve used on standing pilot models.

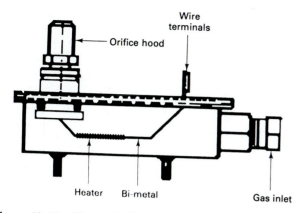

Figure 19–29 Electronically controlled safety valve (non-self-cleaning models). (Courtesy of WCI Major Appliance Group.)

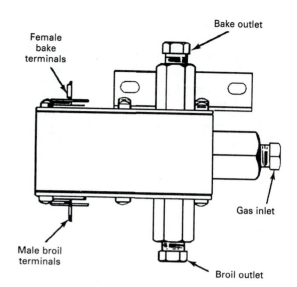

Figure 19–30 Electronically controlled safety valve (self-clean models). (Courtesy of WCI Major Appliance Group.)

The gas safety valve controls the gas flow to the oven burner by means of a diaphragm. The diaphragm is controlled by the flame-responsive element in the pilot assembly. As the flame element is heated by the heater pilot, pressure is built up, opening the diaphragm and allowing gas to flow to the oven burner. See Figure 19–28.

Electronically controlled safety valve (non-self-cleaning models) The following is a description of the electronically controlled safety valve.

The electric safety valve controls the gas flow by means of a bimetal opening a diaphragm. The bimetal is in series with the thermostat glow-bar ignitor. See Figure 19–29.

Caution: Use only the specified electric safety valves and glow-bar ignitors for service replacement. Incorrect safety valves or ignitors may result in ignition failure or premature ignitor burnout.

Electronically controlled safety valve (self-clean oven models) The following is a description of the electronically controlled safety valve.

Self-clean oven models have a single-inlet, dual-outlet, bimetal-operated safety valve. The valve is internally constructed so that it is not possible to open the bake and broil outlets at the same time. To prevent miswiring, the bake terminals are female and the broil terminals are male. See Figure 19–30.

Caution: Use only the specified electric safety valves and glow-bar ignitors for service replacement. Incorrect safety valves or ignitors may result in ignition failure or premature ignitor burnout.

Oven Pilot

The following is a description of the oven pilot assembly.

The oven pilot assembly is mounted to the back of the oven burner. The pilot assembly consists of a single pilot burner with a single gas supply tube which provides a small constant pilot flame. With an increase in gas volume it also provides the heater pilot.

The safety valve sensing element is clamped to the pilot bracket. When properly positioned, the tip of the sensing element will be completely engulfed by the heater pilot flame. A heater pilot flame that is too small may be caused by contamination within the pilot assembly. Disassemble and clean the pilot assembly with compressed air. If necessary, use a piece of stranded copper wire to clean the orifice opening. The orifice must not be distorted or enlarged.

Oven Ignitors

The following is a description of the oven ignitors.

The oven ignitor is made of silicon-carbide material. When voltage is applied to the glow-bar, it heats up to 2000°F. As its temperature increases, its resistance decreases, allowing 2.9 to 3.3 volts ac to flow to the bimetal in the safety valve. The glow-bar ignitor is energized as the oven cycles on and remains on until the oven cycles off. The ignitor may be checked for continuity with an ohmmeter. Its resistance (at 75°F) should be between 50 to 400 ohms. See Figure 19–31.

Caution: Use only the specified glow-bar ignitors and electric safety valves for service replacement. Incorrect ignitors or safety valves may result in ignition failure or premature ignitor burnout.

Oven Burners

The oven burners, both bake and broil, are made of coated steel. The back of the burner fits over the safety valve orifice hood and the front is attached to the oven front with a screw. The air shutter at the back of the burner is adjusted

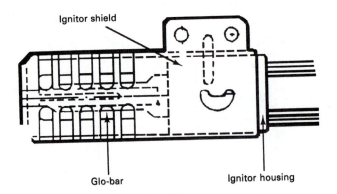

Figure 19–31 Oven ignitor. (Courtesy of WCI Major Appliance Group.)

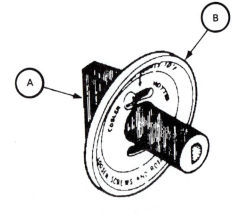

Makes Oven Hotter
(screw moved toward hotter)

Makes Oven Cooler
(screw moved toward cooler)

Figure 19–32 Adjustable oven thermostat knob. (Courtesy of WCI Major Appliance Group.)

by loosening its mounting screw and rotating to obtain the proper flame.

The pilot tube or glow-bar ignitor is mounted to the back left side of the burner.

The bottom of the burner has "carry over" slots cut from side to side to assure ignition of the complete burner. Care should be used in handling and installing the burners as excessive stress may close the carry-over slots and cause failure to ignite the right side of the burner.

Gas Shut-Off Valve

The following is a description of the gas shutoff valve.

All models with an electric control system have a gas shutoff valve in the gas supply line to the safety valve. The shutoff valve is mounted directly in the manifold pipe. The valve must be in its full open position for proper operation of the oven burners. The purpose of the valve is to allow gas to the oven to be turned off in case of an oven control failure.

Adjustable Oven Thermostat Knob

All of the thermostats that have been covered to this point have been adjustable which allows the user to adjust the dial setting to the desired oven temperature. See Figure 19–32.

To adjust the thermostats, use the following procedure:

1. Turn the oven knob to OFF and remove the knob by pulling it straight off the thermostat shaft.

2. On the back of the knob, there is an arrow pointing to the center of the upper screw which indicates the original factory setting. The knob can be adjusted up to 50°F hotter or 50°F cooler, in 10°F increments.

3. Use a screwdriver to loosen the two screws about one turn each.

4. Hold the knob handle (knob A in Figure 19–32) while turning the knob shaft in the desired direction. As the screw is turned clicks should be heard and the notches or teeth should be felt. Each click or notch represents 10°F. The screw can be turned up to five clicks or notches in either direction.

5. After the desired adjustment has been made, retighten both screws and reinstall the knob on the thermostat shaft.

Clock-Timers

The following three types of clock-timers are very popular for use on ovens and ranges:

1. A conventional time-of-day clock with a 4-hour minute timer.

2. A conventional time-of-day clock with a 1-hour minute timer and an hours-to-cook dial.

3. A digital time-of-day clock with a 1-hour minute timer, a start time dial and a stop time dial.

Time of day: All clocks The following is a description of these types of clocks.

The clock is set by pushing in on the minute timer control and turning the control in either direction. If the minute timer hand turns while setting the clock, let the knob come out and return the timer hand to OFF.

Minute timer: Conventional clocks The following is a description of these types of clocks.

The minute timer simply operates a buzzer and serves

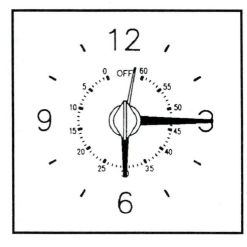

Figure 19–33 Minute timer (conventional clocks). (Courtesy of WCI Major Appliance Group.)

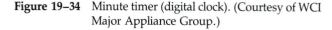

Minute timer
push to set clock

Figure 19–34 Minute timer (digital clock). (Courtesy of WCI Major Appliance Group.)

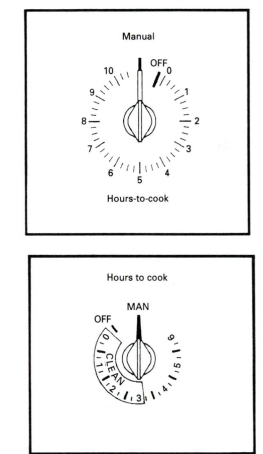

Figure 19–35 Hours-to-cook timer. (Courtesy of WCI Major Appliance Group.)

as a convenient clock-watcher. It does not start or stop cooking. To set the timer, turn the control in a clockwise direction to the desired number of minutes. When the time has elapsed, a buzzer will sound. To stop the buzzer, turn the minute timer control to OFF. See Figure 19–33.

Minute timer: Digital clock The following is a description of these types of clocks.

The minute timer, located to the left of the digital clock, simply operates a buzzer and serves as a convenient clock-watcher. It does not start or stop the oven. To set the minute timer, turn the control in a clockwise direction to the desired number of minutes. When the time has elapsed, a buzzer will sound. To stop the buzzer, turn the control to OFF. See Figure 19–34.

Hours-to-cook dial The automatic timer on the control panel will turn the oven off at any chosen time. To turn the oven on immediately, and have it shut off automatically at the desired serving time use the following procedure:

1. Put the food to be cooked in the oven.
2. Set the oven temperature control to the desired cooking temperature. Cooking begins as soon as the food is placed in the oven.
3. Set the Hours-to-Cook Timer to the desired amount of cooking time. See Figure 19–35.
4. When the cooking is completed and the oven shuts off, turn the oven temperature control to OFF and the Hours-to-Cook control to Manual. If the Hours-to-Cook control is not turned to Manual, the oven will not operate the next time it is turned on.

Automatic Start Time and Stop Time dials Models with these dials have a Delay Start, Cook, and OFF oven setting.

The oven will come on at the selected start time, heat to the selected temperature, and maintain that temperature until the selected stop time has been reached. The oven will then turn off.

To use Delay Start, Cook, and OFF, use the following steps for this control:

1. Select the stop time (push in and turn the knob).
2. Select the start time (push in and turn the knob). Be sure that the start time is set later than the actual time of day, but is set to an earlier time than the stop time. See Figure 19–36.
3. Set the oven temperature control to the desired setting.

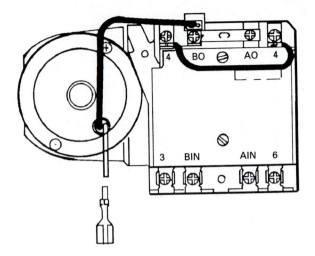

Figure 19–36 Setting delay start, cook, and off timer. (Courtesy of WCI Major Appliance Group.)

When the cooking is completed:

1. Turn the oven temperature control to the OFF position.
2. Push the Stop Time knob in.

Caution: When setting the oven for a delayed start, never allow the food to sit for more than 1 hour before the cooking begins. The room temperature will provide an ideal condition for the growth of harmful bacteria. Be sure that the oven light is turned off, since heat from the bulb will speed up bacteria growth.

To use Instant Start, Cook, and OFF,

1. Select the stop time (push in and turn the knob).
2. Push in and release (do not turn) the Start Timer knob. See Figure 19–36.
3. Select the desired temperature.

When the cooking is completed:

1. Turn the oven temperature control knob to the OFF position.
2. Push the Stop Time knob in. This restores "manual" operation.

Figure 19–37 Self-clean timer. (Courtesy of WCI Major Appliance Group.)

Note: If the Stop Time knob is not pushed in, the oven will not operate the next time it is turned on for a regular bake cycle.

Self-Clean Timer ("C" Line Frigidaire Models)

The clean timer provides all timing functions for the oven cleaning cycle. The motor driven timer has four single-pole, single-throw switches and makes one complete revolution in 200 minutes. See Figure 19–37.

The timer is located in the base of the range and is accessible by removing the service panel. The timer can be manually advanced by turning it with a screwdriver. The switch contacts can be checked for continuity with an ohmmeter in conjunction with the timer chart shown in Figure 19–38.

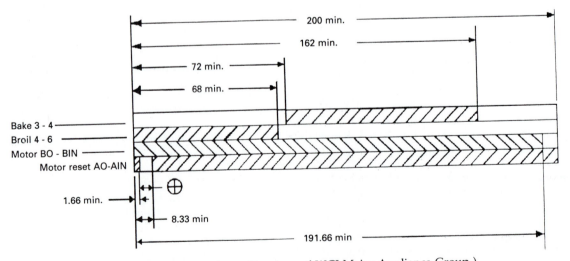

Figure 19–38 Timer chart. (Courtesy of WCI Major Appliance Group.)

Door Latch Switches (Self-Clean Models)

Three door latch switches are mounted to a bracket located directly above the door latch mechanism. See Figure 19–39.

These switches are operated by a sliding switch actuator as the oven door is either latched or unlatched. See Figure 19–40.

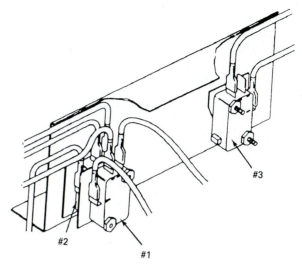

Figure 19–39 Door latch switch location. (Courtesy of WCI Major Appliance Group.)

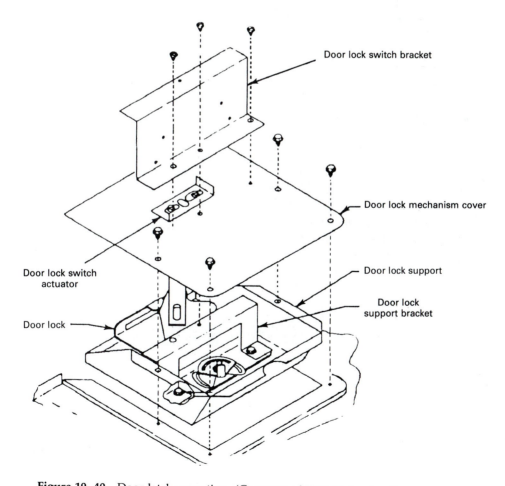

Figure 19–40 Door latch operation. (Courtesy of WCI Major Appliance Group.)

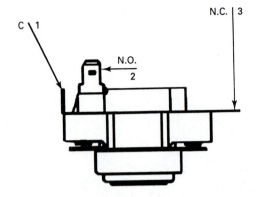

Figure 19–41 Disc thermostat. (Courtesy of WCI Major Appliance Group.)

Disc Thermostat Bake/Broil Selector ("E" Line Self-Clean Frigidaire Models)

The disc thermostat is mounted to the rear of the range and senses the oven temperature. The disc thermostat provides electrical power to either the broil or the bake burner during the clean cycle. During the first part of the clean cycle, the disc thermostat provides power to the broiler circuit. When the temperature in the oven reaches 650°F, the disc thermostat removes power from the broiler circuit and applies it to the bake circuit. See Figure 19–41.

The switches are accessible for checking or replacement by removing the metal cover. Check the switches by using an ohmmeter to check the continuity.

Appendices

Chapter 10: Room Air Conditioning Units

ELECTRICAL DIAGRAMS

INDEX

1990 MODELS	WIRING DIAGRAM	PAGE	1990 MODELS	WIRING DIAGRAM	PAGE
A05LS1N1	A373802	30	AR18NS2F1	A373504	30
A05LE2N1	A373502	29	AR18NS8F1	A373902	29
A05LH5N1	A373502	29	AR22NS5N1	A373504	30
A05LH8N1	A358701	29	AR24NS8F1	A373902	29
A05LE2E3	A233301	29	AR24NS8F3	A373902	29
A05LE2E7	A233301	29	AR25NS5N1	A373902	29
A06LE2N1	A373502	29			
A06LH5N1	A373502	29			
A06LE2L1	A233301	29			
A08LE2N1	A373502	29			
A08LH8N1	A358701	29			
A08LH5L2	A373502	29			
A08ME2E2	A233301	29			
AW08LT5N1	A373504	30			
AR09ME5L1	A373504	30			
AS10ME5L2	A372401	29			
AR10ME5N1	A373504	30			
AR10ME5N2	A373504	30			
AW11MT5N1	A373504	30			
AW11NT6N1	A373504	30			
AHW11NT6N1	A390101	30			
AR12ME8N1	A373902	29			
AR12ME5L1	A373504	30			
AR14ME5L1	A373504	30			
AR18NS2N1	A373504	30			
AR18NS5N1	A373504	30			

ELECTRICAL DIAGRAMS

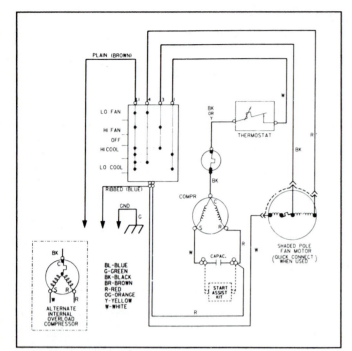

A233301

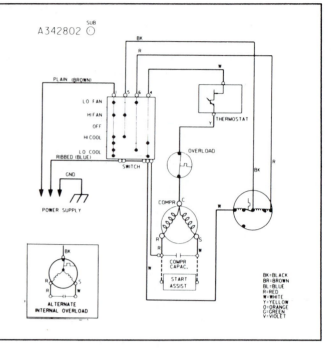

A342802

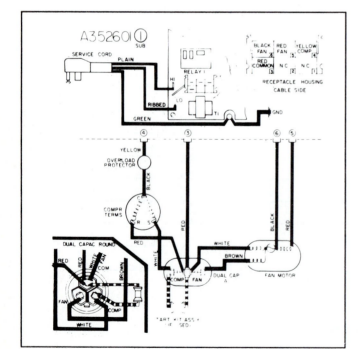

A352601

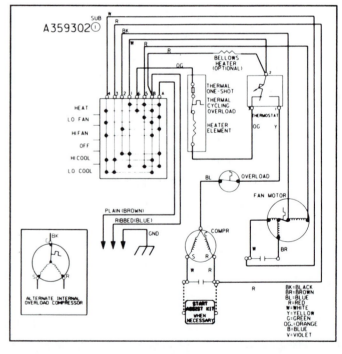

A359302

ELECTRICAL DIAGRAMS

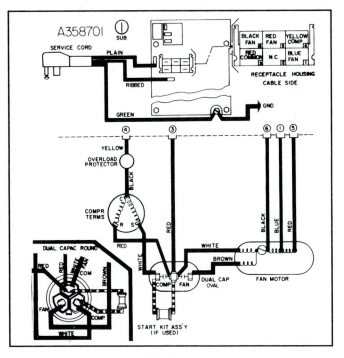

A358701

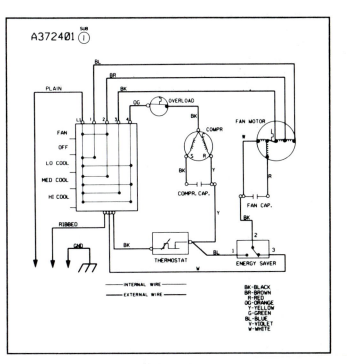

A372401

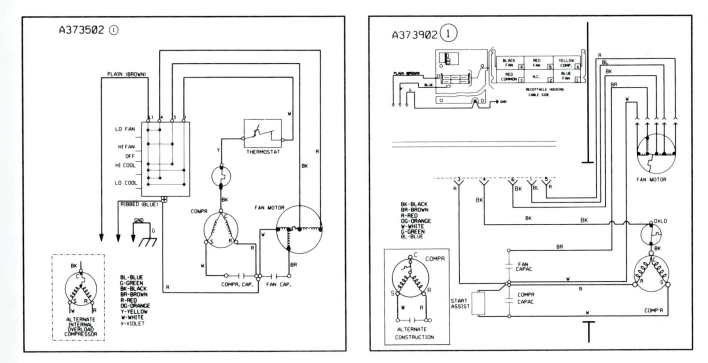

A373502

A373902

ELECTRICAL DIAGRAMS

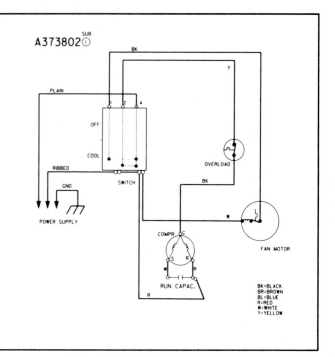

A373504

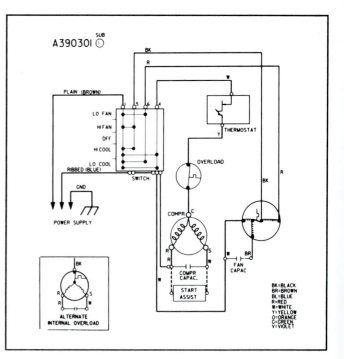

A373802

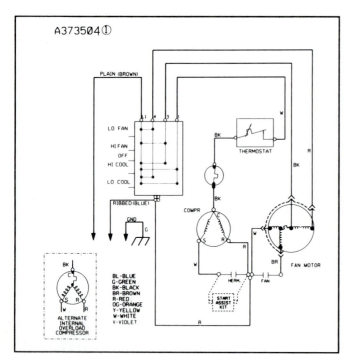

A390101

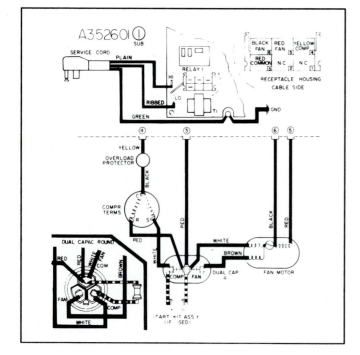

A390301

ELECTRICAL DIAGRAMS

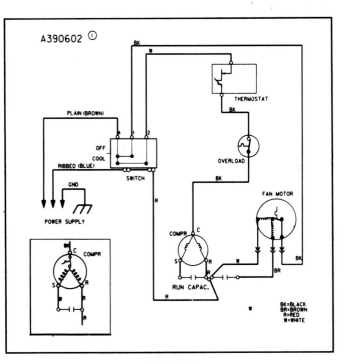

A390602

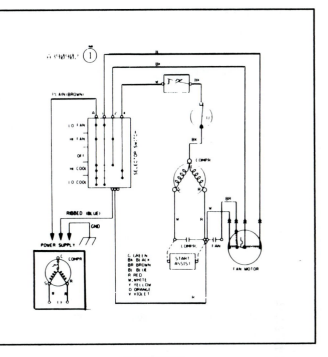

A390902

SECTION G - TROUBLESHOOTING

⚠ CAUTION Review Safe Servicing Practices in front of manual before attempting diagnosis procedures and repairs.

AIR CONDITIONER VOLTAGE LIMITS

NAMEPLATE RATING	MINIMUM	MAXIMUM
115 VAC	103.5 VAC	126.5 VAC
230 VAC	207 VAC	253 VAC
208/230 VAC	197.5 VAC	253 VAC

LOW VOLTAGE

Low voltage is a common cause of trouble in the operation of any room air conditioner. It becomes doubly important, because of the motor size, that the technician check voltages when servicing room air conditioners.

Improper voltage may result in one or more of the following complaints:
1. Unit will not start.
2. Compressor motor cycling on motor protector.
3. Premature failure of motor protector.
4. Blown fuses.
5. Premature failure of compressor or fan motor.
6. Noticeable dimming of lights when air conditioner is running.
7. Evaporator icing. Low voltage may reduce the fan speed resulting in an inadequate air flow over the evaporator, thereby allowing it to ice up.

Often, low voltage can be attributed to extension cords or an inadequately wired circuit, but low voltage into the building and loose fuses or connections in the power supply should not be overlooked. Low voltage may also be a general condition in the area (a responsibility of the power company).

All units will start and run on the minimum voltage stated in the chart above, and will perform satisfactorily if the voltage remains constant. Low voltage caused by defective wiring will not remain constant under load.

Testing for low voltage should be done with a reliable voltmeter with a capacity to measure the required voltage. Measurements should be taken at the electric service entrance and at the electrical outlet serving the air conditioner. Readings should be taken with the unit off, while the unit is starting, and again while the unit is running. The lowest reading should not drop below the lowest value listed in the chart.

HIGH VOLTAGE

High voltage can be equally troublesome by causing motors to overheat, cycle on their protectors, or break down electrically. This problem can only be solved by the power company.

ELECTRONIC CONTROL

This control is not repairable. If any component on the control is defective, the entire control must be replaced.

IMPORTANT NOTE: Repair or replace any malfunctioning line voltage component before testing or replacing the electronic control. Do not assume a service problem is directly caused by the electronic control system. A line voltage component (including power cord and wiring) that has opened, shorted, grounded or otherwise malfunctioned may have created the service problem.

ELECTRONIC CONTROL SERVICE DIAGNOSIS	
SERVICE PROBLEM	**DIAGNOSIS / REMEDY**
Air conditioner will not operate.	1. Check voltage at outlet. Correct if none. 2. Power button must be ON. 3. Set energy saving switch to OFF. 4. Set the delay hours selector to OFF. 5. Temperature control position must be at a setting cooler than room temperature. If fan and compressor do not operate, replace control.

SERVICE PROBLEM	DIAGNOSIS / REMEDY

Fan is running but air conditioner is not cooling.	1. May be normal. If the automatic compressor delay circuit has been activated, the compressor will remain OFF for 3 to 4 minutes.
	
	2. Check for defective relay.
	   a. Remove power supply cord from electric outlet.
	   b. Connect a jumper wire with 1/4" female spade connectors to the relay test terminals "COM" and "NO". Reassemble control panel to the unit. Plug in unit and select HI COOL with energy saving switch OFF and a temperature control setting lower than the room temperature. If compressor starts with the jumper in place, replace electronic control. If compressor does not start when the compressor relay is eliminated (jumpered) from the circuit, remove the jumper wire and check the wiring, capacitor, motor protector and compressor separately.

Compressor runs (cooling) but fan is not operating.	1. Remove power supply cord from electric outlet. 2. Disconnect unit from rear of control. 3. Disconnect fan motor wires and check fan motor. If motor does not have an open or shorted winding, check for binding. 4. Check fan capacitor if used. 5. Check wiring and correct if required. 6. If the above items are found to be satisfactory, replace electronic control.
Unit is cooling but room is not cool.	1. Place a thermometer as close to the temperature thermistor as possible to determine the return air temperature. If the temperature is higher than the thermostat setting, check amps and watts to verify that the unit is operating to its capacity. 2. If unit is not running to its capacity, check sealed refrigeration system. 3. If unit is running to its capacity, determine the heat loss in the area being cooled. If heat loss is excessive, inform the customer of the problem.
Thermistor (sensor) location.	1. Temperature thermistor - Must be located on the lower right side of the evaporator. 2. Freeze prevention thermistor - Must be located on the left side of the evaporator coil. 3. Refer to PART 1, SPECIFICATIONS, TABLE 2 for proper sensor location.

ROOM AIR CONDITIONER SERVICE DIAGNOSIS

SERVICE PROBLEM	POSSIBLE CAUSE	DIAGNOSIS / REMEDY
Fan motor will not run.	1. No power.	1. Check voltage at outlet. Correct if none.
	2. Power supply cord.	2. Check voltage to selector switch. If none, check power supply cord. Replace cord if circuit through cord is open.
	3. Selector switch.	3. Check switch continuity. Refer to wiring diagram for terminal identification. Replace switch if no continuity.
	4. Energy saving switch (if applicable).	4. When the switch is in the ON position, the fan motor will cycle off with the compressor. When the switch is in the OFF position, the fan motor will continue to run when the compressor cycles off. Check switch continuity. Refer to wiring diagram for terminal identification.
	5. Electronic control (if applicable).	5. Replace only if fan motor, capacitor and all line voltage wiring are found to check OK. (Refer to ELECTRONIC CONTROL SERVICE DIAGNOSIS.)
	6. Wire disconnected or connection loose.	6. Connect wire. Refer to wiring diagram for terminal identification. Replace switch if no continuity.
	7. Capacitor. (Discharge capacitor before testing.)	7. Test capacitor. See testing procedure in SECTION E. Replace if not within ±10% of manufacturer's rating. Replace if shorted, open or damaged.
	8. Defective fan motor windings.	8. Test fan motor windings for opens, shorts or grounds. See testing procedure in SECTION D.
	9. Will not rotate. (Motor cycles on overload.)	9a. Fan blade hitting shroud or blower wheel hitting scroll. Realign assembly. NOTE: Slinger ring of the condenser fan must have 1/4" to 5/16" clearance to the base.
		9b. Check fan motor bearings. If motor shaft will not rotate, replace motor.
Fan motor runs intermittently.	1. Cycles on motor protector.	1a. Check voltage. See limits at beginning of this section.

SERVICE PROBLEM	POSSIBLE CAUSE	DIAGNOSIS / REMEDY
Fan motor runs intermittently. (cont.)	1. Cycles on motor protector. (cont.)	1b. If not within limits, consult an electrician.
		1c. Test capacitor. See testing procedure in SECTION E. Check bearings. Does fan blade rotate freely? If not, replace fan motor.
		1d. Note change from high speed to low speed. If speed does not change, replace motor.
Fan motor noisy.	1. Condenser fan blade or evaporator blower wheel.	1. If cracked, out of balance or partially missing, replace.
	2. Loose blower clamp or set screw.	2. Tighten. Note fan blade position in specification tables.
	3. Worn bearings.	3. If knocking sound is heard when running or the shaft is loose, replace motor.
	4. Grommets (if applicable).	4. Check grommets. If worn or missing, replace.
Compressor will not run, but the fan motor runs.	1. Voltage.	1. Check voltage. See limits at beginning of this section.
	2. Wiring.	2. Check wire connections. If loose, repair or replace connection. If wires are disconnected, refer to wiring diagram.
	3. Selector switch.	3. Check for continuity and refer to wiring diagram for terminal identification. Replace switch if circuit is open.
	4. Temperature control.	4a. Check position of control knob. If not at the coldest setting, advance knob to this setting and restart unit.
		4b. Check continuity of control. Replace control if contacts are stuck open.
	5. Capacitor. (Discharge capacitor before testing.)	5. Check capacitor. See testing procedures in SECTION E. Replace if not within ± 10 % of manufacturer's rating. Replace if shorted, open or damaged.
	6. Compressor.	6. Check compressor motor for open circuit or grounded windings. If windings are open, shorted or grounded, replace compressor.
	7. Motor protector (external).	7. If compressor temperature is high, remove overload, cool and check for continuity. Replace if open.

SERVICE PROBLEM	POSSIBLE CAUSE	DIAGNOSIS / REMEDY
Compressor will not run, but the fan motor runs. (cont.)	8. Motor protector (Internal).	8. If compressor temperature is high, allow to cool and test compressor motor windings. (Protector must be below 172°F to close.)
	9. Electronic control (if applicable).	9a. May be normal. If automatic compressor delay circuit has been activated, the compressor will remain off for 3 to 5 minutes.
		9b. Test compressor motor, overload protector and capacitor. See testing procedures in SECTION D and SECTION E.
		9c. Check compressor relay by connecting the NO terminal to the COM with a jumper. If compressor runs with jumper in place, replace control.
	10. Hard starting.	10. Add start assist kit to compressor motor to determine if hard starting is the problem.
Compressor cycles on motor protector.	1. Voltage.	1. Check voltage. See limits at beginning of this section.
	2. Motor protector (external).	2. If compressor temperature is high, remove overload, cool and check for continuity. Replace if open.
	3. Motor protector (internal).	3. If compressor is hot, allow to cool. Test compressor motor windings. (Protector must be below 172°F to close.)
	4. Fan motor.	4. If not running, determine cause. Replace if required.
	5. Condenser air flow restriction.	5. Remove cabinet or sleeve. Inspect the fan side of condenser. If condenser coils are restricted, clean carefully (do not damage or bend fins).
	6. Condenser fins damaged.	6. If condenser fins are collapsed over a large area, head pressure will increase and cause the compressor to cycle. Straighten fins or replace condenser.
	7. Capacitor.	7. Test capacitor. See testing procedure in SECTION E.
	8. Wiring.	8. Check terminals. If loose, repair or replace.
	9. Refrigerant system.	9. Check sealed system for restriction or air in system (low side leak).

SERVICE PROBLEM	POSSIBLE CAUSE	DIAGNOSIS / REMEDY
Insufficient cooling.	1. Low capacity.	1. Check capacity. Refer to capacity test at end of this section.
	2. Air filter.	2. If restricted, clean or replace filter.
	3. Exhaust door open.	3. Close if open. Instruct user.
	4. Unit undersized.	4. Refer to capacity guide to determine if unit is properly sized for area to be cooled.
Excessive noise.	1. Evaporator blower wheel.	1. Check set screw or clamp. If loose or missing, tighten or replace. Refer to blower wheel setting guide in specification tables and reset.
	2. Condenser fan.	2. Check set screw, clamp or clip. If loose or missing, tighten or replace. Refer to condenser fan setting guide in specification tables and reset.
	3. Copper tubing.	3. Remove wrapper or sleeve and carefully realign tubing. Possible tubing contact points: cabinet shell, compressor, shroud, bulkhead, or other tubing.
	4. Compressor internal noise.	4. Replace compressor if abnormally noisy.
	5. Fan motor.	5. Replace fan motor if bearings are worn, causing a knocking or rubbing sound.
Excessive water or condensation.	1. Water hitting condenser fan.	1. Condition will correct itself when humidity is reduced in the area being cooled.
	2. Condensation on room side of unit.	2. Condensation will form on the louvers under extremely high humidity conditions. As the humidity is reduced in the area being cooled, the moisture will disappear.
No cooling.	1. Refrigerant leak.	1. Add leak test charge, find and repair leak, purge and recharge system.
Wattage decreases slowly until abnormally low.	1. Undercharged, restricted strainer or plugged restrictor tube.	1. Install a filter drier. Replace restrictor tube if plugged. Evacuate and recharge.
Wattage decreases immediately.	1. No refrigerant.	1. Leak test and repair. Evacuate and recharge.

SERVICE PROBLEM	POSSIBLE CAUSE	DIAGNOSIS / REMEDY
Wattage decreases immediately. (cont.)	2. Compressor defective.	2. Refer to SECTION H.
Wattage continuously high.	1. Overcharge of refrigerant.	1. Evacuate and recharge.
Evaporator coil partially frosted.	1. System is low on refrigerant.	1. Check for leak and repair. Evacuate and recharge.
Evaporator completely iced.	1. Low outside temperature.	1. Turn to high fan until evaporator defrosts.
No heat. (Heat/Cool models only.)	1. No power.	1. Check voltage at outlet. If none, correct.
	2. Selector switch position.	2. Selector switch must be in the heat mode.
	3. Temperature control position.	3. Check setting. Turn to warmest position.
	4. Fan motor.	4. See fan motor testing in SECTION D if heater is on and there is no air flow.
	5. Heating element.	5. Remove control panel. Check continuity of orange leads. If none, remove cabinet and check heating element, thermal cut-out and fuse link. Replace part that is defective.
	6. Selector switch.	6. Check continuity of switch. Continuity should be between terminals identified on wiring diagram when switch is in the heat mode.
	7. Temperature control.	7. Check continuity of control by removing orange lead from terminal, turn selector switch to warmest setting, and check between common terminal and orange terminal. (NOTE: Room temperature should be 80°F or lower when testing temperature controls.) If no continuity exists, replace control.
	8. Terminals and connectors.	8. Check terminals and connectors for loose connections. Repair or replace.
Fan motor will not rotate during a heat cycle. (Heat/Cool models only.)	1. Thermostatic drain valve. (Water level control, if applicable.)	1. Check valve operation by placing in ice water. Valve should start to open at approximately 40°F.

Chapter 11: Refrigerators

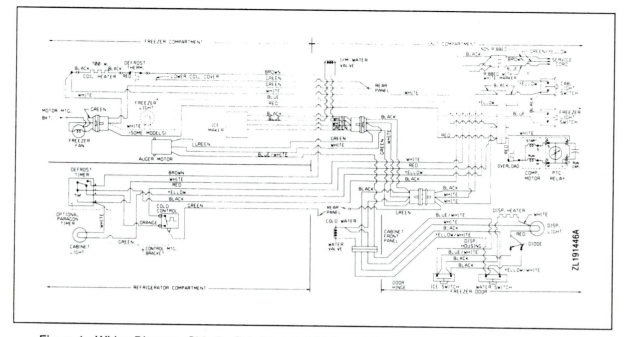

Figure 1 - Wiring Diagram, Side By Side Models With Ice & Water Dispensers and "SG" Compressor

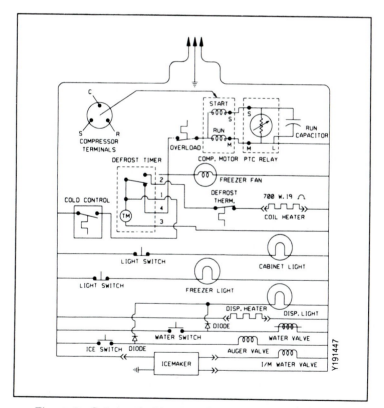

Figure 2 - Schematic Diagram, Side By Side models With
Ice & Water Dispensers and "SG" Compressors

329

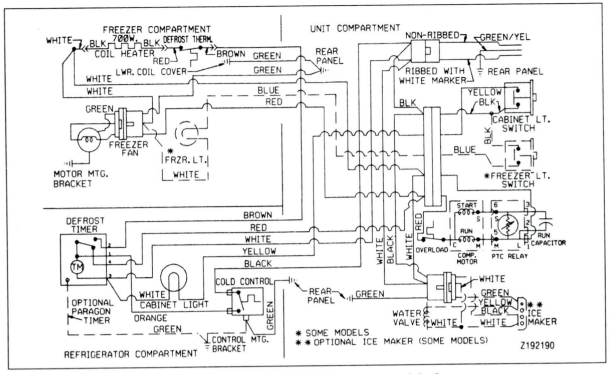

Figure 3 - Wiring Diagram, Side By Side Models With "SG" Compressors

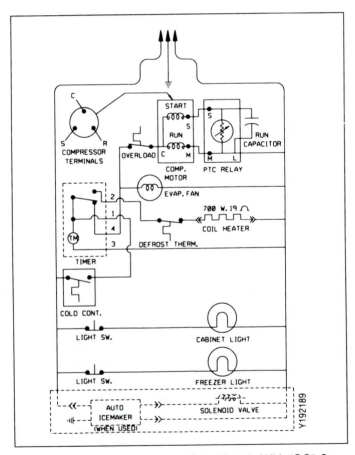

Figure 4 - Schematic Diagram, Side By Side Models With "SG" Compressors

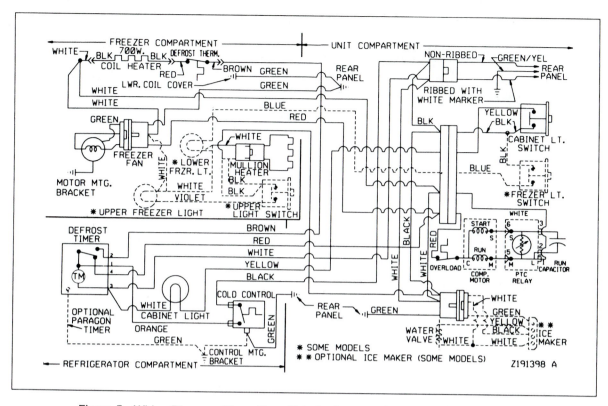

Figure 5 - Wiring Diagram, Three Door Side By Side Models With "SG" Compressors

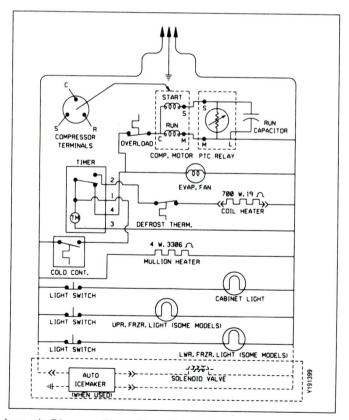

Figure 6 - Schematic Diagram, Three Door Side By Side Models With "SG" Compressors

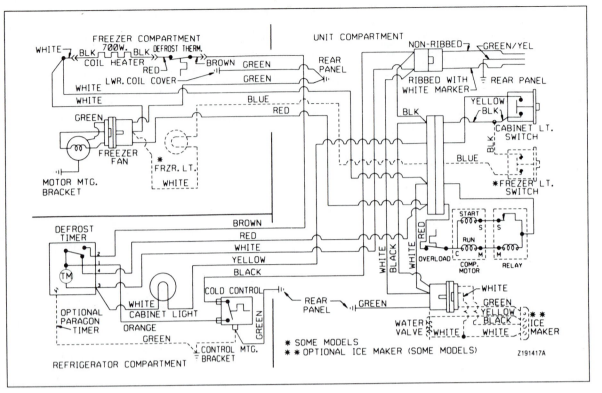

Figure 7 - Wiring Diagram, Side By Side Models

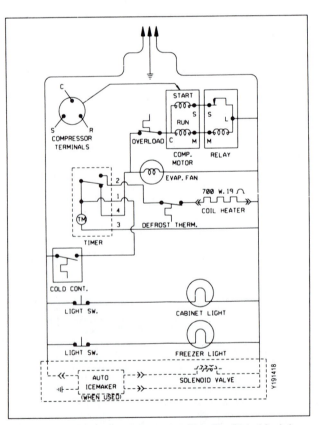

Figure 8 - Schematic Diagram, Side By Side Models

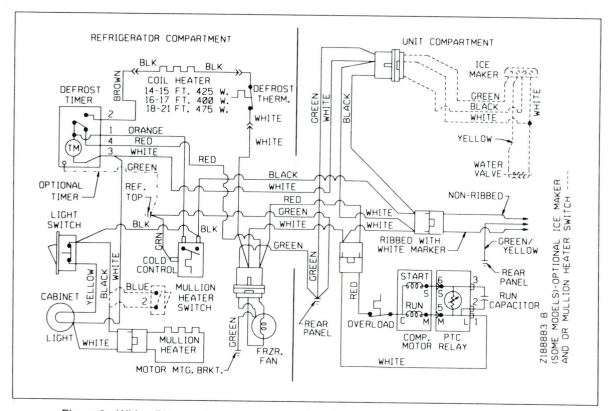

Figure 9 - Wiring Diagram, Top Freezer Frost Proof Models With "SF" & "SG" Compressors

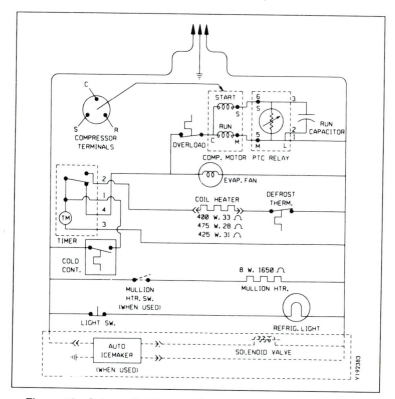

Figure 10 - Schematic Diagram Top Freezer Frost Proof Models
With "SF" & "SG" Compressors

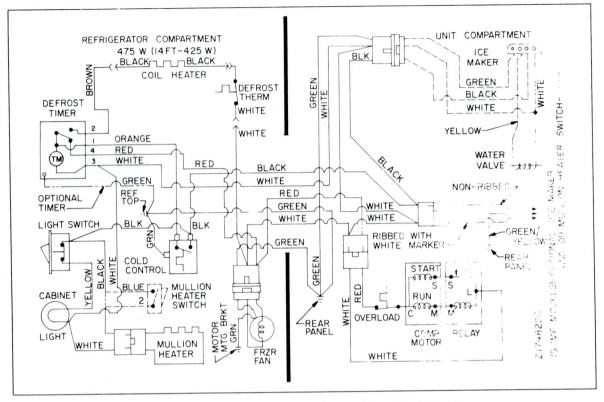

Figure 11 - Wiring Diagram, Top Freezer Frost Proof Models

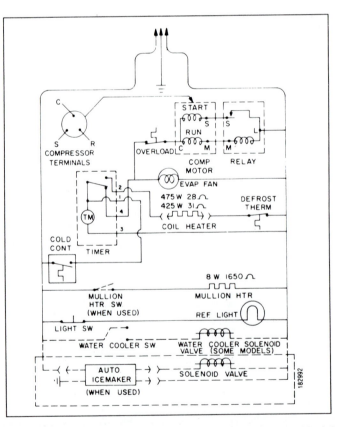

Figure 12 - Schematic Diagram, Top Freezer Frost Proof Models

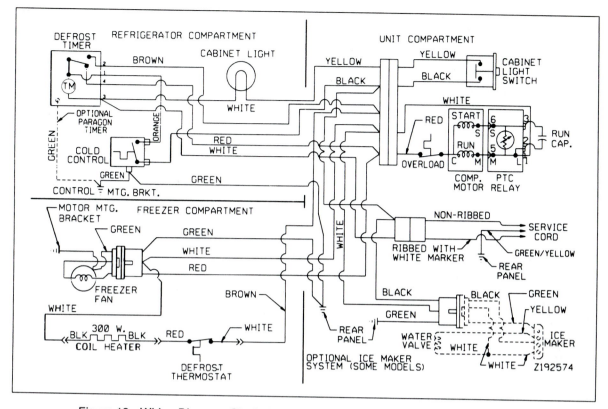

Figure 13 - Wiring Diagram, Single Door Frost Proof Models With "SG" Compressors

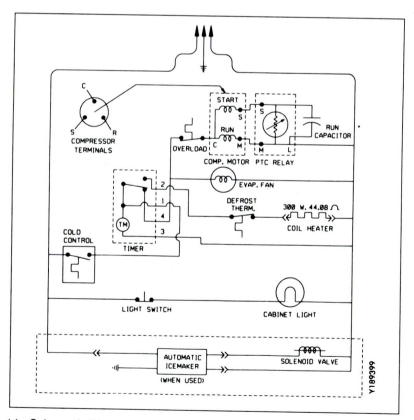

Figure 14 - Schematic Diagram, Single Door Frost Proof Models With "SG" Compressors

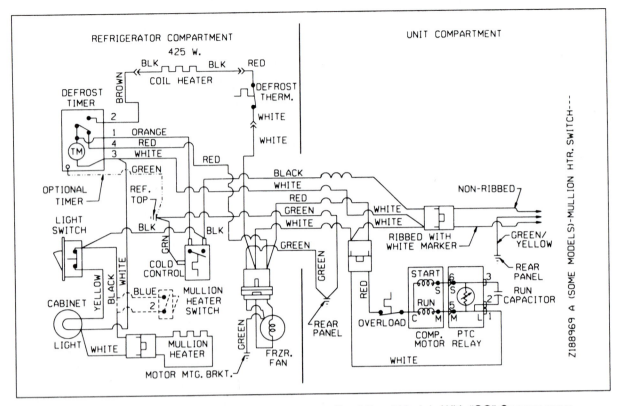

Figure 15 - Wiring Diagram, 12 Cu. Ft. Top Freezer Frost Proof Models With "SG" Compressors

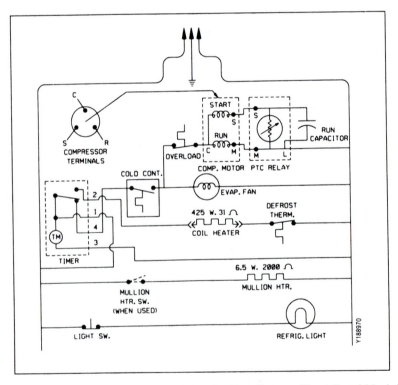

Figure 16 - Schematic Diagram, 12 Cu. Ft. Top Freezer Frost Proof Models
With "SG" Compressors

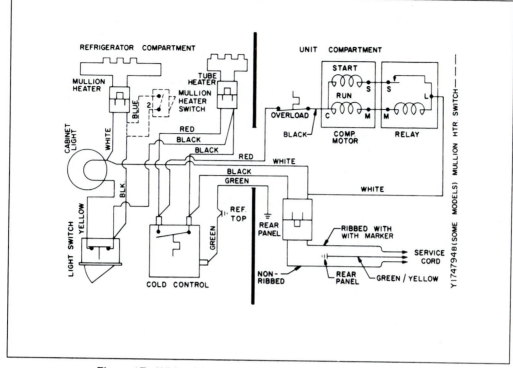

Figure 17 - Wiring Diagram, Top Freezer Cycle Defrost Models

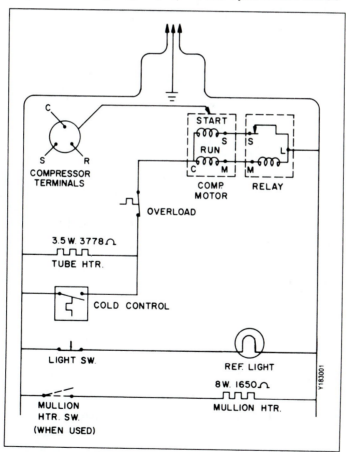

Figure 18 - Schematic Diagram, Top Freezer Frost Proof Models

337

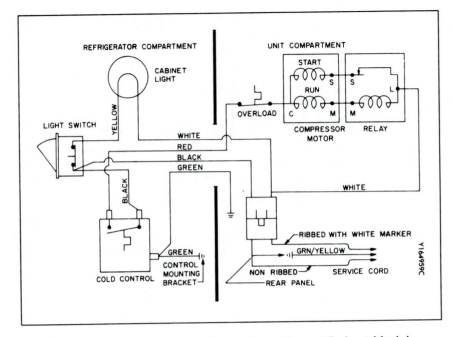

Figure 19 - Wiring Diagram, Single Door Manual Defrost Models

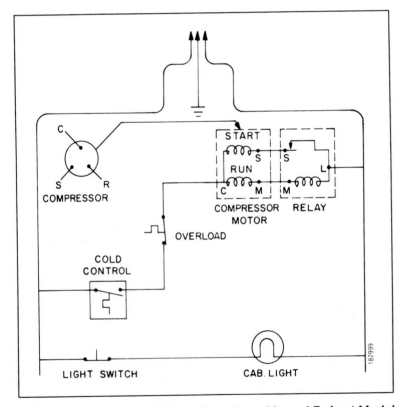

Figure 20 - Schematic Diagram, Single Door Manual Defrost Models

SECTION G - SERVICE DIAGNOSIS CHART

COMPLAINT	CAUSE
Compressor will not run.	1. No voltage at wall receptacle (house fuse blown). 2. Service cord pulled out of wall receptacle - replace. 3. Low voltage causing compressor to cycle on overload. (Voltage fluctuation should not exceed 10% plus or minus from nominal rating.) 4. Control thermostat dial on "Off" position. 5. Inoperative control thermostat - replace. 6. Compressor stuck - replace. 7. Compressor windings open - replace compressor. 8. Defrost timer stuck in defrost. 9. Compressor overload open. 10. Relay lead loose. 11. Relay loose or inoperative. 12. Service cord pulled out of harness. 13. Faulty cabinet wiring - repair.
Compressor runs but no refrigeration.	1. System out of refrigerant - check for leaks. 2. Compressor not pumping - replace. 3. Restricted filter drier - replace. 4. Restricted capillary tube - replace. 5. Moisture in system - clean and recharge.
Compressor short cycles.	1. Erratic control thermostat - replace. 2. Faulty relay - replace. 3. Restricted air flow over condenser. 4. Low voltage - fluctuation exceeds 10%. 5. Compressor draws excessive wattage - replace.
Compressor runs too much or 100%.	1. Erratic control thermostat, or set too cold - replace or reset to normal position. 2. Refrigerator exposed to unusual heat - relocate. 3. Abnormally high room temperature. 4. Low pumping capacity compressor - replace. 5. Door gaskets not sealing. 6. System undercharged or overcharged - correct charge. 7. Interior light stays on - check door switch. 8. Non-condensables in system - evacuate and recharge. 9. Capillary tube kinked or partially restricted. 10. Filter drier partially restricted - replace. 11. Excessive service load. 12. Restricted air flow over condenser.
Noisy.	1. Tubing vibrates - adjust tubing. 2. Internal compressor noise - replace. 3. Compressor vibrating on cabinet frame - adjust. 4. Loose parts, check shelving, kickplate, defrost drain pan. 5. Compressor operating at high head pressure due to restricted air flow over condenser.

COMPLAINT	CAUSE
Freezer compartment too warm.	1. Inoperative fan motor. 2. Improperly positioned fan. 3. Evaporator iced up. 4. Defrost heater inoperative. 5. Inoperative defrost timer. 6. Inoperative defrost thermostat. 7. Wire loose at defrost timer. 8. Fan cover missing. 9. Excessive service load. 10. Abnormally low room temperatures. 11. Freezer compartment or refrigerator compartment door left open. 12. Control thermostat out of calibration. 13. Refrigerator compartment or freezer compartment door gasket not sealing. 14. Control thermostat thermal element improperly positioned. 15. Shortage of refrigerant. 16. Restricted filter drier or capillary tube.
Refrigerator compartment too warm.	1. Inoperative fan motor. 2. Improperly positioned fan. 3. Fan cover missing. 4. Refrigerator compartment air inlet air duct restricted. 5. Refrigerator compartment to freezer compartment return air duct restricted. 6. Abnormally low room temperature. 7. Control thermostat out of calibration. 8. Control thermostat knob set at warm setting. 9. Control thermostat thermal element improperly positioned. 10. Evaporator iced up. 11. Inoperative defrost timer. 12. Inoperative defrost heater. 13. Inoperative defrost thermostat. 14. Wire loose at defrost timer. 15. Excessive service load. 16. Refrigerator compartment or freezer compartment door left open. 17. Inoperative or erratic refrigerator compartment and/or freezer compartment door switch. 18. Shelves covered with foil wrap or paper retarding air circulation. 19. Shortage of refrigerant. 20. Restricted capillary tube or filter drier.
Evaporator blocked with ice.	1. Inoperative defrost timer. 2. Defrost timer terminates too early. 3. Defrost timer incorrectly wired - check wiring. 4. Inoperative fan motor. 5. Inoperative defrost thermostat. 6. Inoperative defrost heater. 7. Refrigerator compartment or freezer compartment door left open. 8. Freezer compartment drain plugged - clean.

340

Chapter 12: Freezers

ELECTRICAL CIRCUIT

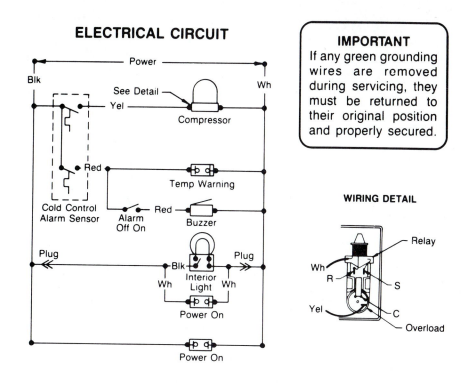

Figure 1 - Wiring Diagram, Compact Chest Models

ELECTRICAL CIRCUIT

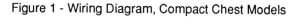

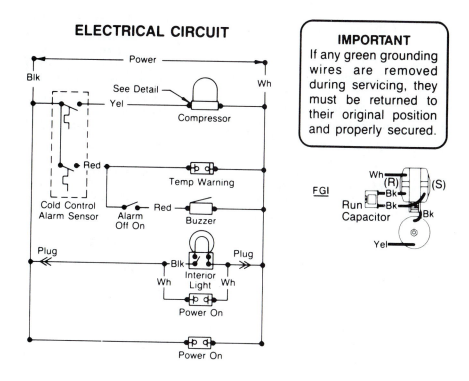

Figure 6 - Wiring Diagram, Standard Chest Models

ELECTRICAL CIRCUIT

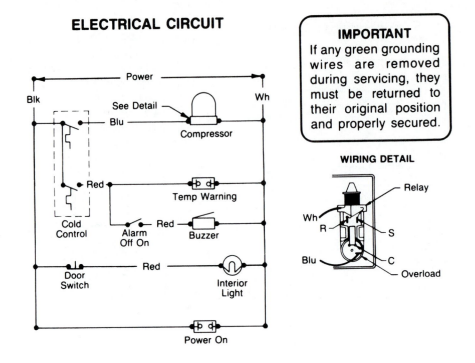

IMPORTANT
If any green grounding wires are removed during servicing, they must be returned to their original position and properly secured.

WIRING DETAIL

Figure 7 - Wiring Diagram, Upright Manual Defrost

ELECTRICAL CIRCUIT

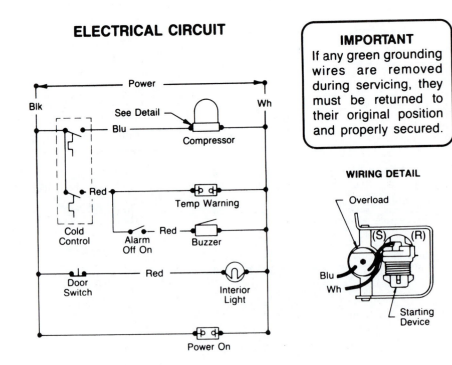

IMPORTANT
If any green grounding wires are removed during servicing, they must be returned to their original position and properly secured.

WIRING DETAIL

Figure 8 - Wiring Diagram, Upright Manual Defrost

ELECTRICAL CIRCUIT

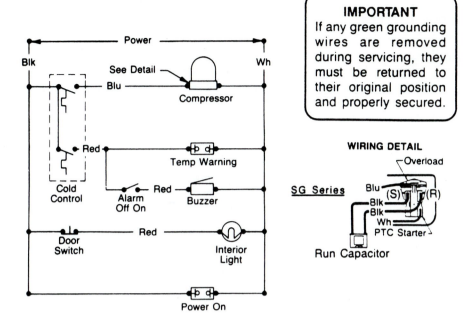

IMPORTANT
If any green grounding wires are removed during servicing, they must be returned to their original position and properly secured.

WIRING DETAIL

Figure 11 - Wiring Diagram, Upright Manual Defrost

ELECTRICAL CIRCUIT

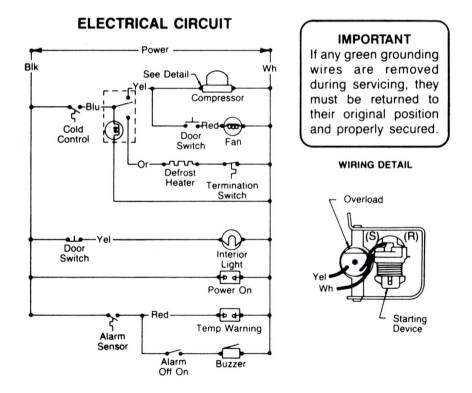

IMPORTANT
If any green grounding wires are removed during servicing, they must be returned to their original position and properly secured.

WIRING DETAIL

Figure 12 - Wiring Diagram, Upright Frost Free

SECTION F - TROUBLESHOOTING CHART

COMPLAINT	CAUSE - REMEDY
Compressor will not run.	1. No voltage at wall receptacle - check circuit breaker or fuse. 2. Service cord pulled out of wall receptacle - replace. 3. Low voltage causing compressor to cycle on overload. (Voltage fluctuation should not exceed 10% plus or minus from nominal rating.) 4. Control thermostat dial in "Off" position - turn control on. 5. Inoperative control thermostat - replace control. 6. Compressor stuck - replace compressor. 7. Compressor windings open - replace compressor. 8. Defrost timer stuck in defrost - replace defrost timer. 9. Compressor overload stuck open - replace overload. 10. Relay lead loose - repair or replace lead. 11. Relay loose or inoperative - replace relay. 12. Service cord pulled out of harness - repair connection. 13. Faulty cabinet wiring - repair wiring.
Compressor runs but no refrigeration.	1. System out of refrigerant - check for leaks. 2. Compressor not pumping - replace compressor. 3. Restricted filter drier - replace filter drier. 4. Restricted capillary tube - replace. (Upright models only) 5. Moisture in system - check for leak in low side.
Compressor short cycles.	1. Erratic control thermostat - replace control. 2. Faulty relay - replace relay. 3. Restricted air flow over condenser - ensure condenser has unobstructed air flow. 4. Low voltage - fluctuation exceeds 10%. (Call qualified electrician.) 5. Compressor draws excessive wattage - replace compressor.
Compressor runs too much or 100%.	1. Erratic control thermostat, or set too cold - replace or reset to normal position. 2. Freezer exposed to unusual heat - relocate freezer. 3. Abnormally high room temperature - advise customer. 4. Low pumping capacity compressor - replace compressor. 5. Door or lid gaskets not sealing - adjust or replace necessary parts. 6. System undercharged - check for leaks. 7. System overcharged - correct charge. 8. Interior light stays on - check door or lid switch. 9. Non-condensables in system - replace filter drier, evacuate, and recharge. 10. Capillary tube kinked or partially restricted - replace heat exchanger. (Upright Models Only) 11. Filter drier partially restricted - replace filter drier. 12. Excessive service load - advise customer. 13. Restricted air flow over condenser - ensure condenser has unobstructed air flow
Noisy.	1. Tubing vibrates - adjust tubing. 2. Internal compressor noise - replace compressor. 3. Compressor vibrating on cabinet frame - adjust compressor. 4. Loose parts - check shelving, kickplate, defrost drain pan. 5. Compressor operating at high head pressure due to restricted air flow over condenser - ensure condenser has unobstructed air flow.

COMPLAINT	CAUSE - REMEDY
Freezer too warm.	1. Inoperative fan motor - check wiring and fan motor. 2. Improperly positioned fan - position blade at end of shaft. 3. Evaporator iced up - check defrost system. 4. Defrost heater inoperative - check wiring and defrost heater. 5. Inoperative defrost timer - check wiring and defrost timer. 6. Inoperative defrost thermostat - check wiring and defrost thermostat. 7. Wire loose at defrost timer - repair wire. 8. Excessive service load - advise customer. 9. Abnormally low room temperatures - advise customer. 10. Freezer door left open - advise customer. 11. Control thermostat out of calibration - replace control. 12. Door or lid gasket not sealing - adjust or replace necessary parts. 13. Control thermostat sensing element improperly positioned - reposition sensing element. 14. Shortage of refrigerant - check for leaks. 15. Restricted filter drier or capillary tube - check for leaks or "burned" compressor windings.
Evaporator blocked with ice.	1. Inoperative defrost timer - check wiring and defrost timer. 2. Defrost thermostat terminates too early - check for correct positioning of defrost thermostat or replace. 3. Defrost timer incorrectly wired - check wiring. 4. Inoperative fan motor - check wiring and fan motor. 5. Inoperative defrost thermostat - check wiring and defrost thermostat. 6. Inoperative defrost heater - check wiring and defrost heater. 7. Freezer door left open - advise customer. 8. Freezer defrost drain plugged - clean drain port.

NOTES

TIMER SEQUENCE CHART (MODEL 61-1170-10-01)

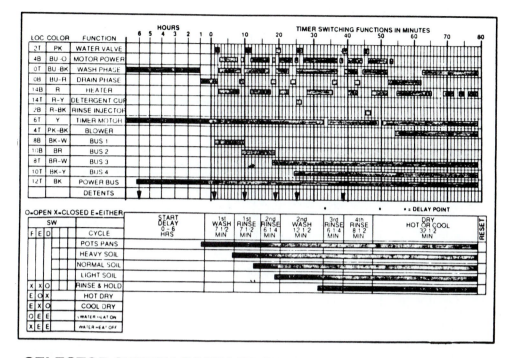

SELECTOR SWITCH CONTACT CHART (MODEL 61-1170-10-01)

SELECTOR SWITCH	6 - 7	7 - 8	9 - 10	1 - 2	1 - 3	1 - 4	1 - 5
POTS AND PANS	/////	/////	/////	OPEN	CLOSED	CLOSED	
HEAVY SOIL	/////	/////	/////	OPEN	CLOSED		OPEN
NORMAL SOIL	/////	/////	/////	OPEN	CLOSED	OPEN	
RINSE HOLD	CLOSED	OPEN	CLOSED		OPEN	OPEN	
WATER HEAT ON	OPEN	OPEN		/////	/////	/////	/////
WATER HEAT OFF	OPEN		CLOSED	/////	/////	/////	/////
HOT DRY	OPEN	CLOSED	/////	/////	/////	/////	/////
COOL DRY	CLOSED	OPEN	/////	/////	/////	/////	/////

■ = CLOSED

▨ = OPEN

▧ = EITHER

351

SELECTOR SWITCH CHART AND WIRING DIAGRAM (MODEL 61-1170-10-01)

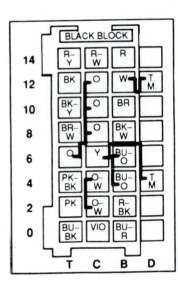

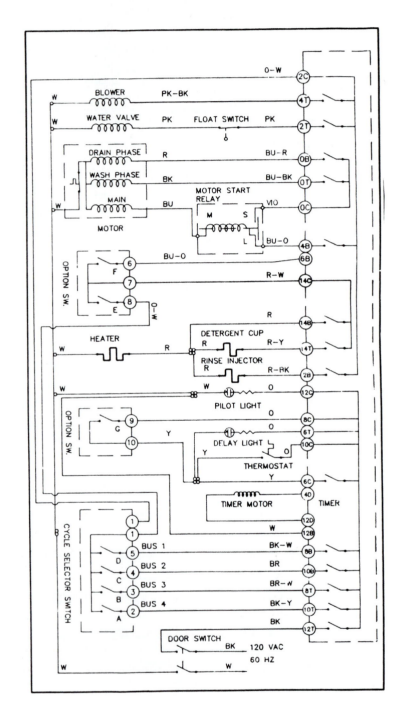

SCHEMATIC DIAGRAM (MODEL 61-1170-10-01)

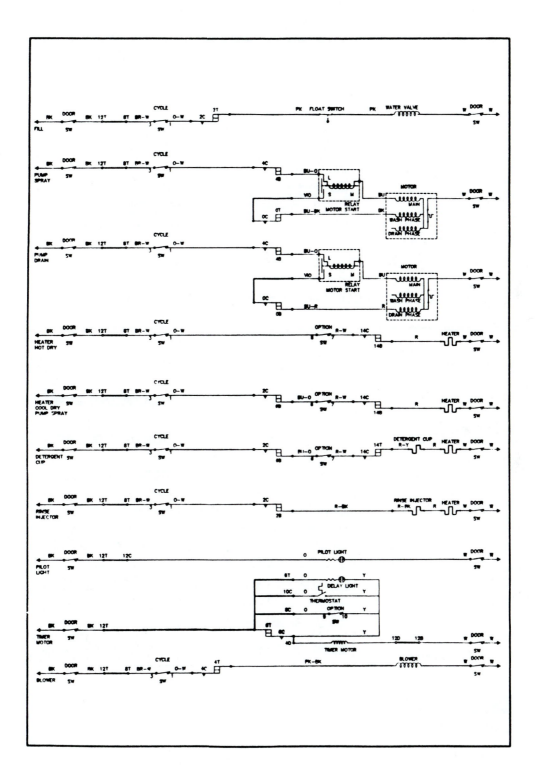

TIMER SEQUENCE CHART (MODEL 61-1160-10-01)

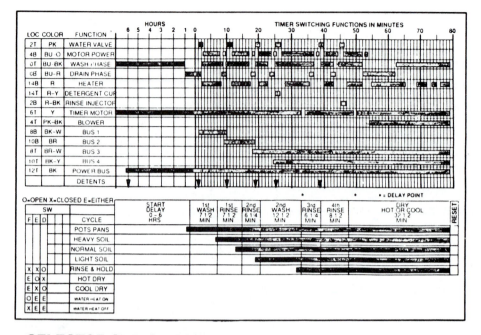

SELECTOR SWITCH CONTACT CHART (MODEL 61-1160-10-01)

SELECTOR SWITCH	6 - 7	7 - 8	9 - 10	1 - 2	1 - 3	1 - 4	1 - 5
POTS AND PANS	EITHER	EITHER	EITHER	OPEN	CLOSED	CLOSED	CLOSED
HEAVY SOIL	EITHER	EITHER	EITHER	OPEN	CLOSED	CLOSED	OPEN
NORMAL SOIL	EITHER	EITHER	EITHER	OPEN	CLOSED	OPEN	OPEN
RINSE HOLD	CLOSED	OPEN	CLOSED	CLOSED	OPEN		
WATER HEAT ON	OPEN	OPEN		EITHER	EITHER	EITHER	EITHER
WATER HEAT OFF	OPEN	OPEN	CLOSED	EITHER	EITHER	EITHER	EITHER
HOT DRY	OPEN	CLOSED	EITHER	EITHER	EITHER	EITHER	EITHER
COOL DRY	CLOSED	OPEN	EITHER	EITHER	EITHER	EITHER	EITHER

= CLOSED

= OPEN

= EITHER

SELECTOR SWITCH CHART AND WIRING DIAGRAM (MODEL 61-1160-10-01)

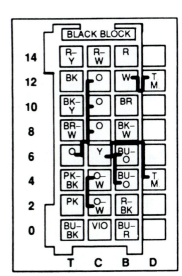

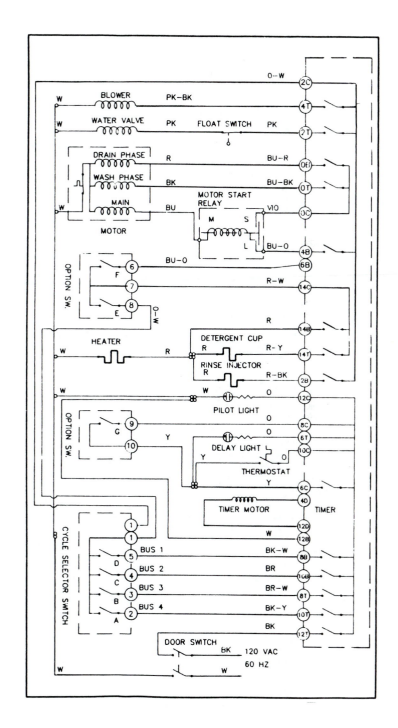

SCHEMATIC DIAGRAM (MODEL 61-1160-10-01)

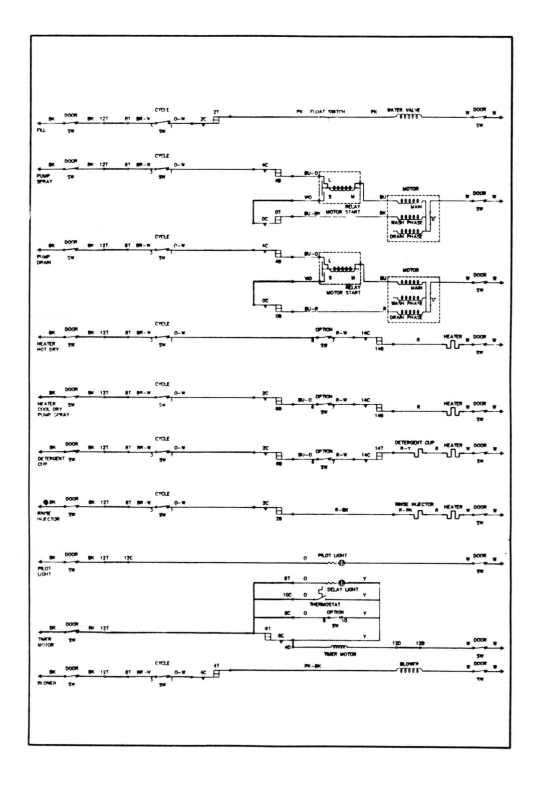

TIMER SEQUENCE CHART (MODEL 61-1140-10-01)

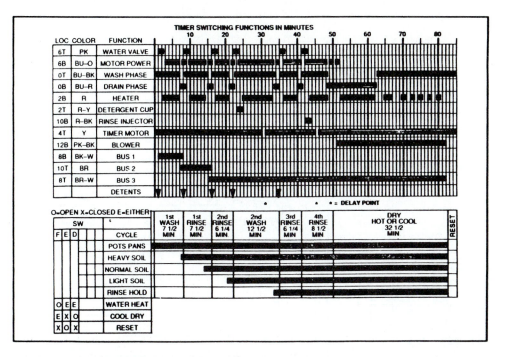

SELECTOR SWITCH CONTACT CHART (MODEL 61-1140-10-01)

SELECTOR SWITCH	5 - 7	5 - 6	3 - 4	1 - 2
POTS AND PANS	EITHER	EITHER	EITHER	CLOSED
NORMAL SOIL	EITHER	EITHER	EITHER	OPEN
WATER HEAT	EITHER	EITHER	OPEN	EITHER
COOL DRY	CLOSED	OPEN	EITHER	EITHER
RESET OPTIONS	OPEN	CLOSED		EITHER

◼ = CLOSED

▦ = OPEN

//////, = EITHER

SELECTOR SWITCH CHART AND WIRING DIAGRAM (MODEL 61-1140-10-01)

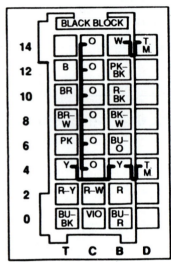

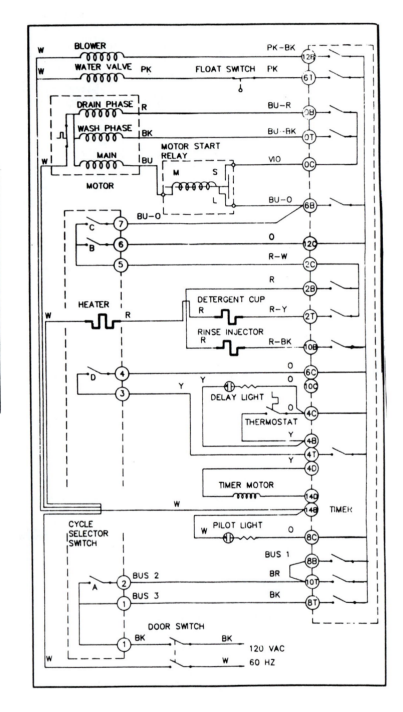

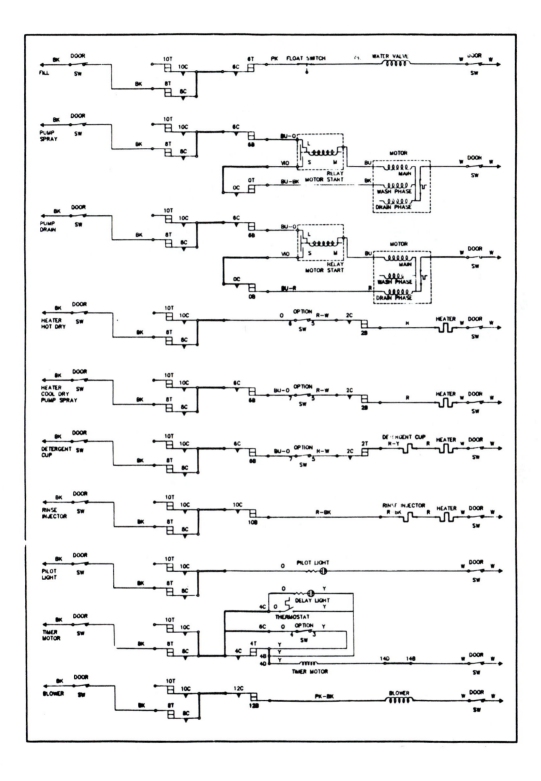

TIMER SEQUENCE CHART (MODEL 61-1120-10-01)

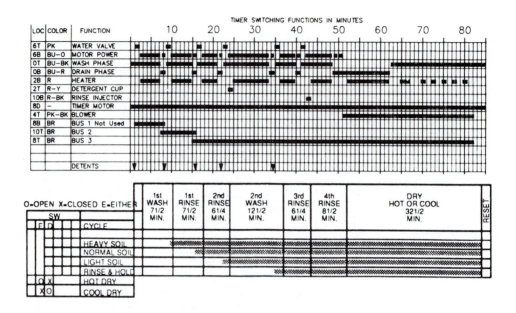

SELECTOR SWITCH CONTACT CHART (MODEL 61-1120-10-01)

SELECTOR SWITCH	1 - 3	2 - 3
COOL DRY	OPEN	CLOSED
HOT DRY	CLOSED	OPEN

■ = CLOSED

▒ = OPEN

SELECTOR SWITCH CHART AND WIRING DIAGRAM (MODEL 61-1120-10-01)

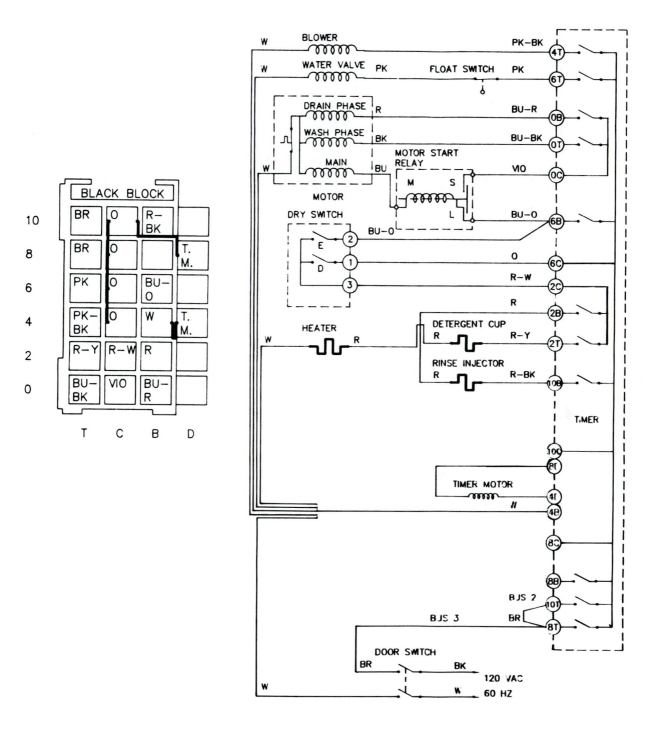

SCHEMATIC DIAGRAM (MODEL 61-1120-10-01)

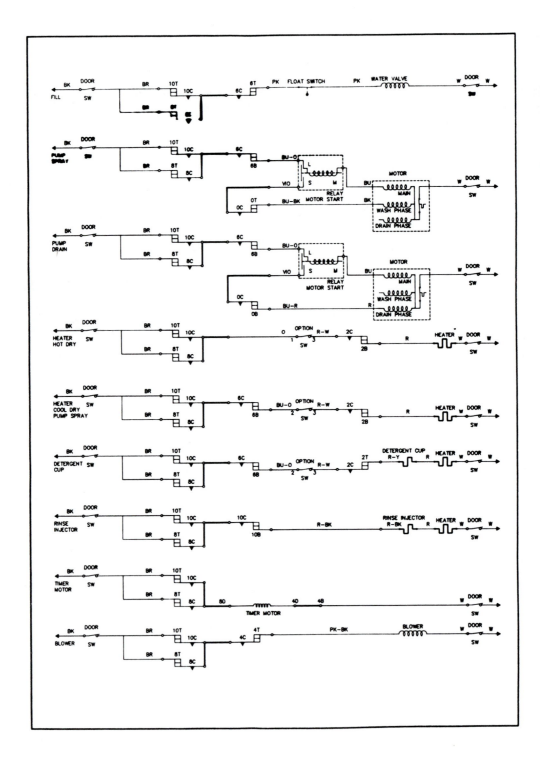

TROUBLESHOOTING

Always turn off the electric power supply before servicing any electrical component, making ohmmeter checks, or replacing any parts.

All voltage checks should be made with a voltmeter having a full scale range of 130 volts or higher.

After service is completed, be sure all safety grounding circuits are complete, all electrical connections are secure, and all access panels are in place.

SYMPTOM   **CHECK**

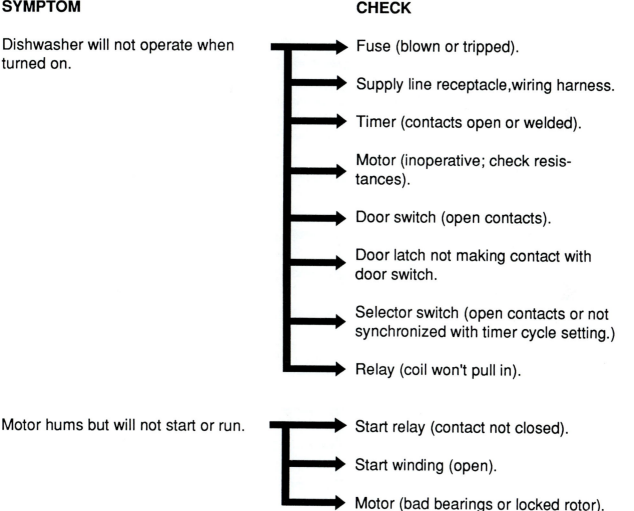

Dishwasher will not operate when turned on.

Fuse (blown or tripped).

Supply line receptacle, wiring harness.

Timer (contacts open or welded).

Motor (inoperative; check resistances).

Door switch (open contacts).

Door latch not making contact with door switch.

Selector switch (open contacts or not synchronized with timer cycle setting.)

Relay (coil won't pull in).

Motor hums but will not start or run.

Start relay (contact not closed).

Start winding (open).

Motor (bad bearings or locked rotor).

SYMPTOM	CHECK
Motor trips out on internal thermal overload protector.	Start relay not dropping out.
	Improper voltage.
	Seal faces binding.
	Motor shaft binding.
	Motor windings shorted.
	Foreign items in pump.
Dishwasher runs but will not heat.	Heat selector switch (not depressed).
	Heater element (open.)
	Timer (contact open or welded).
	Wiring or terminal (welded or broken).
Detergent cup won't latch.	Cam not locking on bi-metal arm.
	Roll pin retainer or shaft broken.
	Broken spring(s).
Detergent cup won't open.	Roll pin retainer or shaft broken.
	Cup binding.
	Defective bi-metal.
	Timer contact (open or welded).
	Wiring or terminal (welded or broken).
	Open heater element.
	Customer used liquid detergent (clean assembly).

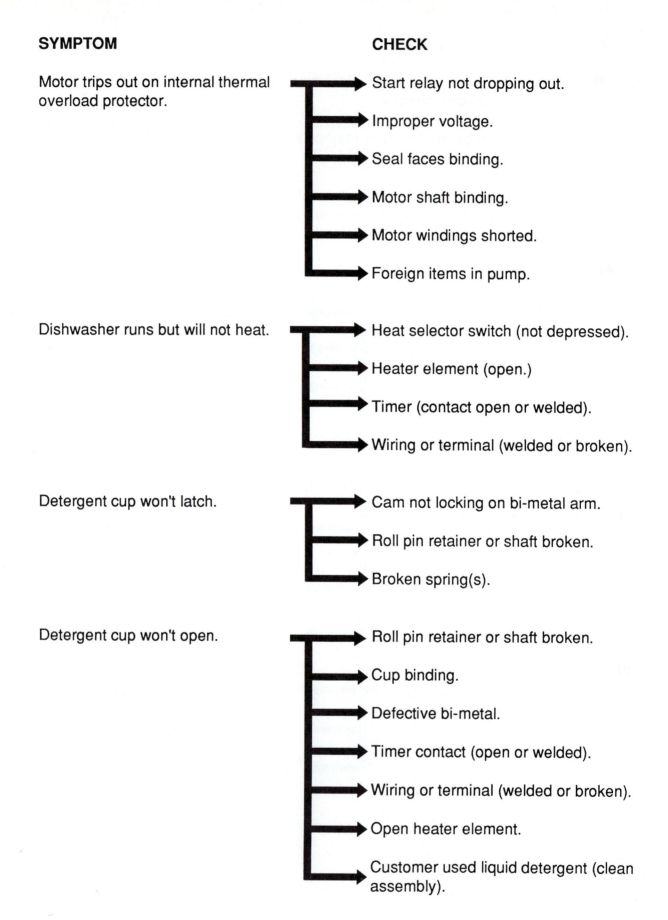

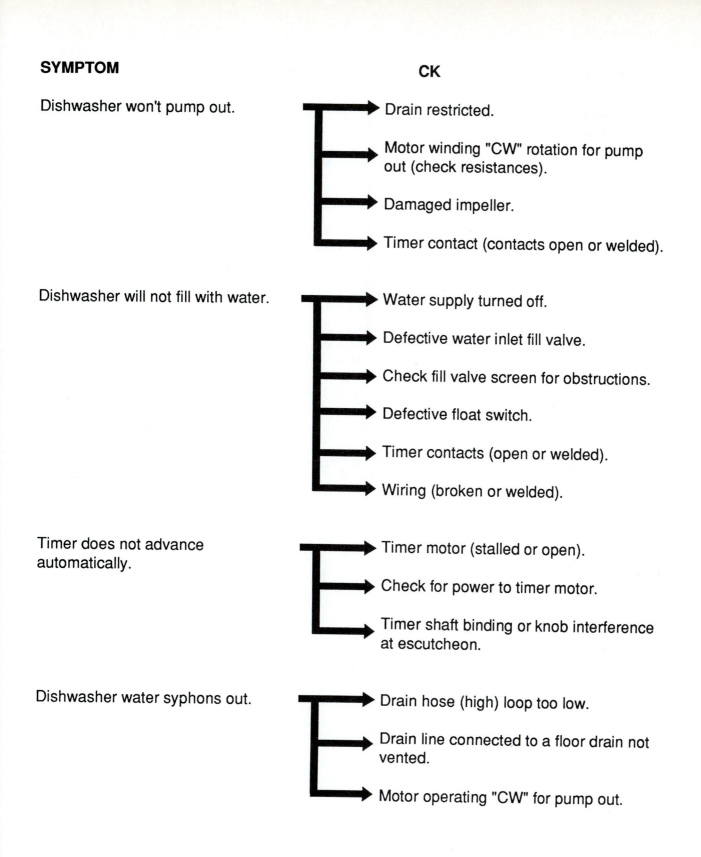

SYMPTOM   **CK**

Dishwasher won't pump out. → Drain restricted.

→ Motor winding "CW" rotation for pump out (check resistances).

→ Damaged impeller.

→ Timer contact (contacts open or welded).

Dishwasher will not fill with water. → Water supply turned off.

→ Defective water inlet fill valve.

→ Check fill valve screen for obstructions.

→ Defective float switch.

→ Timer contacts (open or welded).

→ Wiring (broken or welded).

Timer does not advance automatically. → Timer motor (stalled or open).

→ Check for power to timer motor.

→ Timer shaft binding or knob interference at escutcheon.

Dishwasher water syphons out. → Drain hose (high) loop too low.

→ Drain line connected to a floor drain not vented.

→ Motor operating "CW" for pump out.

SYMPTOM

Poor washability.

CHECK

Improper loading of dishes, pots, pans, and nesting of silverware.

Spray arm not rotating.

Check for proper level of water in tub.

Detergent dispenser inoperative.

Old or insufficient amount of detergent.

Soil separator full.

Chapter 15: Electric and Gas Dryers

ELECTRICAL DIAGRAMS INDEX

MODEL	WIRING DIAGRAM	SCHEMATIC DIAGRAM
	PAGE	
DEF	10	11
DESF	12	13
DEDF	14	15
DEIF	16	17
DECIF	18	19
DEISF	20	21
DGF	22	23
DGSF	24	25
DGDF	26	27
DGIF	28	29
DGCIF	30	31
DGISF	32	33

HOW TO READ TIMER CYCLE CHARTS

• All bar charts represent one complete revolution of the timer shaft.

• Shaded portion of bar charts indicate the proportional times that the internal timer contacts are closed.

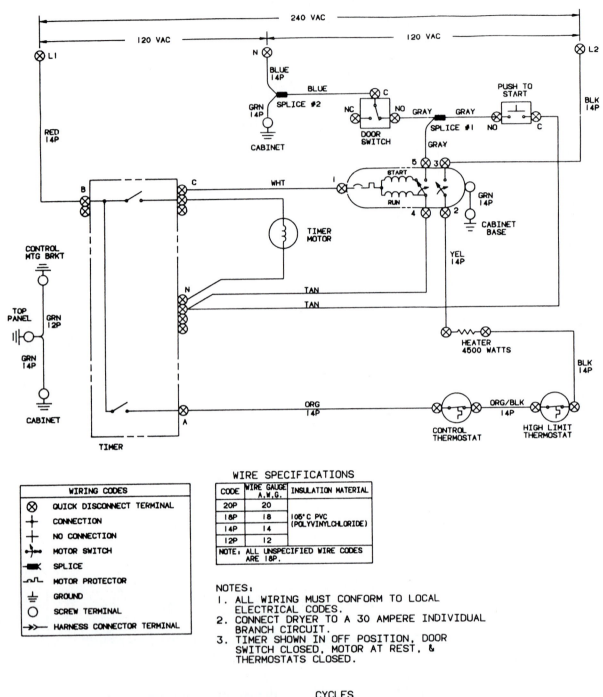

DEF

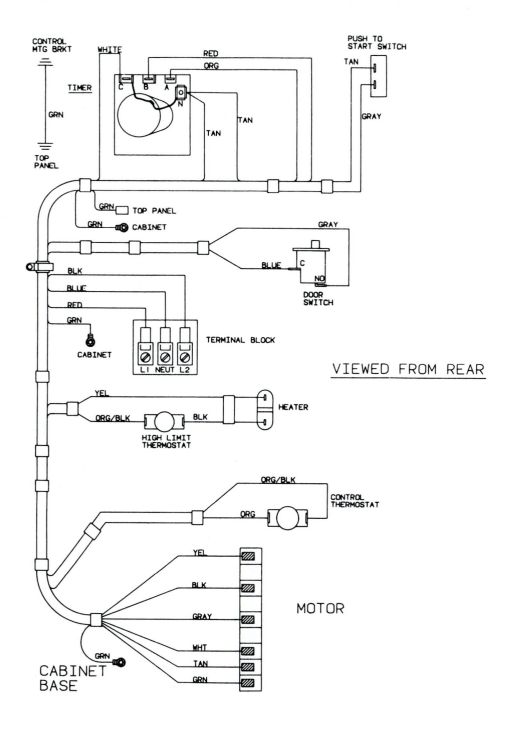

CONTROL
MTG BRKT

WHITE

TIMER

C B A

N

GRN

TOP
PANEL

RED

ORG

TAN

TAN

PUSH TO
START SWITCH

TAN

GRAY

GRN TOP PANEL

GRN CABINET

GRAY

BLUE

C

NO

DOOR
SWITCH

BLK

BLUE

RED

GRN

CABINET

TERMINAL BLOCK

L1 NEUT L2

VIEWED FROM REAR

YEL

ORG/BLK BLK HEATER

HIGH LIMIT
THERMOSTAT

ORG/BLK

ORG

CONTROL
THERMOSTAT

YEL

BLK

GRAY

WHT

TAN

GRN

MOTOR

GRN

CABINET
BASE

27" DRYER, ELECTRIC
"J" ELECTRICAL SYSTEM
WIRING DIAGRAM 142804

DEF

CAUTION: DISCONNECT ELECTRIC CURRENT BEFORE SERVICING DRYER

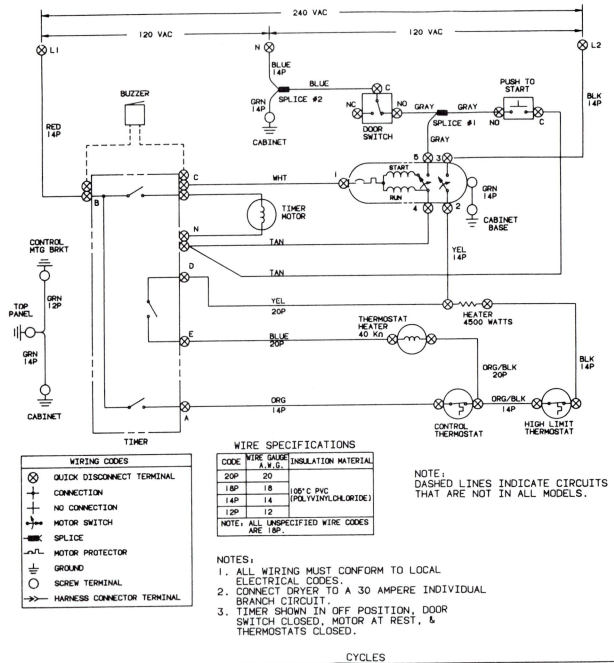

WIRING CODES	
⊗	QUICK DISCONNECT TERMINAL
+	CONNECTION
⊥	NO CONNECTION
⊥•	MOTOR SWITCH
▬K	SPLICE
⌐∿	MOTOR PROTECTOR
⏚	GROUND
○	SCREW TERMINAL
→→	HARNESS CONNECTOR TERMINAL

WIRE SPECIFICATIONS

CODE	WIRE GAUGE A.W.G.	INSULATION MATERIAL
20P	20	
18P	18	105°C PVC (POLYVINYLCHLORIDE)
14P	14	
12P	12	

NOTE: ALL UNSPECIFIED WIRE CODES ARE 18P.

NOTE:
DASHED LINES INDICATE CIRCUITS THAT ARE NOT IN ALL MODELS.

NOTES:
1. ALL WIRING MUST CONFORM TO LOCAL ELECTRICAL CODES.
2. CONNECT DRYER TO A 30 AMPERE INDIVIDUAL BRANCH CIRCUIT.
3. TIMER SHOWN IN OFF POSITION, DOOR SWITCH CLOSED, MOTOR AT REST, & THERMOSTATS CLOSED.

CIRCUIT	REGULAR HEAT CYCLE					OFF	LOW HEAT CYCLE					OFF	AIR FLUFF CYCLE			
	90	75	50	25	0		90	75	50	25	0		60	40	20	0
HEATER B-A																
TIMER & DRIVE MOTOR B-C																
LOW HEAT SW. D-E																

CYCLES

BAR CHART ABOVE REPRESENTS ONE COMPLETE REVOLUTION OF TIMER SHAFT.

DESF

372

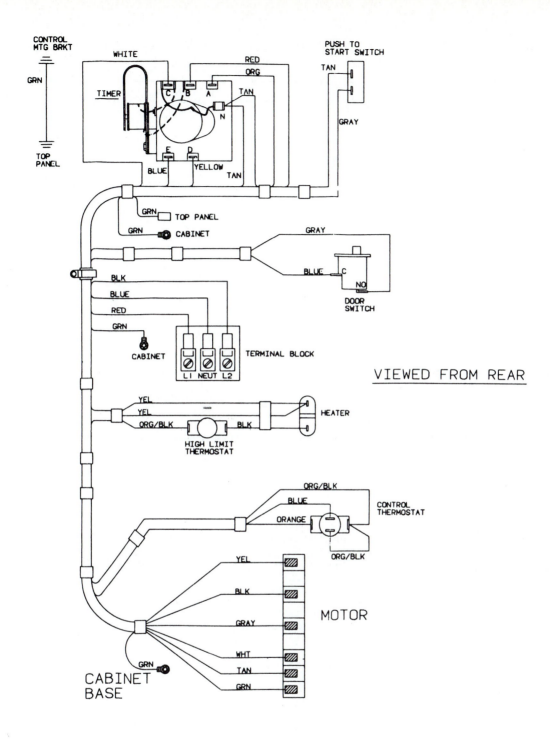

VIEWED FROM REAR

27" DRYER, ELEC.
"K" ELECTRICAL SYSTEM
WIRING DIAGRAM   142829

DESF

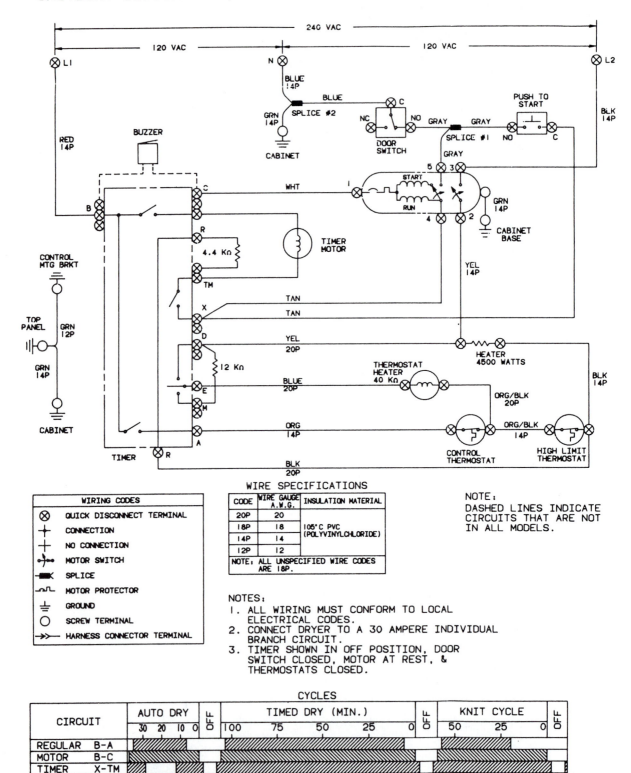

WIRE SPECIFICATIONS

CODE	WIRE GAUGE A.W.G.	INSULATION MATERIAL
20P	20	
18P	18	105°C PVC (POLYVINYLCHLORIDE)
14P	14	
12P	12	

NOTE: ALL UNSPECIFIED WIRE CODES ARE 18P.

WIRING CODES

⊗	QUICK DISCONNECT TERMINAL
+	CONNECTION
+	NO CONNECTION
•	MOTOR SWITCH
▬	SPLICE
ᴖ	MOTOR PROTECTOR
⏚	GROUND
○	SCREW TERMINAL
→→	HARNESS CONNECTOR TERMINAL

NOTE:
DASHED LINES INDICATE CIRCUITS THAT ARE NOT IN ALL MODELS.

NOTES:
1. ALL WIRING MUST CONFORM TO LOCAL ELECTRICAL CODES.
2. CONNECT DRYER TO A 30 AMPERE INDIVIDUAL BRANCH CIRCUIT.
3. TIMER SHOWN IN OFF POSITION, DOOR SWITCH CLOSED, MOTOR AT REST, & THERMOSTATS CLOSED.

CYCLES

CIRCUIT		AUTO DRY				OFF	TIMED DRY (MIN.)					OFF	KNIT CYCLE			OFF
		30	20	10	0		100	75	50	25	0		50	25	0	
REGULAR	B-A															
MOTOR	B-C															
TIMER	X-TM															
MEDIUM	E-M															
LOW	E-D															

DEDF

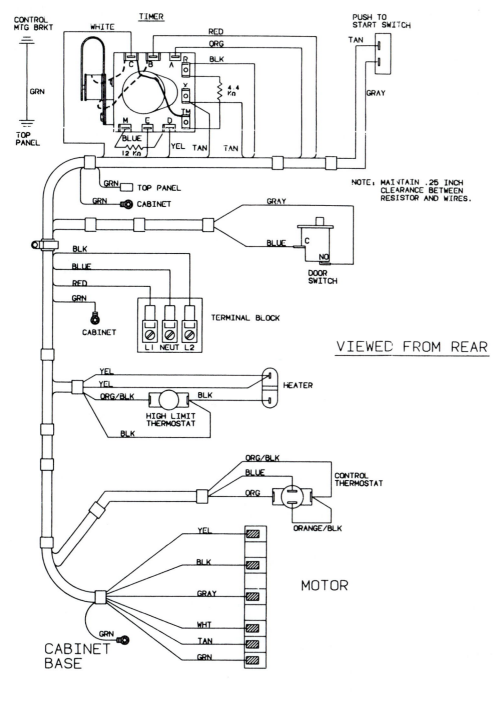

CONTROL
MTG BRKT

WHITE

TIMER

RED

ORG

BLK

PUSH TO
START SWITCH

TAN

GRN

C   B   A   R

Y

4.4
KΩ

GRAY

TOP
PANEL

M   C   D   TM

BLUE

12 KΩ

YEL   TAN   TAN

NOTE: MAINTAIN .25 INCH
CLEARANCE BETWEEN
RESISTOR AND WIRES.

GRN   TOP PANEL

GRN   CABINET

GRAY

BLUE   C

NO

DOOR
SWITCH

BLK

BLUE

RED

GRN

CABINET

L1   NEUT   L2

TERMINAL BLOCK

VIEWED FROM REAR

YEL

YEL

ORG/BLK   BLK

HEATER

HIGH LIMIT
THERMOSTAT

BLK

ORG/BLK

BLUE

ORG

CONTROL
THERMOSTAT

ORANGE/BLK

YEL

BLK

GRAY

MOTOR

WHT

TAN

GRN

GRN

CABINET
BASE

DEDF

27" DRYER, ELECTRIC
"M" ELECTRICAL SYSTEM
WIRING DIAGRAM 142846

CAUTION: DISCONNECT ELECTRIC CURRENT BEFORE SERVICING DRYER

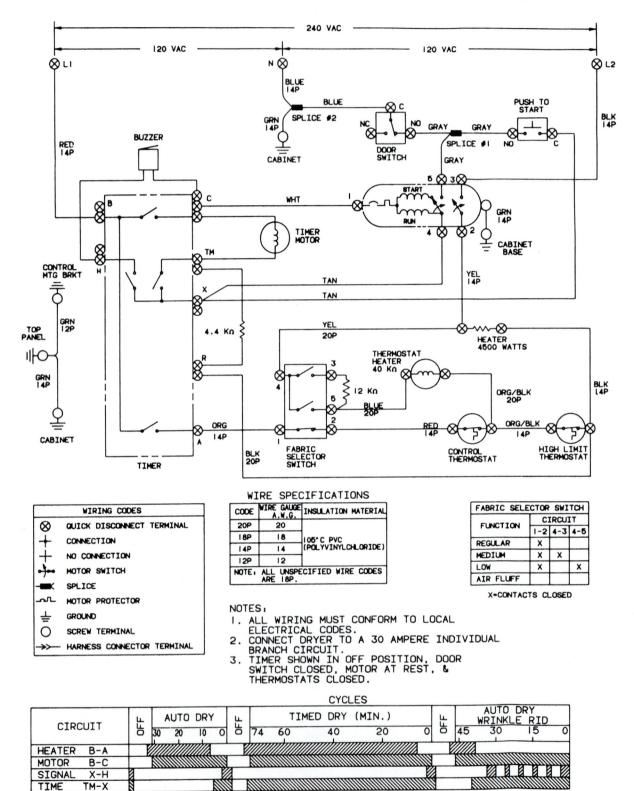

WIRE SPECIFICATIONS

CODE	WIRE GAUGE A.W.G.	INSULATION MATERIAL
20P	20	
18P	18	105°C PVC (POLYVINYLCHLORIDE)
14P	14	
12P	12	
NOTE: ALL UNSPECIFIED WIRE CODES ARE 18P.		

WIRING CODES

Symbol	Description
⊗	QUICK DISCONNECT TERMINAL
┼	CONNECTION
┼	NO CONNECTION
⊶	MOTOR SWITCH
▬	SPLICE
⌁	MOTOR PROTECTOR
⏚	GROUND
○	SCREW TERMINAL
→»	HARNESS CONNECTOR TERMINAL

FABRIC SELECTOR SWITCH

FUNCTION	CIRCUIT		
	1-2	4-3	4-5
REGULAR	X		
MEDIUM	X	X	
LOW	X		X
AIR FLUFF			

X=CONTACTS CLOSED

NOTES:
1. ALL WIRING MUST CONFORM TO LOCAL ELECTRICAL CODES.
2. CONNECT DRYER TO A 30 AMPERE INDIVIDUAL BRANCH CIRCUIT.
3. TIMER SHOWN IN OFF POSITION, DOOR SWITCH CLOSED, MOTOR AT REST, & THERMOSTATS CLOSED.

CYCLES

CIRCUIT	OFF	AUTO DRY			OFF	TIMED DRY (MIN.)				OFF	AUTO DRY WRINKLE RID			
		30	20	10	0	74	60	40	20	0	45	30	15	0
HEATER B-A														
MOTOR B-C														
SIGNAL X-H														
TIME TM-X														

DEIF

376

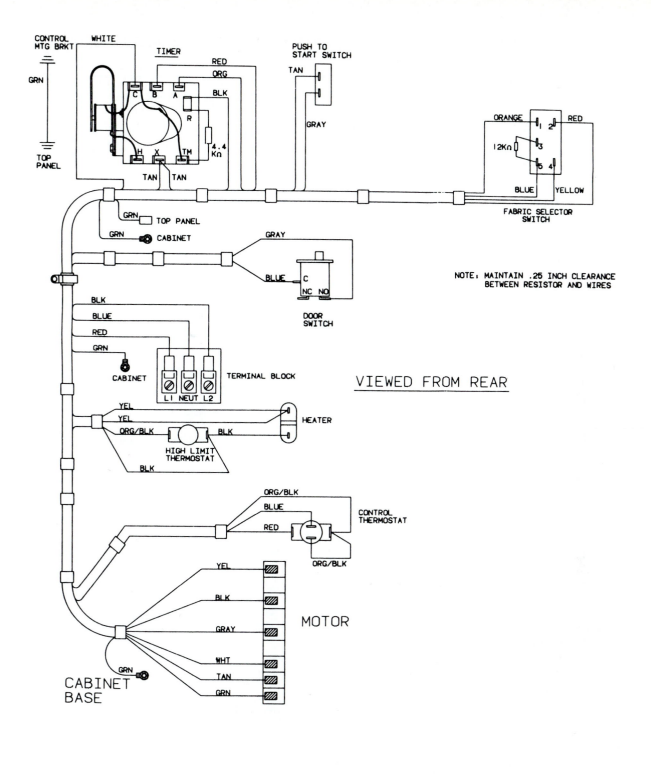

VIEWED FROM REAR

NOTE: MAINTAIN .25 INCH CLEARANCE
BETWEEN RESISTOR AND WIRES

27" DRYER, ELEC.
"P" ELECTRICAL SYSTEM
WIRING DIAGRAM 142841

DEIF

CAUTION: DISCONNECT ELECTRIC CURRENT BEFORE SERVICING DRYER

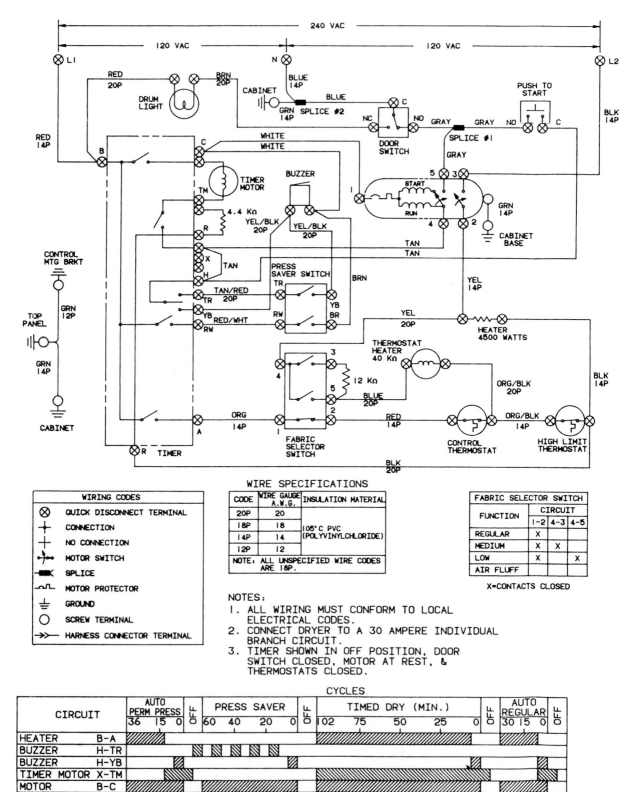

WIRING CODES	
⊗	QUICK DISCONNECT TERMINAL
⊣	CONNECTION
⊥	NO CONNECTION
⟋	MOTOR SWITCH
▬	SPLICE
⏜	MOTOR PROTECTOR
⏚	GROUND
○	SCREW TERMINAL
→→	HARNESS CONNECTOR TERMINAL

WIRE SPECIFICATIONS

CODE	WIRE GAUGE A.W.G.	INSULATION MATERIAL
20P	20	
18P	18	105°C PVC (POLYVINYLCHLORIDE)
14P	14	
12P	12	

NOTE: ALL UNSPECIFIED WIRE CODES ARE 18P.

FABRIC SELECTOR SWITCH

FUNCTION	CIRCUIT		
	1-2	4-3	4-5
REGULAR	X		
MEDIUM	X	X	
LOW	X		X
AIR FLUFF			

X=CONTACTS CLOSED

NOTES:
1. ALL WIRING MUST CONFORM TO LOCAL ELECTRICAL CODES.
2. CONNECT DRYER TO A 30 AMPERE INDIVIDUAL BRANCH CIRCUIT.
3. TIMER SHOWN IN OFF POSITION, DOOR SWITCH CLOSED, MOTOR AT REST, & THERMOSTATS CLOSED.

CYCLES

CIRCUIT		AUTO PERM PRESS			OFF	PRESS SAVER				OFF	TIMED DRY (MIN.)					OFF	AUTO REGULAR			OFF
		36	15	0		60	40	20	0		102	75	50	25	0		30	15	0	
HEATER	B-A																			
BUZZER	H-TR																			
BUZZER	H-YB																			
TIMER MOTOR	X-TM																			
MOTOR	B-C																			
PRESS SAVER	B-RW																			

DECIF

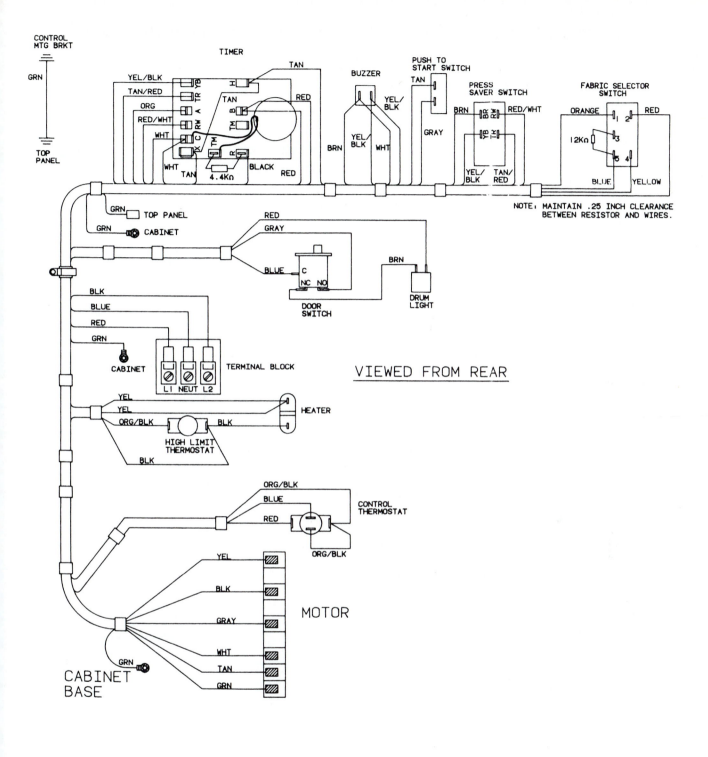

CONTROL MTG BRKT
GRN
TOP PANEL

TIMER
YEL/BLK
TAN/RED
ORG
RED/WHT
WHT
WHT TAN
TAN
TAN
RED
4.4KΩ
BLACK
RED

BUZZER
BRN YEL/BLK WHT

PUSH TO START SWITCH
TAN
GRAY

PRESS SAVER SWITCH
BRN RED/WHT
YEL/BLK TAN/RED

FABRIC SELECTOR SWITCH
ORANGE 2 RED
12KΩ 3
5 4
BLUE YELLOW

NOTE: MAINTAIN .25 INCH CLEARANCE BETWEEN RESISTOR AND WIRES.

GRN TOP PANEL
GRN CABINET

RED
GRAY
BLUE C
NC NO
DOOR SWITCH
BRN
DRUM LIGHT

BLK
BLUE
RED
GRN
CABINET
TERMINAL BLOCK
L1 NEUT L2

VIEWED FROM REAR

YEL
YEL
ORG/BLK BLK
HIGH LIMIT THERMOSTAT
BLK
HEATER

ORG/BLK
BLUE
RED
ORG/BLK
CONTROL THERMOSTAT

YEL
BLK
GRAY
WHT
TAN
GRN
MOTOR

CABINET BASE
GRN

27" DRYER, ELEC.
"T" ELECTRICAL SYSTEM
WIRING DIAGRAM 142848

DECIF

379

CAUTION: DISCONNECT ELECTRIC CURRENT BEFORE SERVICING DRYER

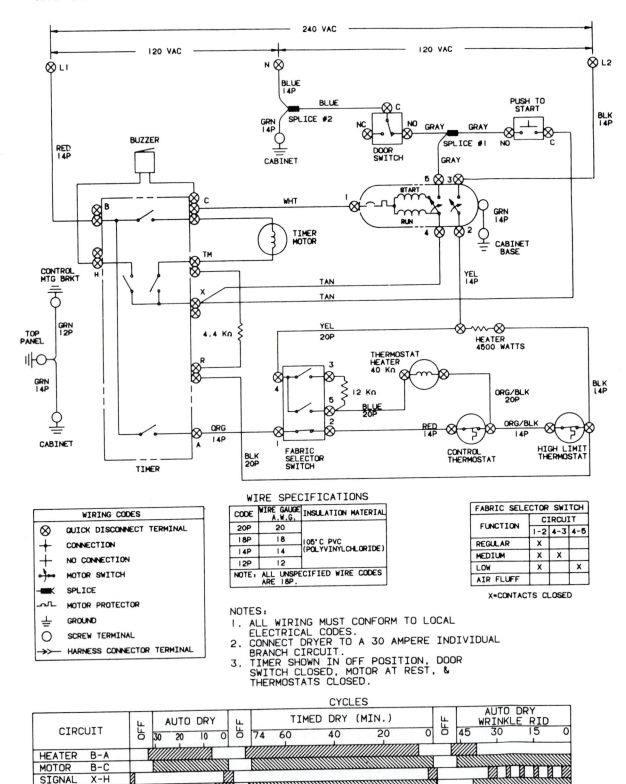

WIRE SPECIFICATIONS

CODE	WIRE GAUGE A.W.G.	INSULATION MATERIAL
20P	20	
18P	18	105°C PVC (POLYVINYLCHLORIDE)
14P	14	
12P	12	

NOTE: ALL UNSPECIFIED WIRE CODES ARE 18P.

WIRING CODES

⊗	QUICK DISCONNECT TERMINAL
+	CONNECTION
+	NO CONNECTION
⊸⊶	MOTOR SWITCH
▬◄	SPLICE
⏝	MOTOR PROTECTOR
⏚	GROUND
○	SCREW TERMINAL
≫	HARNESS CONNECTOR TERMINAL

FABRIC SELECTOR SWITCH

FUNCTION	CIRCUIT		
	1-2	4-3	4-5
REGULAR	X		
MEDIUM	X	X	
LOW	X		X
AIR FLUFF			

X=CONTACTS CLOSED

NOTES:
1. ALL WIRING MUST CONFORM TO LOCAL ELECTRICAL CODES.
2. CONNECT DRYER TO A 30 AMPERE INDIVIDUAL BRANCH CIRCUIT.
3. TIMER SHOWN IN OFF POSITION, DOOR SWITCH CLOSED, MOTOR AT REST, & THERMOSTATS CLOSED.

CYCLES

CIRCUIT	OFF	AUTO DRY			OFF	TIMED DRY (MIN.)				OFF	AUTO DRY WRINKLE RID		
		30	20	10 0		74	60	40	20 0		45	30	15 0
HEATER B-A													
MOTOR B-C													
SIGNAL X-H													
TIME TM-X													

DEISF

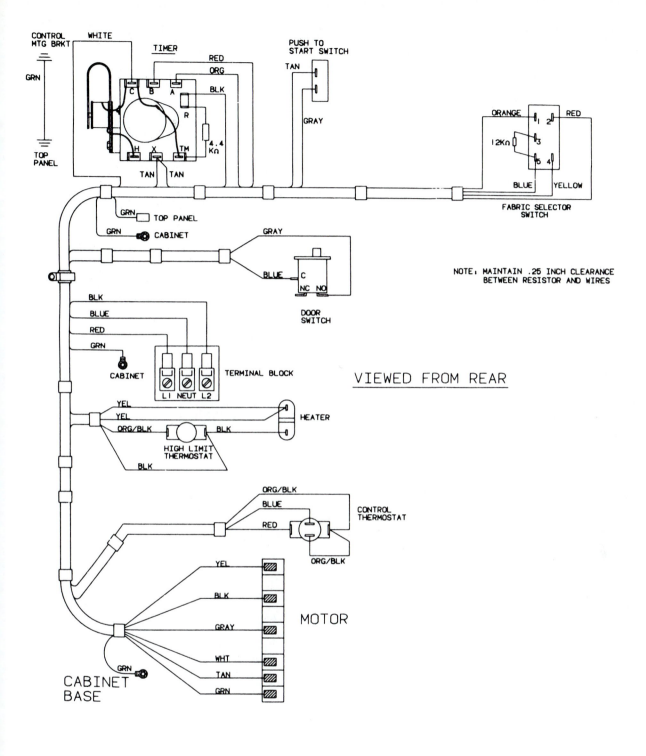

CONTROL MTG BRKT

WHITE

GRN

TOP PANEL

TIMER

RED

ORG

BLK

C   B   A

R

4.4 KΩ

H   X   TM

TAN   TAN

PUSH TO START SWITCH

TAN

GRAY

ORANGE   RED

1   2

12KΩ   3

6   4

BLUE   YELLOW

FABRIC SELECTOR SWITCH

GRN   TOP PANEL

GRN   CABINET

GRAY

BLUE   C

NC   NO

DOOR SWITCH

NOTE: MAINTAIN .25 INCH CLEARANCE BETWEEN RESISTOR AND WIRES

BLK

BLUE

RED

GRN

CABINET

TERMINAL BLOCK

L1   NEUT   L2

VIEWED FROM REAR

YEL

YEL

ORG/BLK   BLK

HEATER

HIGH LIMIT THERMOSTAT

BLK

ORG/BLK

BLUE

RED

ORG/BLK

CONTROL THERMOSTAT

YEL

BLK

GRAY

MOTOR

WHT

TAN

GRN

GRN

CABINET BASE

27" DRYER, ELEC.
"P" ELECTRICAL SYSTEM
WIRING DIAGRAM   142841

DEISF

381

CAUTION: DISCONNECT ELECTRIC CURRENT BEFORE SERVICING DRYER

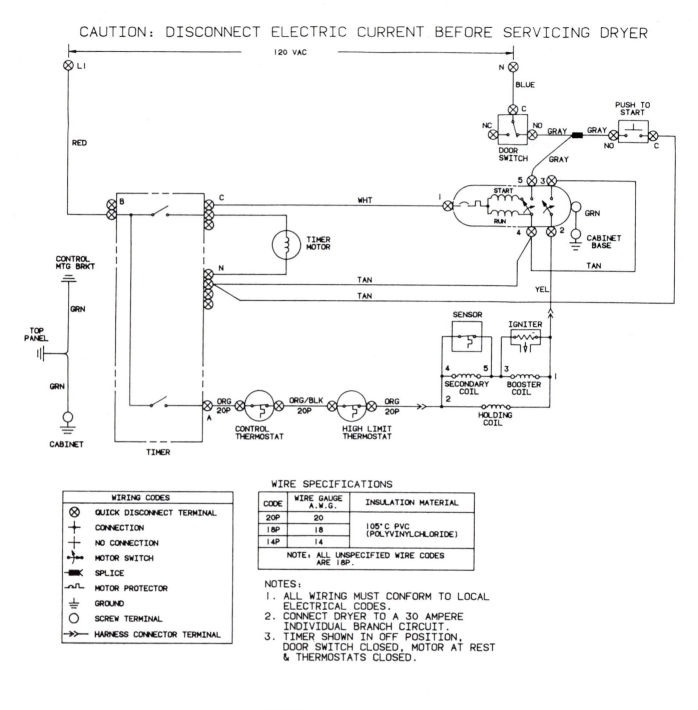

WIRING CODES

⊗	QUICK DISCONNECT TERMINAL
┼	CONNECTION
┼	NO CONNECTION
	MOTOR SWITCH
	SPLICE
	MOTOR PROTECTOR
⏚	GROUND
○	SCREW TERMINAL
→>>	HARNESS CONNECTOR TERMINAL

WIRE SPECIFICATIONS

CODE	WIRE GAUGE A.W.G.	INSULATION MATERIAL
20P	20	
18P	18	105°C PVC (POLYVINYLCHLORIDE)
14P	14	

NOTE: ALL UNSPECIFIED WIRE CODES ARE 18P.

NOTES:
1. ALL WIRING MUST CONFORM TO LOCAL ELECTRICAL CODES.
2. CONNECT DRYER TO A 30 AMPERE INDIVIDUAL BRANCH CIRCUIT.
3. TIMER SHOWN IN OFF POSITION, DOOR SWITCH CLOSED, MOTOR AT REST & THERMOSTATS CLOSED.

CYCLES

CIRCUIT	TIMED DRY (MIN.)					OFF	AIR FLUFF				OFF
	90 75	50	25		0		70	50	25	0	
HEATER B-A											
TIMER MOTOR B-C & DRIVE MOTOR											

DGF

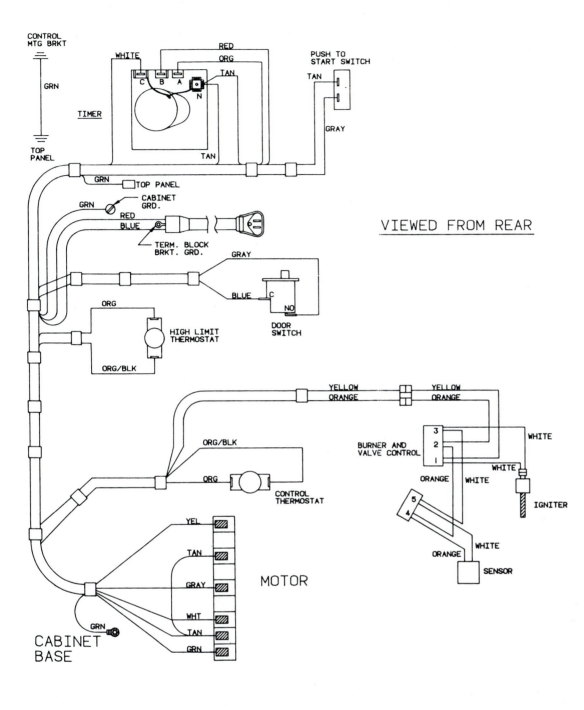

CONTROL
MTG BRKT

GRN

TOP
PANEL

WHITE

TIMER

C B A

RED

ORG

TAN

N

TAN

PUSH TO
START SWITCH

TAN

GRAY

VIEWED FROM REAR

GRN   TOP PANEL

GRN

RED
BLUE

CABINET
GRD.

TERM. BLOCK
BRKT. GRD.

GRAY

BLUE   C

NO

DOOR
SWITCH

ORG

ORG/BLK

HIGH LIMIT
THERMOSTAT

YELLOW
ORANGE

YELLOW
ORANGE

ORG/BLK

ORG

CONTROL
THERMOSTAT

3
2
1

BURNER AND
VALVE CONTROL

WHITE

WHITE

IGNITER

ORANGE   WHITE

5
4

YEL

TAN

GRAY

WHT

TAN

GRN

MOTOR

ORANGE   WHITE

SENSOR

GRN

CABINET
BASE

27" DRYER, GAS
"J" ELECTRICAL SYSTEM
WIRING DIAGRAM 142671

DGF

383

CAUTION: DISCONNECT ELECTRIC CURRENT BEFORE SERVICING DRYER

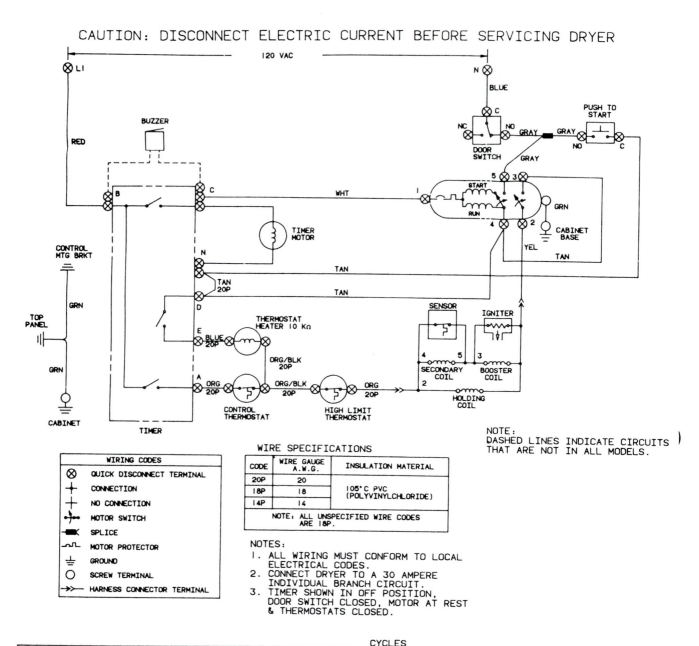

WIRING CODES	
⊗	QUICK DISCONNECT TERMINAL
─•─	CONNECTION
─┼─	NO CONNECTION
─•╲─	MOTOR SWITCH
▬	SPLICE
⌐⌐	MOTOR PROTECTOR
⏚	GROUND
O	SCREW TERMINAL
→»─	HARNESS CONNECTOR TERMINAL

WIRE SPECIFICATIONS

CODE	WIRE GAUGE A.W.G.	INSULATION MATERIAL
20P	20	105°C PVC (POLYVINYLCHLORIDE)
18P	18	
14P	14	

NOTE: ALL UNSPECIFIED WIRE CODES ARE 18P.

NOTES:
1. ALL WIRING MUST CONFORM TO LOCAL ELECTRICAL CODES.
2. CONNECT DRYER TO A 30 AMPERE INDIVIDUAL BRANCH CIRCUIT.
3. TIMER SHOWN IN OFF POSITION, DOOR SWITCH CLOSED, MOTOR AT REST & THERMOSTATS CLOSED.

NOTE:
DASHED LINES INDICATE CIRCUITS THAT ARE NOT IN ALL MODELS.

CYCLES

CIRCUIT	REGULAR HEAT DRY					OFF	LOW HEAT DRY					OFF	AIR FLUFF			OFF
	90	75	50	25	0		90	75	50	25	0		60	40	20	0	
HEATER B-A																
TIMER & DRIVE MOTOR B-C																
LOW HEAT SW. D-E																

BAR CHART ABOVE REPRESENTS ONE COMPLETE REVOLUTION OF TIMER SHAFT.

SHADED PORTION OF BAR CHART INDICATES THE PROPORTIONAL TIMES THAT INTERNAL TIMER CONTACTS ARE CLOSED.

DGSF

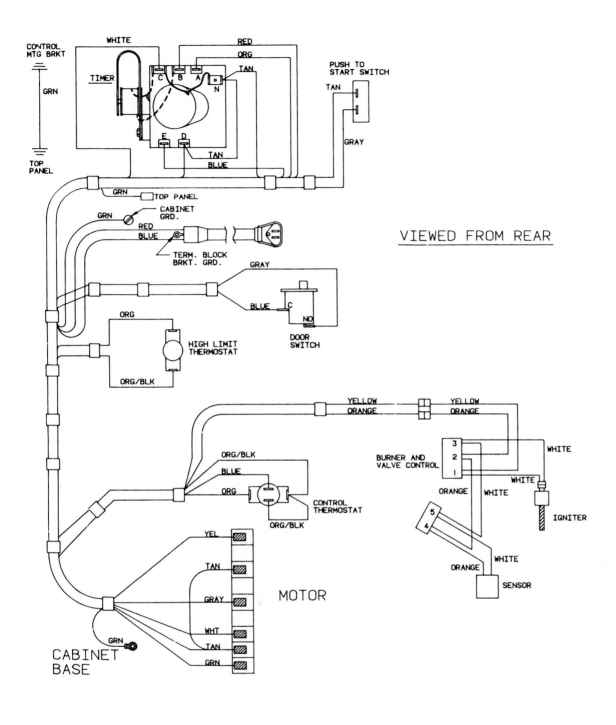

VIEWED FROM REAR

27" DRYER, GAS
"K" ELECTRICAL SYSTEM
WIRING DIAGRAM 142735

DGSF

CAUTION: DISCONNECT ELECTRIC CURRENT BEFORE SERVICING DRYER

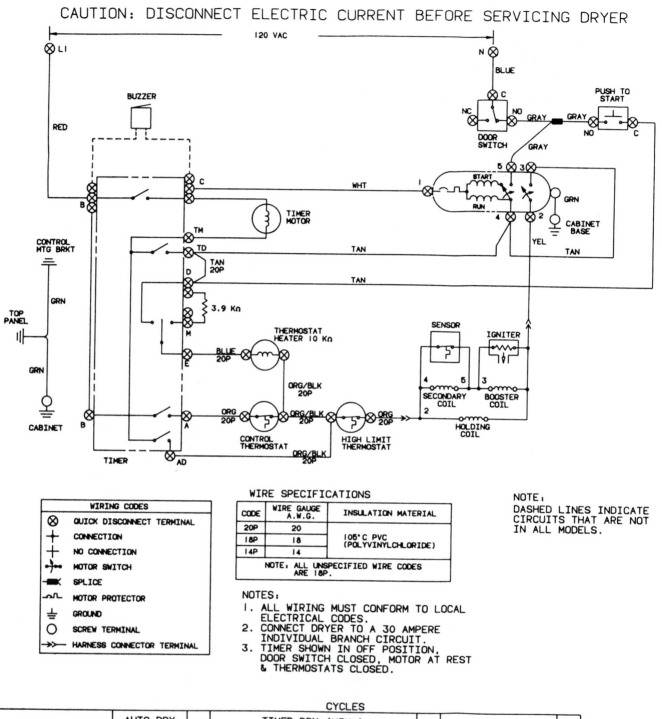

WIRING CODES

⊗	QUICK DISCONNECT TERMINAL
+	CONNECTION
┼	NO CONNECTION
⊶	MOTOR SWITCH
▰	SPLICE
⌁	MOTOR PROTECTOR
⏚	GROUND
○	SCREW TERMINAL
⟫	HARNESS CONNECTOR TERMINAL

WIRE SPECIFICATIONS

CODE	WIRE GAUGE A.W.G.	INSULATION MATERIAL
20P	20	
18P	18	105°C PVC (POLYVINYLCHLORIDE)
14P	14	

NOTE: ALL UNSPECIFIED WIRE CODES ARE 18P.

NOTES:
1. ALL WIRING MUST CONFORM TO LOCAL ELECTRICAL CODES.
2. CONNECT DRYER TO A 30 AMPERE INDIVIDUAL BRANCH CIRCUIT.
3. TIMER SHOWN IN OFF POSITION, DOOR SWITCH CLOSED, MOTOR AT REST & THERMOSTATS CLOSED.

NOTE:
DASHED LINES INDICATE CIRCUITS THAT ARE NOT IN ALL MODELS.

CYCLES

CIRCUIT		AUTO DRY			OFF	TIMED DRY (MIN.)				OFF	KNIT CYCLE			OFF
		30	20	10 0		100	75	50	25 0		50	25 0		
TM MOTOR	TM-TD					▨	▨	▨	▨		▨	▨		
TM MOTOR	TM-AD	▨	▨	▨										
MOTOR	B-C	▨	▨			▨	▨	▨	▨		▨	▨		
HEATER	B-A	▨	▨			▨	▨	▨	▨		▨	▨		
MED. HEAT	E-M	▨	▨	▨										
LOW HEAT	E-D										▨	▨		

DGDF

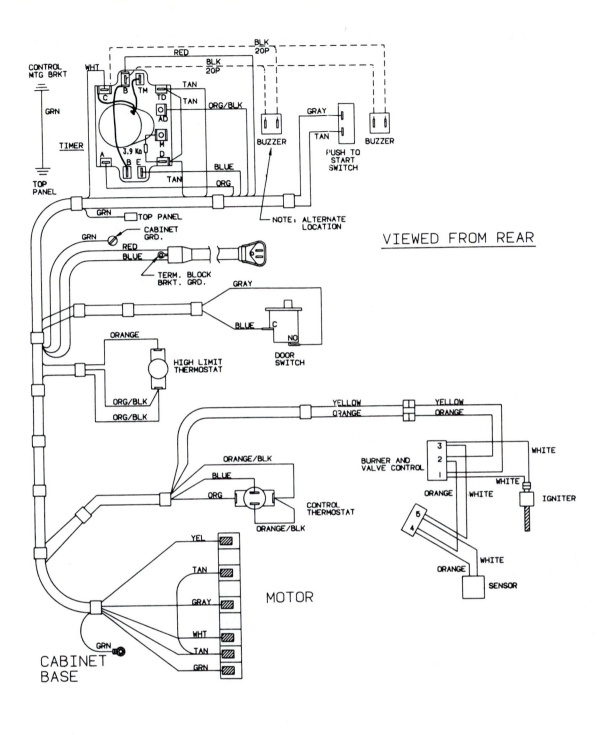

VIEWED FROM REAR

27" DRYER, GAS
"M" ELECTRICAL SYSTEM
WIRING DIAGRAM 142847

DGDF

CAUTION: DISCONNECT ELECTRIC CURRENT BEFORE SERVICING DRYER

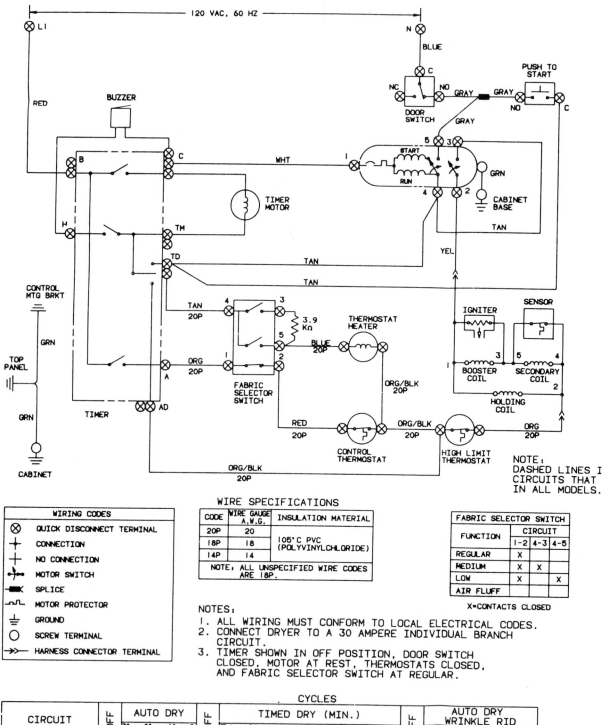

WIRING CODES	
⊗	QUICK DISCONNECT TERMINAL
╪	CONNECTION
╧	NO CONNECTION
	MOTOR SWITCH
■	SPLICE
	MOTOR PROTECTOR
⏚	GROUND
○	SCREW TERMINAL
→»	HARNESS CONNECTOR TERMINAL

WIRE SPECIFICATIONS

CODE	WIRE GAUGE A.W.G.	INSULATION MATERIAL
20P	20	105°C PVC (POLYVINYLCHLORIDE)
18P	18	
14P	14	

NOTE: ALL UNSPECIFIED WIRE CODES ARE 18P.

FABRIC SELECTOR SWITCH			
FUNCTION	CIRCUIT		
	1-2	4-3	4-5
REGULAR	X		
MEDIUM	X	X	
LOW	X		X
AIR FLUFF			

X=CONTACTS CLOSED

NOTES:
1. ALL WIRING MUST CONFORM TO LOCAL ELECTRICAL CODES.
2. CONNECT DRYER TO A 30 AMPERE INDIVIDUAL BRANCH CIRCUIT.
3. TIMER SHOWN IN OFF POSITION, DOOR SWITCH CLOSED, MOTOR AT REST, THERMOSTATS CLOSED, AND FABRIC SELECTOR SWITCH AT REGULAR.

CYCLES

CIRCUIT		OFF	AUTO DRY				OFF	TIMED DRY (MIN.)					OFF	AUTO DRY WRINKLE RID			
			30	20	10	0		74	60	40	20	0		45	30	15	0
HEATER	B-A																
MOTOR	B-C																
SIGNAL	TM-H																
TIMER	TM-TD																
AUTO	TM-AD																

DGIF

388

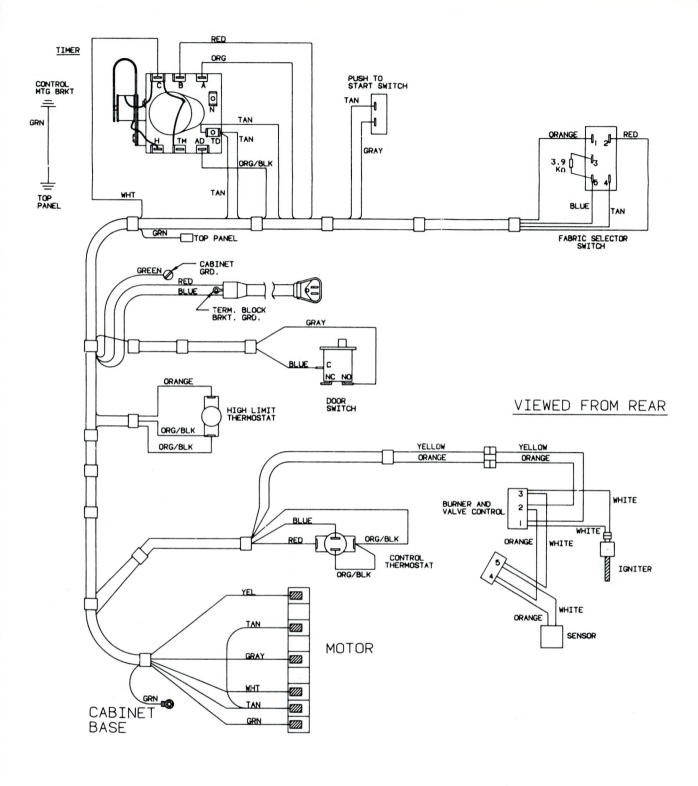

VIEWED FROM REAR

MOTOR

27" DRYER, GAS
"P" ELECTRICAL SYSTEM
WIRING DIAGRAM 142832

DGIF

CAUTION: DISCONNECT ELECTRIC CURRENT BEFORE SERVICING DRYER

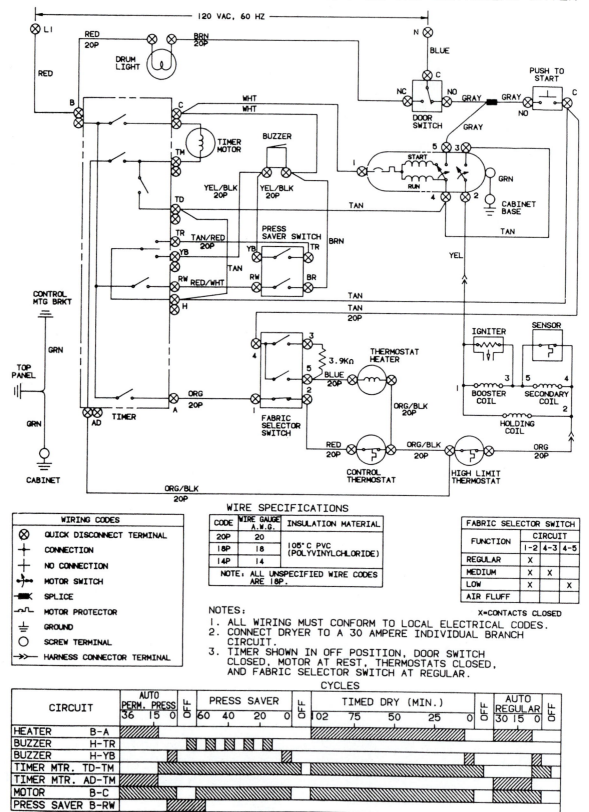

DGCIF

WIRING CODES

⊗	QUICK DISCONNECT TERMINAL
┿	CONNECTION
┬	NO CONNECTION
⊶	MOTOR SWITCH
▬	SPLICE
⏦	MOTOR PROTECTOR
⏚	GROUND
○	SCREW TERMINAL
→≫	HARNESS CONNECTOR TERMINAL

WIRE SPECIFICATIONS

CODE	WIRE GAUGE A.W.G.	INSULATION MATERIAL
20P	20	105°C PVC (POLYVINYLCHLORIDE)
18P	18	
14P	14	

NOTE: ALL UNSPECIFIED WIRE CODES ARE 18P.

FABRIC SELECTOR SWITCH

FUNCTION	CIRCUIT		
	1-2	4-3	4-5
REGULAR	X		
MEDIUM	X	X	
LOW	X		X
AIR FLUFF			

X=CONTACTS CLOSED

NOTES:
1. ALL WIRING MUST CONFORM TO LOCAL ELECTRICAL CODES.
2. CONNECT DRYER TO A 30 AMPERE INDIVIDUAL BRANCH CIRCUIT.
3. TIMER SHOWN IN OFF POSITION, DOOR SWITCH CLOSED, MOTOR AT REST, THERMOSTATS CLOSED, AND FABRIC SELECTOR SWITCH AT REGULAR.

CYCLES

CIRCUIT		AUTO PERM. PRESS			OFF	PRESS SAVER				OFF	TIMED DRY (MIN.)					OFF	AUTO REGULAR			OFF
		36	15	0		60	40	20	0		102	75	50	25	0		30	15	0	
HEATER	B-A																			
BUZZER	H-TR																			
BUZZER	H-YB																			
TIMER MTR.	TD-TM																			
TIMER MTR.	AD-TM																			
MOTOR	B-C																			
PRESS SAVER	B-RW																			

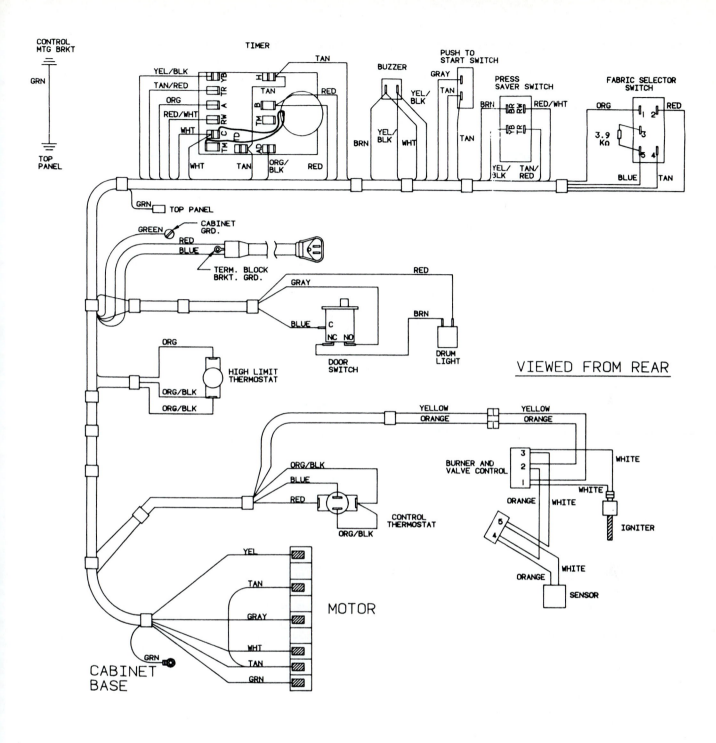

CONTROL
MTG BRKT

GRN

TOP
PANEL

TIMER

YEL/BLK
TAN/RED
ORG
RED/WHT
WHT

WHT   TAN   ORG/   RED
                         BLK

TAN

TAN

RED

BUZZER

YEL/
BLK

BRN   YEL/   WHT
       BLK

PUSH TO
START SWITCH

GRAY

TAN

TAN

PRESS
SAVER SWITCH

BRN   RED/WHT

YEL/   TAN/
BLK   RED

FABRIC SELECTOR
SWITCH

ORG   RED

3.9
KΩ

BLUE   TAN

GRN   TOP PANEL

GREEN   CABINET
         GRD.
RED
BLUE

TERM. BLOCK
BRKT. GRD.

GRAY   RED

BLUE

NC   NO

DOOR
SWITCH

BRN

DRUM
LIGHT

VIEWED FROM REAR

ORG

ORG/BLK

ORG/BLK

HIGH LIMIT
THERMOSTAT

YELLOW   YELLOW
ORANGE   ORANGE

WHITE

ORG/BLK

BLUE

RED

ORG/BLK

CONTROL
THERMOSTAT

BURNER AND
VALVE CONTROL

3
2
1

ORANGE   WHITE

WHITE

WHITE

IGNITER

5
4

ORANGE   WHITE

SENSOR

YEL

TAN

GRAY

WHT

TAN

GRN

MOTOR

CABINET
BASE

GRN

27" DRYER, GAS
"T" ELECTRICAL SYSTEM
WIRING DIAGRAM 142849

DGCIF

391

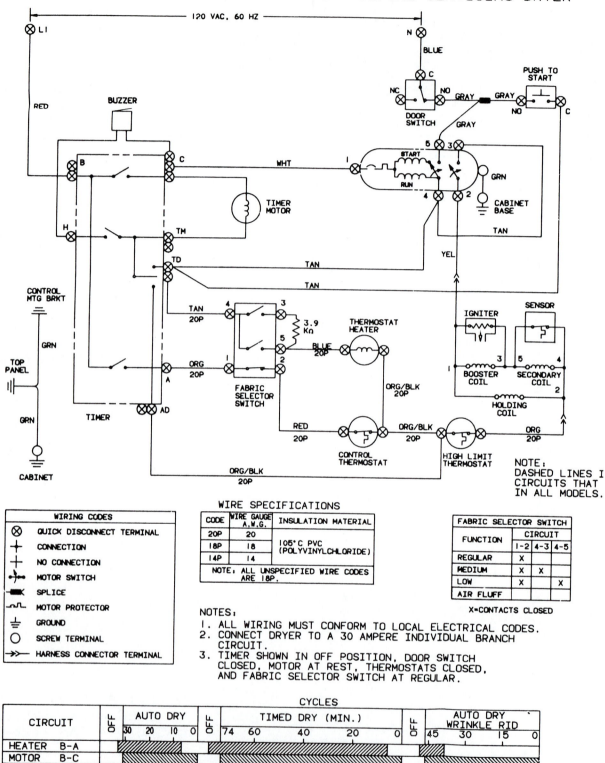

WIRE SPECIFICATIONS

CODE	WIRE GAUGE A.W.G.	INSULATION MATERIAL
20P	20	105°C PVC (POLYVINYLCHLORIDE)
18P	18	
14P	14	

NOTE: ALL UNSPECIFIED WIRE CODES ARE 18P.

WIRING CODES

⊗	QUICK DISCONNECT TERMINAL
╈	CONNECTION
╅	NO CONNECTION
•⌐•	MOTOR SWITCH
▬█▬	SPLICE
⌐⌐	MOTOR PROTECTOR
⏚	GROUND
○	SCREW TERMINAL
→»	HARNESS CONNECTOR TERMINAL

FABRIC SELECTOR SWITCH

FUNCTION	CIRCUIT		
	1-2	4-3	4-5
REGULAR	X		
MEDIUM	X	X	
LOW	X		X
AIR FLUFF			

X=CONTACTS CLOSED

NOTES:
1. ALL WIRING MUST CONFORM TO LOCAL ELECTRICAL CODES.
2. CONNECT DRYER TO A 30 AMPERE INDIVIDUAL BRANCH CIRCUIT.
3. TIMER SHOWN IN OFF POSITION, DOOR SWITCH CLOSED, MOTOR AT REST, THERMOSTATS CLOSED, AND FABRIC SELECTOR SWITCH AT REGULAR.

CYCLES

CIRCUIT	OFF	AUTO DRY				OFF	TIMED DRY (MIN.)					OFF	AUTO DRY WRINKLE RID			
		30	20	10	0		74	60	40	20	0		45	30	15	0
HEATER B-A																
MOTOR B-C																
SIGNAL TM-H																
TIMER TM-TD																
AUTO TM-AD																

DGISF

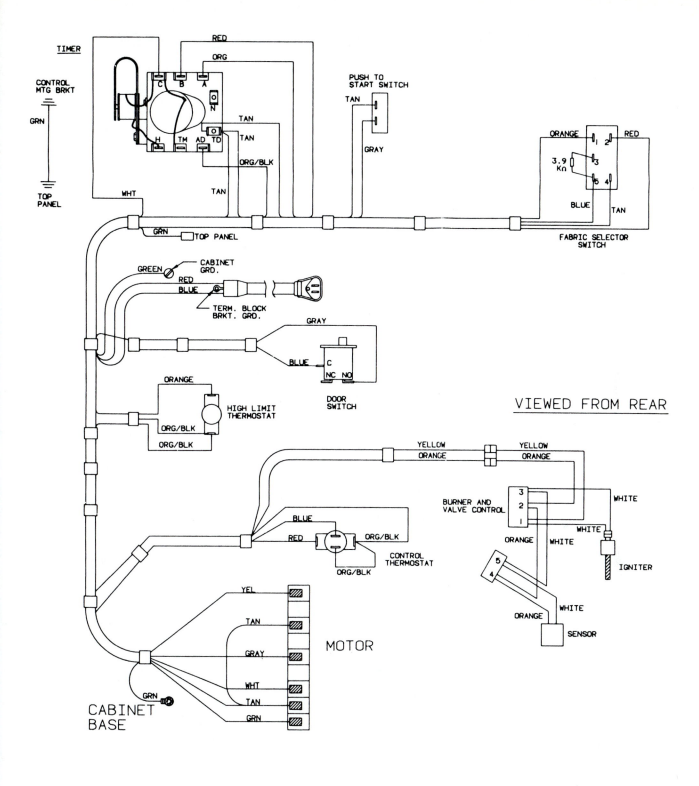

TIMER

CONTROL
MTG BRKT

GRN

TOP
PANEL

RED

ORG

C B A

N

H TM AD TD

TAN

TAN

ORG/BLK

WHT   TAN

GRN   TOP PANEL

GREEN   CABINET
             GRD.
     RED
     BLUE

TERM. BLOCK
BRKT. GRD.

PUSH TO
START SWITCH

TAN

GRAY

ORANGE   RED

1   2

3.9
KΩ   3

5   4

BLUE   TAN

FABRIC SELECTOR
SWITCH

GRAY

BLUE   C

NC NO

DOOR
SWITCH

VIEWED FROM REAR

ORANGE

ORG/BLK

ORG/BLK

HIGH LIMIT
THERMOSTAT

YELLOW   YELLOW

ORANGE   ORANGE

BURNER AND
VALVE CONTROL

3

2

1

WHITE

WHITE

ORANGE   WHITE

IGNITER

5

4

ORANGE   WHITE

SENSOR

BLUE

RED   ORG/BLK

CONTROL
THERMOSTAT

ORG/BLK

MOTOR

YEL

TAN

GRAY

WHT

TAN

GRN

CABINET
BASE

GRN

27" DRYER, GAS
"P" ELECTRICAL SYSTEM
WIRING DIAGRAM 142832

DGISF

SECTION E - EXPLODED VIEWS OF DRYER

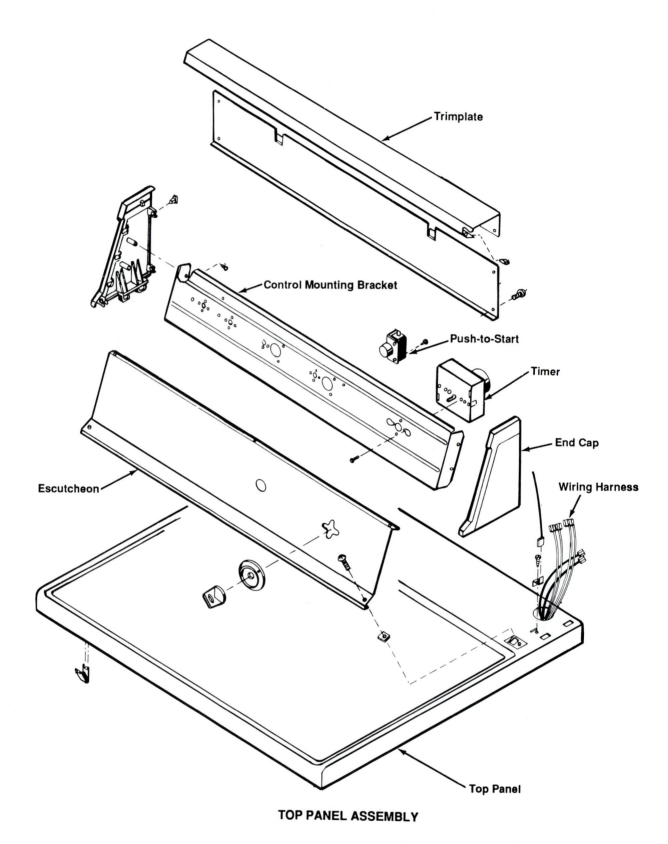

Trimplate

Control Mounting Bracket

Push-to-Start

Timer

End Cap

Escutcheon

Wiring Harness

Top Panel

TOP PANEL ASSEMBLY

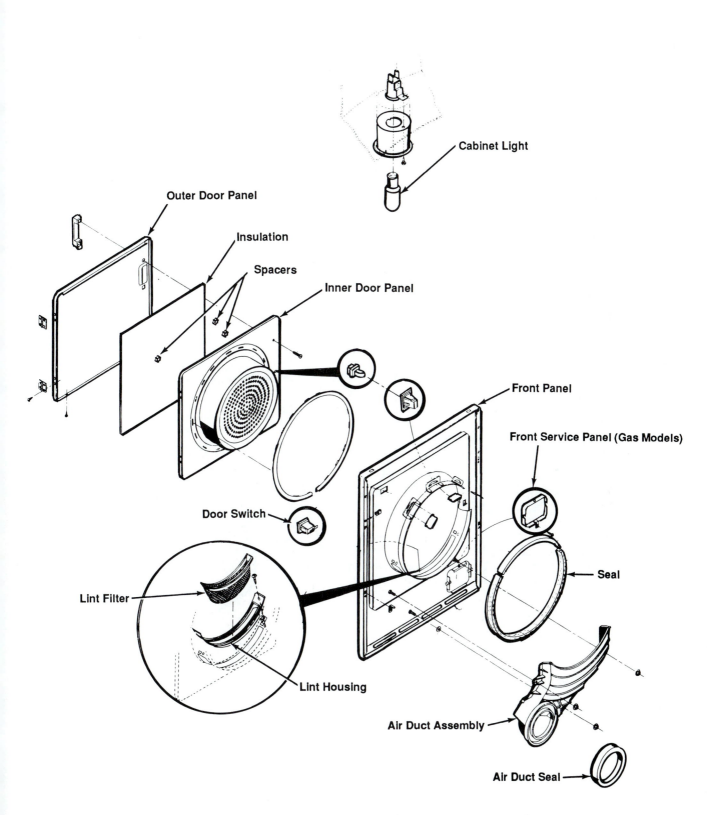

Cabinet Light

Outer Door Panel

Insulation

Spacers

Inner Door Panel

Front Panel

Front Service Panel (Gas Models)

Door Switch

Seal

Lint Filter

Lint Housing

Air Duct Assembly

Air Duct Seal

FRONT PANEL ASSEMBLY

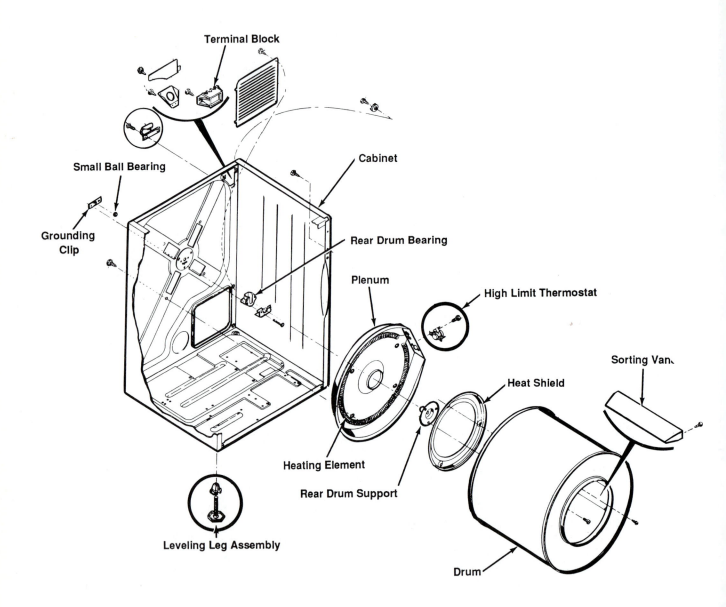

Terminal Block

Cabinet

Small Ball Bearing

Grounding Clip

Rear Drum Bearing

Plenum

High Limit Thermostat

Sorting Vane

Heat Shield

Heating Element

Rear Drum Support

Drum

Leveling Leg Assembly

CABINET ASSEMBLY (ELECTRIC MODELS)

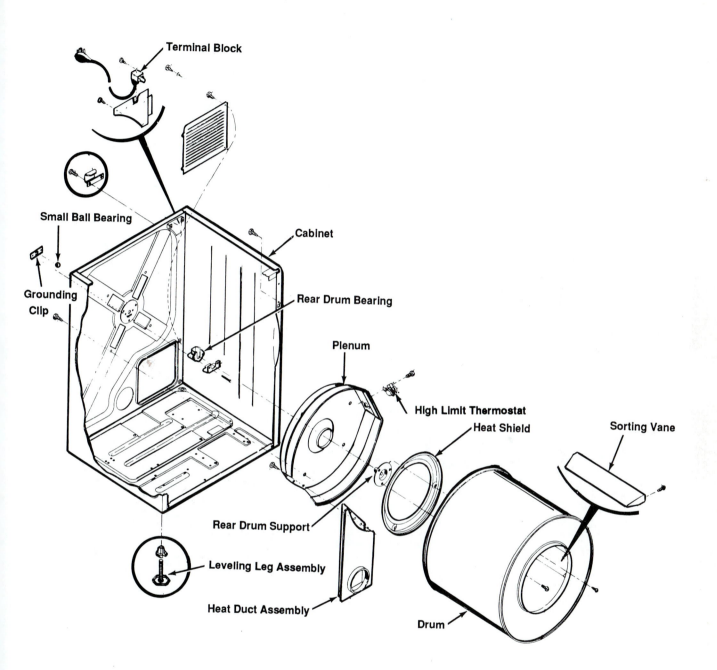

Terminal Block

Small Ball Bearing

Grounding Clip

Cabinet

Rear Drum Bearing

Plenum

High Limit Thermostat

Heat Shield

Sorting Vane

Rear Drum Support

Leveling Leg Assembly

Heat Duct Assembly

Drum

CABINET ASSEMBLY (GAS MODELS)

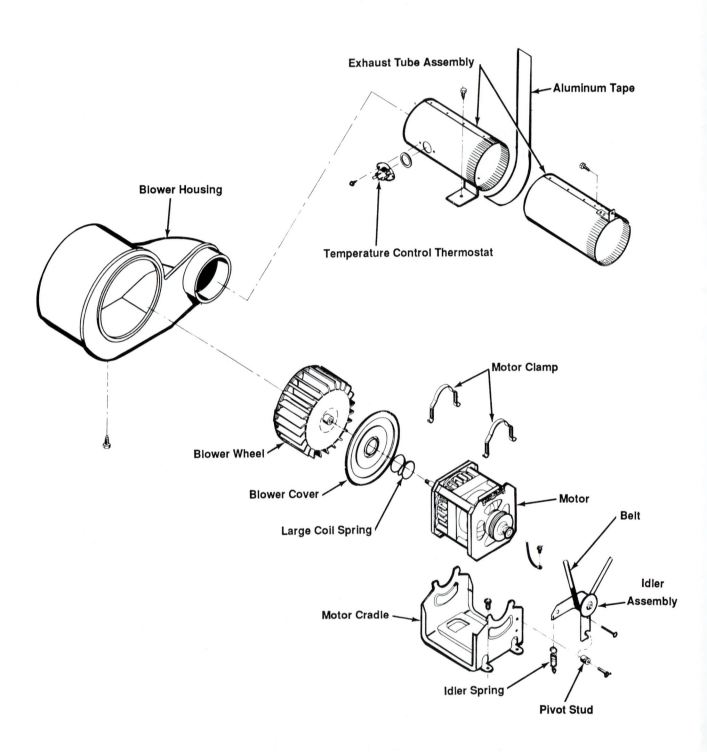

Exhaust Tube Assembly

Aluminum Tape

Blower Housing

Temperature Control Thermostat

Motor Clamp

Blower Wheel

Blower Cover

Large Coil Spring

Motor

Belt

Idler Assembly

Motor Cradle

Idler Spring

Pivot Stud

MOTOR/BLOWER ASSEMBLY

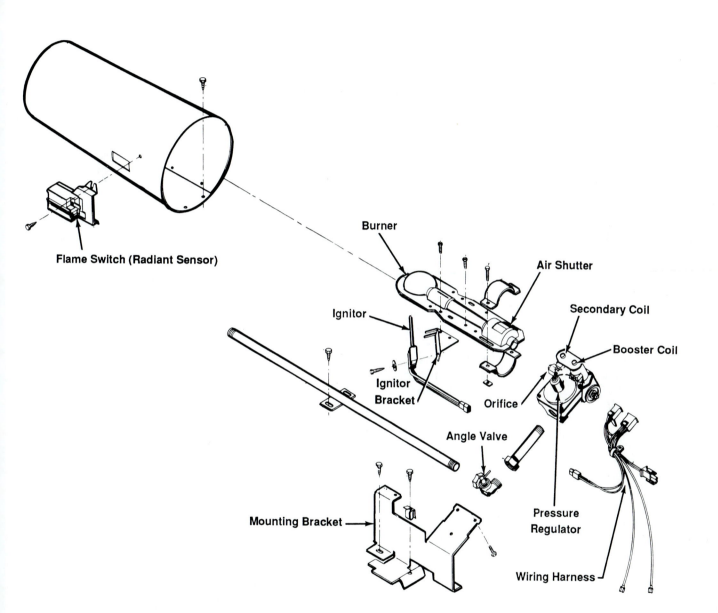

Flame Switch (Radiant Sensor)

Burner

Air Shutter

Ignitor

Secondary Coil

Booster Coil

Ignitor
Bracket

Orifice

Angle Valve

Mounting Bracket

Pressure
Regulator

Wiring Harness

BURNER ASSEMBLY

SECTION F - TROUBLESHOOTING CHARTS

The diagnosis check list is common to both gas and electric heated dryers. They use different type parts to accomplish the same thing and diagnosis will remain similar.

When a problem arises, and the possible cause is listed, follow the testing part outlined in Section C or D. The wiring diagram, timer cycle chart and/or selector switch chart is a necessity. In most cases, an ohmmeter will handle all the tests necessary.

For checking any particular cycle operation, it is absolutely necessary that the cycle be set up as outlined in the instruction manual. Some auto cycles will not operate on time cycle settings or vice-versa. The safety start switch must be pushed to start all operations. Perma Press/Knits Auto-Dry can be checked the same as Auto-Dry. Timed dry operation is quite similar to Auto-Dry knowing that the timer advances continuously from start to off.

SYMPTOMS	POSSIBLE CAUSE	REMEDY
1. Dryer will not run.	1. Blown fuse or tripped breaker.	1. Replace fuse or reset breaker.
	2. Wiring or terminal block burnt.	2. Replace terminal block, re-dress or replace wiring.
	3. Open contacts in timer.	3. Replace timer.
	4. Open door switch.	4. Replace switch.
	5. Open safety start switch	5. Replace switch.
	6. Open selector switch.	6. Replace switch.
	7. Inoperative motor.	7. Repair or replace motor.
2. Dryer runs but will not heat.	1. Open selector switch	1. Replace switch.
	2. Open thermostat.	2. Replace thermostat.
	3. Open switch on motor.	3. Replace switch.
	4. Open ignitor.	4. Replace ignitor.
	5. Open heater coil.	5. Replace heater.
	6. Wire or terminal block burnt off.	6. Replace terminal block, redress or replace wiring.
	7. Inoperative timer.	7. Replace timer.
	8. Inoperative flame switch.	8. Replace flame switch.
	9. Inoperative burner valve.	9. Replace burner assembly.
3. Dryer will not stop.	1. Timer motor inoperative.	1. Replace timer.
	2. Wiring.	2. Check all wiring.
	3. Thermostat not opening (AUTO-DRY gas dryer).	3. Replace thermostat.
	4. Open timer contact.	4. Replace timer.
	5. Wrong thermostat used.	5. Replace with correct thermostat.
4. Buzzer does not operate.	1. Buzzer coil open.	1. Replace buzzer
	2. Open contact in timer.	2. Replace timer.
	3. Loose connection.	3 Check wiring.
5. Drum light does not come on.	1. Lamp burned out.	1. Replace lamp.
	2. Door switch not working.	2. Replace door switch.

SYMPTOMS	POSSIBLE CAUSE	REMEDY
6. Clothes are still wet when dryer stops.	1. Exhaust clogged. 2. Auto-Dry thermostat opened and does not close. 3. Time Dry timer improperly set. 4. Heater failed during operation. 5. Burner failed during operation. 6. Vent pipe too long. 7. Improper burner orifice.	1. Clean exhaust. 2. Replace thermostat. 3. Explain to user. 4. Replace heater. 5. Replace burner assembly. 6. Check installation units. 7. Install proper orifice in burner.
7. Dryer failed to complete and cycle.	1. Motor protector opened. 2. Timer motor failed. 3. Various other parts may have failed in cycle.	1. Check for mechanical over load correct. 2. Replace timer. 3. Check out and replace inoperative part as in Step 1.
8. Motor runs but drum does not turn.	1. Broken or loose belt. 2. Loose motor or idler pulley. 3. Too heavy of a load.	1. Replace belt. 2. Adjust or replace pulleys. 3. Advise customer.
9. Drum operates but is noisy.	1. Drum out of shape. 2. Glides not striking or not in shape. 3. Idler pulley noisy. 4. Metal objects in drum. 5. Belt frayed or squeaking. 6. Drum bearing is worn or needs lubrication. 7. Machine not level.	1. Replace drum. 2. Check and/or replace glides. 3. Adjust or replace pulley 4. Remove. 5. Replace belt. 6. Lubricate or replace drum bearing 7. Level.
10. Motor hums but will not start or run	1. Start switch contacts not closed. 2. Open start winding. 3. Bad bearings or locked motor.	1. Adjust switch contacts or replace switch or terminal board. 2. Replace motor. 3. Replace motor.
11. Dryer smokes.	1. Lint accumulated in dryer. 2. Wire insulation burning. 3. Overheated motor.	1. Advise customer on cleaning dryer. 2. Correct heat source and change wiring. 3. See symptom 10.
12. Clothes too dry or scorched	1. Improper setting. 2. Improper mix of clothes. 3. Detergent left in clothes.	1. Advise customer to follow use care instructions. 2. Same as 1. 3. Rewash clothes.
13. Ignitor does not glow.	1. Flame switch open. 2. Thermostat open. 3. Ignitor broken. 4. Wiring	1. Replace flame switch. 2. Replace thermostat. 3. Replace ignitor. 4. Check and replace connections.
14. Ignitor heats but no flame.	1. No gas. 2. Flame switch doesn't open. 3. Inoperative valve.	1. Turn on gas. 2. Replace flame switch. 3. Replace coils.
15. Burner cycles properly but gas does not flow.	1. Valves may not be opening. 2. Clogged orifice.	1. Check coils for continuity. 2. Check and clean orifice.

Chapter 16: Gas Laundry Centers

Dryer Electrical Diagram
(Models LCG731LW*0 and LCG731LL*0)

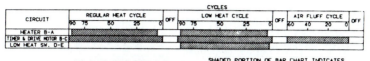

BAR CHART ABOVE REPRESENTS ONE COMPLETE REVOLUTION OF TIMER SHAFT.

SHADED PORTION OF BAR CHART INDICATES THE PROPORTIONAL TIMES THAT INTERNAL TIMER CONTACTS ARE CLOSED.

CAUTION: DISCONNECT ELECTRIC CURRENT BEFORE SERVICING DRYER

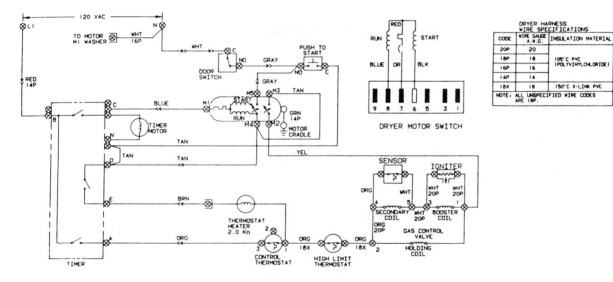

145963-000 WIRING DIAGRAM.

Washer Electrical Diagram
(Models LCG751LW*0 and LCG751LL*0)

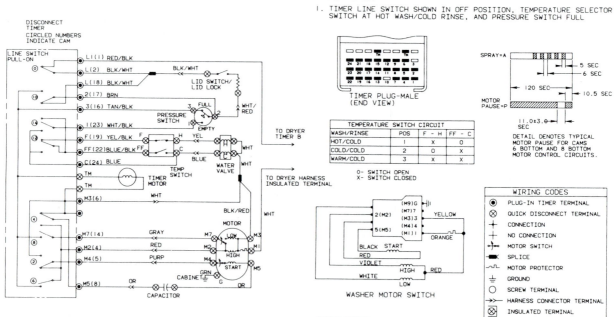

WASHER NOTES:

1. TIMER LINE SWITCH SHOWN IN OFF POSITION, TEMPERATURE SELECTOR SWITCH AT HOT WASH/COLD RINSE, AND PRESSURE SWITCH FULL

TIMER PLUG-MALE (END VIEW)

TEMPERATURE SWITCH CIRCUIT			
WASH/RINSE	POS	F - H	FF - C
HOT/COLD	1	X	O
COLD/COLD	2	O	X
WARM/COLD	3	X	X

O- SWITCH OPEN
X- SWITCH CLOSED

SPRAY=A
MOTOR PAUSE=P

DETAIL DENOTES TYPICAL MOTOR PAUSE FOR CAMS 6 BOTTOM AND 8 BOTTOM MOTOR CONTROL CIRCUITS.

5 SEC
6 SEC
120 SEC
10.5 SEC
11.0±3.0 SEC

WASHER MOTOR SWITCH

WIRING CODES	
⊙	PLUG-IN TIMER TERMINAL
⊗	QUICK DISCONNECT TERMINAL
	CONNECTION
	NO CONNECTION
	MOTOR SWITCH
	SPLICE
	MOTOR PROTECTOR
⏚	GROUND
○	SCREW TERMINAL
>>>	HARNESS CONNECTOR TERMINAL
⊗	INSULATED TERMINAL

DRYER NOTES:

1. ALL WIRING MUST CONFORM TO LOCAL ELECTRICAL CODES.
2. CONNECT DRYER TO A 20 AMPERE INDIVIDUAL BRANCH CIRCUIT.
3. TIMER SHOWN IN OFF POSITION, DOOR SWITCH CLOSED, MOTOR AT REST, AND THERMOSTATS CLOSED.

146564-000 WIRING DIAGRAM.

Dryer Electrical Diagram
(Models LCG751LW*0 and LCG751LL*0)

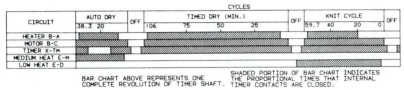

CIRCUIT	AUTO DRY		OFF	TIMED DRY (MIN.)				OFF	KNIT CYCLE			OFF
	38.3	20		106	75	50	25		59.7	40	20	0
HEATER B-A												
MOTOR B-C												
TIMER X-TM												
MEDIUM HEAT E-M												
LOW HEAT E-D												

BAR CHART ABOVE REPRESENTS ONE COMPLETE REVOLUTION OF TIMER SHAFT.

SHADED PORTION OF BAR CHART INDICATES THE PROPORTIONAL TIMES THAT INTERNAL TIMER CONTACTS ARE CLOSED.

CAUTION: DISCONNECT ELECTRIC CURRENT BEFORE SERVICING DRYER

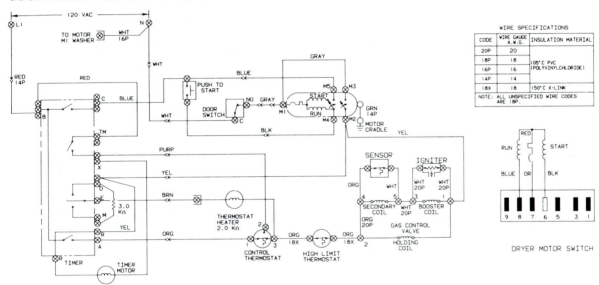

146564-000 WIRING DIAGRAM.

Washer Electrical Diagram
(Models LCG771LW*0 and LCG771LL*0)

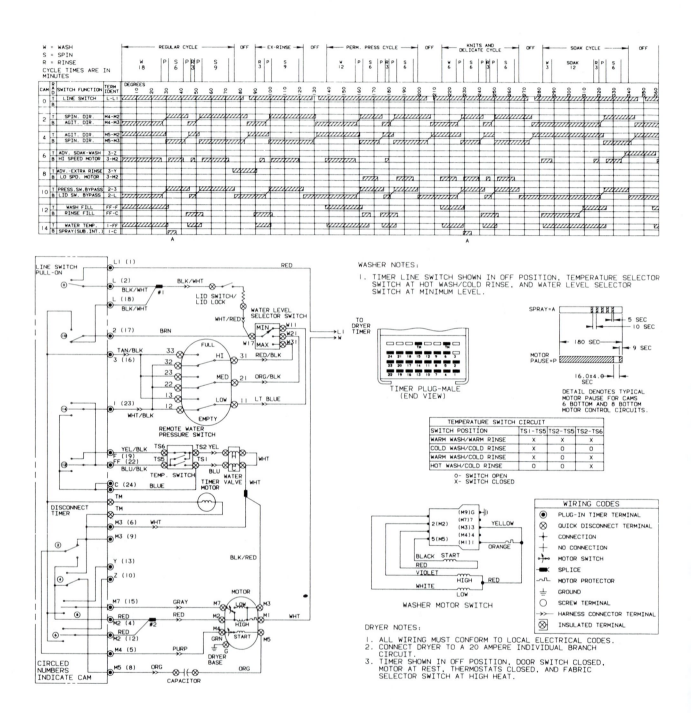

WASHER NOTES:

1. TIMER LINE SWITCH SHOWN IN OFF POSITION, TEMPERATURE SELECTOR SWITCH AT HOT WASH/COLD RINSE, AND WATER LEVEL SELECTOR SWITCH AT MINIMUM LEVEL.

TEMPERATURE SWITCH CIRCUIT			
SWITCH POSITION	TS1-TS5	TS2-TS5	TS2-TS6
WARM WASH/WARM RINSE	X	X	X
COLD WASH/COLD RINSE	X	O	O
WARM WASH/COLD RINSE	X	O	X
HOT WASH/COLD RINSE	O	O	X

O- SWITCH OPEN
X- SWITCH CLOSED

WIRING CODES	
⊙	PLUG-IN TIMER TERMINAL
⊗	QUICK DISCONNECT TERMINAL
+	CONNECTION
⊥	NO CONNECTION
⌒	MOTOR SWITCH
■	SPLICE
⌒	MOTOR PROTECTOR
⏚	GROUND
O	SCREW TERMINAL
⤙	HARNESS CONNECTOR TERMINAL
⊗	INSULATED TERMINAL

DRYER NOTES:

1. ALL WIRING MUST CONFORM TO LOCAL ELECTRICAL CODES.
2. CONNECT DRYER TO A 20 AMPERE INDIVIDUAL BRANCH CIRCUIT.
3. TIMER SHOWN IN OFF POSITION, DOOR SWITCH CLOSED, MOTOR AT REST, THERMOSTATS CLOSED, AND FABRIC SELECTOR SWITCH AT HIGH HEAT.

146547-000 WIRING DIAGRAM.

Dryer Electrical Diagram
(Models LCG771LW*0 and LCG771LL*0)

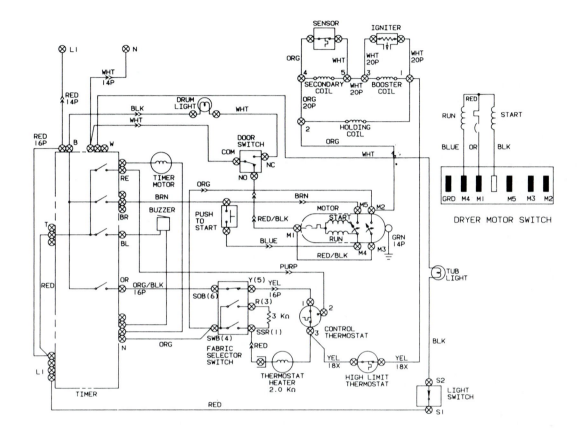

DRYER NOTES:

1. ALL WIRING MUST CONFORM TO LOCAL ELECTRICAL CODES.
2. CONNECT DRYER TO A 20 AMPERE INDIVIDUAL BRANCH CIRCUIT.
3. TIMER SHOWN IN OFF POSITION, DOOR SWITCH CLOSED, MOTOR AT REST, THERMOSTATS CLOSED, AND FABRIC SELECTOR SWITCH AT HIGH HEAT.

SWITCH CLOSED	CYCLES					
	AUTO DRY 36.9 MIN	OFF	TIMED CYCLES (MIN) 99.4 MIN		AUTO DRY WRINKLE RID 55.0 MIN	OFF
HEATER B-OR						
MOTOR B-BR						
TM. MOTOR T-RE						
BUZZER T-BL						

BAR CHART ABOVE REPRESENTS ONE
COMPLETE REVOLUTION OF TIMER SHAFT.

SHADED PORTION OF BAR CHART INDICATES
THE PROPORTIONAL TIMES THAT INTERNAL
TIMER CONTACTS ARE CLOSED.

FABRIC SELECTOR SWITCH

POSITION	SOB-Y	SWB-R	SWB-SSR
NO HEAT	O	O	O
LOW HEAT	X	O	X
MEDIUM HEAT	X	X	O
HIGH HEAT	X	O	O

X=SWITCH CLOSED   O=SWITCH OPEN

DRYER HARNESS WIRE SPECIFICATIONS

CODE	WIRE GAUGE A.W.G.	INSULATION MATERIAL
20P	20	105°C PVC (POLYVINYLCHLORIDE)
18P	18	
16P	16	
14P	14	
18X	18	150°C PVC X-LINK PVC

NOTE: ALL UNSPECIFIED WIRE CODES
ARE 18P.

146547-000 WIRING DIAGRAM.

SECTION K - TROUBLESHOOTING

The following diagnostic information is provided as a guideline to aid in isolating possible service problems. It does not include every situation that may be encountered. The technician must have the ability to read and comprehend schematic and wiring diagrams in conjunction with the information presented.

It is assumed (unless specifically mentioned) that the correct voltage, water supply pressure, and drain facilities are available to the Laundry Center.

⚠WARNING Some of the following test procedures are performed with the Laundry Center connected to power. Ensure that any disassembled component's electrical connections are spaced away from any structural or electrical component to prevent electrical shorts.

The troubleshooting check is common to all Laundry Center models. They use different parts to accomplish the same thing, but diagnosis will remain similar.

When a problem arises, and a possible cause is listed, follow the testing procedure as outlined in the service manual. The schematic and timer cycle charts are a necessity when making electrical checks. In most cases, an ohmmeter will handle all the tests necessary.

For checking a particular cycle of operation, it is absolutely necessary that the cycle be set up as outlined in the product instruction manual.

CLOTHES WASHER

SYMPTOM	CHECK THE FOLLOWING	REMEDY
Washer motor will not run, timer is set to operate and pulled out.	1. Fuse blown or tripped. 2. Supply line receptacle, wiring harness. 3. Timer knob not pulled out or open contacts. 4. Selector switch (open contacts). 5. Drive motor inoperative, check resistances. 6. Drive motor overload protector open. 7. Drive motor start capacitor open. 8. Lid switch (open contacts). 9. Pressure fill switch (inoperative). 10. Pressure fill switch tube air leak. 11. Water valve (loose wiring connections).	1. Replace fuse or reset breaker. 2. Repair or replace. 3. Pull out timer knob or replace timer. 4. Replace selector switch. 5. Replace motor. 6. Replace motor. 7. Replace motor start capacitor. 8. Replace lid switch. 9. Replace pressure fill switch. 10. Repair or replace. 11. Repair or replace.
Washer motor runs but will not wash.	1. Drive belt broken or off drive pulley. 2. Damaged part in transmission. 3. Damaged drive block. 4. Low voltage condition.	1. Repair or replace drive belt. 2. Repair transmission. 3. Repair or replace drive block. 4. Advise homeowner.

CLOTHES WASHER

SYMPTOM	CHECK THE FOLLOWING	REMEDY
Will not pump out water.	1. Pump clogged.	1. Remove obstruction or replace pump.
	2. Kinked or restricted drain hose.	2. Repair.
	3. Pump inpeller shaft broken.	3. Replace pump assembly.
	4. Clogged inlet to pump assembly.	4. Remove obstruction or replace pump.
	5. Drive motor will not run counterclockwise for pump out.	5. Replace drive motor.
Agitator will not oscillate.	1. Drive belt broken or off pulley.	1. Repair or replace drive belt.
	2. Damaged part in transmission.	2. Repair transmission.
	3. Damaged drive block.	3. Repair or replace drive block.
Water will not enter wash basket.	1. Faucet closed or kinked inlet hoses.	1. Repair or replace faucet or hoses.
	2. Water inlet valve screens clogged.	2. Clean screens or remove and use water hose screen /washer.
	3. Loose electrical connection at water inlet valve, pressure fill switch, selector switch, or timer.	3. Tighten connection.
	4. Pressure fill switch inoperative.	4. Replace pressure fill switch.
	5. Pressure fill tube kinked, clogged, or water logged.	5. Repair or replace tube.
	6. Timer inoperative.	6. Replace timer.
	7. Fuse blown or tripped.	7. Replace fuse or reset breaker.
Water will not shut off, water continues to flow.	1. Water inlet valve diaphram bleed hole plugged or diaphragm not seating.	1. Replace water inlet valve.
	2. Water inlet valve armature sticking in guide.	2. Replace water inlet valve.
	3. Timer contacts open or burnt.	3. Replace timer.
	4. Water level control switch contacts open or burnt.	4. Replace water level control switch.
Washer will not stop.	1. Wiring or terminal burnt or broken.	1. Repair wiring or replace terminal.
	2. Timer stalled or runs continuously.	2. Replace timer.
Washer runs with lid open in spin.	1. Damaged lid switch.	1. Replace lid switch.

CLOTHES WASHER

SYMPTOM	CHECK THE FOLLOWING	REMEDY
Washer motor trips on internal overload protector.	1. Low voltage condition. 2. Motor shaft binding. 3. Motor windings shorted. 4. Drive belt too tight. 5. Tight or damaged part in transmission. 6. Overload in wash basket or not enough water.	1. Advise home owner. 2. Replace drive motor. 3. Replace drive motor. 4. Adjust belt idler arm tension. 5. Repair transmission. 6. Adjust load or adjust water level to match load.
Washer incomplete fill.	1. Low water pressure. 2. Clogged water inlet valve screen. 3. Heavy water usage elsewhere in the home. 4. Kinked or restricted fill hoses.	1. Minimum water pressure of 20 P.S.I. (Advise home owner). 2. Clean water inlet screens. 3. Use washer when water usage is at a minimum. 4. Repair or replace hoses.
Water leaks.	1. Seal on tub cover damaged or out of position. 2. Over sudsing. 3. Water level control switch inoperative, causing an overfill. 4. Loose hose connections at the faucet and/or at the water inlet fill valve. 5. Water pump seal. 6. Water seal leak between transmission and outer tub, caused by a damaged seal or insufficient pressure on seal. 7. Loose hose clamp on internal fill or drain hoses. 8. Inlet fill hose at tub cover out of position.	1. Repair or replace seal. 2. Advise user to follow package directions or use low sudsing detergent. 3. Repair or replace water level control switch. 4. Tighten connections. 5. Replace water pump assembly. 6. Repair or replace tub seal. 7. Tighten clamps. 8. Reposition inlet fill hose.
Agitates but will not spin, or does not spin at full speed; motor overload protector opens.	1. Damaged bearing in tub bearing housing. 2. Belt tension too tight. 3. Low voltage condition. 4. Damaged or broken small LGS spring.	1. Repair or replace tub housing bearing. 2. Adjust belt idler arm tension. 3. Advise home owner. 4. Repair transmission.

CLOTHES WASHER

SYMPTOM	CHECK THE FOLLOWING	REMEDY
Wash water temperature not as selected.	1. Inlet water fill hoses reversed at faucet or at inlet fill valve.	1. Position hoses correctly.
	2. Water supply faucets not turned fully on.	2. Turn faucets fully on.
	3. Water inlet fill valve inoperative, or clogged inlet screens.	3. Clean or replace screens, repair or replace water inlet valve.
	4. Timer and/or selector switch miswired or inoperative.	4. Check wiring, or replace defective timer and/or selector switch.
Noise during agitation and/or spin cycle.	1. Loose agitator, cap or drive lug.	1. Tighten agitator/cap/lug.
	2. Loose set screw in drive motor pulley.	2. Tighten set screw.
	3. Worn idler arm pulley.	3. Replace idler arm pulley.
	4. Drive belt damaged or worn.	4. Replace drive belt.
	5. Transmission gears damaged or worn.	5. Repair transmission.
	6. Product not level, floor not solid enough to support unit.	6. Level product and/or place on solid floor.
	7. Out of balance load.	7. Readjust load.
	8. Snubber damaged.	8. Replace snubber.
	9. Motor mounting bolts loose.	9. Tighten bolts.
	10. Water seal under inner wash basket damaged or worn.	10. Repair or replace water seal.

CLOTHES DRYER

SYMPTOM	CHECK THE FOLLOWING	REMEDY
Dryer will not operate.	1. Fuse blown or tripped. 2. Open door safety switch. 3. Open push-to-start switch. 4. Open contacts in timer. 5. Open fabric temperature selector switch. 6. Inoperative drive motor. 7. Wiring or terminal block burned.	1. Replace fuse or reset breaker. 2. Replace door safety switch. 3. Replace push-to-start switch. 4. Replace dryer timer. 5. Replace fabric temperature selector switch. 6. Replace drive motor. 7. Repair wiring or replace terminal block.
Dryer runs but will not heat.	1. Open fabric temperature selector switch. 2. Open heat control thermostat. 3. Open safety thermostat. 4. Open switch on drive motor. 5. Open ignitor. 6. Inoperative dryer timer. 7. Wiring or terminal block burned. 8. Inoperative flame switch. 9. Inoperative burner valve or coil.	1. Replace fabric temperature selector switch. 2. Replace heat control thermostat. 3. Replace safety thermostat. 4. Replace switch on drive motor. 5. Replace ignitor. 6. Replace dryer timer. 7. Repair wiring or replace terminal block. 8. Replace flame switch. 9. Replace burner valve or coil.
Dryer will not stop.	1. Timer motor inoperative. 2. Inoperative dryer timer. 3. Wiring damaged. 4. Thermostat not opening. 5. Wrong heat control thermostat used.	1. Replace timer. 2. Replace timer. 3. Check and repair wiring. 4. Replace heat control thermostat. 5. Replace with correct heat control thermostat.
Buzzer does not operate.	1. Buzzer coil open. 2. Open contact in dryer timer. 3. Loose connection.	1. Replace buzzer. 2. Replace dryer timer. 3. Check and repair wiring.
Dryer fails to complete cycle.	1. Dryer drive motor overload protector open. 2. Timer motor failed. 3. Various other components may have failed in cycle.	1. Replace dryer drive motor. 2. Replace dryer timer. 3. Check out and replace inoperative components. (See Dryer will not operate.)
Clothes are still wet when dryer stops.	1. Exhaust vent clogged. 2. Heat control thermostat open. 3. Time dry cycle improperly set. 4. Incorrect gas. 5. Exhaust vent system is too long. 6. Temperature selector switch set on Air Fluff.	1. Clean exhaust vent. 2. Replace heat control thermostat. 3. Explain drying time to user. 4. Units are pre-set for natural gas. If L.P., conversion is necessary. 5. Check installation venting chart. 6. Change setting.

CLOTHES DRYER

SYMPTOM	CHECK THE FOLLOWING	REMEDY
Dryer motor runs but drum does not turn.	1. Broken or loose drum drive belt. 2. Loose drive motor pulley or idler arm pulley. 3. Too heavy a load.	1. Replace drive belt. 2. Replace pulleys or adjust idler arm tension spring. 3. Advise user.
Dryer drum rotates but is noisy.	1. Drum out of shape. 2. Idler arm pulley noisy. 3. Metal objects in drum. 4. Drive belt frayed or squeaking. 5. Drum rear support dry (no lube). 6. Drum front seal worn or torn. 7. Product not level.	1. Replace dryer drum assembly. 2. Replace idler arm pulley or assembly. 3. Remove objects from drum. 4. Replace drive belt. 5. Repair rear drum support. 6. Replace front drum seal. 7. Level product for proper operation.
Dryer motor hums but will not start or run.	1. Drive motor start switch contacts stuck. 2. Open start windings. 3. Bad bearings or locked motor.	1. Replace drive motor. 2. Replace drive motor. 3. Replace drive motor.
Dryer smokes.	1. Lint accumulated in dyer. 2. Loose terminal connection. 3. Overheated dryer drive motor.	1. Advise user to follow use and care instructions. 2. Tighten connections. 3. Replace drive motor.
Clothes too dry or scorched.	1. Improper temperature setting. 2. Improper mix of clothes. 3. Detergent left in clothes. 4. Inoperative heat control thermostat.	1. Advise user to follow use and care instructions. 2. Advise user to follow use and care instructions. 3. Rewash clothes. 4. Replace heat control thermostat.
Clothing damaged.	1. Foreign objects in dryer drum. 2. Rough drum or sorting vane surfaces. (Check with old nylon stocking or equivalent.) 3. Rough surfaces on rear plenum, lint housing, or lint screen filter. 4. Clothing previously damaged by improper use of chlorine bleach or other chemicals before drying.	1. Remove objects from drum. 2. Sand or replace rough surfaced vanes or drum assembly. 3. Sand or replace. 4. Review use and laundering procedure for damaged articles.
Ignitor does not glow.	1. Open flame sensor switch. 2. Open heat control thermostat. 3. Broken ignitor. 4. Wiring damaged.	1. Replace flame senor switch. 2. Replace heat control thermostat. 3. Replace ignitor. 4. Repair or replace wiring.

CLOTHES DRYER

SYMPTOM	CHECK THE FOLLOWING	REMEDY
Ignitor glows, but no flame.	1. No gas. 2. Flame sensor switch doesn't open. 3. Inoperative gas valve.	1. Turn on gas. 2. Replace flame sensor switch. 3. Test coil and replace as necessary.
Ignitor and flame switch cycle properly, but no flame.	1. Valves may not be opening. 2. Clogged orifice.	1. Check coils for continuity. 2. Check and clean orifice.

FABRIC CARE GUIDE

FIBER	WASHING CARE	DRYING CARE
Natural Fibers		
Cotton and Linen	Hot water for whites and sturdy colorfast cottons and linens. Follow Label directions.	Can be dried at high heat.
Silk	Warm water wash. Mild detergent. No chlorine bleach. Follow label directions; some are "dry-clean only".	Hang to dry. Press while still damp.
Wool	If machine washable, use warm or cold water. No chlorine bleach. Excessive agitation can cause felting or shrinkage.	If machine dry, use low heat, remove from dryer while still damp. Dry flat.
Man-Made Fibers (synthetics)		
Acetate	If washable, use warm or cold water wash.	Dry at low heat.
Acrylic and modacrylic	Wash in warm water. Dry-clean deep pile garments.	Dry at low heat. Some acrylics need tumble drying to maintain garment size.
Nylon	Wash in warm water. Rinse in cold water. Whites pick up color easily; wash separately.	Dry at low heat.
Polyester	Wash in small loads with high water level to prevent wrinkling. Wash colors in warm water. Wash whites in hot water. Rinse in cold water.	Dry at medium or low heat.
Rayon	Check label. If washable, use warm water wash.	Follow label directions. Dry at medium heat.
Spandex	Wash in warm water, slow speed. Never use chlorine bleach.	Dry at low heat.
Triacetate	Wash in warm water, slow speed.	Air dry permanently pleated garments. Dry medium to low heat.

Always follow label directions.

STAIN REMOVAL PROCEDURES

Read instructions on all products and keep them out of children's reach. Keep products in original labeled containers. Thoroughly wash any utensil used.

For successful stain removal:
1. Remove stains promptly.
2. Determine the type of stain:
   a) Greasy stain: Use safe drycleaning fluid, pre-wash spray or liquid laundry detergent.
   b) Non-greasy stain: Sponge in cold water.
3. Use cold or warm water on unknown stains. Hot water can set stains.
4. Before removing stain, check fabric and finish. Read the garment care label. Test bleach or spot remover on inside seam before using.

STAIN	TREATMENT
Adhesive tape, chewing gum, rubber cement	Apply ice or cold water to hardened surface; scrape with a dull knife. Place stain face down on paper towels. Saturate with pre-wash stain remover or cleaning fluid. Rinse, then launder.
Baby formula	Use a product containing enzymes to pretreat or soak stains. Soak for at least 30 minutes, or several hours for aged stains, then launder.
Beverages, (coffee, tea, soft drinks, wine, and alcoholic beverages)	Spong or soak stain in cold water. Pretreat with prewash stain remover, liquid laundry detergent or paste of granular laundry product and water. Wash using chlorine bleach, if safe for fabric, or an oxygen bleach. Older stains may respond to pretreating or soaking in product containing enzymes, followed by laundering.
Blood	If stain is fresh, rinse and soak in cold water. Rub with bar soap. For dried stains, pretreat or soak in warm water with a product containing enzymes, then launder. If stain remains, rewash using a bleach safe for fabric.
Brown, yellow discoloration from rust, iron, manganese	Use a rust remover recommended for fabrics; launder. **Do not use a chlorine bleach to remove rust stains because it may intensify discoloration.** For a rusty water problem, use a non-precipitating water conditioner in both wash and rinse water. For severe problems, install an iron filter in the water system.
Candle wax	Scrape off surface wax with a dull knife. Place stain between clean paper towels frequently to absorb more wax and press with a warm iron from the back of the fabric to avoid transferring stains. Place stain face down on clean paper towels. Sponge remaining stain with pre-wash stain remover or cleaning fluid; blot with paper towels, then launder. If any color remains, rewash using a bleach safe for fabric.

STAIN	TREATMENT
Carbon paper	Rub detergent into dampened stain; rinse well. If stain is not removed, put a few drops of ammonia on the stain and repeat the treatment with detergent; rinse well. Repeat if necessary.
Chalk	Brush stain with a moderately soft brush, being careful not to abrade fabric. Wash with laundry detergent. If stain remains, rewash using a bleach safe for fabric.
Chocolate	Pretreat or prewash in warm water with a product containing enzymes or a prewash stain remover, then launder. If stain remains, rewash using a bleach safe for fabric.
Collar, cuff soil	Pretreat with prewash stain remover, liquid laundry detergent or paste of granular detergent and water, then launder.
Cosmetics	Pretreat with prewash stain remover, liquid laundry detergent, paste of granular detergent or laundry additive and water, or rub with bar soap, then launder.
Crayon	For a few spots, treat the same as candle wax or dampen the stain and rub with bar soap. Launder using hottest water safe for fabric. For a full load of clothes, wash with hot water using a laundry soap and a cup of baking soda. If color stain remains, launder using chlorine bleach, if safe for fabric. Otherwise, pretreat or soak in a product containing enzymes or an oxygen bleach using hottest water safe for fabric, then launder.
Dairy products (milk, cream, yogurt, cream soups)	Treat as baby formula. Pretreat or soak in a product containing enzymes, then launder.
Deodorant	Pretreat with liquid detergent, then launder. For heavy stains, pretreat with prewash stain remover. Allow to stand 5 to 10 minutes. Launder using oxygen bleach.
Dye transfer	You can attempt restoration of white fabrics that have picked up color from other fabrics by using a packaged color remover. Follow label directions, then launder. If dye stain remains, launder again using a chlorine bleach or soak in oxygen bleach. This type of stain can be prevented if proper sorting and laundry procedures are followed.
Egg	Treat as baby formula. Pretreat or soak in a product containing enzymes, then launder.
Fabric softener	Dampen the stain and rub with bar soap. Rinse out, then launder.
Fruits or juices	Wash with bleach safe for fabric.

STAIN	TREATMENT
Grass	Pretreat by soaking or prewashing in warm water using a detergent containing enzymes, then launder. If stains remain, launder using a bleach safe for fabric.
Grease, oil, butter fats, salad dressing, cooking oils, car grease, and motor oils	Pretreat with prewash stain remover, liquid laundry detergent, or liquid detergent booster. For heavy stains, place stain face down on clean paper towels. Apply dry cleaning fluid to back of stain. Replace paper towels under stain frequently. Let dry; rinse. Launder using hottest water safe for fabric.
Ink	Some inks - ballpoint, felt tip and liquid - may be impossible to remove. Laundering may set some types of ink. Try a pretreatment method using a prewash stain remover, denatured alcohol, or dry cleaning fluid. First sponge the area around the stain with the stain remover before applying it directly on the stain. Place stain face down on clean paper towels. Apply denatured alcohol or dry cleaning fluid to back of stain. Replace towels under stain frequently. Rinse thoroughly, then launder.
Mildew	Badly mildewed fabrics may be damaged beyond repair. Launder stained items using a chlorine bleach safe for fabric. If not safe, soak in oxygen bleach and hot water, then launder.
Mud	When dry, brush off as much as possible. Pretreat with a paste of granular detergent and water, or liquid detergent, then launder. For heavy stains, pretreat or presoak with detergent or a product containing enzymes.
Mustard	Pretreat with prewash stain remover. Launder using a bleach safe for fabric.
Nail polish	May be impossible to remove. Try nail polish remover, but do not use on acetate or triacetate fabrics. Place stain face down on clean paper towels. Apply nail polish remover to back of stain. Replace paper towels under stain frequently. Repeat until stain disappears.
Paint	Water Based: Rinse fabric in cold water while paint stains are wet, then launder. Once paint is dry, it cannot be removed. Oil Based and Varnish: Use the same solvent the label on the can advises as a thinner. If label is not available, use turpentine, and rinse. Pretreat with prewash stain remover, bar soap, or laundry detergent. Rinse and launder.
Perspiration	Use a prewash stain remover or rub with bar soap. If perspiration has changed the color of the fabric, apply either ammonia to fresh stains or white vinegar to old stains and rinse. Launder using hottest water safe for fabric. Stubborn stains may respond to washing in an enyzme product or oxygen bleach in hottest water safe for fabric.

STAIN	TREATMENT
Perfume	Pretreat with prewash stain remover or liquid laundry detergent, then launder.
Pine Resin	Sponge cleaning fluid onto stain; let dry. Mix liquid laundry detergent and ammonia; soak stain in solution. Launder using liquid laundry detergent. **Note:** Choose a liquid laundry detergent that does not contain bleach. Mixing bleach and ammonia can cause dangerous fumes.
Rust	For rust spots, use rust stain remover following instructions. **Note: If rust remover contains hydrofluoric acid, do not use near washer as it will damage washer finish.** For rust discoloration of an entire load, launder using a phosphate detergent and an oxygen bleach.
Scorch	Treat as for mildew. Launder using bleach safe for fabric and hot water.
Shoe Polish	Liquid: Pretreat with a paste of granular detergent and water, then launder. Paste: Scrape residue from fabric with a dull knife. Pretreat with a stain remover or dry cleaning fluid, then rinse. Rub detergent into dampened area. Launder using bleach safe for fabric.
Tar	Scrape residue from fabric. Place stain face down on clean paper towels. Sponge with dry cleaning fluid. Replace towels frequently to absorb more tar and to avoid transferring stains. Launder in hottest water safe for fabric.
Tobacco	Dampen stain and rub with bar soap. Rinse. Pretreat or soak in a product containing enzymes, then launder. If stains remain, launder again using a bleach safe for fabric.
Typewriter correction fluid	Let stain dry thoroughly. Gently brush excess off with a clothes brush. Send to professional dry cleaner and mention type of stain.
Urine, vomit, mucus, feces	Pretreat or soak in a product containing enzymes. Launder using bleach safe for fabric.

FABRIC CARE GUIDE

FIBER	WASHING CARE	DRYING CARE
Natural Fibers		
Cotton and Linen	Hot water for whites and sturdy colorfast cottons and linens. Follow Label directions.	Can be dried at high heat.
Silk	Warm water wash. Mild detergent. No chlorine bleach. Follow label directions; some are "dry-clean only".	Hang to dry. Press while still damp.
Wool	If machine washable, use warm or cold water. No chlorine bleach. Excessive agitation can cause felting or shrinkage.	If machine dry, use low heat, remove from dryer while still damp. Dry flat.
Man-Made Fibers (synthetics)		
Acetate	If washable, use warm or cold water wash.	Dry at low heat.
Acrylic and modacrylic	Wash in warm water. Dry-clean deep pile garments.	Dry at low heat. Some acrylics need tumble drying to maintain garment size.
Nylon	Wash in warm water. Rinse in cold water. Whites pick up color easily; wash separately.	Dry at low heat.
Polyester	Wash in small loads with high water level to prevent wrinkling. Wash colors in warm water. Wash whites in hot water. Rinse in cold water.	Dry at medium or low heat.
Rayon	Check label. If washable, use warm water wash.	Follow label directions. Dry at medium heat.
Spandex	Wash in warm water, slow speed. Never use chlorine bleach.	Dry at low heat.
Triacetate	Wash in warm water, slow speed.	Air dry permanently pleated garments. Dry medium to low heat.

Always follow label directions.

422

Control Setting Chart

Always consult fabric care labels before drying. For items without labels, follow the suggestions below. The suggested drying times are approximate. Actual time needed will vary depending on size of load, weight and type of fabrics.

Dryers operating on a 208 Volt circuit instead of 240 Volt circuit require 20 percent more drying time than the chart listing.

For more helpful Information , see the Total Laundry Care book.

Type of Load	Temperature Setting	Cycle Setting/Time
Articles containing elastic	Low	Auto Dry Regular
Articles to be fluffed, freshened or dusted	Air Fluff	Time Dry 15-20 Minutes
Bedspreads, chenille	Regular	Time Dry 50-60 Minutes
Blankets		
Electric	See manufacturer's care label	See Total Laundry Care book
Synthetic	Regular	Time Dry 30 Minutes; check frequently
Wool	Regular	Partially dry on Time Dry 10 Minutes, then air dry on flat surface
Cottons, Linens		
Full Dry	Regular	Auto Dry Regular
Damp Dry, for ironing	Regular	Time Dry 20-30 Minutes
Curtains		
Cotton or cotton blends	Medium or Low	Auto Cycle Delicates/Perm Press
Sheer synthetic	Low	Auto Cycle Delicates/Perm Press
Heavy synthetic	Low	Auto Cycle Regular
To freshen all types	Air Fluff	Time Dry 15-20 Minutes
Delicate sheer lingerie and blouses, silk, rayon and nylon	Low	Time Dry 15-20 Minutes, check frequently
Knits		
Cotton or cotton blends	Medium or Low	Auto Cycle Delicates/Perm Press
Synthetics	Low	Auto Cycle Delicates/Perm Press
Wool	If machine dryable, Low	Time Dry 20 Minutes to damp dry, then air dry on a flat surface
Glass Fiber	Do not dry glass fiber items in a dryer unless recommended by the manufacturer. Small glass particles left in the drum can be picked up with the next load. These articles may cause skin irritation.	
Permanent Press		
100% Cotton	Medium	Auto Cycle Delicates/Perm Press
100% Synthetics & blends	Medium or Low	Auto Cycle Delicates/Perm Press
Pillows, feather or down	Regular	Time Dry 1-2 hours or longer. Timer may need to be reset

DO NOT DRY FOAM PILLOWS IN DRYER

Type of Load	Temperature Setting	Cycle Setting/Time
Plastics	To avoid fire hazards, never dry plastics, foam rubber or similarly textured, rubber-like materials in a heated dryer. Use the AIR FLUFF (No Heat) setting only.	

Rugs (4'x6')
    Cottons
    Nylon, polyester blends
	Regular
	Low
		Auto Dry Regular
		Auto Dry Regular

Slipcovers	Select temperature and time according to fabric; remove when slightly damp.	
Table linens, damp dry for ironing	Regular	Time Dry 20-30 Minutes
Tennis Shoes	Always consult care labels before washing or drying tennis shoes. Improper care can permanently damage some shoes.	
Wrinkled, but not soiled, permanent press or knit garments	Medium	Time Dry 15-20 Minutes

DRYER OPERATION AND REVIEW

DRYER	
SORTING	• Dry items of similar fabric, weight, and construction in the same load. • As a general rule, clothes that can be washed together can be dried together • Separate lint givers (towels, sweaters and flannels) from lint receivers (corduroys, dark cottons, man-made fabrics and permanent press items). • Sort Dark colored items from light ones. Dry non-colorfast items separately. • Turn pockets inside out for faster drying. Fasten hooks, close zippers and button buttons to prevent snagging.
LOADING	• For best results, do not overload the dryer. Clothes need enough room to tumble freely. They will dry faster, more evenly, and wrinkle less under these conditions. • Dry only 2 or 3 large items at one time. Fill out the load with medium or small items. Dry only 1 very large item, such as a bedspread or rug, at one time to ensure good tumbling and fluffing of the article. • For better tumbling action, add 2 lint-free towels to very small or delicate loads. • Man-made knits and permanent press items need plenty of room to tumble.
USING FABRIC SOFTENERS	• Use dryer fabric softeners according to manufacturer's instructions. • Fabric softeners make fabrics softer, fluffier and reduce static electricity. • If a fabric softener is used regularly in your dryer, more than normal lint release from fabrics will occur. Be sure to clean the lint screen after each load and occasionally wash in soapy water. Also, inspect and clean the dryer and venting system periodically.
DRYER FEATURES	• Safety door switch stops the dryer automatically if the door is opened during a cycle. Close the door and press the start button for the cycle to continue. • The safety door latch allows the dryer to be opened from the inside. • The automatic temperature safety switch shuts off the heat if overheating occurs. • The motor safety switch shuts off the heat if the motor stops. • The automatic motor protector stops the motor if it overheats and resets automatically after the motor cools. Press the start button to continue the cycle.

DRYING TIMES	• Many factors affect drying times. The dryer will operate most efficiently when fabrics of similar weights are dried together. Times may vary according to:
	  * Amount of moisture in clothes.
	  * Size of load. Overloading lengthens drying time.
	  * Type of load. Heavy, absorbent fabrics like bath towels will take longer to dry than sheer man-made fabrics.
	  * Amount of lint on screen. Lint build-up will lengthen drying time.
	  * Clean the lint screen after each load.
	  * Heat and moisture in the room.
	  * Type of installation (length of exhaust, number of elbows, and type of vent pipe.)
	  * Electrical voltage.

DRYER	
DRYING TEMPERATURES	• Select the temperature setting according to the fabric content. See the **Control Settings Chart** and the **"Important Laundry Tips"** book for more information.
DRYING HINTS FOR BEST RESULTS	• Remove clothes from the dryer as soon as tumbling stops to help prevent wrinkling. Immediately place permanent press garments on hangers or fold items for storage. • Do not overdry items. Overdrying can cause wrinkling, shrinkage and the build-up of static electricity. • Remove knits and corduroys while still damp for better shape retention. • Use the AIR FLUFF setting to freshen garments, draperies or pillows.
CYCLE SELECTIONS	• **Auto Dry:** Provides drying at a medium temperature. When the items are dry, the heat automatically shuts off and the load is tumbled for 5 minutes without heat. This cool-down period helps reduce wrinkling. **When using the Auto Dry cycle for the first time**, set the timer about halfway between MORE DRY and LESS DRY. Experience will determine if adjustments are needed for future loads. This cycle works best with items of similar fabric, weight and construction. Do not use Auto Dry for fabrics requiring low heat. **Note**: During Auto Dry, the timer temperature selector does not move constantly toward off during the cycle (as it does during Time Dry). • **Time Dry**: Provides up to 100 minutes of drying time at a regular temperature. The last few minutes of the cycle is a cool-down period. • **Delicate Dry**: Provides low heat drying for delicate items such as knits and man-made fabrics. The last 20 minutes of the cycle is Air Fluff. • **Air Fluff**: May also be selected to provide a no heat drying period for plastics, fluffing pillows, curtains or rubber like materials. • **For best results, follow label directions on specific items to be dried. Check the Control Settings chart and the "Important Laundry Tips" book if the fabric care label is not available.**

SECTION M - GLOSSARY

LENGTH CONVERSION

1 Millimeter	=	0.0394	Inch
1 Centimeter	=	0.394	Inch
1 Meter	=	39.4	Inches
1 Meter	=	3.2808	Feet
1 Meter	=	1.0936	Yards
1 Inch	=	25.4	Millimeters
1 Inch	=	2.54	Centimeters
1 Foot	=	304.8	Millimeters
1 Foot	=	0.3048	Meter
1 Yard	=	0.9144	Meter

ENERGY CONVERSION

1 Horsepower	=	746 Watts = .746	Kilowatts
1 Foot-pound	=	0.138	Kilogram-meter
1 Foot-pound	=	0.001	B. T. U.
1 Watt	=	0.00134	Horsepower
1 Kilogram-meter	=	7.23	Foot-pounds
1 B. T. U.	=	778	Foot-pounds

TEMPERATURE CONVERSION

$$Fahrenheit = \frac{9 \times C°}{5} + 32°$$

$$Centigrade = \frac{5 \times (F° - 32°)}{9}$$

AREA or SQUARE CONVERSION

1 Sq. Centimeter	=	0.155	Sq. Inch
1 Sq. Meter	=	10.764	Sq. Feet
1 Sq. Meter	=	1.196	Sq. Yards
1 Sq. Inch	=	6.452	Sq. Centimeters
1 Sq. Foot	=	0.0929	Sq. Meter
1 Sq. Yard	=	0.836	Sq. Meter

VOLUME CONVERSION

1 Cu. Meter	=	35.314 Cu. Ft. = 1.308 Cu. Yd.
1 Cu. Meter	=	264.2 U.S. Gallons
1 Cu. Centimeter	=	0.061 Cu. Inch
1 Liter	=	0.0353 Cu. Ft. = 61.023 Cu. In.
1 Liter	=	0.2642 U.S. Gal.=1.0567 U.S. Qt.
1 Cu. Yard	=	0.7645 Cu. Meter
1 Cu. Foot	=	0.02832 Cu. Meter =26.317 Liters
1 Cu. Inch	=	16.38716 Cu. Centimeters
1 U.S. Gallon	=	3.785 Liters
1 U.S. Quart	=	0.946 Liters

WEIGHT CONVERSION (Avoirdupois*)

1 Gram	=	0.03527	Ounce
1 Gram	=	0.002046	Pound
1 Pound	=	453.6	Grams
1 Ounce	=	28.35	Grams

* Based on 16 oz. = 1 lb.

PRESSURE CONVERSION

1 Pound per Square Inch	=	0.0703	Kilogram per Square Centimeter
1 Kilogram per Square Centimeter	=	14.22	Pounds per Square Inch

DEFINITION OF TERMS

Ampere (a) - The measurement of a unit of electrical current produced by one volt applied across a circuit with a resistance of one ohm. (Amps = Volts + Ohms)

Bimetal - Two dissimilar strips of metal, bonded together that expand differently when heated.

British Thermal Unit (B.T.U.) - The measurement of a unit of energy defined as the quantity of heat required to raise the temperature of one pound of water 1°F at or near its point of maximum density.

Centigrade (C°) - A unit of temperature defined as a measurement on a thermometric scale on which the interval between the freezing point of water and the boiling point of water is divided into 100 degrees with the 0° representing the freezing point and 100° the boiling point.

Centimeter (cm) - The measurement of a unit of length defined as one hundredth of 1 meter.

Diode (Rectifier) - A device which has two terminals and has a high resistance to current flow in one direction and a low resistance to current flow in the other direction. (Rectifies or converts alternating current into pulsating direct current.)

Energy - Measured in many different forms, energy is defined as the capacity for performing work.

Fahrenheit (F°) - A unit of temperature defined as a measurement on a thermometric scale on which (under standard atmospheric pressure) the boiling point of water is 212 degrees above the zero of the scale, the freezing point 32 degrees above the zero, and the zero point approximates the temperature produced by mixing equal quantities by weight of snow and common salt.

Foot-pound (ft/lb) - The measurement of a unit of energy defined as the work done in raising one pound (avoirdupois) against the force of gravity the height of one foot.

Gram (g) - The measurement of a unit of weight defined as the weight of one cubic centimeter of distilled water at 4°C.

Horsepower (hp) - The measurement of a unit of power defined as numerically equal to a rate of 33,000 foot-pounds of work per minute (550 foot-pounds per second).

Kilogram (kg) - The measurement of a unit of weight defined as one thousand grams.

Kilogram-meter (kg/m) - The measurement of a unit of energy defined as the amount of work expended in raising one kilogram through the height of one meter, in the latitude of Paris, France.

Kilometer (km) - The measurement of a unit of length defined as one thousand meters.

Liter (l) - The measurement of a unit of volume defined as the volume of one kilogram of distilled water at 4°C.

Meter (m) - A measurement of a unit of length defined as 1,650,763.73 times the wavelength of the orange-red line of Krypton-86 under specified conditions.

Millimeter (mm) - The measurement of a unit of length defined as one thousandth of 1 meter.

Shunt Switch - A switch that carries the majority of current when closed and placed in parallel with a component.

Ohm (Ω) - The measurement of a unit of electrical resistance equal to the resistance of a circuit in which a potential difference of one volt produces a current of one ampere. (Ohms = Volts + Amps)

Volt (v) - The measurement of a unit of electromotive force equal to the potential difference across a circuit with a resistance of one ohm with one ampere of current flowing. (Volts = Amps x Ohms)

Watt (w) - The measurement of a unit of electrical energy defined as the rate of work represented by a current of one ampere under a pressure of one volt.

CONVERSION CHART FOR DETERMINING AMPERES, OHMS, VOLTS, OR WATTS
(Amperes = a, Ohms = Ω, Volts = v, Watts = w)

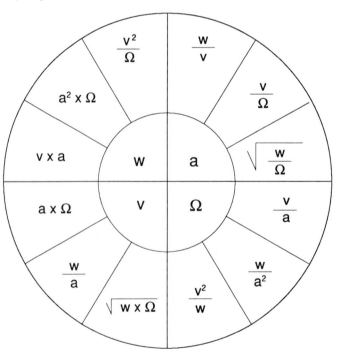

428

Chapter 17: Microwave Ovens

SAFE SERVICING PRACTICES
FOR MICROWAVE OVENS

To avoid the possibility of personal injury and/or property damage, it is important that safe servicing practices be observed. The following are examples, but without limitation, of such practices:

1. Always disconnect the microwave oven from the electric power source when preparing for any test or repair called for in this manual. Make all connections for the test and check them for tightness before reconnecting power. Do not reach into the component area while the oven is connected to power.

2. Always discharge the capacitor with an insulated-handle screwdriver before working in the electrical component compartment. Some types of failures may leave a high voltage charge in the capacitor and accidental discharge could cause injury.

3. When the oven is operating, there is HIGH VOLTAGE present near the transformer, capacitor, and magnetron. Keep the area clear and free of anything that could possibly come into contact with the high voltage.

4. A microwave oven should never be operated with any of its components removed and/or by-passed; or when any of the safety interlocks are found to be defective.

5. Do not, for any reason, defeat the interlock switches, or operate any electrical component on a test cord. To do so may expose you to dangerous levels of microwave energy.

6. USE ONLY REPLACEMENT PARTS CATALOGED FOR EACH RESPECTIVE MODEL OVEN. SUBSTITUTION MAY DEFEAT COMPLIANCE WITH SAFETY STANDARDS SET FOR MICROWAVE OVENS.

7. GROUNDING: The standard color coding for safety ground wires is GREEN or GREEN with YELLOW STRIPES. Ground leads are not to be used as current carrying conductors. IT IS EXTREMELY IMPORTANT THAT THE SERVICE TECHNICIAN RE-ESTABLISH ALL SAFETY GROUNDS PRIOR TO COMPLETION OF SERVICE. FAILURE TO DO SO WOULD CREATE A POTENTIAL HAZARD.

8. Prior to returning the product to service ensure that:
   • all electrical connections are correctly and securely connected; all uninsulated electrical terminals, connectors, etc. have adequate spacing from metal parts and panels; and all electrical leads are properly dressed and secured away from sharp edges, high-temperature components and moving parts.
   • all safety grounds (both internal and external to the product) are correctly and securely connected.
   • all panels are properly and securely reassembled.
   • IMPORTANT: Check for proper interlock switch action and test for microwave leakage to ensure that the oven meets the Bureau of Radiological Health Guidelines for microwave leakage.

MICROWAVE ENERGY REGULATIONS

The basic design of the Microwave oven makes it an inherently safe device to use and service. However, there are some precautions which should be followed to maintain this safety.

PRECAUTIONS TO BE OBSERVED BEFORE AND DURING SERVICING TO AVOID POSSIBLE EXPOSURE TO EXCESSIVE MICROWAVE ENERGY

a. Do not operate or allow the oven to be operated with the door open.

b. Make the following safety checks, on all ovens to be serviced, before activating the magnetron or other microwave source, and make repairs as necessary: (1) interlock operation, (2) proper door closing, (3) seal and sealing surfaces (arcing, wear, and other damage), (4) damage to or loosening of hinges and latches, (5) evidence of dropping or abuse.

c. Before turning on microwave power for any service test or inspection within the microwave generating compartments, check the magnetron, wave guide or transmission line, and cavity for proper alignment, integrity, and connections.

d. Any defective or misadjusted components in the interlock, monitor, door seal, and microwave generation and transmission systems, shall be repaired, replaced, or adjusted by procedures described in this manual before the oven is released to the owner.

e. A microwave leakage check, to verify compliance with the FEDERAL PERFORMANCE STANDARD, should be performed on each oven prior to release to the owner.

SERVICING PROCEDURES

a. If the oven is operative prior to servicing, a microwave emission check should be performed prior to servicing the oven.

b. Check for microwave leakage, also referred to as Radio Frequency (R.F.) leakage, after each servicing. Should the leakage be more than $4mW/cm^2$, inform the WCI Region Service Office, immediately. After repairing or replacing any radiation and/or safety device, keep a written record of the maximum microwave leakage for future reference, as required by DHHS regulations. In addition, the microwave leakage reading must be recorded on the service repair invoice.

c. If the oven operates with the door open:

   1. Tell user not to operate the oven.

   2. Immediately contact your WCI Region Service Office, and the Office of Compliance, Center for Devices and Radiological Health, 8757 Georgia Avenue, Silver Spring, Md. 20910.

432

MICROWAVE LEAKAGE TEST - ALL MODELS

To perform a microwave leakage test:

1. Fill a glass measuring container with 275 milliters (9 ounces) of tap water at 20° ± 5 °C (68° ± 9°F).

2. Place the container in the center of the oven cavity.

3. Perform the proper adjustment and operation of the microwave energy leakage meter. (Refer to the operating instructions supplied with the meter.)

4. Set the microwave oven controls for a cook operation of at least three minutes with a power setting of 100%.

5. Check for microwave leakage as follows: The receiver head of the meter probe must be held perpendicular to the surface or opening being checked and positioned approximately two (2) inches from the point of measurement. This is accomplished by the use of a two (2) inch plastic spacer normally attached to the probe. Check the oven by moving the probe along the juncture of the oven door and the cabinet, across all oven vent openings, all cabinet openings, and all non-metallic surfaces of the oven. Move the probe no faster than one (1) inch per second. Highest reading must be less than 4mW/cm².

Acceptable microwave leakage measurement instruments are available from the following manufacturers:

NARDA Microwave Corporation
Plainview, New York 11803

Holaday Industries, Inc.
14825 Martin Drive
Eden Praire, Minnesota 55344

Simpson Electric Co.
Elgin, Illinois 60120

OVEN SPECIFICATIONS

POWER SOURCE

120 Volts Nominal
15 Amperes (Single Circuit)
60 Hertz
Single Phase, 3 Wire Ground

POWER OUTPUT - VARIABLE

See Features Chart For Output Power
Off Condition ———————— None
Idle Condition ———————— 100 Watts Nominal
Cook Condition ———————— 1500 Watts
Current Draw ———————— 12.5 Amps Approx.

DIMENSIONS

Cabinet:

Width ———— 23 7/16"
Height ———— 15 1/2"
Depth ———— 17 3/4"

Cooking Cavity:

Width ———— 15 1/2"
Height ———— 9 3/8"
Depth ———— 14 1/2"

Shipping Weight: 68 pounds

SECTION C - ELECTRICAL DIAGRAMS

ELECTRICAL DIAGRAM INDEX

MODEL	ENGINEERING CHANGE CODE	WIRING DIAGRAM	SCHEMATIC DIAGRAM
56-9139	1	C-2	C-3
56-9339	1	C-4	C-5
56-9439	1	C-6	C-7

RESISTANCE CHART	
COMPONENT	RESISTANCE
Fan motor	15 to 25Ω
Stirrer motor	2300 to 2400Ω
Magnetron filament	Less than 1Ω
Capacitor	1,000,000+Ω
Transformer (primary)	Less than 1Ω
Transformer (HV)	50 to 100Ω
Transformer (filament)	Less than 1Ω

WIRING COLOR CODE											
BK	Black	BR	Brown	O	Orange	V	Violet	T	Tan	Y	Yellow
C	Copper	G	Green	BL	Blue	GY	Gray	P	Pink	R	Red
W	White	PR	Purple	G/Y	Green with Yellow Stripe						

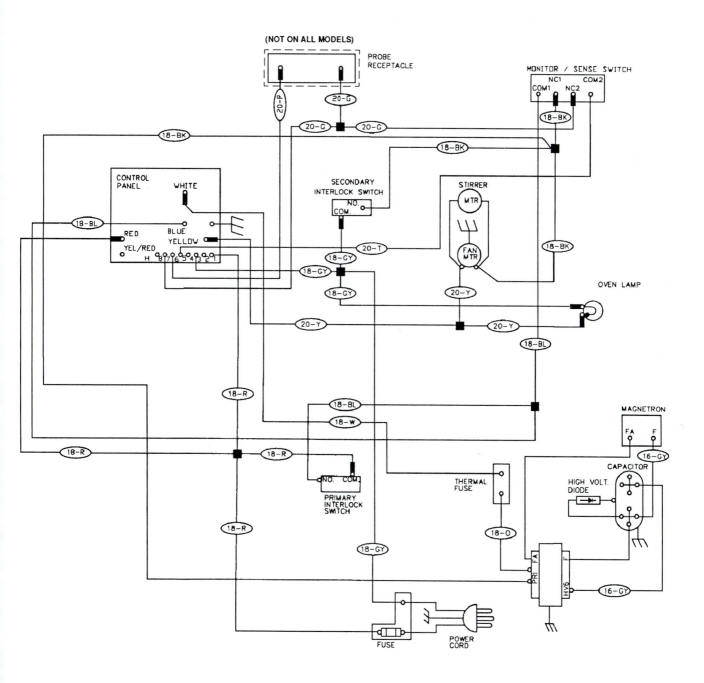

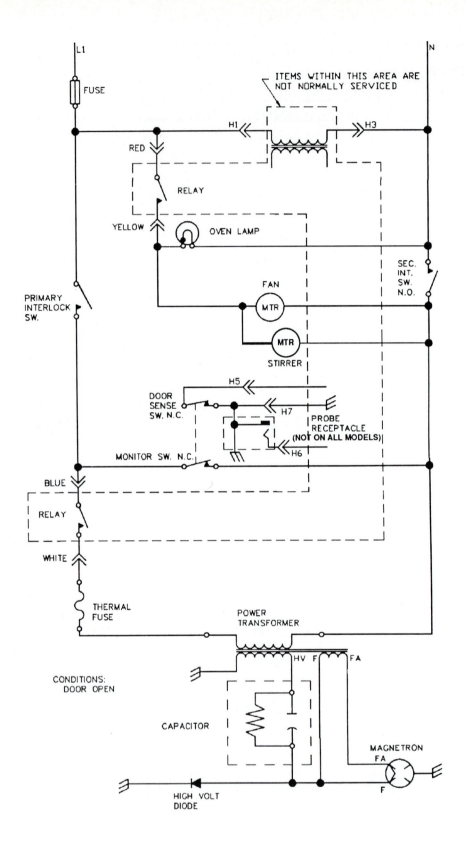

WIRING COLOR CODE									
BK	Black	BR	Brown	O	Orange	V	Violet	T	Tan	Y	Yellow
C	Copper	G	Green	BL	Blue	GY	Gray	P	Pink	R	Red
W	White	PR	Purple	G/Y	Green with Yellow Stripe						

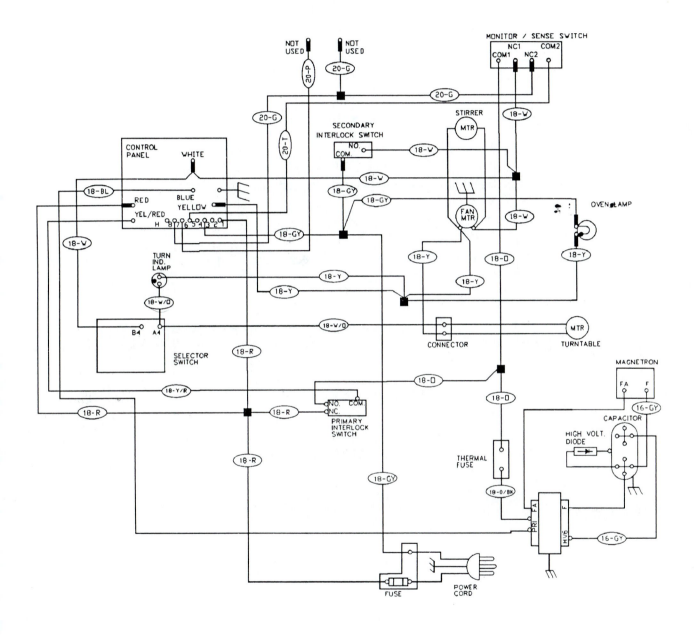

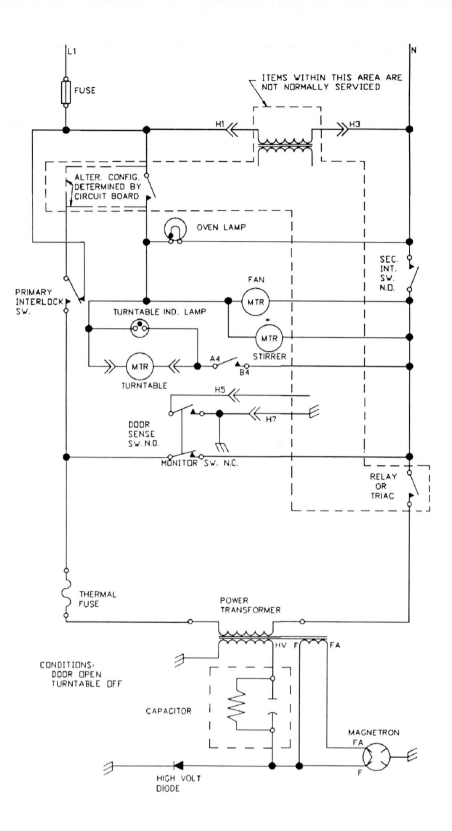

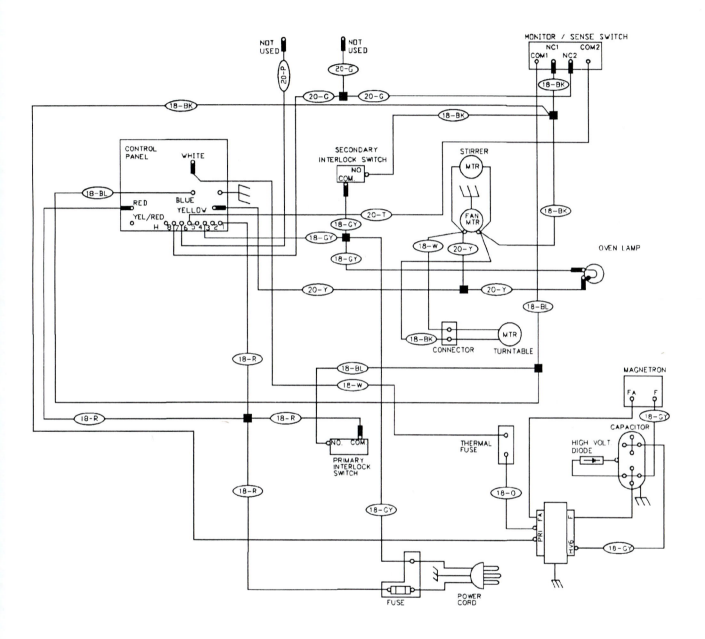

WIRING COLOR CODE									
BK	Black	BR	Brown	O	Orange	V	Violet	T	Tan	Y	Yellow
C	Copper	G	Green	BL	Blue	GY	Gray	P	Pink	R	Red
W	White	PR	Purple	G/Y	Green with Yellow Stripe						

439

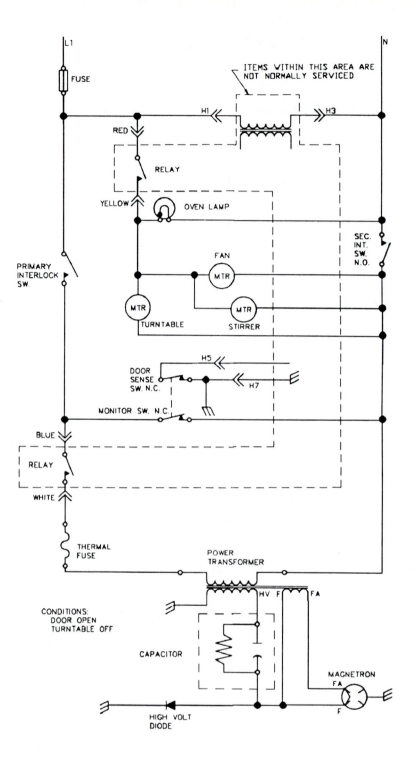

440

FLOW CHART SYMBOLS

What is a flow chart? A diagram that shows step-by-step progression through a procedure using connecting lines and a set of symbols.

Why use flow charts? To expedite the diagnosis of a product. A flow chart gives the proper test procedure (with the least amount of written information) that will eventually lead to a faulty component or system failure. (The flow charts used in this manual lead to a single component failure.)

Why are symbols used in a flow chart? Flow chart symbols have definite meanings. Each symbol is a key to that particular step in the diagnostic procedure. Once these symbols are understood, the diagnosis goes much faster. The following symbols and their meanings are used in this manual.

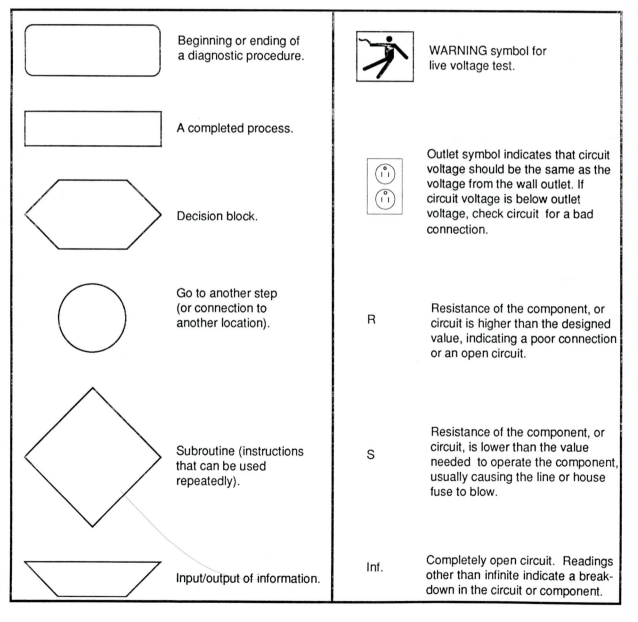

	Beginning or ending of a diagnostic procedure.
	A completed process.
	Decision block.
	Go to another step (or connection to another location).
	Subroutine (instructions that can be used repeatedly).
	Input/output of information.
	WARNING symbol for live voltage test.
	Outlet symbol indicates that circuit voltage should be the same as the voltage from the wall outlet. If circuit voltage is below outlet voltage, check circuit for a bad connection.
R	Resistance of the component, or circuit is higher than the designed value, indicating a poor connection or an open circuit.
S	Resistance of the component, or circuit, is lower than the value needed to operate the component, usually causing the line or house fuse to blow.
Inf.	Completely open circuit. Readings other than infinite indicate a breakdown in the circuit or component.

FLOW CHART ENTRY POINTS

PROBLEM	MODEL NUMBERS		
	56-9139 LOT 1	56-9339 LOT 1	56-9439 LOT 1
Display does not illuminate.	EP 1 Pg D-5	EP 2 Pg D-6	EP 4 Pg D-9
Display is illuminated, but control board will not program.	EP 5 Pg D-10	EP 5 Pg D-10	EP 5 Pg D-10
With the oven door closed, oven light, and cooling fan operate without the control board being programmed.	EP 6 Pg D-10	EP 6 Pg D-10	EP 6 Pg D-10
Display is illuminated, but the oven lamp will not glow during cooking program.	EP 7 Pg D-11	EP 7 Pg D-11	EP 7 Pg D-11
Display is illuminated, and the oven lamp glows when the oven is in cook, but does not glow when the oven door is open.	EP 15 Pg D-15	N/A	EP 15 Pg D-15
Display is illuminated, but the cooling fan does not run during cooking program.	EP 8 Pg D-11	EP 8 Pg D-11	EP8 Pg D-11
Display is illuminated, oven light glows, cooling fan runs, but power transformer does not operate during cooking program.	EP 9 Pg D-12	EP 9 Pg D-12	EP 9 Pg D-12
The oven sounds and acts normal, but does not cook.	EP 11 Pg D-13	EP 11 Pg D-13	EP13 Pg D-13
Power transformer makes a loud noise when the oven is placed in cook.	EP 12 Pg D-14	EP 12 Pg D-14	EP 12 Pg D-14
Display is illuminated, but the turntable does not turn during cooking program.	N/A	N/A	EP 10 Pg D-12
Display is illuminated, but stirrer does not rotate during cooking program.	EP 14 Pg D-15	EP 14 Pg D-15	EP 14 Pg D-15
Control board does not stop counting down when oven door is open.	EP 13 Pg D-15	EP 13 Pg D-15	EP 13 Pg D-15
Oven cooks slow.	EP 16 Pg D-16	EP 16 Pg D-16	EP 16 Pg D-16
Oven light glows, but turntable does not turn.	N/A	EP 17 Pg D-18	N/A
Turntable turns, but turntable ind. light does not glow.	N/A	EP 18 Pg D-18	N/A
Oven light glows when the oven is in cook, but does not glow when the oven is off, and the door is open.	N/A	EP 3 Pg D-8	N/A

Items listed under the problem column are indicators that may be observed without disassembling the appliance, and should be used as a troubleshooting starting point.

N/A = Not applicable.

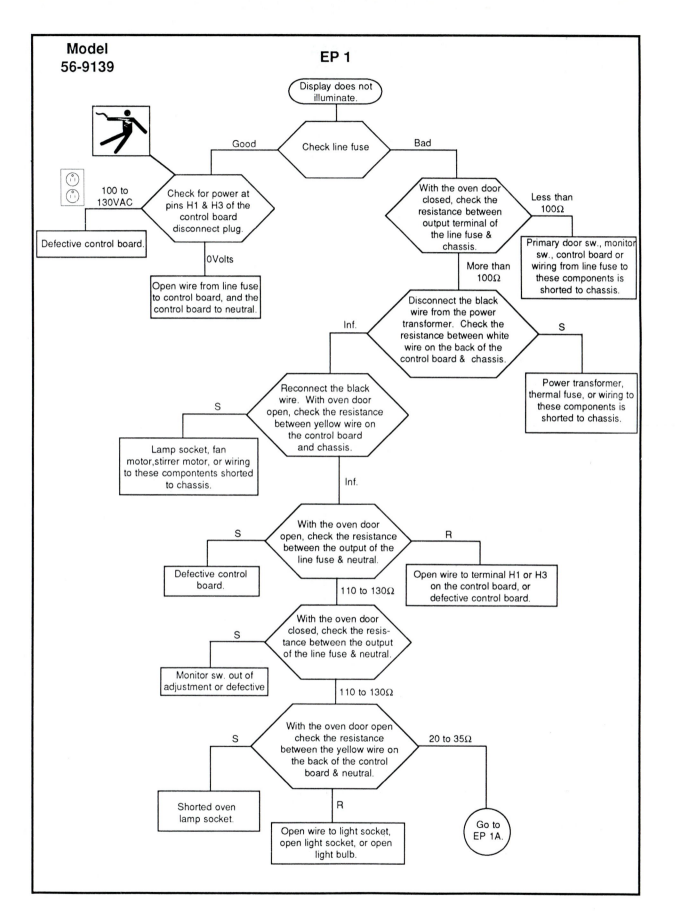

Model 56-9139

EP 1

Display does not illuminate.

Check line fuse

Good / Bad

100 to 130VAC

Check for power at pins H1 & H3 of the control board disconnect plug.

Defective control board.

0 Volts

Open wire from line fuse to control board, and the control board to neutral.

With the oven door closed, check the resistance between output terminal of the line fuse & chassis.

Less than 100Ω

Primary door sw., monitor sw., control board or wiring from line fuse to these components is shorted to chassis.

More than 100Ω

Disconnect the black wire from the power transformer. Check the resistance between white wire on the back of the control board & chassis.

Inf. / S

Power transformer, thermal fuse, or wiring to these components is shorted to chassis.

Reconnect the black wire. With oven door open, check the resistance between yellow wire on the control board and chassis.

S

Lamp socket, fan motor, stirrer motor, or wiring to these compontents shorted to chassis.

Inf.

With the oven door open, check the resistance between the output of the line fuse & neutral.

S / R

Defective control board.

Open wire to terminal H1 or H3 on the control board, or defective control board.

110 to 130Ω

With the oven door closed, check the resis- tance between the output of the line fuse & neutral.

S

Monitor sw. out of adjustment or defective

110 to 130Ω

With the oven door open check the resistance between the yellow wire on the back of the control board & neutral.

S / 20 to 35Ω

Shorted oven lamp socket.

R

Open wire to light socket, open light socket, or open light bulb.

Go to EP 1A.

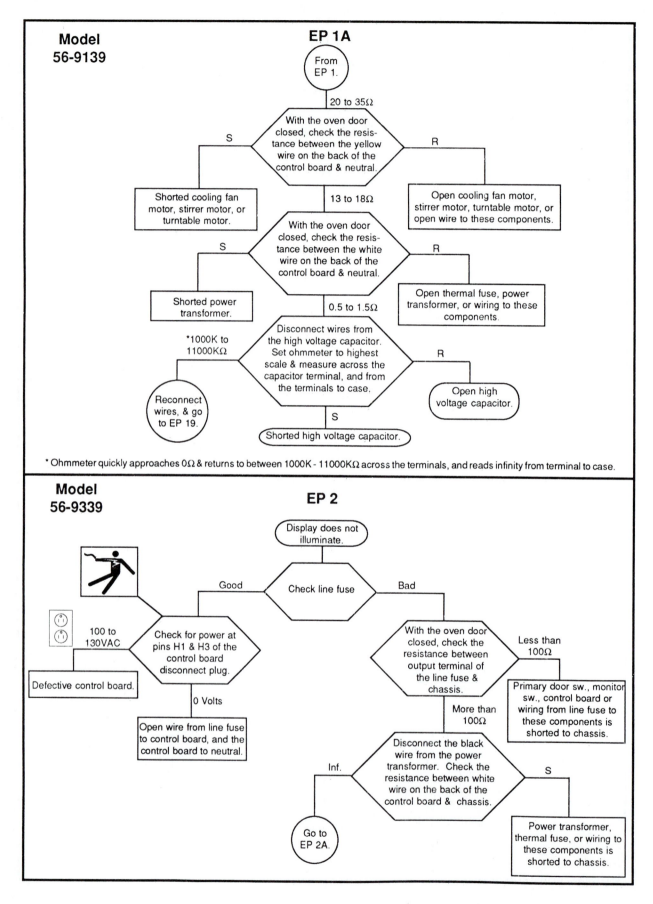

Model 56-9139

EP 1A

From EP 1.

20 to 35Ω

With the oven door closed, check the resistance between the yellow wire on the back of the control board & neutral.

S → Shorted cooling fan motor, stirrer motor, or turntable motor.

R → Open cooling fan motor, stirrer motor, turntable motor, or open wire to these components.

13 to 18Ω

With the oven door closed, check the resistance between the white wire on the back of the control board & neutral.

S → Shorted power transformer.

R → Open thermal fuse, power transformer, or wiring to these components.

0.5 to 1.5Ω

Disconnect wires from the high voltage capacitor. Set ohmmeter to highest scale & measure across the capacitor terminal, and from the terminals to case.

*1000K to 11000KΩ → Reconnect wires, & go to EP 19.

R → Open high voltage capacitor.

S → Shorted high voltage capacitor.

* Ohmmeter quickly approaches 0Ω & returns to between 1000K - 11000KΩ across the terminals, and reads infinity from terminal to case.

Model 56-9339

EP 2

Display does not illuminate.

Check line fuse

Good → Check for power at pins H1 & H3 of the control board disconnect plug.

100 to 130VAC → Defective control board.

0 Volts → Open wire from line fuse to control board, and the control board to neutral.

Bad → With the oven door closed, check the resistance between output terminal of the line fuse & chassis.

Less than 100Ω → Primary door sw., monitor sw., control board or wiring from line fuse to these components is shorted to chassis.

More than 100Ω → Disconnect the black wire from the power transformer. Check the resistance between white wire on the back of the control board & chassis.

Inf. → Go to EP 2A.

S → Power transformer, thermal fuse, or wiring to these components is shorted to chassis.

444

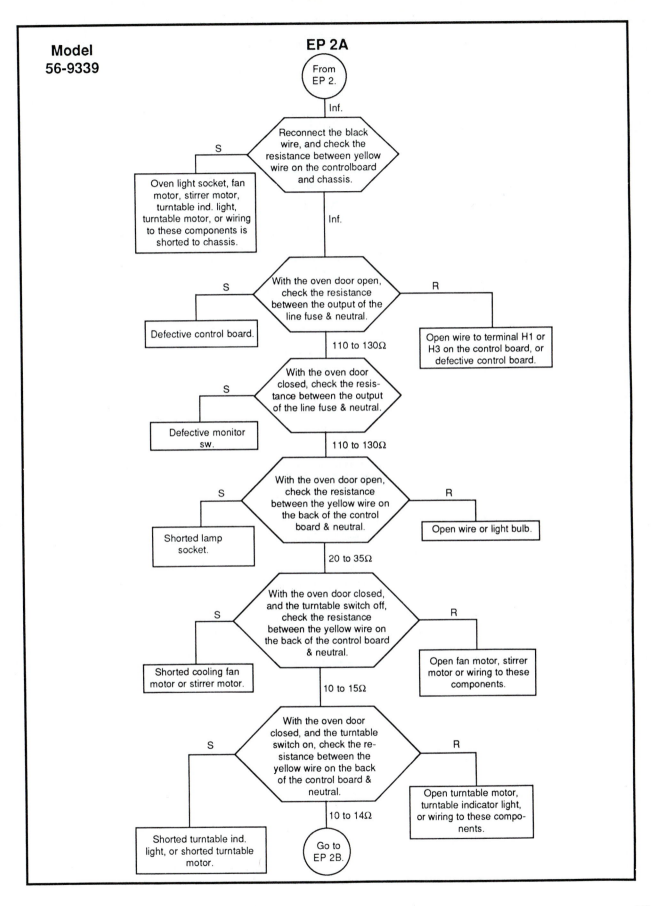

Model 56-9339

EP 2A

From EP 2.

Inf.

Reconnect the black wire, and check the resistance between yellow wire on the controlboard and chassis.

S → Oven light socket, fan motor, stirrer motor, turntable ind. light, turntable motor, or wiring to these components is shorted to chassis.

Inf.

With the oven door open, check the resistance between the output of the line fuse & neutral.

S → Defective control board.

R → Open wire to terminal H1 or H3 on the control board, or defective control board.

110 to 130Ω

With the oven door closed, check the resistance between the output of the line fuse & neutral.

S → Defective monitor sw.

110 to 130Ω

With the oven door open, check the resistance between the yellow wire on the back of the control board & neutral.

S → Shorted lamp socket.

R → Open wire or light bulb.

20 to 35Ω

With the oven door closed, and the turntable switch off, check the resistance between the yellow wire on the back of the control board & neutral.

S → Shorted cooling fan motor or stirrer motor.

R → Open fan motor, stirrer motor or wiring to these components.

10 to 15Ω

With the oven door closed, and the turntable switch on, check the resistance between the yellow wire on the back of the control board & neutral.

S → Shorted turntable ind. light, or shorted turntable motor.

R → Open turntable motor, turntable indicator light, or wiring to these components.

10 to 14Ω

Go to EP 2B.

445

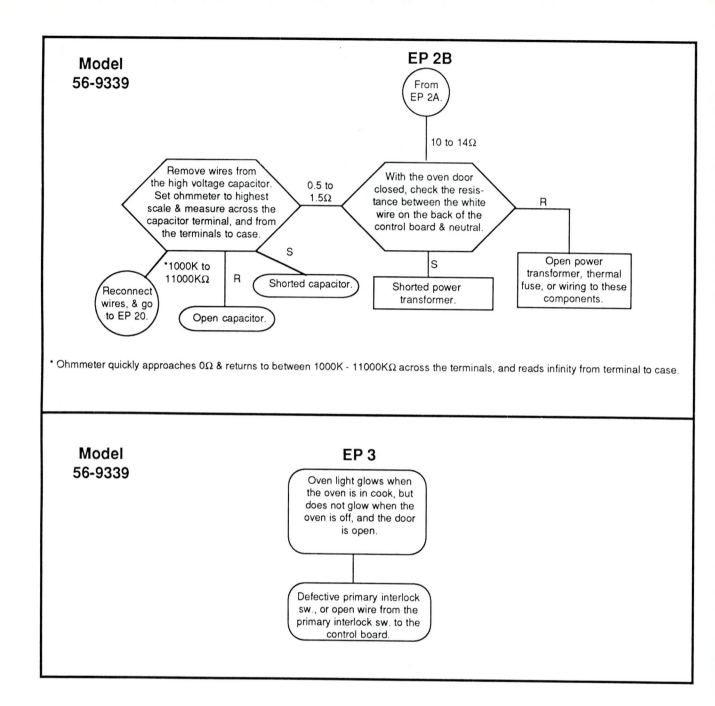

**Model
56-9339**

EP 2B

From EP 2A.

10 to 14Ω

Remove wires from the high voltage capacitor. Set ohmmeter to highest scale & measure across the capacitor terminal, and from the terminals to case.

0.5 to 1.5Ω

With the oven door closed, check the resistance between the white wire on the back of the control board & neutral.

R

Open power transformer, thermal fuse, or wiring to these components.

*1000K to 11000KΩ

R

S

S

Shorted power transformer.

Reconnect wires, & go to EP 20.

Shorted capacitor.

Open capacitor.

* Ohmmeter quickly approaches 0Ω & returns to between 1000K - 11000KΩ across the terminals, and reads infinity from terminal to case.

**Model
56-9339**

EP 3

Oven light glows when the oven is in cook, but does not glow when the oven is off, and the door is open.

Defective primary interlock sw., or open wire from the primary interlock sw. to the control board.

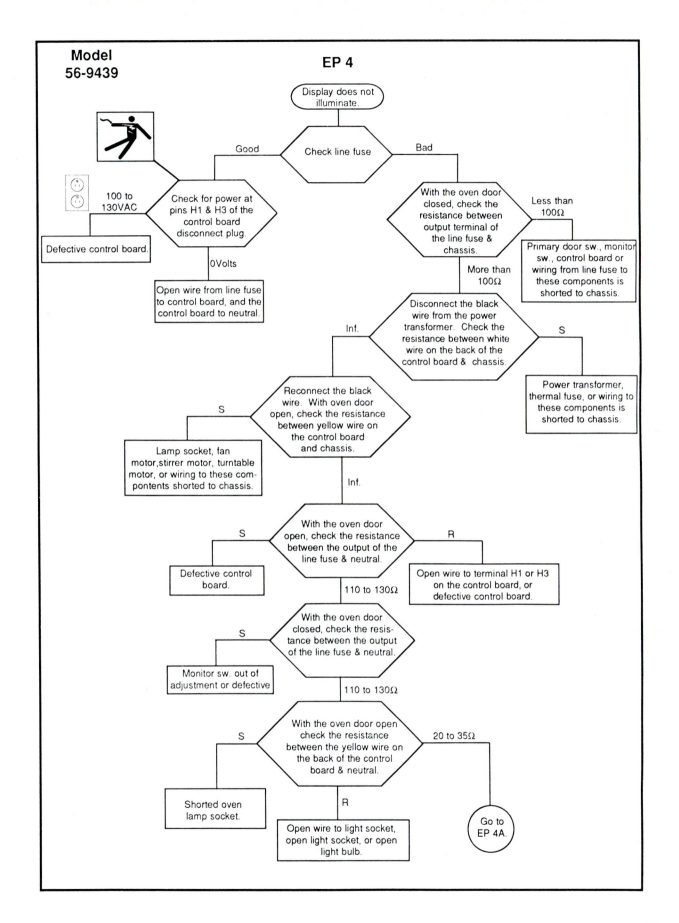

Model 56-9439

EP 4

Display does not illuminate.

Check line fuse

Good / Bad

100 to 130VAC

Check for power at pins H1 & H3 of the control board disconnect plug.

Defective control board.

0 Volts

Open wire from line fuse to control board, and the control board to neutral.

With the oven door closed, check the resistance between output terminal of the line fuse & chassis.

Less than 100Ω

Primary door sw., monitor sw., control board or wiring from line fuse to these components is shorted to chassis.

More than 100Ω

Disconnect the black wire from the power transformer. Check the resistance between white wire on the back of the control board & chassis.

Inf.

S

Power transformer, thermal fuse, or wiring to these components is shorted to chassis.

Reconnect the black wire. With oven door open, check the resistance between yellow wire on the control board and chassis.

S

Lamp socket, fan motor, stirrer motor, turntable motor, or wiring to these compontents shorted to chassis.

Inf.

With the oven door open, check the resistance between the output of the line fuse & neutral.

S

Defective control board.

R

Open wire to terminal H1 or H3 on the control board, or defective control board.

110 to 130Ω

With the oven door closed, check the resistance between the output of the line fuse & neutral.

S

Monitor sw. out of adjustment or defective

110 to 130Ω

With the oven door open check the resistance between the yellow wire on the back of the control board & neutral.

S

Shorted oven lamp socket.

R

Open wire to light socket, open light socket, or open light bulb.

20 to 35Ω

Go to EP 4A.

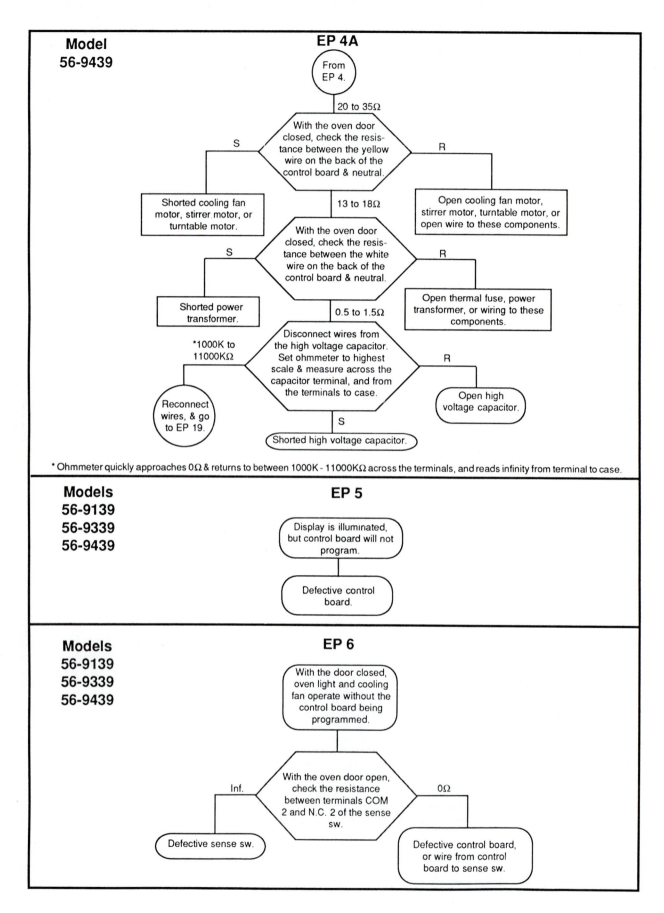

Model
56-9439

EP 4A

From EP 4.

20 to 35Ω

With the oven door closed, check the resistance between the yellow wire on the back of the control board & neutral.

S — Shorted cooling fan motor, stirrer motor, or turntable motor.

R — Open cooling fan motor, stirrer motor, turntable motor, or open wire to these components.

13 to 18Ω

With the oven door closed, check the resistance between the white wire on the back of the control board & neutral.

S — Shorted power transformer.

R — Open thermal fuse, power transformer, or wiring to these components.

0.5 to 1.5Ω

Disconnect wires from the high voltage capacitor. Set ohmmeter to highest scale & measure across the capacitor terminal, and from the terminals to case.

*1000K to 11000KΩ — Reconnect wires, & go to EP 19.

R — Open high voltage capacitor.

S — Shorted high voltage capacitor.

* Ohmmeter quickly approaches 0Ω & returns to between 1000K - 11000KΩ across the terminals, and reads infinity from terminal to case.

Models
56-9139
56-9339
56-9439

EP 5

Display is illuminated, but control board will not program.

Defective control board.

Models
56-9139
56-9339
56-9439

EP 6

With the door closed, oven light and cooling fan operate without the control board being programmed.

With the oven door open, check the resistance between terminals COM 2 and N.C. 2 of the sense sw.

Inf. — Defective sense sw.

0Ω — Defective control board, or wire from control board to sense sw.

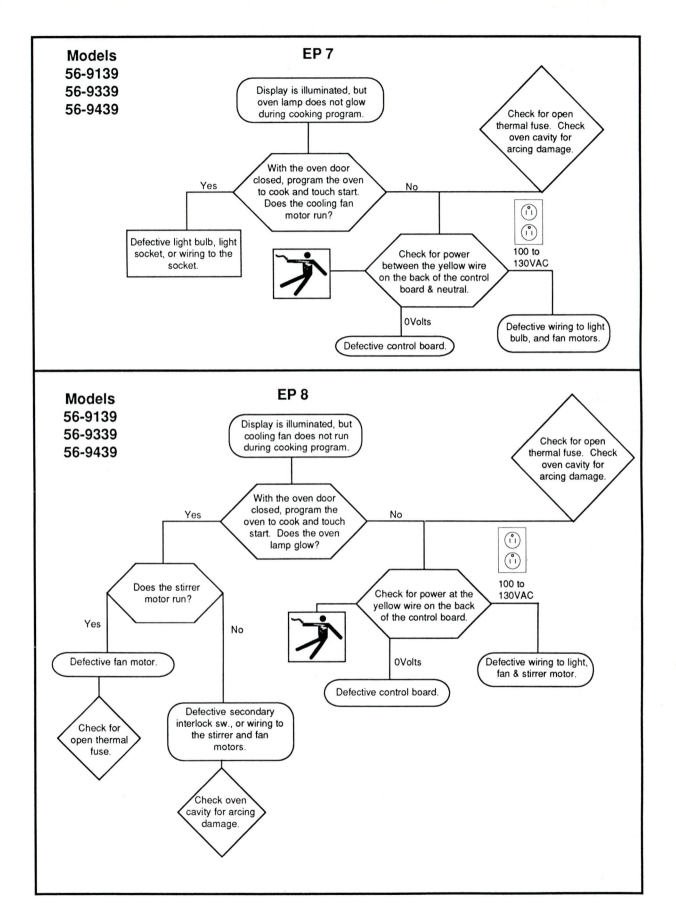

Models
56-9139
56-9339
56-9439

EP 7

Display is illuminated, but oven lamp does not glow during cooking program.

With the oven door closed, program the oven to cook and touch start. Does the cooling fan motor run?

Yes

No

Defective light bulb, light socket, or wiring to the socket.

Check for open thermal fuse. Check oven cavity for arcing damage.

100 to 130VAC

Check for power between the yellow wire on the back of the control board & neutral.

0Volts

Defective control board.

Defective wiring to light bulb, and fan motors.

Models
56-9139
56-9339
56-9439

EP 8

Display is illuminated, but cooling fan does not run during cooking program.

With the oven door closed, program the oven to cook and touch start. Does the oven lamp glow?

Yes

No

Does the stirrer motor run?

Yes

No

Check for open thermal fuse. Check oven cavity for arcing damage.

100 to 130VAC

Check for power at the yellow wire on the back of the control board.

0Volts

Defective control board.

Defective wiring to light, fan & stirrer motor.

Defective fan motor.

Defective secondary interlock sw., or wiring to the stirrer and fan motors.

Check for open thermal fuse.

Check oven cavity for arcing damage.

449

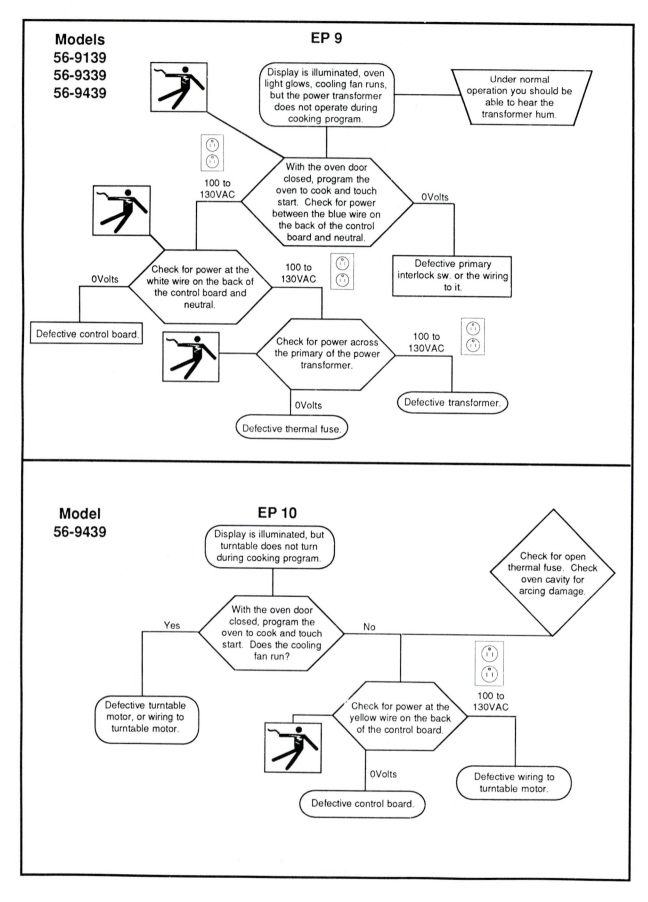

Models
56-9139
56-9339
56-9439

EP 9

Display is illuminated, oven light glows, cooling fan runs, but the power transformer does not operate during cooking program.

Under normal operation you should be able to hear the transformer hum.

With the oven door closed, program the oven to cook and touch start. Check for power between the blue wire on the back of the control board and neutral.

100 to 130VAC

0Volts

Defective primary interlock sw. or the wiring to it.

Check for power at the white wire on the back of the control board and neutral.

0Volts

100 to 130VAC

Defective control board.

Check for power across the primary of the power transformer.

100 to 130VAC

Defective transformer.

0Volts

Defective thermal fuse.

Model
56-9439

EP 10

Display is illuminated, but turntable does not turn during cooking program.

Check for open thermal fuse. Check oven cavity for arcing damage.

With the oven door closed, program the oven to cook and touch start. Does the cooling fan run?

Yes

No

Defective turntable motor, or wiring to turntable motor.

Check for power at the yellow wire on the back of the control board.

100 to 130VAC

0Volts

Defective wiring to turntable motor.

Defective control board.

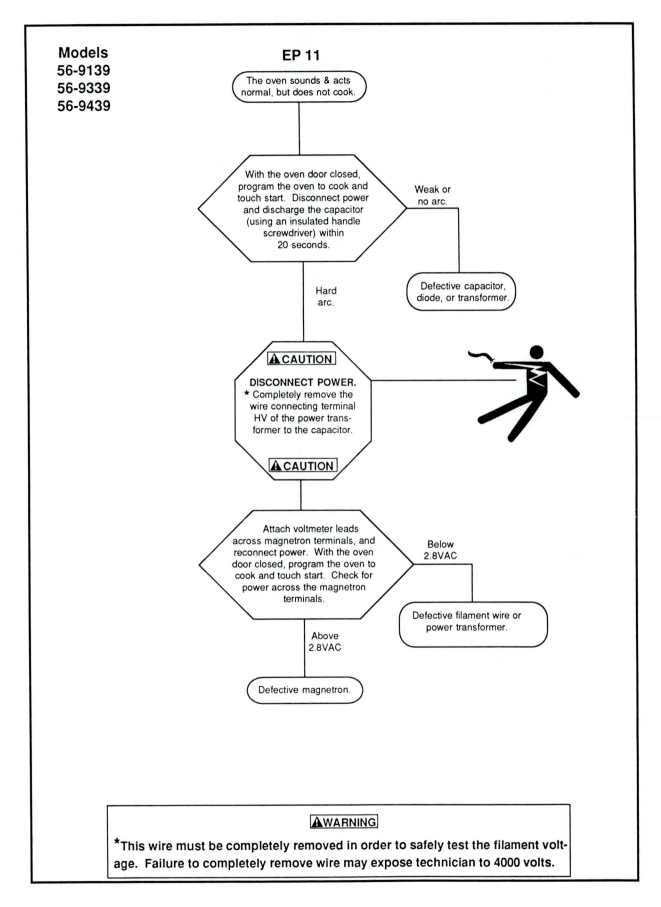

Models
56-9139
56-9339
56-9439

EP 11

The oven sounds & acts normal, but does not cook.

With the oven door closed, program the oven to cook and touch start. Disconnect power and discharge the capacitor (using an insulated handle screwdriver) within 20 seconds.

Weak or no arc.

Defective capacitor, diode, or transformer.

Hard arc.

⚠CAUTION

DISCONNECT POWER.
* Completely remove the wire connecting terminal HV of the power transformer to the capacitor.

⚠CAUTION

Attach voltmeter leads across magnetron terminals, and reconnect power. With the oven door closed, program the oven to cook and touch start. Check for power across the magnetron terminals.

Below 2.8VAC

Defective filament wire or power transformer.

Above 2.8VAC

Defective magnetron.

⚠WARNING

*This wire must be completely removed in order to safely test the filament voltage. Failure to completely remove wire may expose technician to 4000 volts.

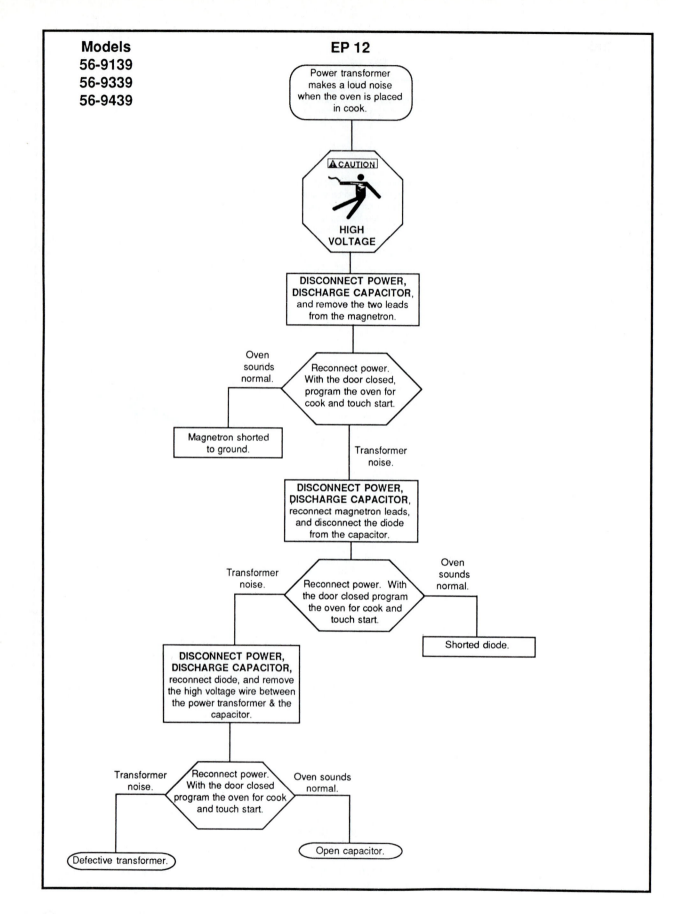

Models
56-9139
56-9339
56-9439

EP 12

Power transformer makes a loud noise when the oven is placed in cook.

⚠ CAUTION

HIGH VOLTAGE

DISCONNECT POWER, DISCHARGE CAPACITOR, and remove the two leads from the magnetron.

Reconnect power. With the door closed, program the oven for cook and touch start.

Oven sounds normal. → Magnetron shorted to ground.

Transformer noise.

DISCONNECT POWER, DISCHARGE CAPACITOR, reconnect magnetron leads, and disconnect the diode from the capacitor.

Reconnect power. With the door closed program the oven for cook and touch start.

Transformer noise.

Oven sounds normal. → Shorted diode.

DISCONNECT POWER, DISCHARGE CAPACITOR, reconnect diode, and remove the high voltage wire between the power transformer & the capacitor.

Reconnect power. With the door closed program the oven for cook and touch start.

Transformer noise. → Defective transformer.

Oven sounds normal. → Open capacitor.

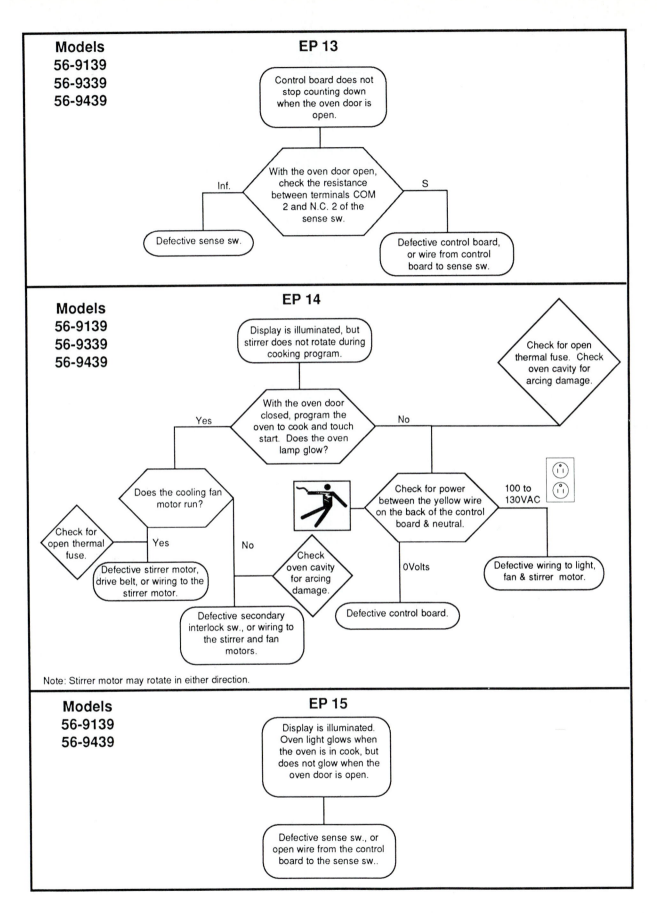

Models
56-9139
56-9339
56-9439

EP 13

Control board does not stop counting down when the oven door is open.

With the oven door open, check the resistance between terminals COM 2 and N.C. 2 of the sense sw.

Inf.

Defective sense sw.

S

Defective control board, or wire from control board to sense sw.

Models
56-9139
56-9339
56-9439

EP 14

Display is illuminated, but stirrer does not rotate during cooking program.

With the oven door closed, program the oven to cook and touch start. Does the oven lamp glow?

Yes

Does the cooling fan motor run?

Check for open thermal fuse.

Yes

Defective stirrer motor, drive belt, or wiring to the stirrer motor.

No

Defective secondary interlock sw., or wiring to the stirrer and fan motors.

No

Check oven cavity for arcing damage.

Check for power between the yellow wire on the back of the control board & neutral.

100 to 130VAC

0Volts

Defective control board.

Defective wiring to light, fan & stirrer motor.

Check for open thermal fuse. Check oven cavity for arcing damage.

Note: Stirrer motor may rotate in either direction.

Models
56-9139
56-9439

EP 15

Display is illuminated. Oven light glows when the oven is in cook, but does not glow when the oven door is open.

Defective sense sw., or open wire from the control board to the sense sw..

Models
56-9139
56-9339
56-9439

EP 16

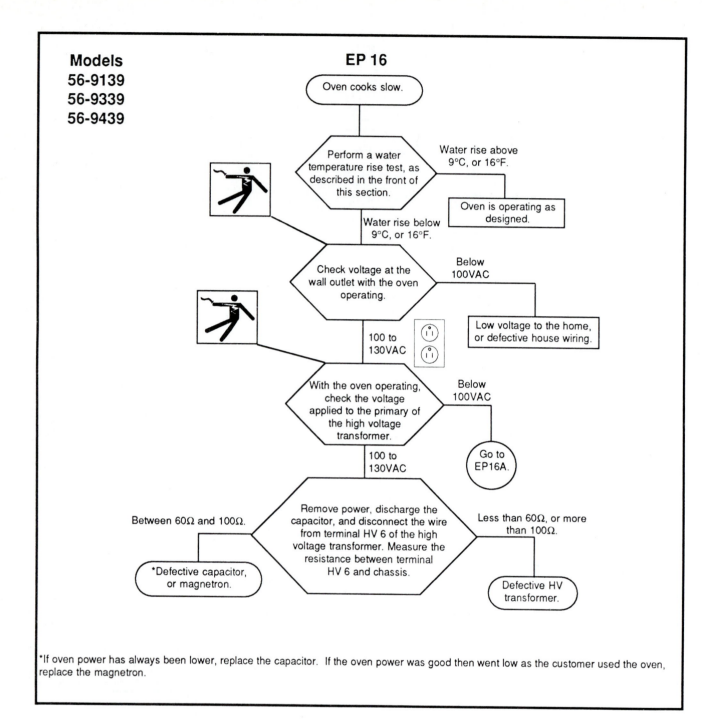

Oven cooks slow.

Perform a water temperature rise test, as described in the front of this section.

Water rise above 9°C, or 16°F.

Oven is operating as designed.

Water rise below 9°C, or 16°F.

Check voltage at the wall outlet with the oven operating.

Below 100VAC

Low voltage to the home, or defective house wiring.

100 to 130VAC

With the oven operating, check the voltage applied to the primary of the high voltage transformer.

Below 100VAC

Go to EP16A.

100 to 130VAC

Between 60Ω and 100Ω.

Remove power, discharge the capacitor, and disconnect the wire from terminal HV 6 of the high voltage transformer. Measure the resistance between terminal HV 6 and chassis.

Less than 60Ω, or more than 100Ω.

*Defective capacitor, or magnetron.

Defective HV transformer.

*If oven power has always been lower, replace the capacitor. If the oven power was good then went low as the customer used the oven, replace the magnetron.

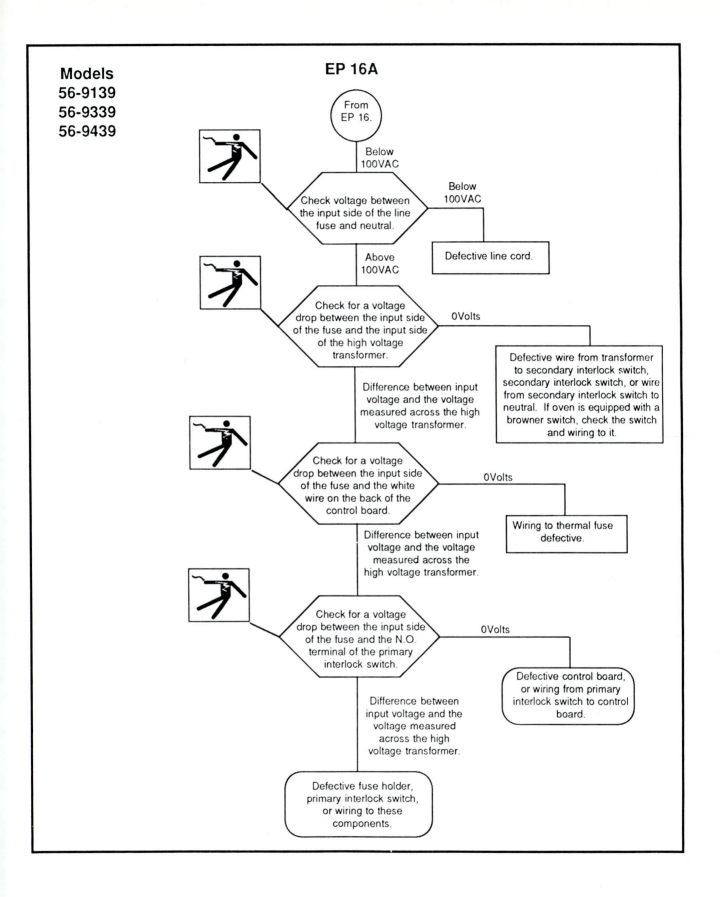

Models
56-9139
56-9339
56-9439

EP 16A

From EP 16.

Below 100VAC

Check voltage between the input side of the line fuse and neutral. — Below 100VAC → Defective line cord.

Above 100VAC

Check for a voltage drop between the input side of the fuse and the input side of the high voltage transformer. — 0Volts → Defective wire from transformer to secondary interlock switch, secondary interlock switch, or wire from secondary interlock switch to neutral. If oven is equipped with a browner switch, check the switch and wiring to it.

Difference between input voltage and the voltage measured across the high voltage transformer.

Check for a voltage drop between the input side of the fuse and the white wire on the back of the control board. — 0Volts → Wiring to thermal fuse defective.

Difference between input voltage and the voltage measured across the high voltage transformer.

Check for a voltage drop between the input side of the fuse and the N.O. terminal of the primary interlock switch. — 0Volts → Defective control board, or wiring from primary interlock switch to control board.

Difference between input voltage and the voltage measured across the high voltage transformer.

Defective fuse holder, primary interlock switch, or wiring to these components.

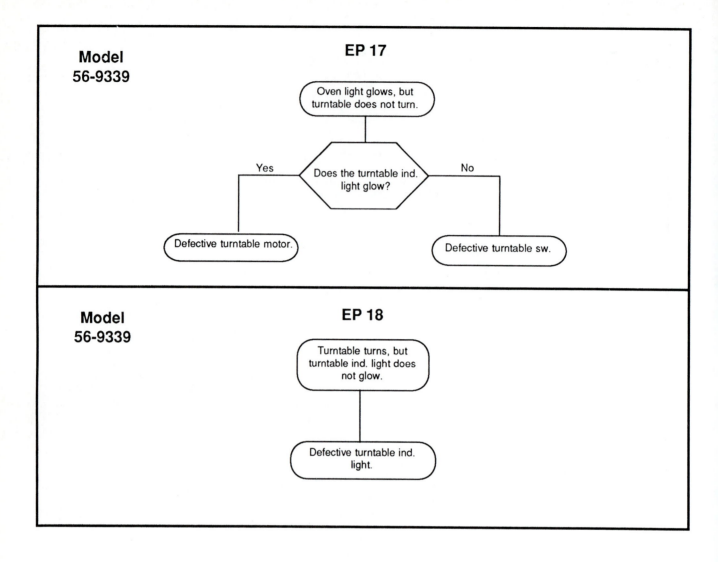

**Model
56-9339**

EP 17

Oven light glows, but turntable does not turn.

Does the turntable ind. light glow?

Yes

No

Defective turntable motor.

Defective turntable sw.

**Model
56-9339**

EP 18

Turntable turns, but turntable ind. light does not glow.

Defective turntable ind. light.

Models
56-9139
56-9439

EP 19

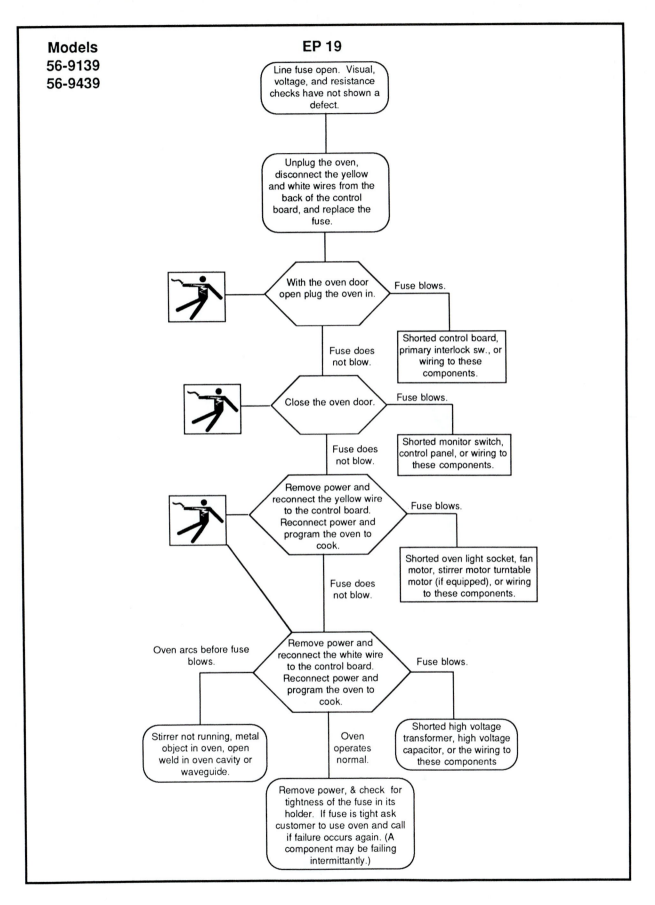

Line fuse open. Visual, voltage, and resistance checks have not shown a defect.

Unplug the oven, disconnect the yellow and white wires from the back of the control board, and replace the fuse.

With the oven door open plug the oven in.

Fuse blows.

Shorted control board, primary interlock sw., or wiring to these components.

Fuse does not blow.

Close the oven door.

Fuse blows.

Shorted monitor switch, control panel, or wiring to these components.

Fuse does not blow.

Remove power and reconnect the yellow wire to the control board. Reconnect power and program the oven to cook.

Fuse blows.

Shorted oven light socket, fan motor, stirrer motor turntable motor (if equipped), or wiring to these components.

Fuse does not blow.

Oven arcs before fuse blows.

Remove power and reconnect the white wire to the control board. Reconnect power and program the oven to cook.

Fuse blows.

Stirrer not running, metal object in oven, open weld in oven cavity or waveguide.

Oven operates normal.

Shorted high voltage transformer, high voltage capacitor, or the wiring to these components

Remove power, & check for tightness of the fuse in its holder. If fuse is tight ask customer to use oven and call if failure occurs again. (A component may be failing intermittantly.)

Model
56-9339

EP 20

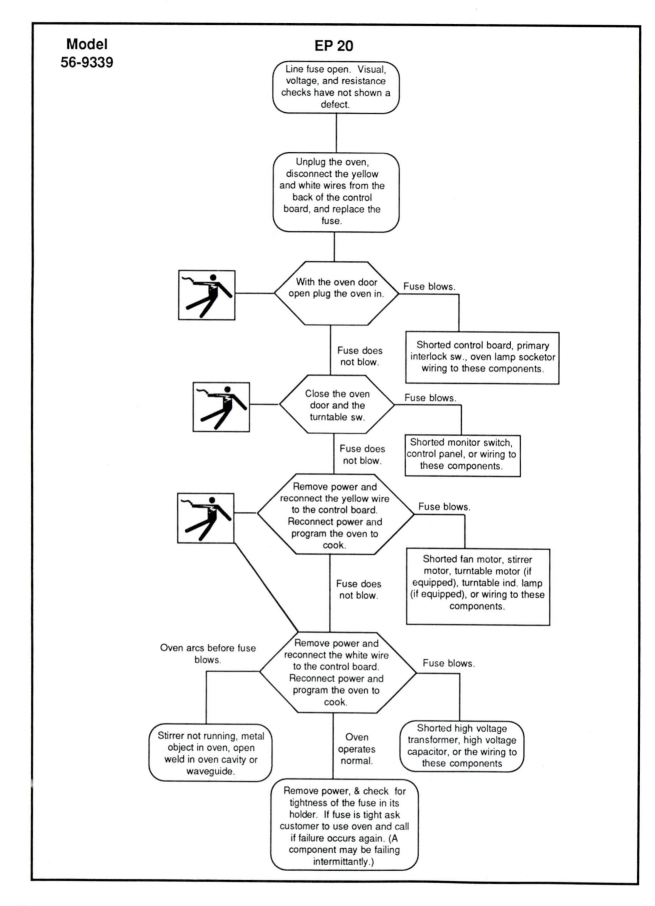

Line fuse open. Visual, voltage, and resistance checks have not shown a defect.

Unplug the oven, disconnect the yellow and white wires from the back of the control board, and replace the fuse.

With the oven door open plug the oven in.

Fuse blows.

Shorted control board, primary interlock sw., oven lamp socket or wiring to these components.

Fuse does not blow.

Close the oven door and the turntable sw.

Fuse blows.

Shorted monitor switch, control panel, or wiring to these components.

Fuse does not blow.

Remove power and reconnect the yellow wire to the control board. Reconnect power and program the oven to cook.

Fuse blows.

Shorted fan motor, stirrer motor, turntable motor (if equipped), turntable ind. lamp (if equipped), or wiring to these components.

Fuse does not blow.

Oven arcs before fuse blows.

Remove power and reconnect the white wire to the control board. Reconnect power and program the oven to cook.

Fuse blows.

Stirrer not running, metal object in oven, open weld in oven cavity or waveguide.

Oven operates normal.

Shorted high voltage transformer, high voltage capacitor, or the wiring to these components

Remove power, & check for tightness of the fuse in its holder. If fuse is tight ask customer to use oven and call if failure occurs again. (A component may be failing intermittantly.)

Chapter 18: Electric Wall Ovens

ELECTRICAL SPECIFICATIONS

ELECTRICAL SPECIFICATIONS												
Model	REG77BF*0		REG75WF*0		REG74BF*0		RG74BF*0		REG94BF*0		RG94BF*0	
Voltage	240	208	240	208	240	208	240	208	240	208	240	208
Component												
Broil Element												
Watts	3400	2560	3400	2560	3400	2560	3400	2560	3400	2560	3400	2560
Resistance (ohms)	16.9	16.9	16.9	16.9	16.9	16.9	16.9	16.9	16.9	16.9	16.9	16.9
Amps	14.1	12.3	14.1	12.3	14.1	12.3	14.1	12.3	14.1	12.3	14.1	12.3
Bake Element												
Watts	2100	1578	2100	1578	2100	1578	2100	1578	2100	1578	2100	1578
Resistance (ohms)	27.4	27.4	27.4	27.4	27.4	27.4	27.4	27.4	27.4	27.4	27.4	27.4
Amps	8.75	7.6	8.75	7.6	8.75	7.6	8.75	7.6	8.75	7.6	8.75	7.6
Electrical Rating												
Kilowatts	7.4	5.8	5.6	4.2	5.6	4.2	3.9	3.2	3.9	3.2	3.9	3.2
Amps	30.8	27.8	23.3	20.2	23.3	20.2	16.9	15.3	16.9	15.3	16.9	15.3

NOTE: Model REG77BF*0 has two ovens with identical elements.

ELECTRICAL DIAGRAMS INDEX

MODELS	WIRING PAGE NUMBER	SCHEMATIC PAGE NUMBER
REG77BF*0	8	9
REG75WF*0	10	11
REG74BF*0	10	11
RG74BF*0	12	13
REG94BF*0	14	15
RG94BF*0	12	13

> **NOTICE**
> When servicing a product that does not have an engineering code or lot number listed in this service manual, do not reference the technical specifications or electrical diagrams contained herein. Use the technical information supplied with the product. However, the repair procedures in this manual can be referenced.

MODEL/SERIAL NUMBERING SYSTEM

Model Number Coding System

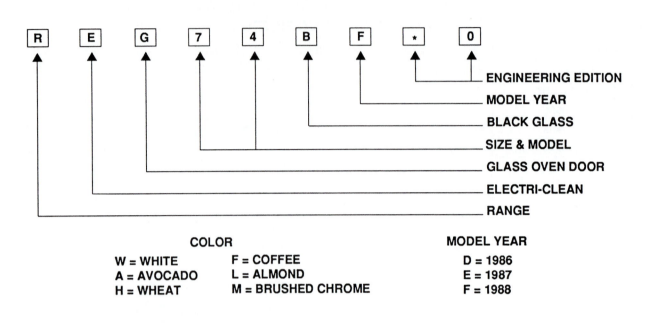

| R | E | G | 7 | 4 | B | F | * | 0 |

— ENGINEERING EDITION
— MODEL YEAR
— BLACK GLASS
— SIZE & MODEL
— GLASS OVEN DOOR
— ELECTRI-CLEAN
— RANGE

COLOR

W = WHITE   F = COFFEE
A = AVOCADO   L = ALMOND
H = WHEAT   M = BRUSHED CHROME

MODEL YEAR

D = 1986
E = 1987
F = 1988

Serial Number Coding System Before January 1, 1989

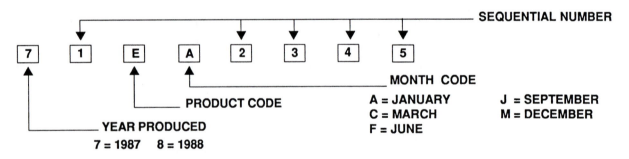

| 7 | 1 | E | A | 2 | 3 | 4 | 5 |

— SEQUENTIAL NUMBER
— MONTH CODE
— PRODUCT CODE
— YEAR PRODUCED

7 = 1987   8 = 1988

MONTH CODE

A = JANUARY   J = SEPTEMBER
C = MARCH   M = DECEMBER
F = JUNE

Serial Number Coding System After January 1, 1989

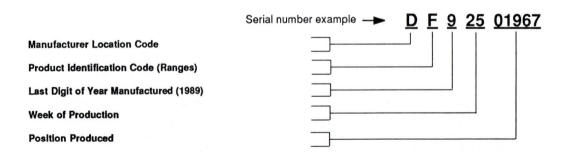

Serial number example → **D F 9 25 01967**

Manufacturer Location Code
Product Identification Code (Ranges)
Last Digit of Year Manufactured (1989)
Week of Production
Position Produced

WIRING DIAGRAM

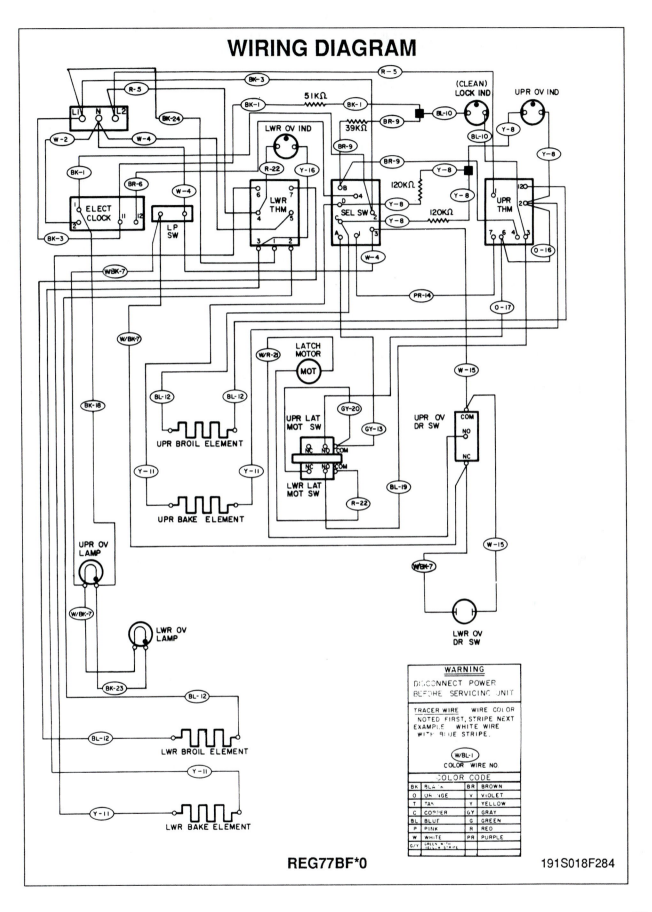

REG77BF*0

191S018F284

463

SCHEMATIC DIAGRAM

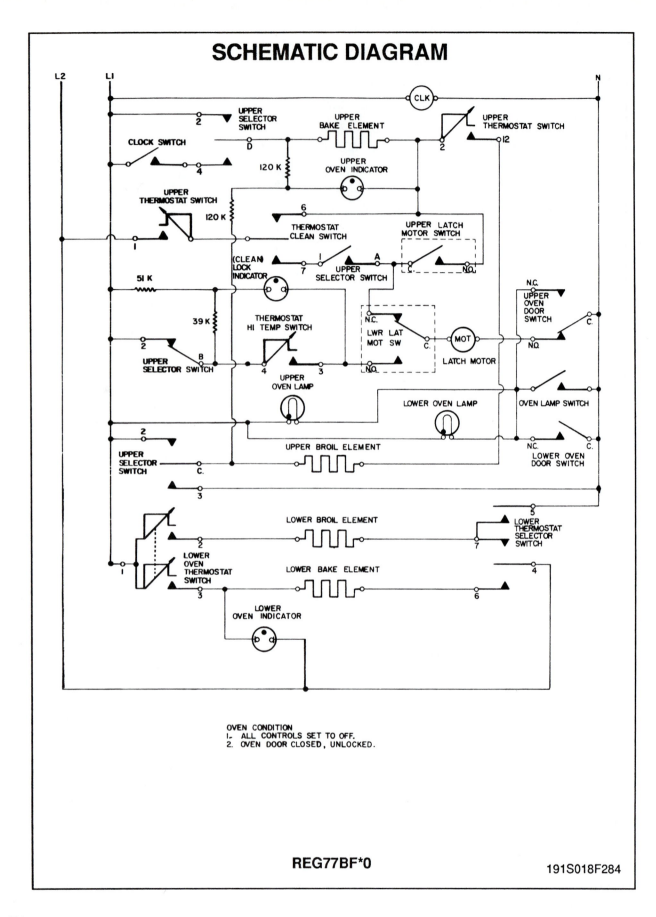

OVEN CONDITION
1. ALL CONTROLS SET TO OFF.
2. OVEN DOOR CLOSED, UNLOCKED.

REG77BF*0

191S018F284

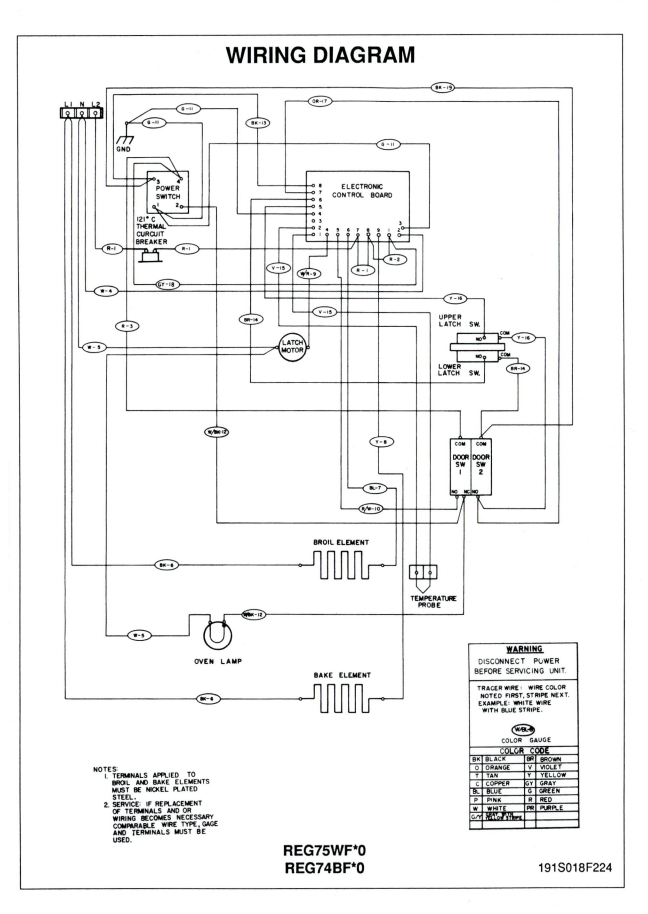

WIRING DIAGRAM

REG75WF*0
REG74BF*0

191S018F224

465

SCHEMATIC DIAGRAM

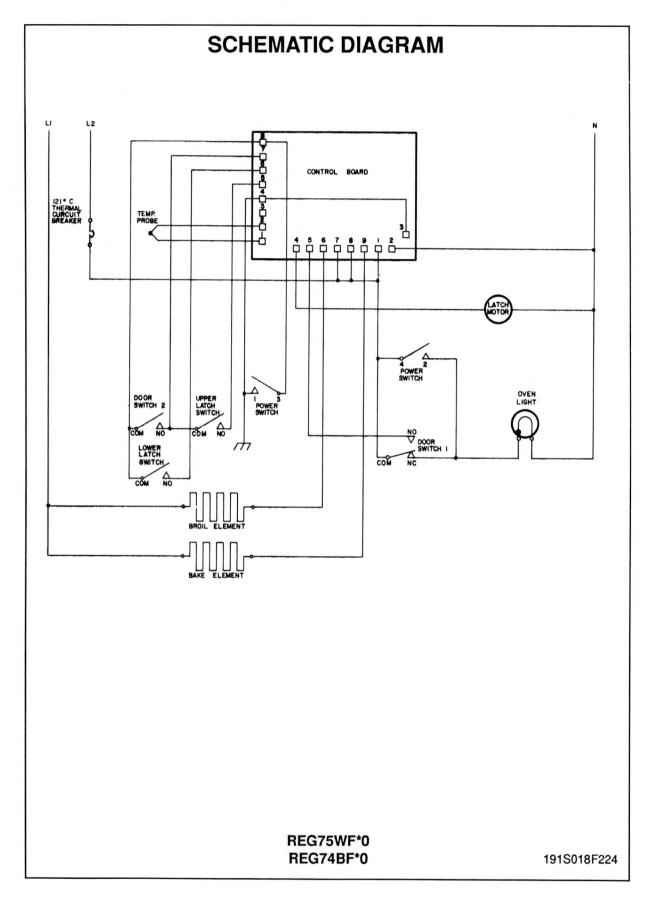

REG75WF*0
REG74BF*0

191S018F224

WIRING DIAGRAM

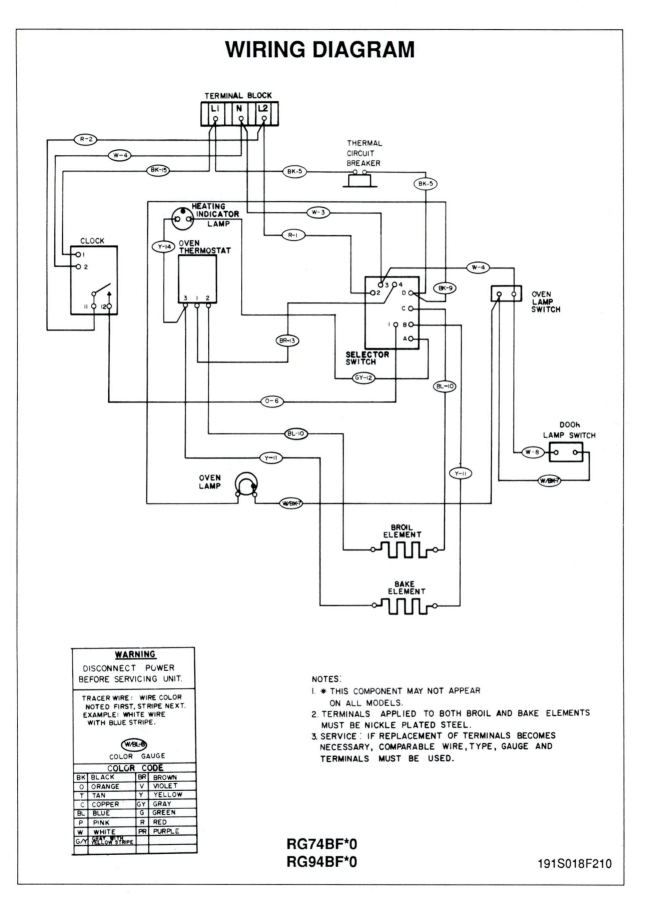

SCHEMATIC DIAGRAM

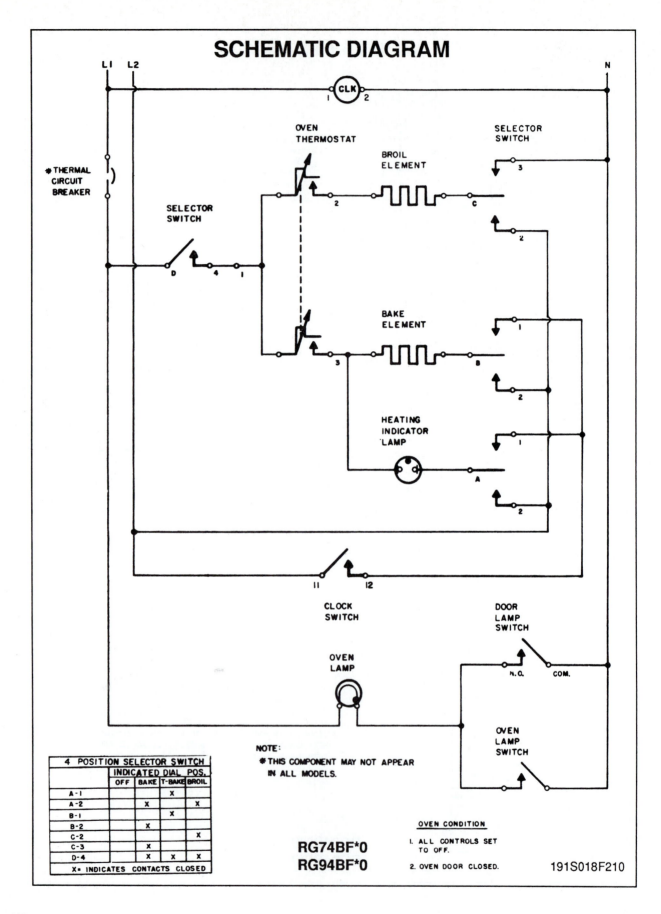

4 POSITION SELECTOR SWITCH				
	INDICATED DIAL POS.			
	OFF	BAKE	T-BAKE	BROIL
A-1			X	
A-2		X		X
B-1			X	
B-2		X		
C-2				X
C-3		X		
D-4		X	X	X
X= INDICATES CONTACTS CLOSED				

NOTE:

＊THIS COMPONENT MAY NOT APPEAR IN ALL MODELS.

RG74BF*0
RG94BF*0

OVEN CONDITION

1. ALL CONTROLS SET TO OFF.

2. OVEN DOOR CLOSED.

191S018F210

468

WIRING DIAGRAM

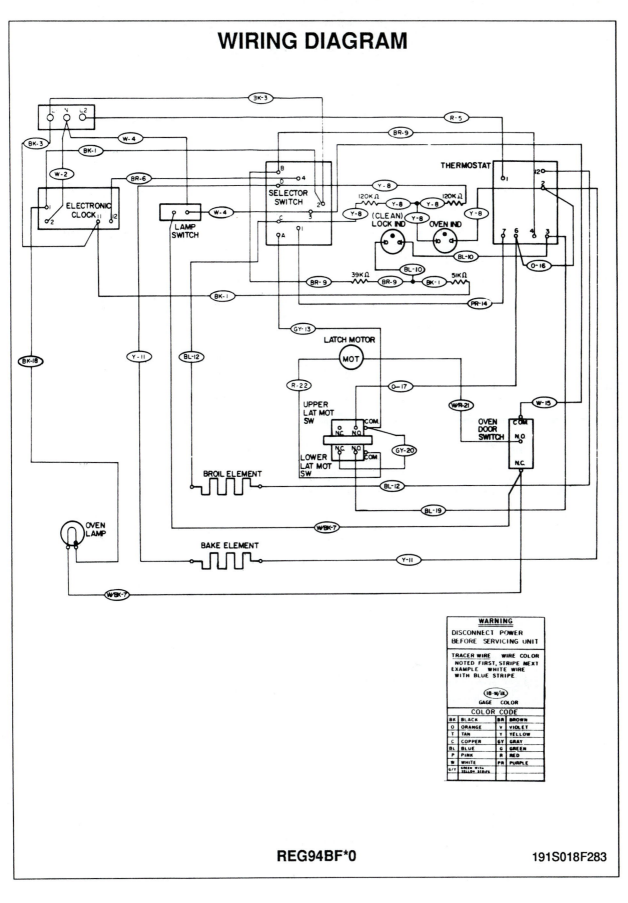

REG94BF*0

191S018F283

SCHEMATIC DIAGRAM

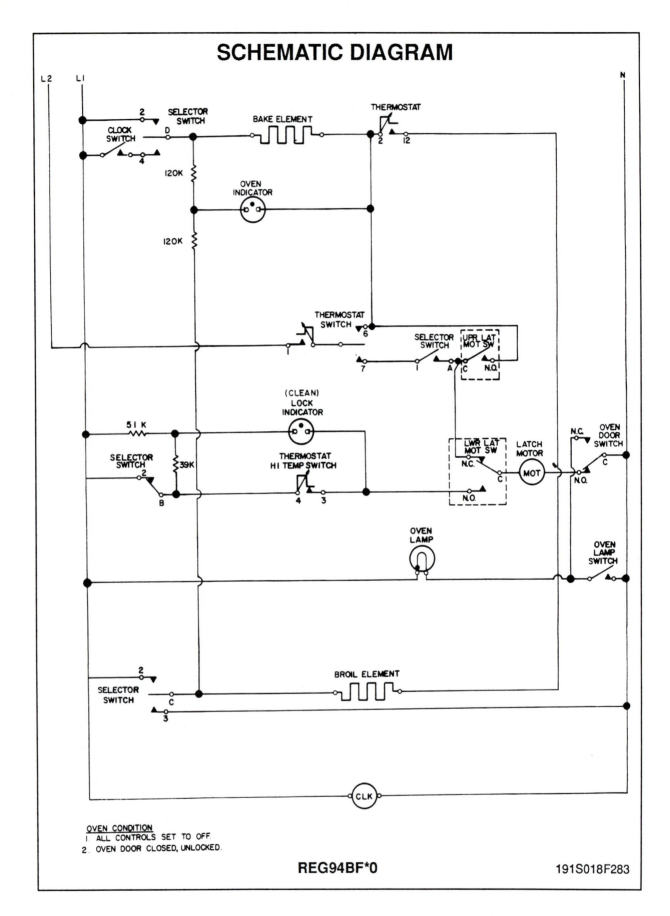

REG94BF*0   191S018F283

470

SECTION D - TROUBLESHOOTING

HIGH VOLTAGE IS PRESENT DURING TESTING!

1.   THE VOLTAGE VALUES STATED IN THE FOLLOWING FLOW CHARTS MAY VARY DEPENDING ON THE INCOMING POWER SUPPLY CONNECTED TO THE WALL OVEN.

2.   IT IS ASSUMED THAT THE INCOMING POWER SUPPLY IS PROPERLY POLARIZED, GROUNDED, AND WITHIN THE VOLTAGE RATING PRINTED ON THE MODEL/SERIAL PLATE.

⚠ WARNING
TESTS IN THIS SECTION SHOULD ONLY BE
PERFORMED BY A QUALIFIED TECHNICIAN.

FLOW CHART SYMBOLS

What is a flow chart? A diagram that shows step-by-step progression through a procedure using connecting lines and a set of symbols.

Why use flow charts? To expedite the diagnosis of a product. A flow chart gives the proper test procedure (with the least amount of written information) that will eventually lead to a faulty component or system failure. (The flow charts used in this manual lead to a single component failure.)

Why are symbols used in a flow chart? Flow chart symbols have definite meanings. Each symbol is a key to that particular step in the diagnostic procedure. Once these symbols are understood, the diagnosis goes much faster. The following symbols and their meanings are used in this manual.

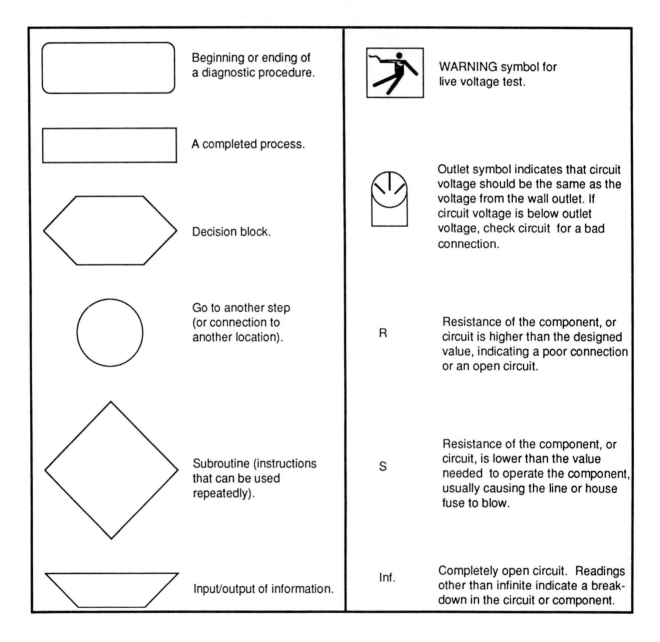

(rounded rectangle)	Beginning or ending of a diagnostic procedure.
(rectangle)	A completed process.
(hexagon)	Decision block.
(circle)	Go to another step (or connection to another location).
(diamond)	Subroutine (instructions that can be used repeatedly).
(trapezoid)	Input/output of information.
(warning figure)	WARNING symbol for live voltage test.
(outlet symbol)	Outlet symbol indicates that circuit voltage should be the same as the voltage from the wall outlet. If circuit voltage is below outlet voltage, check circuit for a bad connection.
R	Resistance of the component, or circuit is higher than the designed value, indicating a poor connection or an open circuit.
S	Resistance of the component, or circuit, is lower than the value needed to operate the component, usually causing the line or house fuse to blow.
Inf.	Completely open circuit. Readings other than infinite indicate a breakdown in the circuit or component.

FLOW CHART ENTRY POINTS (EP)						
PROBLEM	MODEL NUMBERS					
	REG77BF*0	REG75WF*0	REG74BF*0	RG74BF*0	REG94BF*0	RG94BF*0
Broil element does not heat.	EP 1 Pg D4	EP23 Pg D20	EP23 Pg 20	EP 12 Pg D14	EP 11 Pg D13	EP 12 Pg D14
Bake element not heating or insufficient heat.	EP 2 Pg D5	EP24 Pg D21	EP24 Pg D21	EP 13 Pg D15	EP 8 Pg D10	EP 13 Pg D15
Timed bake feature not working.	EP 4 Pg D6	EP25 Pg D22	EP25 Pg D22	EP 14 Pg D16	EP 9 Pg D11	EP 14 Pg D16
Self-clean feature not working.	EP 5 Pg D7	EP26 Pg D23	EP26 Pg D23	NA	EP 10 Pg D12	NA
Lower oven's broil element not heating.	EP 6 Pg D8	NA	NA	NA	NA	NA
Lower oven's bake element not heating or insufficient heat.	EP 7 Pg D9	NA	NA	NA	NA	NA
Displayed fault conditions.	NA	EP20 Pg D17	EP20 Pg D17	NA	NA	NA
Quick check: Electronic Oven Control	NA	EP21 Pg D18	EP21 Pg D18	NA	NA	NA
No word 'Lock' in display.	NA	EP22 Pg D19	EP22 Pg D19	NA	NA	NA
No display in EOC.	NA	EP27 Pg D24	EP27 Pg D24	NA	NA	NA
Door switch lock circuit.	NA	EP28 Pg D25	EP28 Pg D25	NA	NA	NA
Checking RTD.	NA	EP29 Pg D25	EP29 Pg D25	NA	NA	NA
Calibrating the EOC.	NA	EP30 Pg D26	EP30 Pg D26	NA	NA	NA

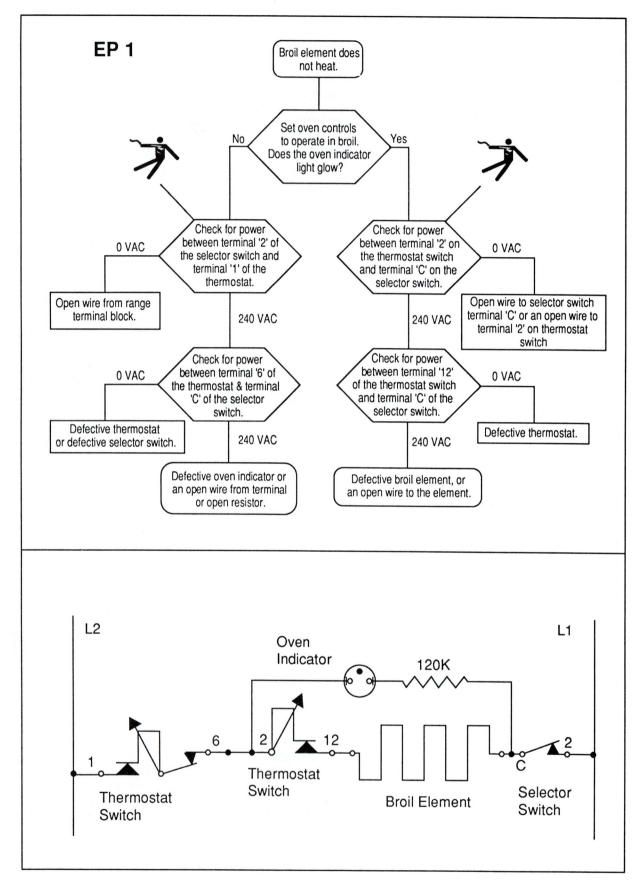

EP 1

Broil element does not heat.

Set oven controls to operate in broil. Does the oven indicator light glow?

No — Check for power between terminal '2' of the selector switch and terminal '1' of the thermostat.

Yes — Check for power between terminal '2' on the thermostat switch and terminal 'C' on the selector switch.

0 VAC — Open wire from range terminal block.

240 VAC — Check for power between terminal '6' of the thermostat & terminal 'C' of the selector switch.

0 VAC — Defective thermostat or defective selector switch.

240 VAC — Defective oven indicator or an open wire from terminal or open resistor.

0 VAC — Open wire to selector switch terminal 'C' or an open wire to terminal '2' on thermostat switch

240 VAC — Check for power between terminal '12' of the thermostat switch and terminal 'C' of the selector switch.

0 VAC — Defective thermostat.

240 VAC — Defective broil element, or an open wire to the element.

L2   Oven Indicator   120K   L1

1   Thermostat Switch   6   2   Thermostat Switch   12   Broil Element   C   Selector Switch   2

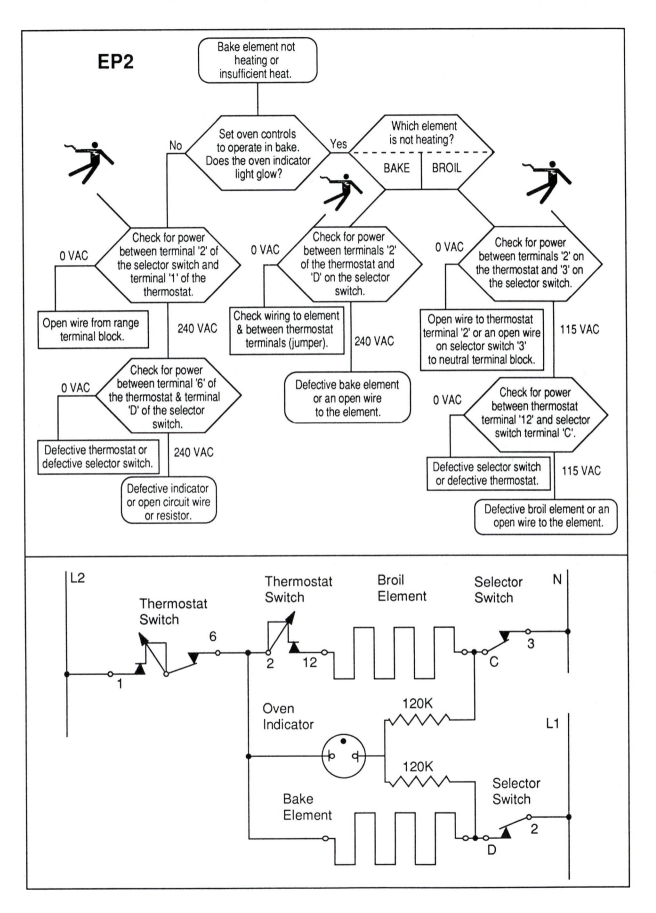

EP2

Bake element not heating or insufficient heat.

Set oven controls to operate in bake. Does the oven indicator light glow?

No / Yes

Which element is not heating?

BAKE | BROIL

Check for power between terminal '2' of the selector switch and terminal '1' of the thermostat.

0 VAC

Open wire from range terminal block.

240 VAC

Check for power between terminal '6' of the thermostat & terminal 'D' of the selector switch.

0 VAC

Defective thermostat or defective selector switch.

240 VAC

Defective indicator or open circuit wire or resistor.

Check for power between terminals '2' of the thermostat and 'D' on the selector switch.

0 VAC

Check wiring to element & between thermostat terminals (jumper).

240 VAC

Defective bake element or an open wire to the element.

Check for power between terminals '2' on the thermostat and '3' on the selector switch.

0 VAC

Open wire to thermostat terminal '2' or an open wire on selector switch '3' to neutral terminal block.

115 VAC

Check for power between thermostat terminal '12' and selector switch terminal 'C'.

0 VAC

Defective selector switch or defective thermostat.

115 VAC

Defective broil element or an open wire to the element.

L2

Thermostat Switch

Thermostat Switch

Broil Element

Selector Switch

N

6

2   12

C

3

Oven Indicator

120K

L1

120K

Bake Element

Selector Switch

1

D

2

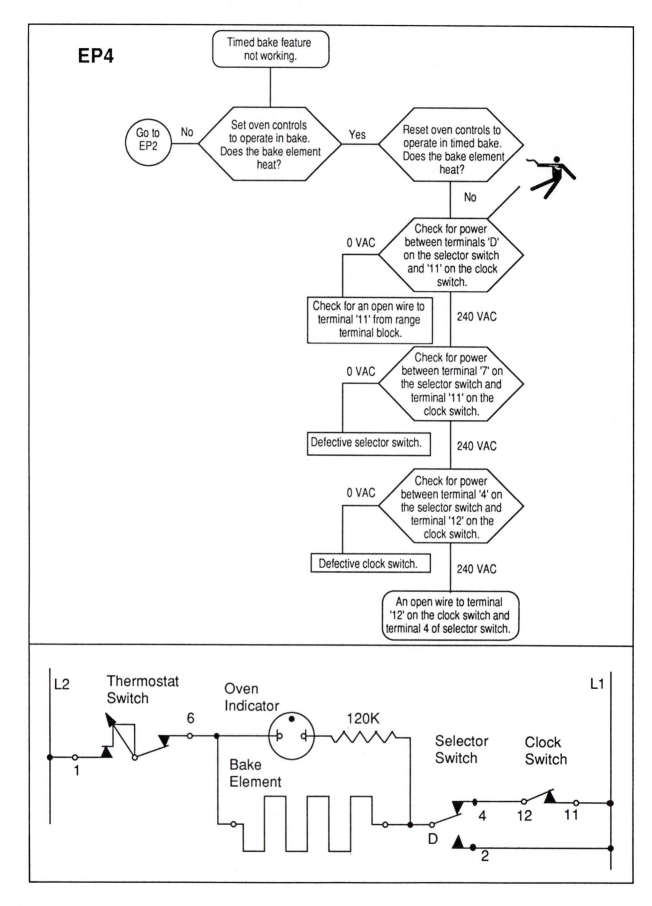

EP4

Timed bake feature not working.

Go to EP2 ← No — Set oven controls to operate in bake. Does the bake element heat? — Yes → Reset oven controls to operate in timed bake. Does the bake element heat?

No

Check for power between terminals 'D' on the selector switch and '11' on the clock switch.

0 VAC → Check for an open wire to terminal '11' from range terminal block.

240 VAC

Check for power between terminal '7' on the selector switch and terminal '11' on the clock switch.

0 VAC → Defective selector switch.

240 VAC

Check for power between terminal '4' on the selector switch and terminal '12' on the clock switch.

0 VAC → Defective clock switch.

240 VAC

An open wire to terminal '12' on the clock switch and terminal 4 of selector switch.

L2   Thermostat Switch   Oven Indicator   120K   Selector Switch   Clock Switch   L1

6   Bake Element   4   12   11

1   D   2

476

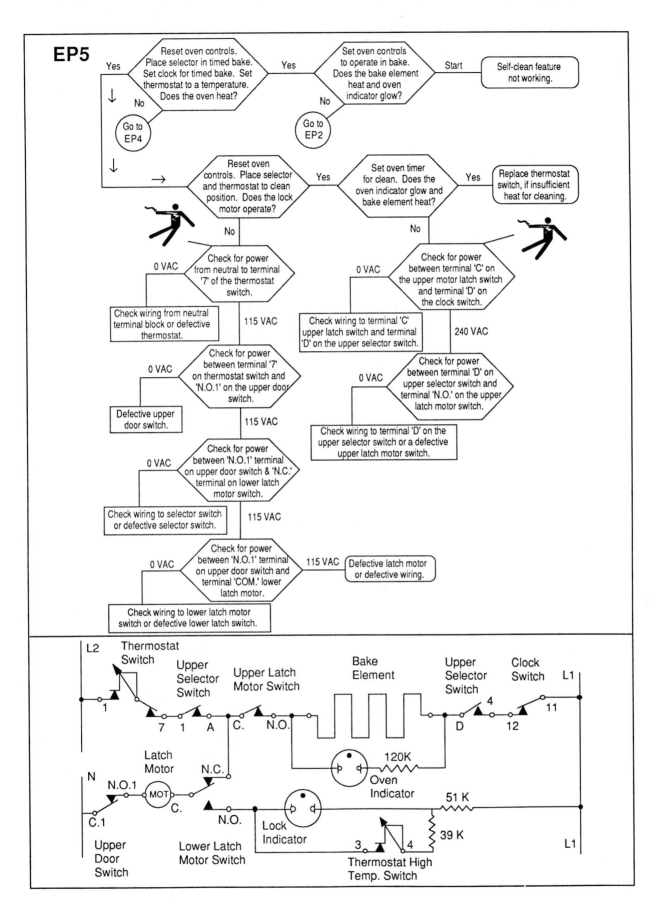

EP5

Self-clean feature not working. **Start** → Set oven controls to operate in bake. Does the bake element heat and oven indicator glow?

— **Yes** → Reset oven controls. Place selector in timed bake. Set clock for timed bake. Set thermostat to a temperature. Does the oven heat?

— **No** → Go to EP2

Yes ↓

No → Go to EP4

↓ →

Reset oven controls. Place selector and thermostat to clean position. Does the lock motor operate? — **Yes** → Set oven timer for clean. Does the oven indicator glow and bake element heat? — **Yes** → Replace thermostat switch, if insufficient heat for cleaning.

No

Check for power from neutral to terminal '7' of the thermostat switch.
- **0 VAC** → Check wiring from neutral terminal block or defective thermostat.
- **115 VAC**

Check for power between terminal '7' on thermostat switch and 'N.O.1' on the upper door switch.
- **0 VAC** → Defective upper door switch.
- **115 VAC**

Check for power between 'N.O.1' terminal on upper door switch & 'N.C.' terminal on lower latch motor switch.
- **0 VAC** → Check wiring to selector switch or defective selector switch.
- **115 VAC**

Check for power between 'N.O.1' terminal on upper door switch and terminal 'COM.' lower latch motor.
- **0 VAC** → Check wiring to lower latch motor switch or defective lower latch switch.
- **115 VAC** → Defective latch motor or defective wiring.

No

Check for power between terminal 'C' on the upper motor latch switch and terminal 'D' on the clock switch.
- **0 VAC** → Check wiring to terminal 'C' upper latch switch and terminal 'D' on the upper selector switch.
- **240 VAC**

Check for power between terminal 'D' on upper selector switch and terminal 'N.O.' on the upper latch motor switch.
- **0 VAC** → Check wiring to terminal 'D' on the upper selector switch or a defective upper latch motor switch.

L2 | Thermostat Switch | Upper Selector Switch | Upper Latch Motor Switch | Bake Element | Upper Selector Switch | Clock Switch | L1

1 · 7 1 A C. N.O. · · D 4 12 11

Latch Motor · N.C.

N · N.O.1 · MOT · C. · N.O. · 120K · Oven Indicator

C.1 · Lock Indicator · 51 K

Upper Door Switch · Lower Latch Motor Switch · 3 4 · 39 K · L1

Thermostat High Temp. Switch

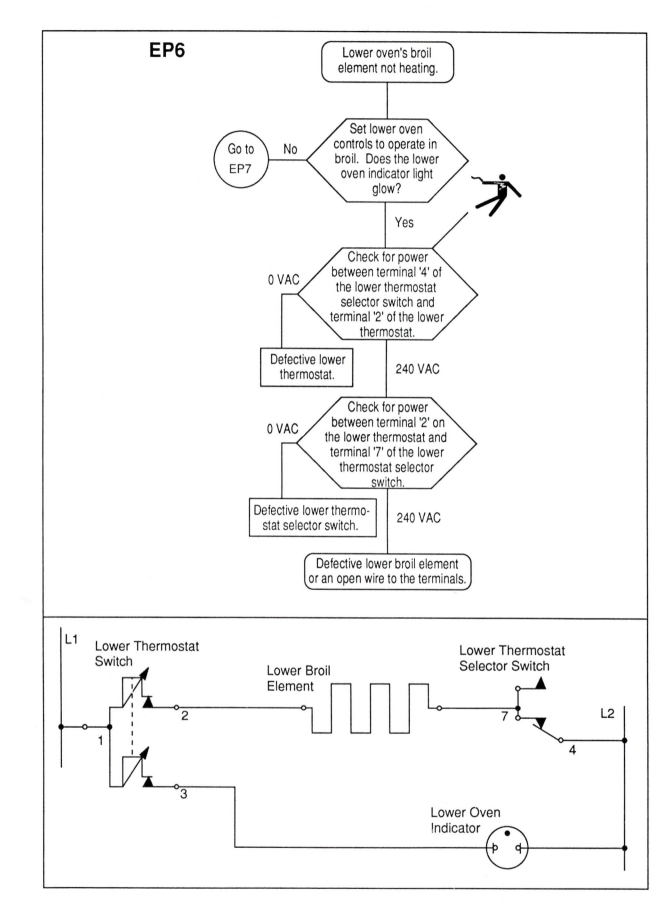

EP6

Lower oven's broil element not heating.

Set lower oven controls to operate in broil. Does the lower oven indicator light glow?

No → Go to EP7

Yes

Check for power between terminal '4' of the lower thermostat selector switch and terminal '2' of the lower thermostat.

0 VAC → Defective lower thermostat.

240 VAC

Check for power between terminal '2' on the lower thermostat and terminal '7' of the lower thermostat selector switch.

0 VAC → Defective lower thermostat selector switch.

240 VAC

Defective lower broil element or an open wire to the terminals.

L1

Lower Thermostat Switch

Lower Broil Element

Lower Thermostat Selector Switch

L2

2

1

3

7

4

Lower Oven Indicator

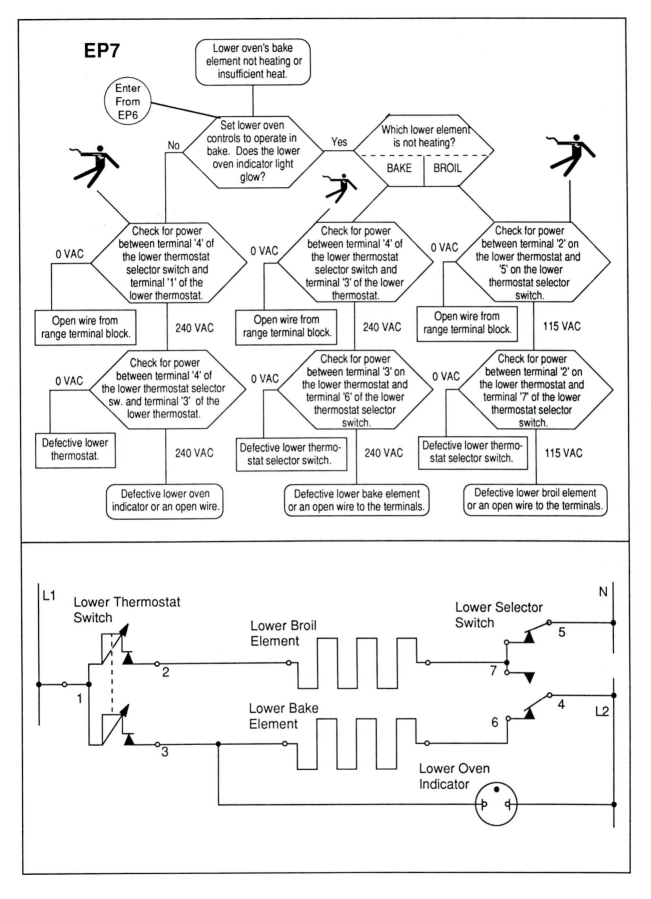

EP7

Lower oven's bake element not heating or insufficient heat.

Enter From EP6

Set lower oven controls to operate in bake. Does the lower oven indicator light glow?

No / Yes

Which lower element is not heating?

BAKE / BROIL

Check for power between terminal '4' of the lower thermostat selector switch and terminal '1' of the lower thermostat.

0 VAC

Open wire from range terminal block.

240 VAC

Check for power between terminal '4' of the lower thermostat selector switch and terminal '3' of the lower thermostat.

0 VAC

Open wire from range terminal block.

240 VAC

Check for power between terminal '2' on the lower thermostat and '5' on the lower thermostat selector switch.

0 VAC

Open wire from range terminal block.

115 VAC

Check for power between terminal '4' of the lower thermostat selector sw. and terminal '3' of the lower thermostat.

0 VAC

Defective lower thermostat.

240 VAC

Check for power between terminal '3' on the lower thermostat and terminal '6' of the lower thermostat selector switch.

0 VAC

Defective lower thermostat selector switch.

240 VAC

Check for power between terminal '2' on the lower thermostat and terminal '7' of the lower thermostat selector switch.

0 VAC

Defective lower thermostat selector switch.

115 VAC

Defective lower oven indicator or an open wire.

Defective lower bake element or an open wire to the terminals.

Defective lower broil element or an open wire to the terminals.

L1

Lower Thermostat Switch

Lower Broil Element

Lower Selector Switch

N

5

2

7

1

Lower Bake Element

6

4

L2

3

Lower Oven Indicator

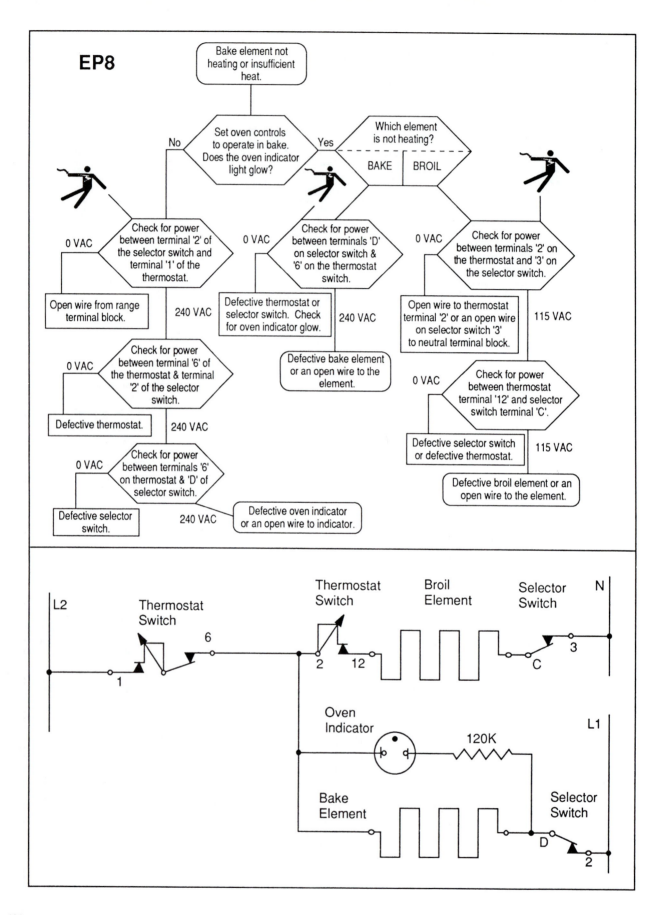

EP8

Bake element not heating or insufficient heat.

Set oven controls to operate in bake. Does the oven indicator light glow?

No / **Yes**

Which element is not heating?

BAKE | BROIL

Check for power between terminal '2' of the selector switch and terminal '1' of the thermostat.

0 VAC — Open wire from range terminal block.

240 VAC

Check for power between terminals 'D' on selector switch & '6' on the thermostat switch.

0 VAC — Defective thermostat or selector switch. Check for oven indicator glow.

240 VAC

Check for power between terminals '2' on the thermostat and '3' on the selector switch.

0 VAC — Open wire to thermostat terminal '2' or an open wire on selector switch '3' to neutral terminal block.

115 VAC

Check for power between terminal '6' of the thermostat & terminal '2' of the selector switch.

0 VAC — Defective thermostat.

240 VAC

Defective bake element or an open wire to the element.

Check for power between thermostat terminal '12' and selector switch terminal 'C'.

0 VAC — Defective selector switch or defective thermostat.

115 VAC

Check for power between terminals '6' on thermostat & 'D' of selector switch.

0 VAC — Defective selector switch.

240 VAC — Defective oven indicator or an open wire to indicator.

Defective broil element or an open wire to the element.

L2

Thermostat Switch

6

Thermostat Switch

Broil Element

Selector Switch

N

2 12

C 3

1

Oven Indicator

120K

L1

Bake Element

Selector Switch

D

2

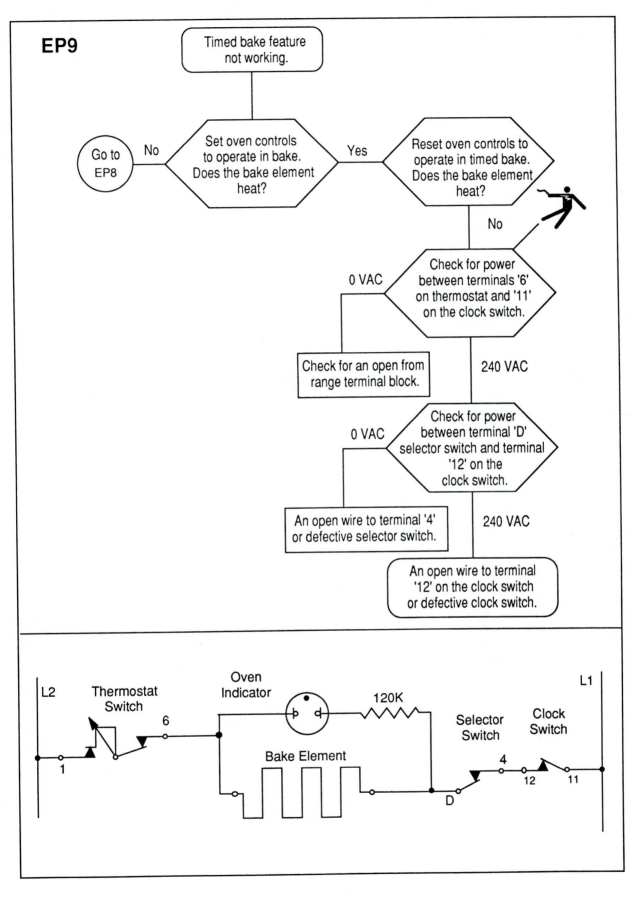

EP9

Timed bake feature not working.

Go to EP8 ←No— Set oven controls to operate in bake. Does the bake element heat? —Yes→ Reset oven controls to operate in timed bake. Does the bake element heat?

No

Check for power between terminals '6' on thermostat and '11' on the clock switch.

0 VAC → Check for an open from range terminal block.

240 VAC

Check for power between terminal 'D' selector switch and terminal '12' on the clock switch.

0 VAC → An open wire to terminal '4' or defective selector switch.

240 VAC

An open wire to terminal '12' on the clock switch or defective clock switch.

L2

Thermostat Switch

Oven Indicator

120K

6

Bake Element

Selector Switch

Clock Switch

1

4

12

11

D

L1

481

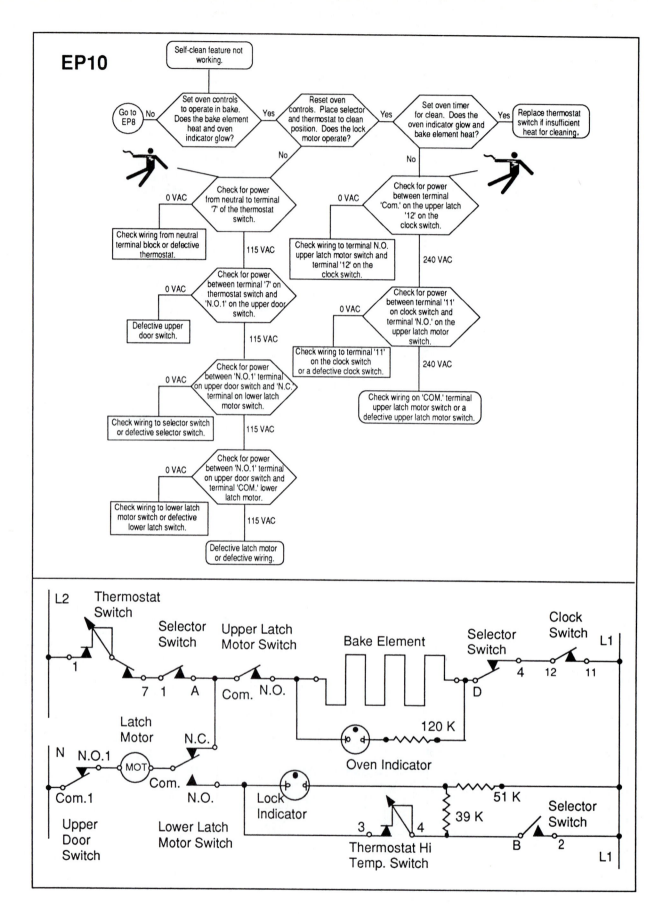

EP10

Self-clean feature not working.

Set oven controls to operate in bake. Does the bake element heat and oven indicator glow? — No → Go to EP8

Yes ↓

Reset oven controls. Place selector and thermostat to clean position. Does the lock motor operate? — Yes → Set oven timer for clean. Does the oven indicator glow and bake element heat? — Yes → Replace thermostat switch if insufficient heat for cleaning.

No ↓ (left) Check for power from neutral to terminal '7' of the thermostat switch.
- 0 VAC → Check wiring from neutral terminal block or defective thermostat.
- 115 VAC ↓

Check for power between terminal '7' on thermostat switch and 'N.O.1' on the upper door switch.
- 0 VAC → Defective upper door switch.
- 115 VAC ↓

Check for power between 'N.O.1' terminal on upper door switch and 'N.C.' terminal on lower latch motor switch.
- 0 VAC → Check wiring to selector switch or defective selector switch.
- 115 VAC ↓

Check for power between 'N.O.1' terminal on upper door switch and terminal 'COM.' lower latch motor.
- 0 VAC → Check wiring to lower latch motor switch or defective lower latch switch.
- 115 VAC ↓
Defective latch motor or defective wiring.

No ↓ (right) Check for power between terminal 'Com.' on the upper latch '12' on the clock switch.
- 0 VAC → Check wiring to terminal N.O. upper latch motor switch and terminal '12' on the clock switch.
- 240 VAC ↓

Check for power between terminal '11' on clock switch and terminal 'N.O.' on the upper latch motor switch.
- 0 VAC → Check wiring to terminal '11' on the clock switch or a defective clock switch.
- 240 VAC ↓
Check wiring on 'COM.' terminal upper latch motor switch or a defective upper latch motor switch.

Thermostat Switch — L2
Selector Switch
Upper Latch Motor Switch
Bake Element
Selector Switch
Clock Switch — L1
1   7 1 A Com. N.O.   D   4 12 11

Latch Motor
N.C.
120 K
Oven Indicator

N N.O.1
Com.1
Upper Door Switch

MOT Com. N.O.
Lower Latch Motor Switch

Lock Indicator

3   4
Thermostat Hi Temp. Switch

39 K   51 K

Selector Switch
B   2
L1

482

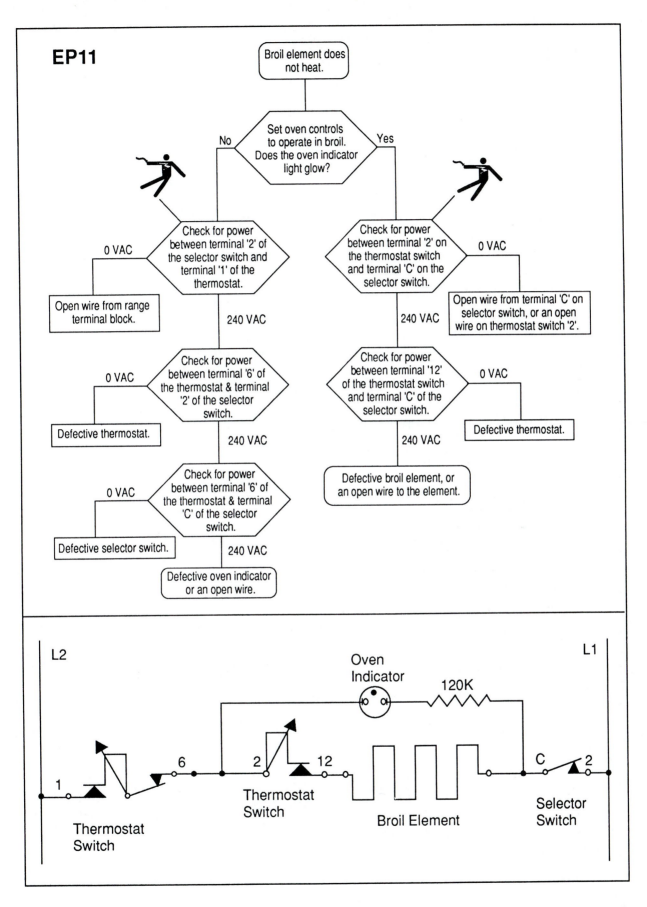

EP11

Broil element does not heat.

Set oven controls to operate in broil. Does the oven indicator light glow?

No / Yes

No branch:

Check for power between terminal '2' of the selector switch and terminal '1' of the thermostat.

0 VAC → Open wire from range terminal block.

240 VAC →

Check for power between terminal '6' of the thermostat & terminal '2' of the selector switch.

0 VAC → Defective thermostat.

240 VAC →

Check for power between terminal '6' of the thermostat & terminal 'C' of the selector switch.

0 VAC → Defective selector switch.

240 VAC → Defective oven indicator or an open wire.

Yes branch:

Check for power between terminal '2' on the thermostat switch and terminal 'C' on the selector switch.

0 VAC → Open wire from terminal 'C' on selector switch, or an open wire on thermostat switch '2'.

240 VAC →

Check for power between terminal '12' of the thermostat switch and terminal 'C' of the selector switch.

0 VAC → Defective thermostat.

240 VAC → Defective broil element, or an open wire to the element.

L2   L1

Oven Indicator   120K

1   Thermostat Switch   6   2   Thermostat Switch   12   Broil Element   C   2   Selector Switch

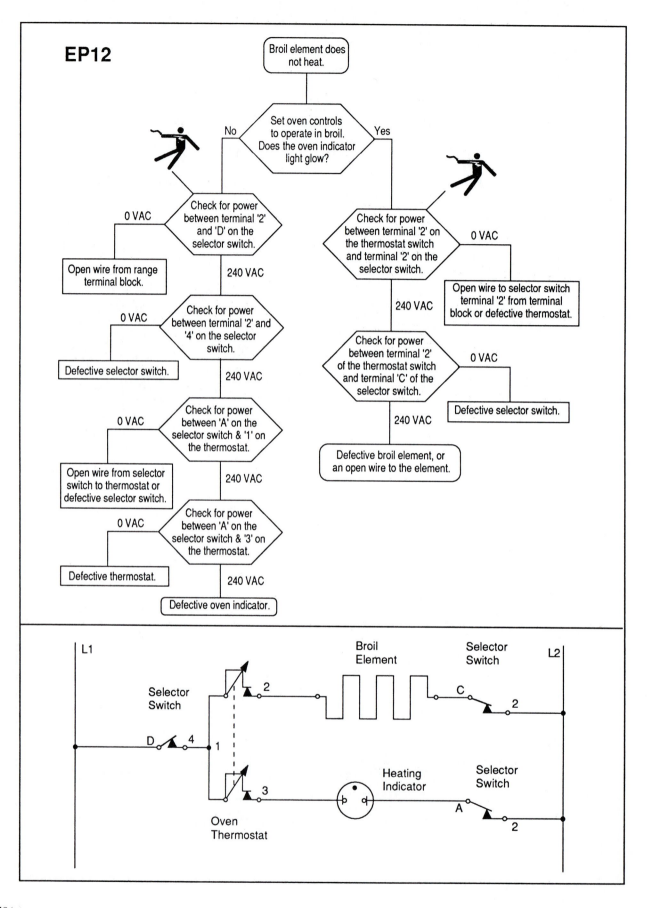

EP12

Broil element does not heat.

Set oven controls to operate in broil. Does the oven indicator light glow?

No — Check for power between terminal '2' and 'D' on the selector switch.

0 VAC — Open wire from range terminal block.

240 VAC — Check for power between terminal '2' and '4' on the selector switch.

0 VAC — Defective selector switch.

240 VAC — Check for power between 'A' on the selector switch & '1' on the thermostat.

0 VAC — Open wire from selector switch to thermostat or defective selector switch.

240 VAC — Check for power between 'A' on the selector switch & '3' on the thermostat.

0 VAC — Defective thermostat.

240 VAC — Defective oven indicator.

Yes — Check for power between terminal '2' on the thermostat switch and terminal '2' on the selector switch.

0 VAC — Open wire to selector switch terminal '2' from terminal block or defective thermostat.

240 VAC — Check for power between terminal '2' of the thermostat switch and terminal 'C' of the selector switch.

0 VAC — Defective selector switch.

240 VAC — Defective broil element, or an open wire to the element.

L1

Selector Switch

Broil Element

Selector Switch

L2

D 4 1 2

C 2

Oven Thermostat

Heating Indicator

Selector Switch

3

A 2

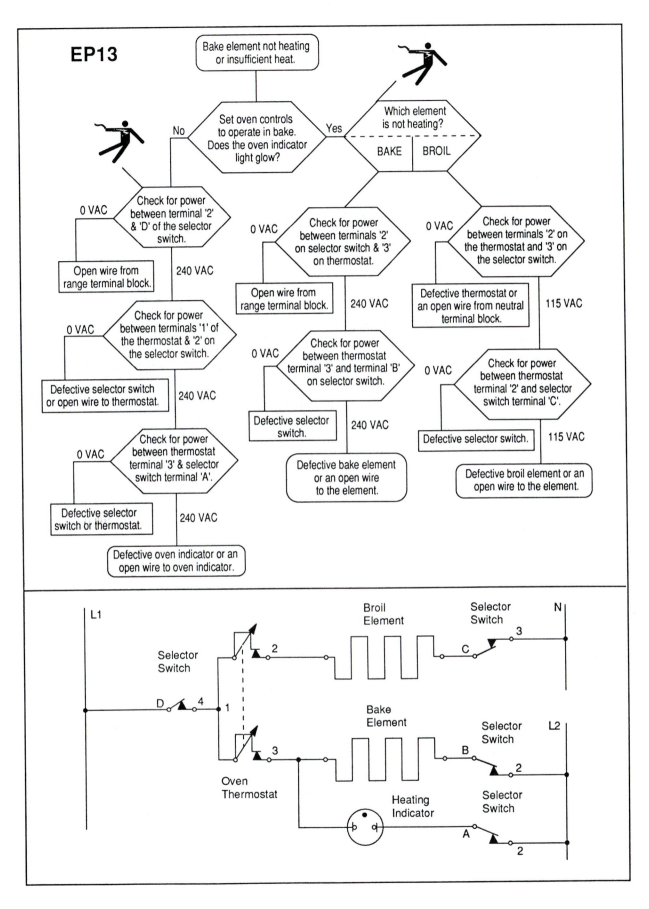

EP13

Bake element not heating or insufficient heat.

Set oven controls to operate in bake. Does the oven indicator light glow?

No / **Yes**

Which element is not heating?

BAKE | BROIL

Check for power between terminal '2' & 'D' of the selector switch.
- 0 VAC → Open wire from range terminal block.
- 240 VAC →

Check for power between terminals '1' of the thermostat & '2' on the selector switch.
- 0 VAC → Defective selector switch or open wire to thermostat.
- 240 VAC →

Check for power between thermostat terminal '3' & selector switch terminal 'A'.
- 0 VAC → Defective selector switch or thermostat.
- 240 VAC → Defective oven indicator or an open wire to oven indicator.

Check for power between terminals '2' on selector switch & '3' on thermostat.
- 0 VAC → Open wire from range terminal block.
- 240 VAC →

Check for power between thermostat terminal '3' and terminal 'B' on selector switch.
- 0 VAC → Defective selector switch.
- 240 VAC → Defective bake element or an open wire to the element.

Check for power between terminals '2' on the thermostat and '3' on the selector switch.
- 0 VAC → Defective thermostat or an open wire from neutral terminal block.
- 115 VAC →

Check for power between thermostat terminal '2' and selector switch terminal 'C'.
- 0 VAC → Defective selector switch.
- 115 VAC → Defective broil element or an open wire to the element.

L1

Selector Switch

D — 4 — 1

Oven Thermostat

Broil Element

Selector Switch

2

C — 3 — N

Bake Element

Selector Switch

B

2 — L2

Heating Indicator

Selector Switch

A

2

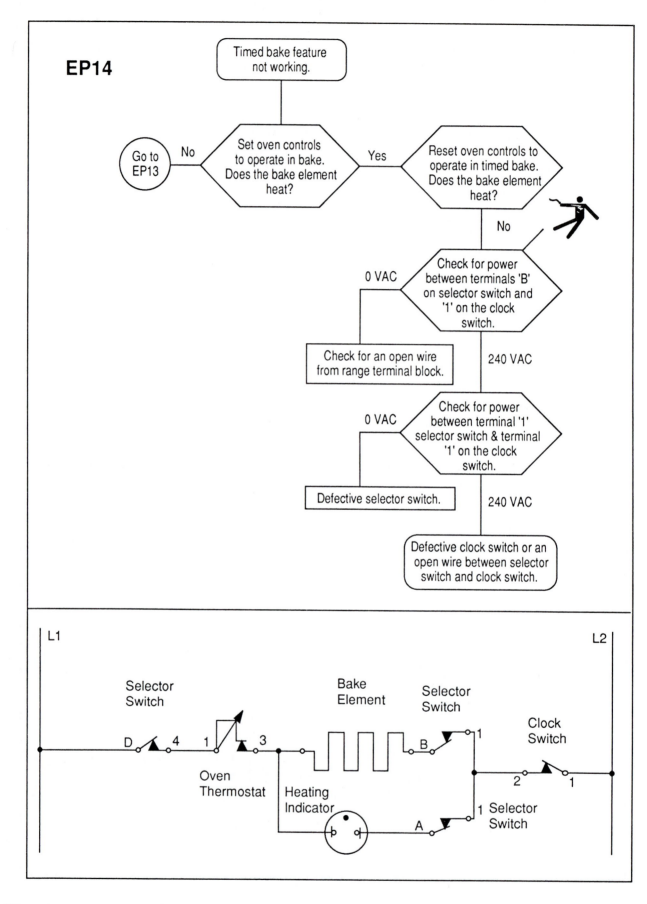

EP14

Timed bake feature not working.

Go to EP13 — No — Set oven controls to operate in bake. Does the bake element heat? — Yes — Reset oven controls to operate in timed bake. Does the bake element heat?

No

Check for power between terminals 'B' on selector switch and '1' on the clock switch. — 0 VAC — Check for an open wire from range terminal block.

240 VAC

Check for power between terminal '1' selector switch & terminal '1' on the clock switch. — 0 VAC — Defective selector switch.

240 VAC

Defective clock switch or an open wire between selector switch and clock switch.

L1   L2

Selector Switch

Bake Element

Selector Switch

Clock Switch

D   4   1   3

Oven Thermostat

Heating Indicator

B   1

2   1

A   1   Selector Switch

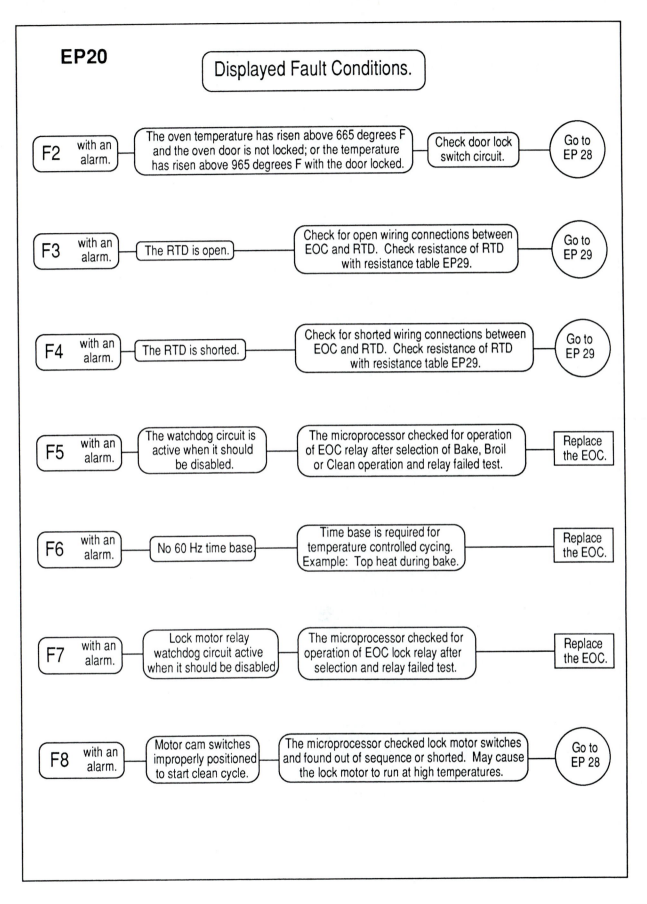

EP20

Displayed Fault Conditions.

F2 with an alarm. → The oven temperature has risen above 665 degrees F and the oven door is not locked; or the temperature has risen above 965 degrees F with the door locked. → Check door lock switch circuit. → Go to EP 28

F3 with an alarm. → The RTD is open. → Check for open wiring connections between EOC and RTD. Check resistance of RTD with resistance table EP29. → Go to EP 29

F4 with an alarm. → The RTD is shorted. → Check for shorted wiring connections between EOC and RTD. Check resistance of RTD with resistance table EP29. → Go to EP 29

F5 with an alarm. → The watchdog circuit is active when it should be disabled. → The microprocessor checked for operation of EOC relay after selection of Bake, Broil or Clean operation and relay failed test. → Replace the EOC.

F6 with an alarm. → No 60 Hz time base → Time base is required for temperature controlled cycing. Example: Top heat during bake. → Replace the EOC.

F7 with an alarm. → Lock motor relay watchdog circuit active when it should be disabled. → The microprocessor checked for operation of EOC lock relay after selection and relay failed test. → Replace the EOC.

F8 with an alarm. → Motor cam switches improperly positioned to start clean cycle. → The microprocessor checked lock motor switches and found out of sequence or shorted. May cause the lock motor to run at high temperatures. → Go to EP 28

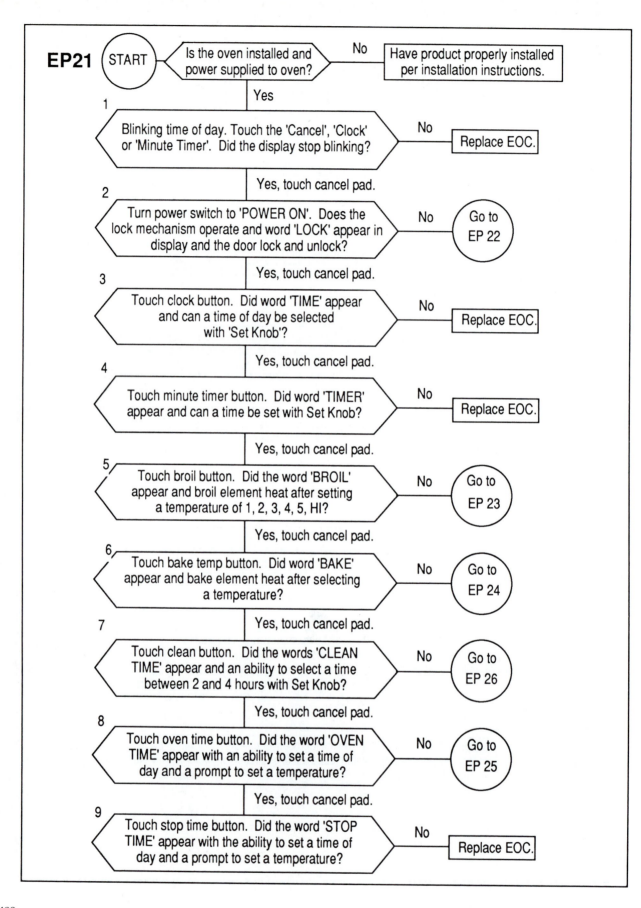

EP21 START — Is the oven installed and power supplied to oven? — **No** → Have product properly installed per installation instructions.

Yes

1. Blinking time of day. Touch the 'Cancel', 'Clock' or 'Minute Timer'. Did the display stop blinking? — **No** → Replace EOC.

Yes, touch cancel pad.

2. Turn power switch to 'POWER ON'. Does the lock mechanism operate and word 'LOCK' appear in display and the door lock and unlock? — **No** → Go to EP 22

Yes, touch cancel pad.

3. Touch clock button. Did word 'TIME' appear and can a time of day be selected with 'Set Knob'? — **No** → Replace EOC.

Yes, touch cancel pad.

4. Touch minute timer button. Did word 'TIMER' appear and can a time be set with Set Knob? — **No** → Replace EOC.

Yes, touch cancel pad.

5. Touch broil button. Did the word 'BROIL' appear and broil element heat after setting a temperature of 1, 2, 3, 4, 5, HI? — **No** → Go to EP 23

Yes, touch cancel pad.

6. Touch bake temp button. Did word 'BAKE' appear and bake element heat after selecting a temperature? — **No** → Go to EP 24

Yes, touch cancel pad.

7. Touch clean button. Did the words 'CLEAN TIME' appear and an ability to select a time between 2 and 4 hours with Set Knob? — **No** → Go to EP 26

Yes, touch cancel pad.

8. Touch oven time button. Did the word 'OVEN TIME' appear with an ability to set a time of day and a prompt to set a temperature? — **No** → Go to EP 25

Yes, touch cancel pad.

9. Touch stop time button. Did the word 'STOP TIME' appear with the ability to set a time of day and a prompt to set a temperature? — **No** → Replace EOC.

EP22

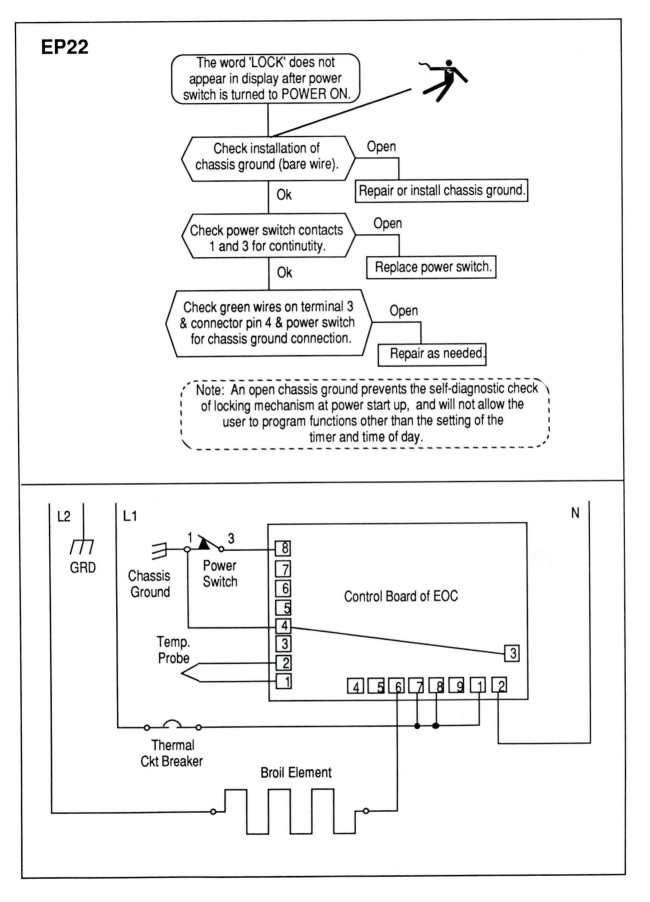

The word 'LOCK' does not appear in display after power switch is turned to POWER ON.

Check installation of chassis ground (bare wire).
— Open → Repair or install chassis ground.
— Ok ↓

Check power switch contacts 1 and 3 for continutity.
— Open → Replace power switch.
— Ok ↓

Check green wires on terminal 3 & connector pin 4 & power switch for chassis ground connection.
— Open → Repair as needed.

Note: An open chassis ground prevents the self-diagnostic check of locking mechanism at power start up, and will not allow the user to program functions other than the setting of the timer and time of day.

L2   L1   N

GRD

Chassis Ground   Power Switch   1   3

8
7
6
5
4
3
2
1

Control Board of EOC

3

Temp. Probe

4 5 6 7 8 9 1 2

Thermal Ckt Breaker

Broil Element

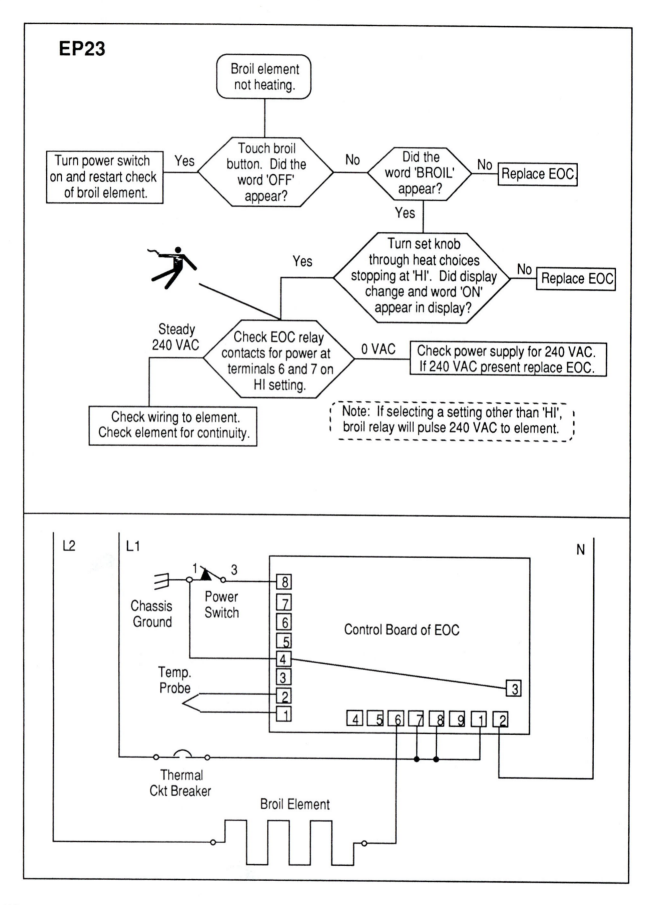

EP23

Broil element not heating.

Touch broil button. Did the word 'OFF' appear?

Yes → Turn power switch on and restart check of broil element.

No → Did the word 'BROIL' appear?

No → Replace EOC.

Yes → Turn set knob through heat choices stopping at 'HI'. Did display change and word 'ON' appear in display?

No → Replace EOC

Yes → Check EOC relay contacts for power at terminals 6 and 7 on HI setting.

Steady 240 VAC → Check wiring to element. Check element for continuity.

0 VAC → Check power supply for 240 VAC. If 240 VAC present replace EOC.

Note: If selecting a setting other than 'HI', broil relay will pulse 240 VAC to element.

L2   L1   N

Chassis Ground

Power Switch   1   3

8
7
6
5
4
3
2
1

Control Board of EOC

3

Temp. Probe

Thermal Ckt Breaker

Broil Element

4 5 6 7 8 9 1 2

490

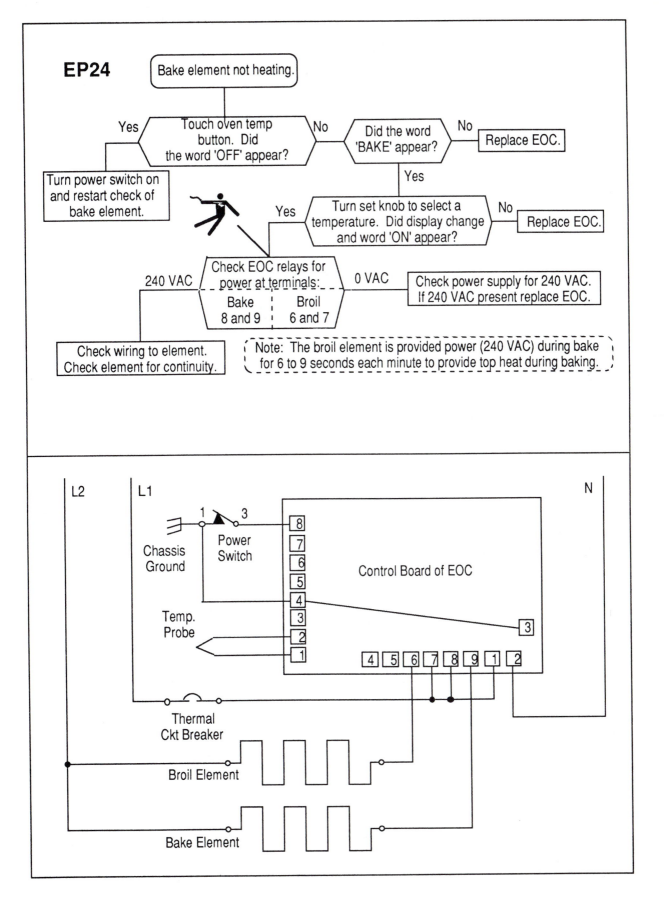

EP24

Bake element not heating.

Touch oven temp button. Did the word 'OFF' appear?
- Yes → Turn power switch on and restart check of bake element.
- No → Did the word 'BAKE' appear?
  - No → Replace EOC.
  - Yes → Turn set knob to select a temperature. Did display change and word 'ON' appear?
    - No → Replace EOC.
    - Yes → Check EOC relays for power at terminals:

Check EOC relays for power at terminals:
Bake 8 and 9 | Broil 6 and 7
- 240 VAC → Check wiring to element. Check element for continuity.
- 0 VAC → Check power supply for 240 VAC. If 240 VAC present replace EOC.

Note: The broil element is provided power (240 VAC) during bake for 6 to 9 seconds each minute to provide top heat during baking.

L2   L1   N

Chassis Ground

Power Switch   1   3   8

7
6
5
4   Control Board of EOC   3
3
Temp. Probe   2
1

4 5 6 7 8 9 1 2

Thermal Ckt Breaker

Broil Element

Bake Element

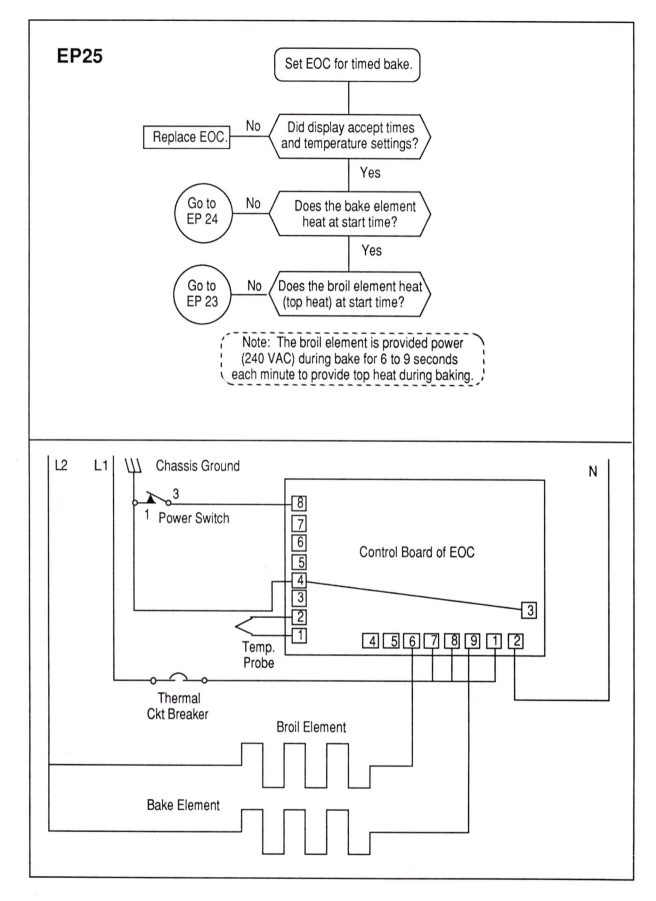

EP25

Set EOC for timed bake.

Did display accept times and temperature settings? — No → Replace EOC.

Yes

Does the bake element heat at start time? — No → Go to EP 24

Yes

Does the broil element heat (top heat) at start time? — No → Go to EP 23

Note: The broil element is provided power (240 VAC) during bake for 6 to 9 seconds each minute to provide top heat during baking.

L2 L1 Chassis Ground   N

3
1 Power Switch   8
                                   7
                                   6
                                   5   Control Board of EOC
                                   4   3
                                   3
                                   2
                                   1
Temp.
Probe
                        4 5 6 7 8 9 1 2

Thermal
Ckt Breaker

Broil Element

Bake Element

EP26

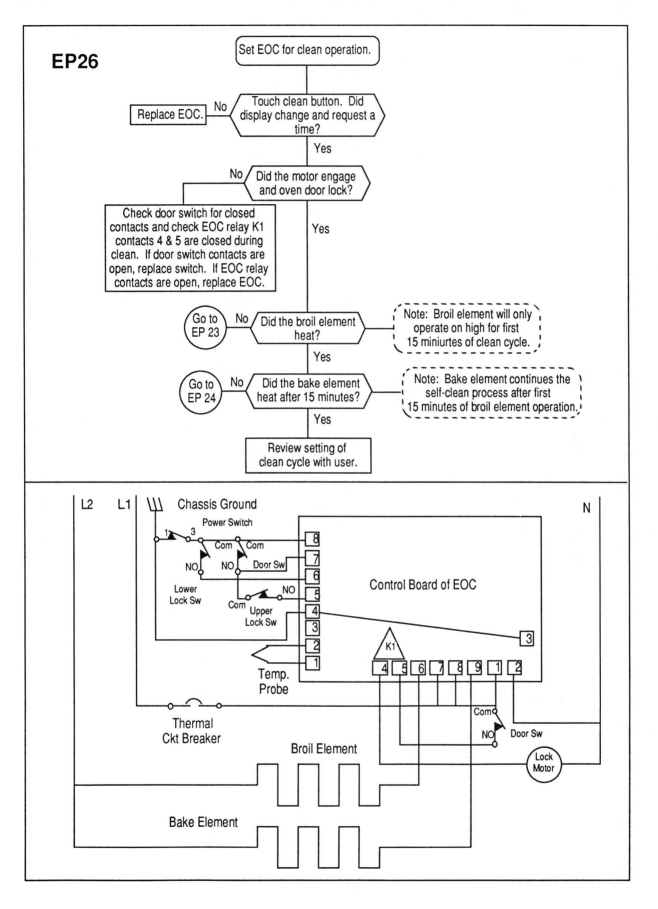

Set EOC for clean operation.

Replace EOC. ← No — Touch clean button. Did display change and request a time?

Yes

No ← Did the motor engage and oven door lock?

Check door switch for closed contacts and check EOC relay K1 contacts 4 & 5 are closed during clean. If door switch contacts are open, replace switch. If EOC relay contacts are open, replace EOC.

Yes

Go to EP 23 ← No — Did the broil element heat?

Note: Broil element will only operate on high for first 15 miniurtes of clean cycle.

Yes

Go to EP 24 ← No — Did the bake element heat after 15 minutes?

Note: Bake element continues the self-clean process after first 15 minutes of broil element operation.

Yes

Review setting of clean cycle with user.

L2 L1 \|/ Chassis Ground   N

Power Switch

Control Board of EOC

Lower Lock Sw

Upper Lock Sw

Temp. Probe

K1

Thermal Ckt Breaker

Broil Element

Bake Element

Door Sw

Lock Motor

493

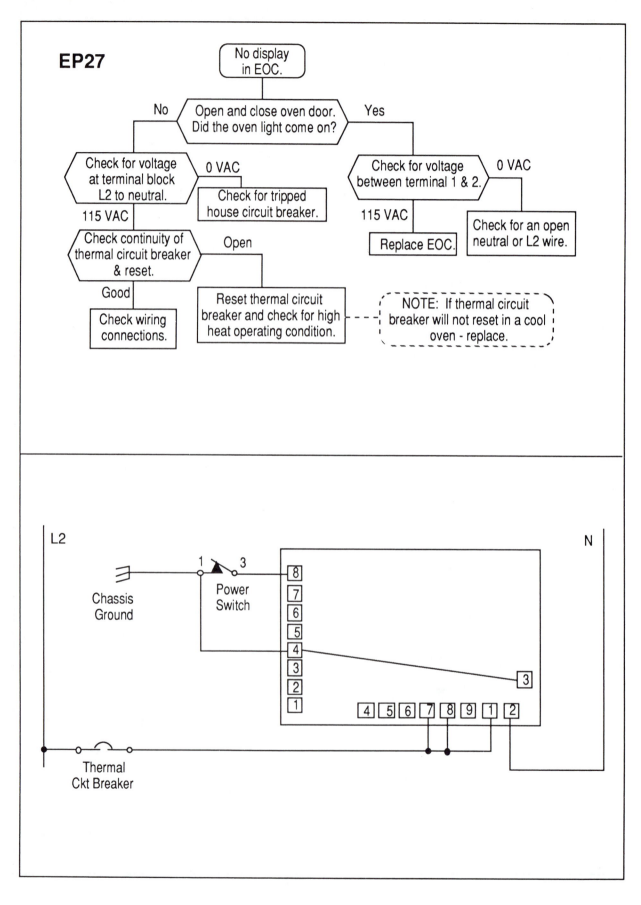

EP27

No display in EOC.

Open and close oven door. Did the oven light come on?

No → Check for voltage at terminal block L2 to neutral.

0 VAC → Check for tripped house circuit breaker.

115 VAC → Check continuity of thermal circuit breaker & reset.

Open → Reset thermal circuit breaker and check for high heat operating condition.

Good → Check wiring connections.

Yes → Check for voltage between terminal 1 & 2.

0 VAC → Check for an open neutral or L2 wire.

115 VAC → Replace EOC.

NOTE: If thermal circuit breaker will not reset in a cool oven - replace.

L2

Chassis Ground

1   3
Power Switch

8
7
6
5
4
3
2
1

4 5 6 7 8 9 1 2

3

N

Thermal Ckt Breaker

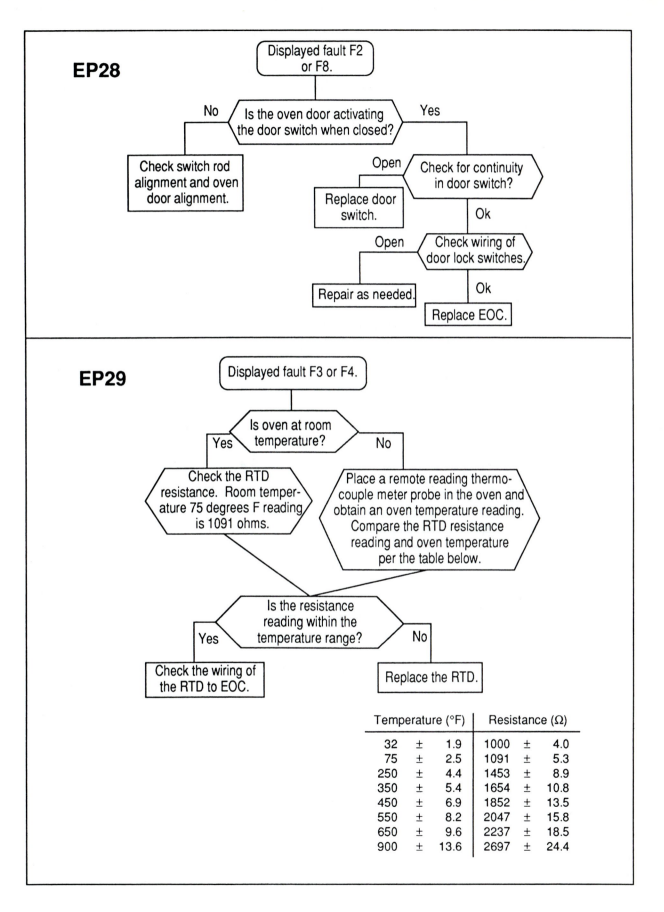

EP28

Displayed fault F2 or F8.

No ← Is the oven door activating the door switch when closed? → Yes

Check switch rod alignment and oven door alignment.

Open → Check for continuity in door switch?

Replace door switch.

Ok ↓

Open → Check wiring of door lock switches.

Repair as needed.

Ok ↓

Replace EOC.

EP29

Displayed fault F3 or F4.

Is oven at room temperature?

Yes ← → No

Check the RTD resistance. Room temperature 75 degrees F reading is 1091 ohms.

Place a remote reading thermo-couple meter probe in the oven and obtain an oven temperature reading. Compare the RTD resistance reading and oven temperature per the table below.

Is the resistance reading within the temperature range?

Yes ← → No

Check the wiring of the RTD to EOC.

Replace the RTD.

Temperature (°F)			Resistance (Ω)		
32	±	1.9	1000	±	4.0
75	±	2.5	1091	±	5.3
250	±	4.4	1453	±	8.9
350	±	5.4	1654	±	10.8
450	±	6.9	1852	±	13.5
550	±	8.2	2047	±	15.8
650	±	9.6	2237	±	18.5
900	±	13.6	2697	±	24.4

EP30

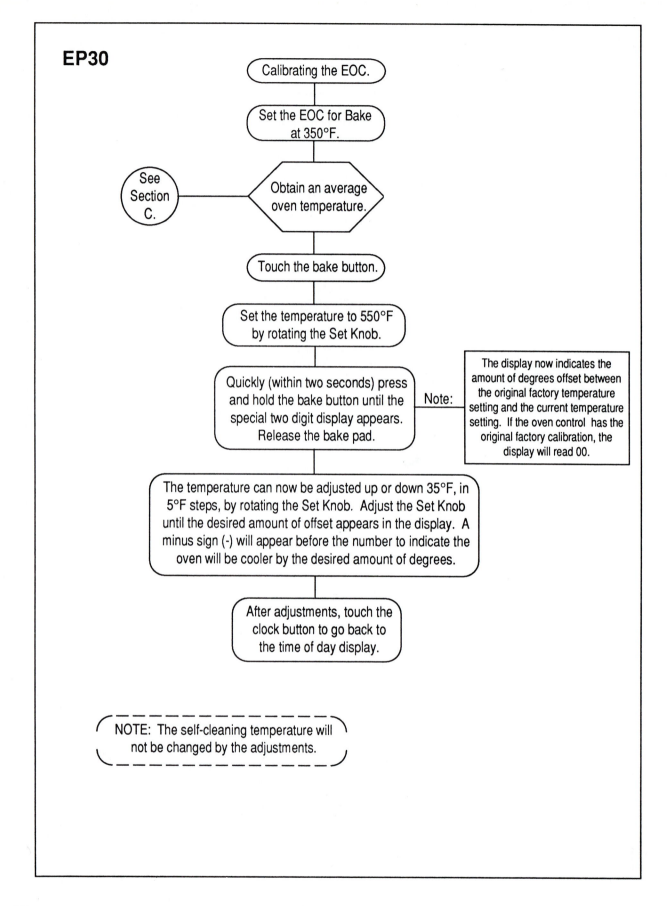

Calibrating the EOC.

Set the EOC for Bake at 350°F.

See Section C.

Obtain an average oven temperature.

Touch the bake button.

Set the temperature to 550°F by rotating the Set Knob.

Quickly (within two seconds) press and hold the bake button until the special two digit display appears. Release the bake pad.

Note: The display now indicates the amount of degrees offset between the original factory temperature setting and the current temperature setting. If the oven control has the original factory calibration, the display will read 00.

The temperature can now be adjusted up or down 35°F, in 5°F steps, by rotating the Set Knob. Adjust the Set Knob until the desired amount of offset appears in the display. A minus sign (-) will appear before the number to indicate the oven will be cooler by the desired amount of degrees.

After adjustments, touch the clock button to go back to the time of day display.

NOTE: The self-cleaning temperature will not be changed by the adjustments.

SECTION C - ELECTRICAL DIAGRAMS

ELECTRICAL DIAGRAM INDEX

MODEL	ENGINEERING CHANGE CODE†	ELECTRICAL DIAGRAM
G30BC†	0, 1, & 2	Page C-5
G30C†	0, 1, 2, & 3	Page C-5
G30PC†	0, 1, & 2	N/A
G31BF†	0 & 1	Page C-5
G32BC†	0, 1, 2, 3, & 4	Page C-5
G32BPC†	0, 1, & 2	Page C-4
GC32BC†	0, 1, & 2	Page C-5
GC32C†	0, 1, 2, 3, & 4	Page C-5
GG32BC†	0, 1, 2, 3, 4, & 5	Page C-5
GG32BPC†	0, 1, & 2	Page C-4

ELECTRICAL DIAGRAM INDEX

MODEL	ENGINEERING CHANGE CODE†	ELECTRICAL DIAGRAM
GG32C†	0, 1, 2, & 3	Page C-5
GG32PC†	0, 1, 2, 3, & 4	Page C-4
GP32BC†	0, 1, & 2	Pages C-6 & C-7
GP32BE†	0, 1, & 2	Pages C-9 & C-10
GCG34BC†	0, 1, 2, & 3	Page C-12
GG34BC†	0, 1, 2, 3, 4, & 5	Page C-12
GP34BC†	0, 1, & 2	Pages C-7 & C-13
GP34BE†	0, 1, & 2	Pages C-9 & C-10
GCG38BC†	0, 1, 2, & 3	Page C-16
GP38BC†	0, 1, & 2	Pages C-7 & C-14
GPG38BC†	0, 1, & 2	Pages C-7 & C-15

ELECTRICAL DIAGRAM INDEX

MODEL	ENGINEERING CHANGE CODE†	ELECTRICAL DIAGRAM
GPG38BE†	0, 1, 2, 3, & 4	Pages C-10 & C-11
GG46C†	0, 1, 2, 3, & 4	Page C-8

NOTE: COMPONENTS SHOWN MAY NOT APPEAR ON ALL PRODUCTS.

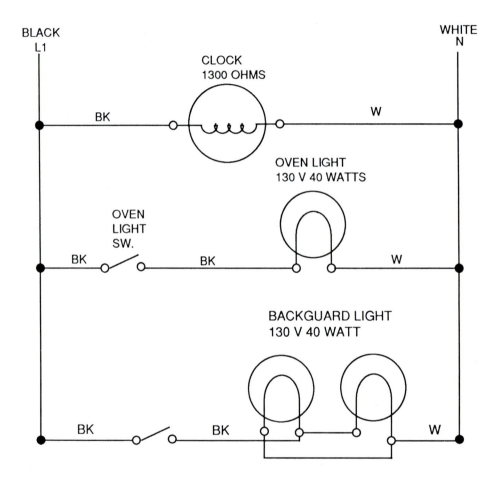

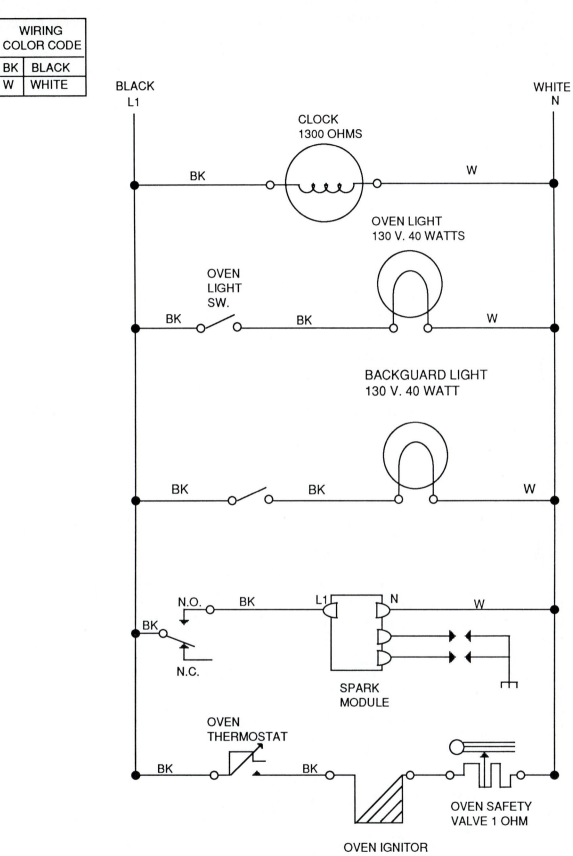

503

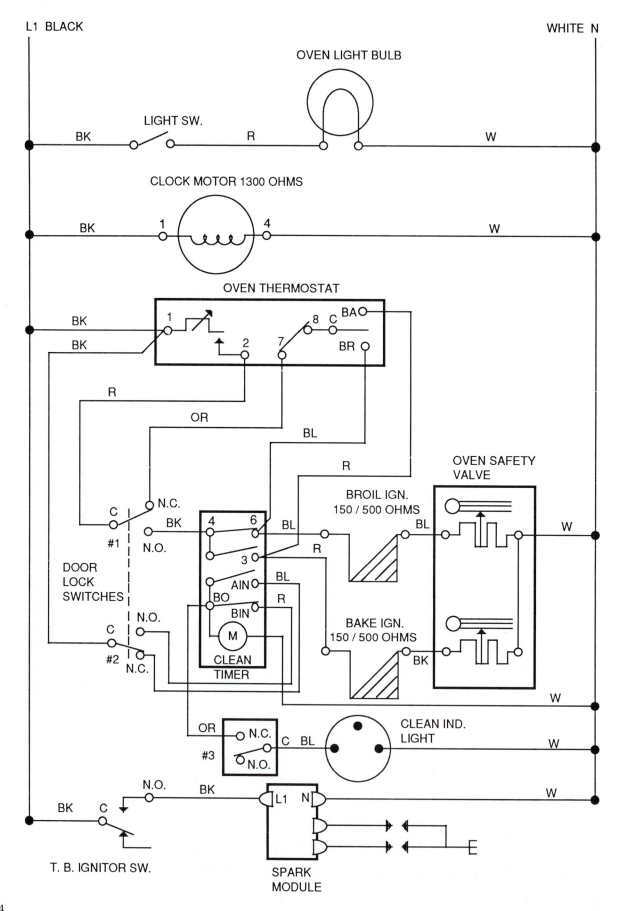

L1 BLACK   WHITE N

OVEN LIGHT BULB

BK   LIGHT SW.   R   W

CLOCK MOTOR 1300 OHMS

BK   1   4   W

OVEN THERMOSTAT

BK   1   8 C BA
BK   2   7   BR

R

OR

BL

R   OVEN SAFETY
                                     VALVE

BROIL IGN.
150 / 500 OHMS

C   N.C.   BK   4   6   BL   BL   W

#1   R
N.O.   3

DOOR
LOCK   AIN   BL
SWITCHES
BO   BIN   R

C   N.O.   BAKE IGN.
                   M   150 / 500 OHMS
#2
N.C.   CLEAN   BK
         TIMER

                                                              W

OR   N.C.   CLEAN IND.
                C BL   LIGHT   W
#3   N.O.

N.O.   BK   W
BK   C   L1 N

T. B. IGNITOR SW.   SPARK
                         MODULE

                                                        E

504

| CONTACT POSITION | OVEN THERMOSTAT | | | | DOOR LOCK SWITCHES | | | | |
| | | | | | No. 1 | | No. 2 | | No. 3 |
	1-2	7-8	C-BA	C-BR	C-NO	C-NC	C-NO	C-NC	C-NC
THERMOSTAT SET TO BROIL.	X	X	O	X	O	X	O	X	X
THERMOSTAT SET TO BAKE.	X	X	X	O	O	X	O	X	X
THERMO. SET TO CLEAN. LEVER IN LOCK POSITION. OVEN TEMP. BELOW 550°F. CLEAN TIMER RUN FROM 0 TO 8 MIN.	X	O	O	O	X	O	X	O	O
THERMO. SET TO CLEAN. OVEN DOOR LOCKED. OVEN TEMP. BELOW 550°F. CLEAN TIMER 0 TO 8 MIN.	X	O	O	O	X	O	X	O	O
THERMO. SET TO CLEAN. OVEN DOOR LOCKED. OVEN TEMP. 550° to 920° F. CLEAN TIMER RUN 72 TO 162 MIN.	X	O	O	O	X	O	X	O	O
THERMO. SET TO CLEAN. OVEN DOOR LOCKED. OVEN TEMP. 920°F THERMO. CYCLED OFF. CLEAN TIMER RUN 72 TO 162 .	O	O	O	O	X	O	X	O	O
THERMO. SET TO CLEAN. OVEN DOOR LOCKED. OVEN TEMP. 550°TO 920°F. CLEAN TIMER RUN 162 TO 191 MIN.	X	O	O	O	X	O	X	O	O
THERMO. SET TO CLEAN. OVEN DOOR LOCKED. OVEN TEMP. LESS THAN 550°F. CLEAN TIMER RUN 191 TO 200 MIN.	X	O	O	O	X	O	X	O	O
THERMO. SET TO CLEAN. LEVEL IN UNLOCKED POSITION. CLEAN TIMER RUN 191 TO 202 MIN.	X	O	O	O	O	X	O	X	X
THERMOSTAT OFF. LEVER IN UNLOCKED POSITION. TIMER RUN 202 MIN.	O	O	O	O	O	X	O	X	X

| CONTACT POSITION | CLEAN TIMER | | | |
	BO-BIN	BO-AIN	4-3	4-6
THERMOSTAT SET TO BROIL.	X	O	O	X
THERMOSTAT SET TO BAKE.	X	O	O	X
THERMO. SET TO CLEAN. LEVER IN LOCK POSITION. OVEN TEMP. BELOW 550°F. CLEAN TIMER RUN FROM 0 T0 8 MIN.	X	O	O	X
THERMO. SET TO CLEAN. OVEN DOOR LOCKED. OVEN TEMP. BELOW 550°F. CLEAN TIMER 0 TO 8 MIN.	X	X	O	X
THERMO. SET TO CLEAN. OVEN DOOR LOCKED. OVEN TEMP. 550° to 920° F. CLEAN TIMER RUN 72 TO 162 MIN.	X	X	X	O
THERMO. SET TO CLEAN. OVEN DOOR LOCKED. OVEN TEMP. 920°F THERMO. CYCLED OFF. CLEAN TIMER RUN 72 TO 162 .	X	X	X	O
THERMO. SET TO CLEAN. OVEN DOOR LOCKED. OVEN TEMP. 550°TO 920°F. CLEAN TIMER RUN 162 TO 191 MIN.	X	X	O	O
THERMO. SET TO CLEAN. OVEN DOOR LOCKED. OVEN TEMP. LESS THAN 550°F. CLEAN TIMER RUN 191 TO 200 MIN.	O	X	O	O
THERMO. SET TO CLEAN. LEVEL IN UNLOCKED POSITION. CLEAN TIMER RUN 191 TO 202 MIN.	O	X	O	O
THERMOSTAT OFF. LEVER IN UNLOCKED POSITION. TIMER RUN 202 MIN.	X	O	O	X

WIRING COLOR CODE	
BK	BLACK
W	WHITE
C	COPPER
BR	BROWN
G	GREEN
PR	PURPLE
O	ORANGE
BL	BLUE
V	VIOLET
GY	GRAY
T	TAN
P	PINK
Y	YELLOW
R	RED
G/Y	GREEN WITH YELLOW STRIPE

WIRING COLOR CODE	
BK	BLACK
W	WHITE

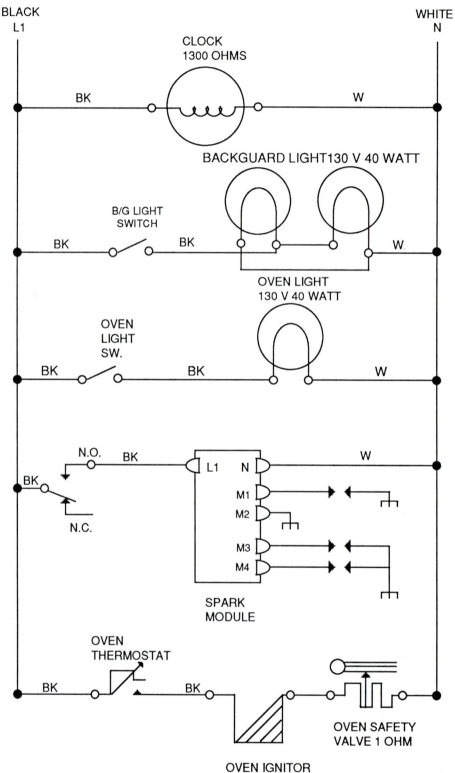

506

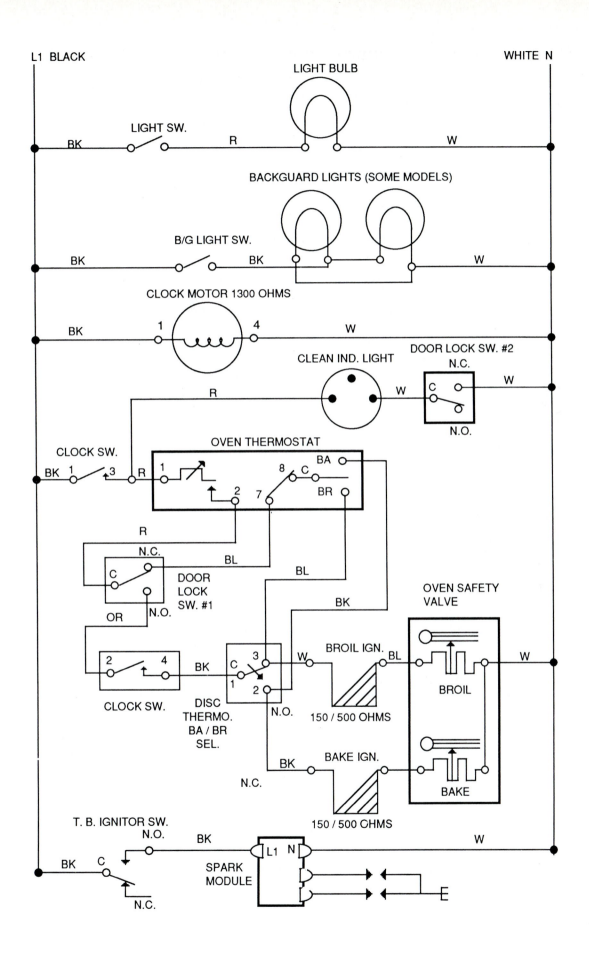

L1 BLACK

WHITE N

LIGHT BULB

LIGHT SW.

BK   R   W

BACKGUARD LIGHTS (SOME MODELS)

B/G LIGHT SW.

BK   BK   W

CLOCK MOTOR 1300 OHMS

BK   1   4   W

CLEAN IND. LIGHT

DOOR LOCK SW. #2
N.C.

R   W   C   W

N.O.

CLOCK SW.   OVEN THERMOSTAT

BK 1   3   R   1   BA

                        8   C

                    2   7   BR

R

N.C.   BL

C   BL

DOOR
LOCK
SW. #1   BK

OR   N.O.

OVEN SAFETY
VALVE

2   4   BK   C   3   W   BROIL IGN.   BL   W

                    1

                    2   BROIL

CLOCK SW.   DISC   N.O.
              THERMO.   150 / 500 OHMS
              BA / BR
              SEL.

                    BK   BAKE IGN.

              N.C.   BAKE

                    150 / 500 OHMS

T. B. IGNITOR SW.
N.O.   BK   W

BK   C   L1 N

              SPARK
              MODULE   E

N.C.

507

WIRING COLOR CODE									
BK	Black	BR	Brown	O	Orange	V	Violet	T	Tan	Y	Yellow
C	Copper	G	Green	BL	Blue	GY	Gray	P	Pink	R	Red
W	White	PR	Purple	G/Y	Green with Yellow Stripe						

CLOCK SWITCH FUNCTIONS				
TERMINALS	MANUAL	DELAY	COOK	OFF
1 - 3	CLOSED	OPEN	CLOSED	OPEN
2 - 4	OPEN	OPEN	*	OPEN

*CLOSED BETWEEN 30 MINUTES & 3 HOURS

CONTACT POSITION	OVEN THERMOSTAT				DOOR LOCK SWITCHES NO. 1		NO. 2		DISC THERMO	
	1-2	7-8	C-BA	C-BR	C-NO	C-NC	C-NO	C-NC	C-NC	C-NO
THERMOSTAT SET TO BROIL.	X	X	O	X	O	X	X	O	X	O
THERMOSTAT SET TO BAKE.	X	X	X	O	O	X	X	O	X	O
THERMO. SET TO CLEAN. LEVER IN LOCK POSITION. OVEN TEMP. BELOW 650°F.	X	X	O	O	X	O	O	X	X	O
THERMO. SET TO CLEAN. OVEN DOOR LOCKED. OVEN TEMP. ABOVE 650°F.	X	X	O	O	X	O	O	X	O	X
THERMO. SET TO CLEAN. OVEN DOOR LOCKED. OVEN TEMP. 920°F. THERMO. CYCLE OFF.	O	X	O	O	X	O	O	X	O	X
THERMO. SET TO CLEAN. LEVER IN LOCKED POSITION. OVEN TEMP. LESS THAN 550°F.	X	X	O	O	X	O	O	X	X	O
THERMOSTAT SET TO CLEAN. LEVER IN UNLOCKED POSITION.	X	X	O	O	X	O	O	X	X	O
THERMOSTAT OFF. LEVER IN UNLOCKED POSITION.	O	O	O	O	O	X	X	O	X	O

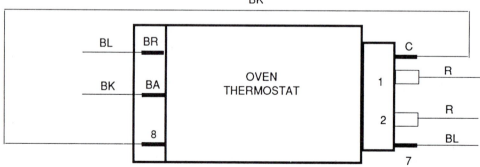

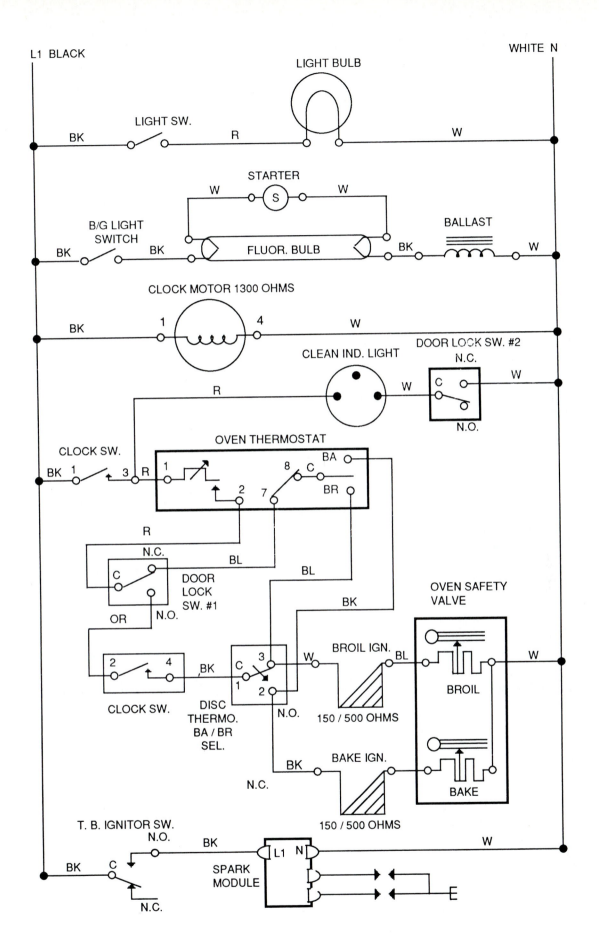

L1 BLACK

WHITE N

LIGHT BULB

LIGHT SW.

BK R W

STARTER

S

W W

B/G LIGHT
SWITCH

BK BK

FLUOR. BULB

BALLAST

BK W

CLOCK MOTOR 1300 OHMS

BK 1 4 W

CLEAN IND. LIGHT

DOOR LOCK SW. #2
N.C.

C W

N.O.

R W

CLOCK SW.

OVEN THERMOSTAT

BK 1 3 R 1

2 7

8 C BA

BR

R

DOOR
LOCK
SW. #1

N.C.

C

OR N.O.

BL

BL

BK

OVEN SAFETY
VALVE

CLOCK SW.

2 4 BK

DISC
THERMO.
BA / BR
SEL.

C 3

1 2

N.O.

W BROIL IGN. BL

150 / 500 OHMS

BROIL

W

BK BAKE IGN.

N.C.

150 / 500 OHMS

BAKE

T. B. IGNITOR SW.
N.O.

BK C

N.C.

BK

SPARK
MODULE

L1 N

W

E

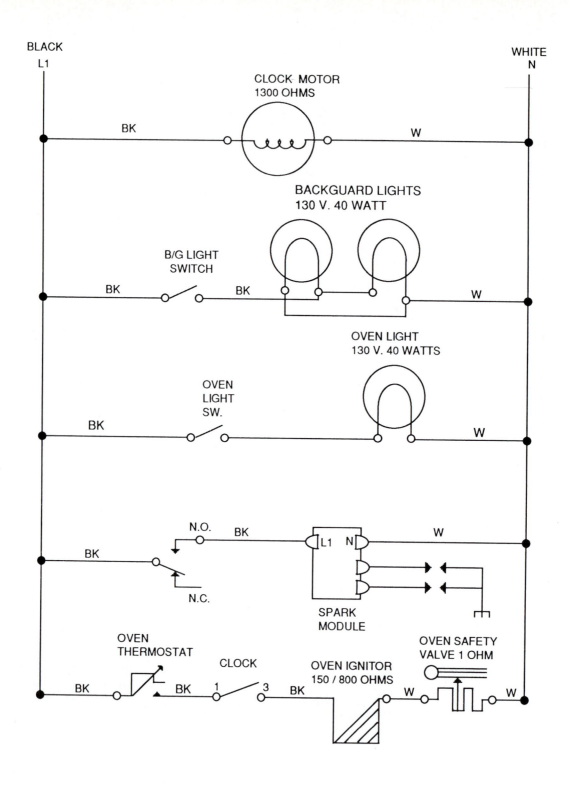

BLACK
L1

WHITE
N

CLOCK MOTOR
1300 OHMS

BK

W

BACKGUARD LIGHTS
130 V. 40 WATT

B/G LIGHT
SWITCH

BK

BK

W

OVEN LIGHT
130 V. 40 WATTS

OVEN
LIGHT
SW.

BK

W

N.O.

BK

W

BK

L1   N

N.C.

SPARK
MODULE

OVEN
THERMOSTAT

CLOCK

OVEN IGNITOR
150 / 800 OHMS

OVEN SAFETY
VALVE 1 OHM

BK

BK   1   3

BK

W   W

WIRING COLOR CODE	
BK	BLACK
W	WHITE
R	RED

510

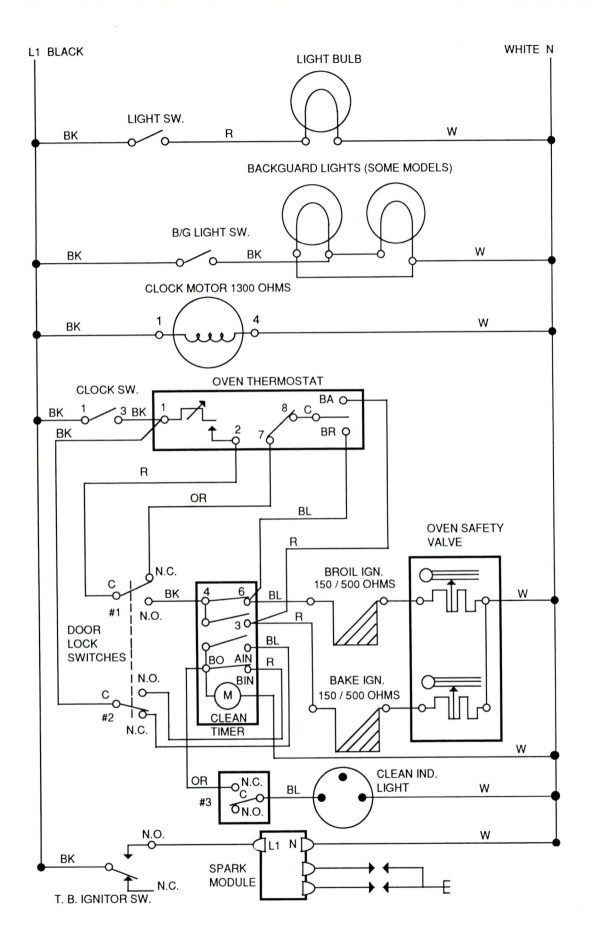

L1 BLACK

WHITE N

LIGHT BULB

LIGHT SW.

BK   R   W

BACKGUARD LIGHTS (SOME MODELS)

B/G LIGHT SW.

BK   BK   W

CLOCK MOTOR 1300 OHMS

BK   1   4   W

OVEN THERMOSTAT

CLOCK SW.

BK 1   3 BK   1

BK

BA

8   C

BR

2   7

R

OR

BL

R

OVEN SAFETY
VALVE

N.C.

C

#1   N.O.

BK   4   6   BL

3

R

BROIL IGN.
150 / 500 OHMS

W

DOOR
LOCK
SWITCHES

N.O.

C

#2   N.C.

BO   AIN

BIN

M

CLEAN
TIMER

BL

R

BAKE IGN.
150 / 500 OHMS

W

OR

#3

N.C.

C

N.O.

BL

CLEAN IND.
LIGHT

W

W

N.O.

L1 N

W

BK

N.C.

T. B. IGNITOR SW.

SPARK
MODULE

E

511

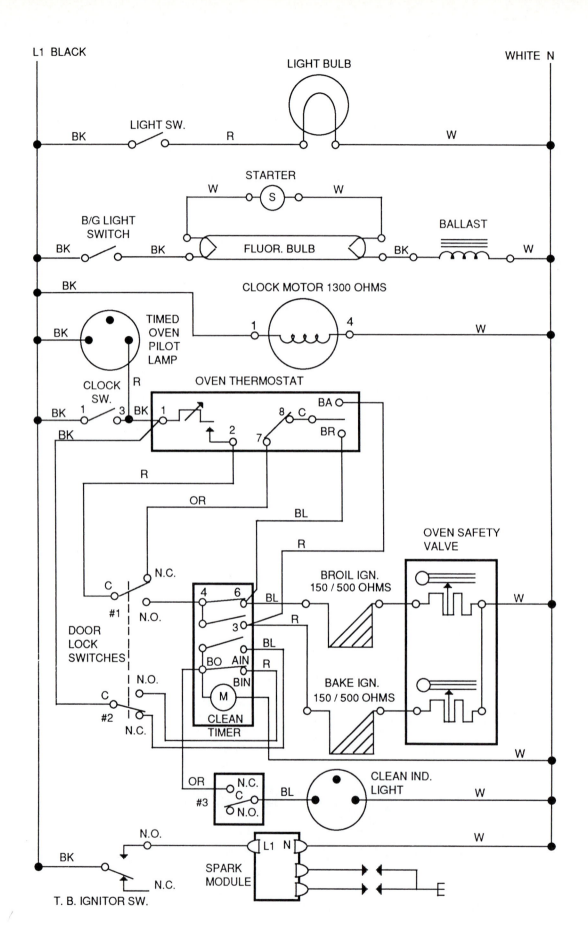

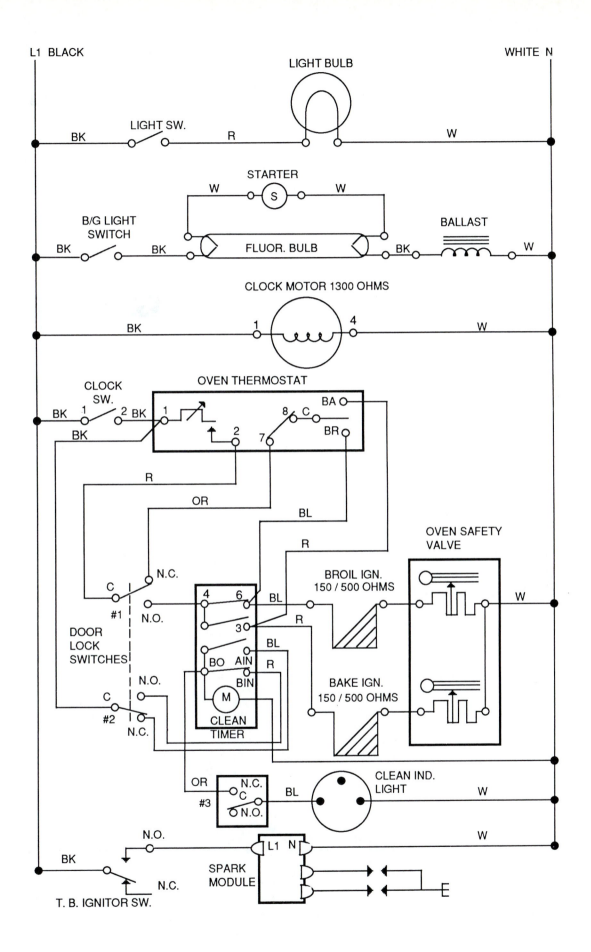

L1 BLACK

WHITE N

LIGHT BULB

BK LIGHT SW. R W

STARTER
W S W

B/G LIGHT SWITCH
BK BK FLUOR. BULB BK BALLAST W

CLOCK MOTOR 1300 OHMS
BK 1 4 W

OVEN THERMOSTAT
CLOCK SW.
BK 1 2 BK 1 8 C BA
BK 2 7 BR

R
OR
BL
R

OVEN SAFETY VALVE

C N.C.
#1

DOOR LOCK SWITCHES

4 6 BL BROIL IGN. 150 / 500 OHMS
3 R
BO AIN BL
R BIN
N.O.
C
#2
N.C.
M CLEAN TIMER
BAKE IGN. 150 / 500 OHMS

OR N.C. BL CLEAN IND. LIGHT W
#3 C N.O.

W

N.O.
BK L1 N W
SPARK MODULE
N.C. E
T. B. IGNITOR SW.

513

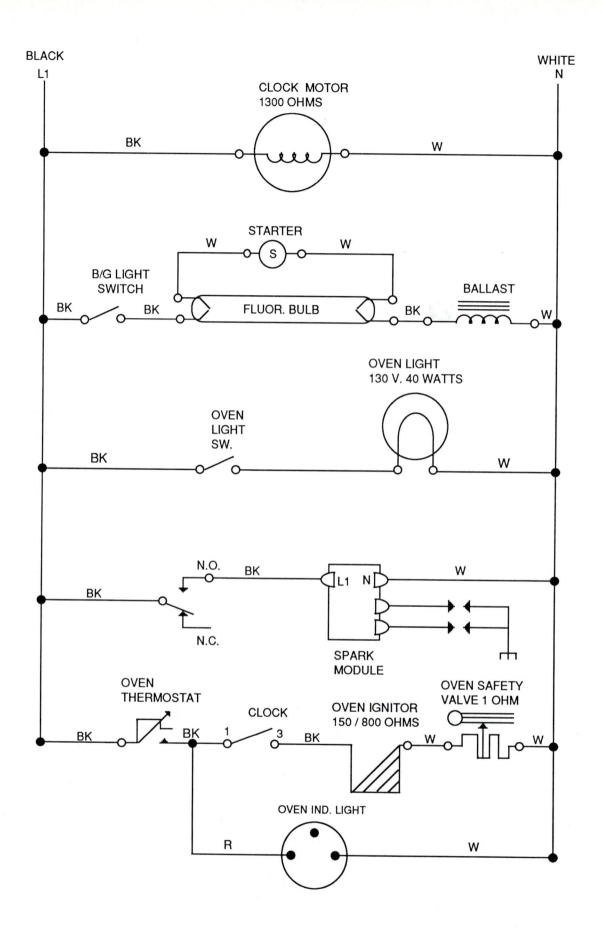

BLACK
L1

WHITE
N

CLOCK MOTOR
1300 OHMS

BK

W

STARTER
S

W W

B/G LIGHT
SWITCH

BK

BK

FLUOR. BULB

BK

BALLAST

W

OVEN LIGHT
130 V. 40 WATTS

OVEN
LIGHT
SW.

BK

W

N.O.

BK

L1 N

W

N.C.

SPARK
MODULE

BK

OVEN
THERMOSTAT

CLOCK

OVEN IGNITOR
150 / 800 OHMS

OVEN SAFETY
VALVE 1 OHM

BK

BK

1 3

BK

W

W

OVEN IND. LIGHT

R

W

514

SECTION D - TROUBLESHOOTING

⚠️WARNING

LINE TO NEUTRAL VOLTAGE MAY BE PRESENT DURING TESTING!

1. **REMOVE POWER BEFORE TOUCHING ANY ELECTRICAL COMPONENT.**

2. **NEVER CHECK FOR A GAS LEAK WITH A FLAME.**

3. **ALWAYS TURN THE GAS OFF BEFORE DISCONNECTING ANY GAS CARRYING COMPONENT.**

4. **READING OBTAINED DURING TESTING MAY VARY FROM VALUES STATED IN FLOW CHARTS. ALLOWABLE VARIANCE IS ±10% AT 77° F. (RESISTANCE VALUES GIVEN ON THE FLOW CHARTS WERE MEASURED WITH A SIMPSON 260 VOLT/OHMMETER.)**

5. **THESE FLOW CHARTS ASSUME THAT THE POWER SUPPLY AVAILABLE TO THE RANGE IS WITHIN SPECIFICATIONS, PROPERLY POLARIZED, AND GROUNDED.**

6. **THESE FLOW CHARTS ASSUME THE GAS SUPPLY TO THE RANGE HAS PROPER PRESSURE AND VOLUME.**

⚠️WARNING

TESTS IN THIS SECTION SHOULD ONLY BE PERFORMED BY A QUALIFIED TECHNICIAN.

FLOW CHART SYMBOLS

What is a flow chart? A diagram that shows step-by-step progression through a procedure using connecting lines and a set of symbols.

Why use flow charts? To expedite the diagnosis of a product. A flow chart gives the proper test procedure (with the least amount of written information) that will eventually lead to a faulty component or system failure. (The flow charts used in this manual lead to a single component failure.)

Why are symbols used in a flow chart? Flow chart symbols have definite meanings. Each symbol is a key to that particular step in the diagnostic procedure. Once these symbols are understood, the diagnosis goes much faster. The following symbols and their meanings are used in this manual.

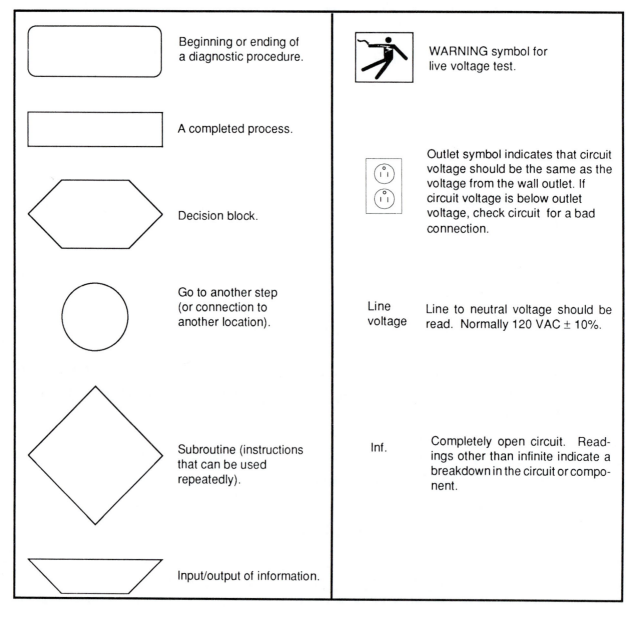

	Beginning or ending of a diagnostic procedure.
	A completed process.
	Decision block.
	Go to another step (or connection to another location).
	Subroutine (instructions that can be used repeatedly).
	Input/output of information.
	WARNING symbol for live voltage test.
	Outlet symbol indicates that circuit voltage should be the same as the voltage from the wall outlet. If circuit voltage is below outlet voltage, check circuit for a bad connection.
Line voltage	Line to neutral voltage should be read. Normally 120 VAC ± 10%.
Inf.	Completely open circuit. Readings other than infinite indicate a breakdown in the circuit or component.

THERMOSTAT CALIBRATION

The following procedures are recommended for obtaining the oven thermostat average temperature for calibration purposes.

1. Use thermocouple type temperature tester.

2. Place the sensor as close to the center of the oven cavity using an empty oven rack.

3. Set the thermostat to 350°F.

4. After the initial off cycle, allow a minimum of three cut-in and three cut-out cycles to permit all oven cavity surfaces to stabilize at temperature.

5. Record the last cycle and calculate the average oven temperature using the following formula:

$$\text{Av. Temp.} = \frac{\text{Cut-in Temp.} + \text{Cut-out Temp.}}{2}$$

Example: Last cycle recorded was 335°F cut-in and 385°F cut-out.

$$\text{Av. Temp.} = \frac{\text{Cut-in T} + \text{Cut-out T}}{2} = \frac{335°F + 385°F}{2}$$

$$\text{Av. Temp.} = \frac{720°F}{2} = 360°F$$

6. If average oven temperature is 50°F above or below the selected setting, the thermostat should be replaced.

7. If average oven temperature is more than 25°F, but less than 50°F from selected setting, the thermostat should be calibrated.*

*NOTE: Self-clean thermostats do not have any calibration adjustment screws. Only the control knob and skirt can be adjusted.

FLOW CHART ENTRY POINTS

PROBLEM	MODEL NUMBERS				
	G30BC0-2	G30C0-3	G30PC0-2	G31BF0-1	G32BC0-4
Top burner pilots go out.	N/A	N/A	EP 1 Pg D-9	N/A	N/A
Clock does not operate.	N/A	N/A	N/A	EP 2 Pg D-9	EP 2 Pg D-9
Top burners will not ignite.	EP 5 Pg D-11	EP 5 Pg D-11	EP 3 Pg D-10	EP 5 Pg D-11	EP 5 Pg D-11
Top burner has yellow flame.	EP 4 Pg D-10	EP 4 Pg D-10	EP 4 Pg D-10	EP 4 Pg D-10	EP 4 Pg D-10
Oven pilot goes out.	N/A	N/A	EP 6 Pg D-11	N/A	N/A
Backguard light does not illuminate.	N/A	N/A	N/A	N/A	N/A
Oven light does not illuminate.	N/A	N/A	N/A	EP 7 Pg D-12	EP 7 Pg D-12
Oven will not come on.	EP 12 Pg D-14	EP 12 Pg D-14	EP 8 Pg D-12	EP 12 Pg D-14	EP 12 Pg D-14
Oven flame will not cycle off unless the thermostat is turned off.	N/A	N/A	EP 9 Pg D-13	N/A	N/A
Oven temperature out of calibration.	EP 10 Pg D-13	EP 10 Pg D-13	EP 10 Pg D-13	EP 10 Pg D-13	EP 10 Pg D-13
Oven will not cycle off.	EP 13 Pg D-15	EP 13 Pg D-15	N/A	EP 13 Pg D-15	EP 13 Pg D-15
Broil burner will operate, but bake burner will not come on.	N/A	N/A	N/A	N/A	N/A
Top burner ignitors do not spark.	EP 15 Pg D-16	EP 15 Pg D-16	N/A	EP 15 Pg D-16	EP 15 Pg D-16
Bake burner will operate, but broil burner will not come on.	N/A	N/A	N/A	N/A	N/A
Neither bake or broil burners will come on.	N/A	N/A	N/A	N/A	N/A
Oven operates in bake & broil, but does not heat in clean.	N/A	N/A	N/A	N/A	N/A
Oven operates in bake & broil, but does not start to heat until 8 minutes into the clean cycle.	N/A	N/A	N/A	N/A	N/A
Oven operates in bake & broil, heats the first 8 minutes of cleaning cycle then stops heating.	N/A	N/A	N/A	N/A	N/A
Clean light does not illuminate during the clean cycle.	N/A	N/A	N/A	N/A	N/A
Clean light glows, but clean timer does not advance.	N/A	N/A	N/A	N/A	N/A
Oven is not clean at the end of the clean cycle.	N/A	N/A	N/A	N/A	N/A
Oven operates, but oven indicator light does not glow.	N/A	N/A	N/A	N/A	N/A
Timed oven pilot lamp does not glow when timed bake is over.	N/A	N/A	N/A	N/A	N/A

Items listed under the problem column are indicators that may be observed without disassembling the appliance, and should be used as a troubleshooting starting point.
N/A = Not applicable.

PROBLEM	MODEL NUMBERS				
	G32BPC0-2	GC32BC0-2	GC32C0-4	GG32BC0-5	GG32BPC0-2
Top burner pilots go out.	EP 1 Pg D-9	N/A	N/A	N/A	EP 1 Pg D-9
Clock does not operate.	EP 2 Pg D-9	EP 2 Pg D-9	EP 2 Pg D-9	EP 2 Pg D-9	EP 2 Pg D-9
Top burners will not ignite.	EP 3 Pg D-10	EP 5 Pg D-11	EP 5 Pg D-11	EP 5 Pg D-11	EP 3 Pg D-10
Top burner has yellow flame.	EP 4 Pg D-10	EP 4 Pg D-10	EP 4 Pg D-10	EP 4 Pg D-10	EP 4 Pg D-10
Oven pilot goes out.	EP 6 Pg D-11	N/A	N/A	N/A	EP 6 Pg D-11
Backguard light does not illuminate.	N/A	N/A	N/A	N/A	N/A
Oven light does not illuminate.	EP 7 Pg D-12	N/A	EP 7 Pg D-12	EP 7 Pg D-12	EP 7 Pg D-12
Oven will not come on.	EP 8 Pg D-12	EP 12 Pg D-14	EP 12 Pg D-14	EP 12 Pg D-14	EP 8 Pg D-12
Oven flame will not cycle off unless the thermostat is turned off.	EP 9 Pg D-13	N/A	N/A	N/A	EP 9 Pg D-13
Oven temperature out of calibration.	EP 10 Pg D-13	EP 10 Pg D-13	EP 10 Pg D-13	EP 10 Pg D-13	EP 10 Pg D-13
Oven will not cycle off.	N/A	EP 13 Pg D-15	EP 13 Pg D-15	EP 13 Pg D-15	N/A
Broil burner will operate, but bake burner will not come on.	N/A	N/A	N/A	N/A	N/A
Top burner ignitors do not spark.	N/A	EP 15 Pg D-16	EP 15 Pg D-16	EP 15 Pg D-16	N/A
Bake burner will operate, but broil burner will not come on.	N/A	N/A	N/A	N/A	N/A
Neither bake or broil burners will come on.	N/A	N/A	N/A	N/A	N/A
Oven operates in bake & broil, but does not heat in clean.	N/A	N/A	N/A	N/A	N/A
Oven operates in bake & broil, but does not start to heat until 8 minutes into the clean cycle.	N/A	N/A	N/A	N/A	N/A
Oven operates in bake & broil, heats the first 8 minutes of cleaning cycle then stops heating.	N/A	N/A	N/A	N/A	N/A
Clean light does not illuminate during the clean cycle.	N/A	N/A	N/A	N/A	N/A
Clean light glows, but clean timer does not advance.	N/A	N/A	N/A	N/A	N/A
Oven is not clean at the end of the clean cycle.	N/A	N/A	N/A	N/A	N/A
Oven operates, but oven indicator light does not glow.	N/A	N/A	N/A	N/A	N/A
Timed oven pilot lamp does not glow when timed bake is over.	N/A	N/A	N/A	N/A	N/A

Items listed under the problem column are indicators that may be observed without disassembling the appliance, and should be used as a troubleshooting starting point. N/A = Not applicable.

FLOW CHART ENTRY POINTS

PROBLEM	MODEL NUMBERS				
	GG32C0-3	GG32PC0-4	GP32BC0-2	GP32BE0-2	GCG34BC0-3
Top burner pilots go out.	N/A	EP 1 Pg D-9	N/A	N/A	N/A
Clock does not operate.	EP 2 Pg D-9	EP 2 Pg D-9	EP 2 Pg D-9	EP 2 Pg D-9	EP 2 Pg D-9
Top burners will not ignite.	EP 5 Pg D-11	EP 3 Pg D-10	EP 5 Pg D-11	EP 5 Pg D-11	EP 5 Pg D-11
Top burner has yellow flame.	EP 4 Pg D-10	EP 4 Pg D-10	EP 4 Pg D-10	EP 4 Pg D-10	EP 4 Pg D-10
Oven pilot goes out.	N/A	EP 6 Pg D-11	N/A	N/A	N/A
Backguard light does not illuminate.	N/A	N/A	EP 17 Pg D-17	EP 17 Pg D-17	EP 17 Pg D-17
Oven light does not illuminate.	EP 7 Pg D-12	EP 7 Pg D-12	EP 7 Pg D-12	EP 7 Pg D-12	EP 7 Pg D-12
Oven will not come on.	EP 12 Pg D-14	EP 8 Pg D-12	N/A	N/A	EP 30 Pg D-24
Oven flame will not cycle off unless the thermostat is turned off.	N/A	EP 9 Pg D-13	N/A	N/A	N/A
Oven temperature out of calibration.	EP 10 Pg D-13	EP 10 Pg D-13	EP 11 Pg D-13	EP 11 Pg D-13	EP 10 Pg D-13
Oven will not cycle off.	EP 13 Pg D-15	N/A	EP 13 Pg D-15	EP 13 Pg D-15	EP 13 Pg D-15
Broil burner will operate, but bake burner will not come on.	N/A	N/A	EP 14 Pg D-15	EP 14 Pg D-15	N/A
Top burner ignitors do not spark.	EP 15 Pg D-16	N/A	EP 15 Pg D-16	EP 15 Pg D-16	EP 15 Pg D-16
Bake burner will operate, but broil burner will not come on.	N/A	N/A	EP 16 Pg D-17	EP 16 Pg D-17	N/A
Neither bake or broil burners will come on.	N/A	N/A	EP 32 Pg D-26	EP 19 Pg D-19	EP 18 Pg D-17
Oven operates in bake & broil, but does not heat in clean.	N/A	N/A	EP 23 Pg D-20	EP 27 Pg D-22	N/A
Oven operates in bake & broil, but does not start to heat until 8 minutes into the clean cycle.	N/A	N/A	EP 21 Pg D-20	N/A	N/A
Oven operates in bake & broil, heats the first 8 minutes of cleaning cycle then stops heating.	N/A	N/A	EP 22 Pg D-20	N/A	N/A
Clean light does not illuminate during the clean cycle.	N/A	N/A	EP 24 Pg D-21	EP 28 Pg D-23	N/A
Clean light glows, but clean timer does not advance.	N/A	N/A	EP 25 Pg D-21	N/A	N/A
Oven is not clean at the end of the clean cycle.	N/A	N/A	EP 26 Pg D-22	EP 29 Pg D-23	N/A
Oven operates, but oven indicator light does not glow.	N/A	N/A	N/A	N/A	N/A
Timed oven pilot lamp does not glow when timed bake is over.	N/A	N/A	N/A	N/A	N/A

Items listed under the problem column are indicators that may be observed without disassembling the appliance, and should be used as a troubleshooting starting point. N/A = Not applicable.

FLOW CHART ENTRY POINTS

PROBLEM	MODEL NUMBERS				
	GG34BC0-5	GP34BC0-2	GP34BE0-2	GCG38BC0-3	GP38BC0-2
Top burner pilots go out.	N/A	N/A	N/A	N/A	N/A
Clock does not operate.	EP 2 Pg D-9	EP 2 Pg D-9	EP 2 Pg D-9	EP 2 Pg D-9	EP 2 Pg D-9
Top burners will not ignite.	EP 5 Pg D-11	EP 5 Pg D-11	EP 5 Pg D-11	EP 5 Pg D-11	EP 5 Pg D-11
Top burner has yellow flame.	EP 4 Pg D-10	EP 4 Pg D-10	EP 4 Pg D-10	EP 4 Pg D-10	EP 4 Pg D-10
Oven pilot goes out.	N/A	N/A	N/A	N/A	N/A
Backguard light does not illuminate.	EP 17 Pg D-17	EP 17 Pg D-17	EP 17 Pg D-17	EP 31 Pg D-25	EP 17 Pg D-17
Oven light does not illuminate.	EP 7 Pg D-12	EP 7 Pg D-12	EP 7 Pg D-12	EP 7 Pg D-12	EP 7 Pg D-12
Oven will not come on.	EP 30 Pg D-24	N/A	N/A	EP 30 Pg D-24	N/A
Oven flame will not cycle off unless the thermostat is turned off.	N/A	N/A	N/A	N/A	N/A
Oven temperature out of calibration.	EP 10 Pg D-13	EP 11 Pg D-13	EP 11 Pg D-13	EP 10 Pg D-13	EP 11 Pg D-13
Oven will not cycle off.	EP 13 Pg D-15	EP 13 Pg D-15	EP 13 Pg D-15	EP 13 Pg D-15	EP 13 Pg D-15
Broil burner will operate, but bake burner will not come on.	N/A	EP 14 Pg D-15	EP 14 Pg D-15	N/A	EP 14 Pg D-15
Top burner ignitors do not spark.	EP 15 Pg D-16	EP 15 Pg D-16	EP 15 Pg D-16	EP 15 Pg D-16	EP 15 Pg D-16
Bake burner will operate, but broil burner will not come on.	N/A	EP 16 Pg D-17	EP 16 Pg D-17	N/A	EP 16 Pg D-17
Neither bake or broil burners will come on.	N/A	EP 19 Pg D-19	EP 19 Pg D-19	N/A	EP 19 Pg D-19
Oven operates in bake & broil, but does not heat in clean.	N/A	EP 20 Pg D-20	EP 27 Pg D-22	N/A	EP 20 Pg D-20
Oven operates in bake & broil, but does not start to heat until 8 minutes into the clean cycle.	N/A	EP 21 Pg D-20	N/A	N/A	EP 21 Pg D-20
Oven operates in bake & broil, heats the first 8 minutes of cleaning cycle then stops heating.	N/A	EP 22 Pg D-20	N/A	N/A	EP 22 Pg D-20
Clean light does not illuminate during the clean cycle.	N/A	EP 24 Pg D-21	EP 28 Pg D-23	N/A	EP 24 Pg D-21
Clean light glows, but clean timer does not advance.	N/A	EP 25 Pg D-21	N/A	N/A	EP 25 Pg D-21
Oven is not clean at the end of the clean cycle.	N/A	EP 26 Pg D-22	EP 29 Pg D-23	N/A	EP 26 Pg D-22
Oven operates, but oven indicator light does not glow.	N/A	N/A	N/A	EP 33 Pg D-27	N/A
Timed oven pilot lamp does not glow when timed bake is over.	N/A	N/A	N/A	N/A	EP 34 Pg D-27

Items listed under the problem column are indicators that may be observed without disassembling the appliance, and should be used as a troubleshooting starting point.   N/A = Not applicable.

FLOW CHART ENTRY POINTS

PROBLEM	MODEL NUMBERS		
	GPG38BC0-2	GPG38BE0-4	GG46C0-4
Top burner pilots go out.	N/A	N/A	N/A
Clock does not operate.	EP 2 Pg D-9	EP 2 Pg D-9	EP 2 Pg D-9
Top burners will not ignite.	EP 5 Pg D-11	EP 5 Pg D-11	EP 5 Pg D-11
Top burner has yellow flame.	EP 4 Pg D-10	EP 4 Pg D-10	EP 4 Pg D-10
Oven pilot goes out.	N/A	N/A	N/A
Backguard light does not illuminate.	EP 31 Pg D-25	EP 31 Pg D-25	N/A
Oven light does not illuminate.	EP 7 Pg D-12	EP 7 Pg D-12	EP 7 Pg D-12
Oven will not come on.	N/A	N/A	EP 12 Pg D-14
Oven flame will not cycle off unless the thermostat is turned off.	N/A	N/A	N/A
Oven temperature out of calibration.	EP 11 Pg D-13	EP 11 Pg D-13	EP 10 Pg D-13
Oven will not cycle off.	EP 13 Pg D-15	EP 13 Pg D-15	EP 13 Pg D-15
Broil burner will operate, but bake burner will not come on.	EP 14 Pg D-15	EP 14 Pg D-15	N/A
Top burner ignitors do not spark.	EP 15 Pg D-16	EP 15 Pg D-16	EP 15 Pg D-16
Bake burner will operate, but broil burner will not come on.	EP 16 Pg D-16	EP 16 Pg D-16	N/A
Neither bake or broil burners will come on.	EP 18 Pg D-18	EP 19 Pg D-19	N/A
Oven operates in bake & broil, but does not heat in clean.	EP 20 Pg D-20	EP 27 Pg D-22	N/A
Oven operates in bake & broil, but does not start to heat until 8 minutes into the clean cycle.	EP 21 Pg D-20	N/A	N/A
Oven operates in bake & broil, heats the first 8 minutes of cleaning cycle then stops heating.	EP 22 Pg D-20	N/A	N/A
Clean light does not illuminate during the clean cycle.	EP 24 Pg D-21	EP 28 Pg D-23	N/A
Clean light glows, but clean timer does not advance.	EP 25 Pg D-21	N/A	N/A
Oven is not clean at the end of the clean cycle.	EP 26 Pg D-22	EP 29 Pg D-23	N/A
Oven operates, but oven indicator light does not glow.	N/A	N/A	N/A
Timed oven pilot lamp does not glow when timed bake is over.	N/A	N/A	N/A

Items listed under the problem column are indicators that may be observed without disassembling the appliance, and should be used as a troubleshooting starting point.   N/A = Not applicable.

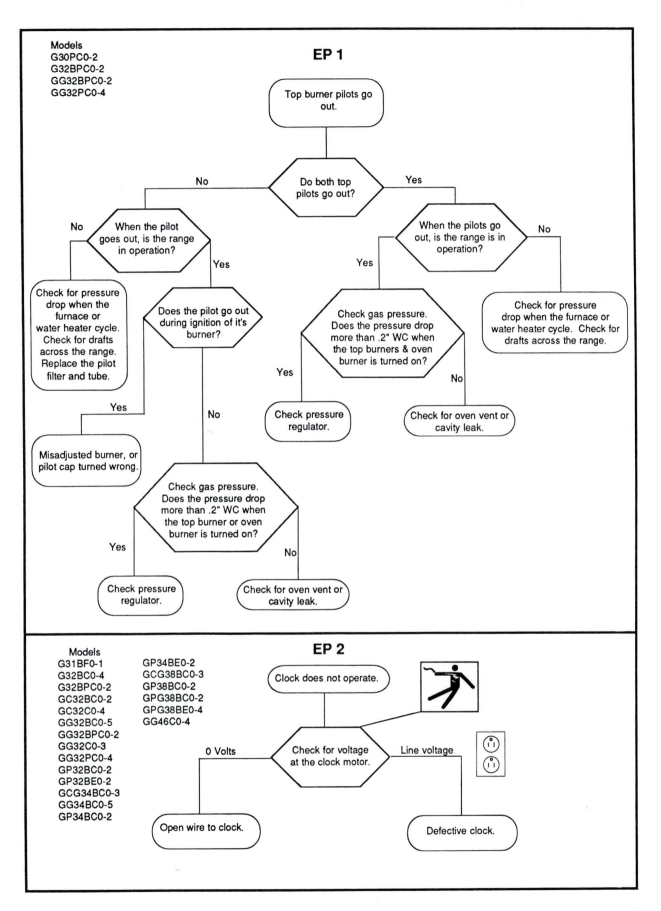

Models
G30PC0-2
G32BPC0-2
GG32BPC0-2
GG32PC0-4

EP 1

Top burner pilots go out.

Do both top pilots go out?

No — When the pilot goes out, is the range in operation?

Yes — When the pilots go out, is the range is in operation?

No — Check for pressure drop when the furnace or water heater cycle. Check for drafts across the range. Replace the pilot filter and tube.

Yes — Does the pilot go out during ignition of it's burner?

Yes — Check gas pressure. Does the pressure drop more than .2" WC when the top burners & oven burner is turned on?

No — Check for pressure drop when the furnace or water heater cycle. Check for drafts across the range.

Yes — Check pressure regulator.

No — Check for oven vent or cavity leak.

Yes — Misadjusted burner, or pilot cap turned wrong.

No — Check gas pressure. Does the pressure drop more than .2" WC when the top burner or oven burner is turned on?

Yes — Check pressure regulator.

No — Check for oven vent or cavity leak.

Models
GG31BF0-1
G32BC0-4
G32BPC0-2
GC32BC0-2
GC32C0-4
GG32BC0-5
GG32BPC0-2
GG32C0-3
GG32PC0-4
GP32BC0-2
GP32BE0-2
GCG34BC0-3
GG34BC0-5
GP34BC0-2

GP34BE0-2
GCG38BC0-3
GP38BC0-2
GPG38BC0-2
GPG38BE0-4
GG46C0-4

EP 2

Clock does not operate.

Check for voltage at the clock motor.

0 Volts — Open wire to clock.

Line voltage — Defective clock.

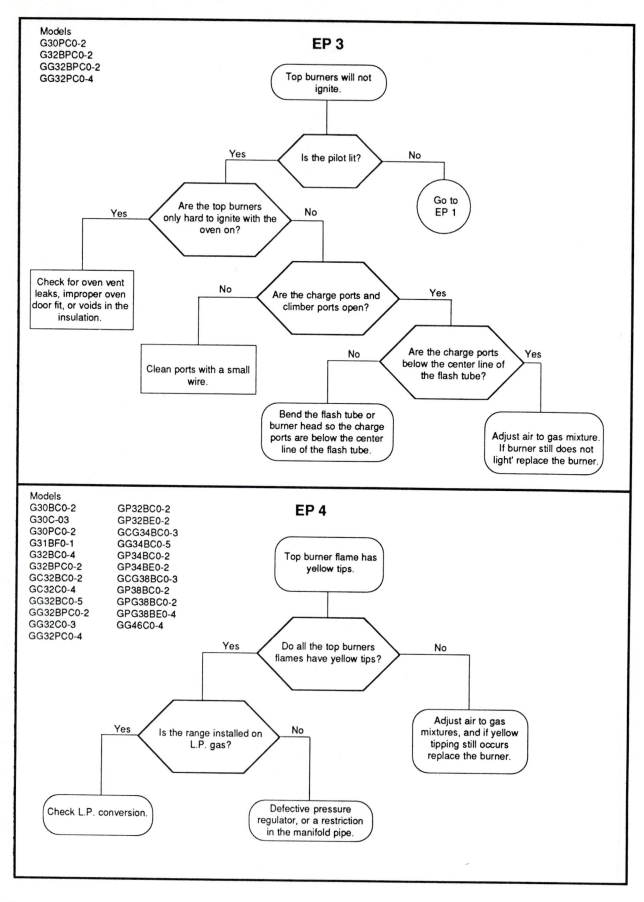

Models
G30PC0-2
G32BPC0-2
GG32BPC0-2
GG32PC0-4

EP 3

Top burners will not ignite.

Is the pilot lit?

Yes

No

Go to EP 1

Are the top burners only hard to ignite with the oven on?

Yes

No

Check for oven vent leaks, improper oven door fit, or voids in the insulation.

Are the charge ports and climber ports open?

No

Yes

Clean ports with a small wire.

Are the charge ports below the center line of the flash tube?

No

Yes

Bend the flash tube or burner head so the charge ports are below the center line of the flash tube.

Adjust air to gas mixture. If burner still does not light' replace the burner.

Models
G30BC0-2   GP32BC0-2
G30C-03   GP32BE0-2
G30PC0-2   GCG34BC0-3
G31BF0-1   GG34BC0-5
G32BC0-4   GP34BC0-2
G32BPC0-2   GP34BE0-2
GC32BC0-2   GCG38BC0-3
GC32C0-4   GP38BC0-2
GG32BC0-5   GPG38BC0-2
GG32BPC0-2   GPG38BE0-4
GG32C0-3   GG46C0-4
GG32PC0-4

EP 4

Top burner flame has yellow tips.

Do all the top burners flames have yellow tips?

Yes

No

Is the range installed on L.P. gas?

Adjust air to gas mixtures, and if yellow tipping still occurs replace the burner.

Yes

No

Check L.P. conversion.

Defective pressure regulator, or a restriction in the manifold pipe.

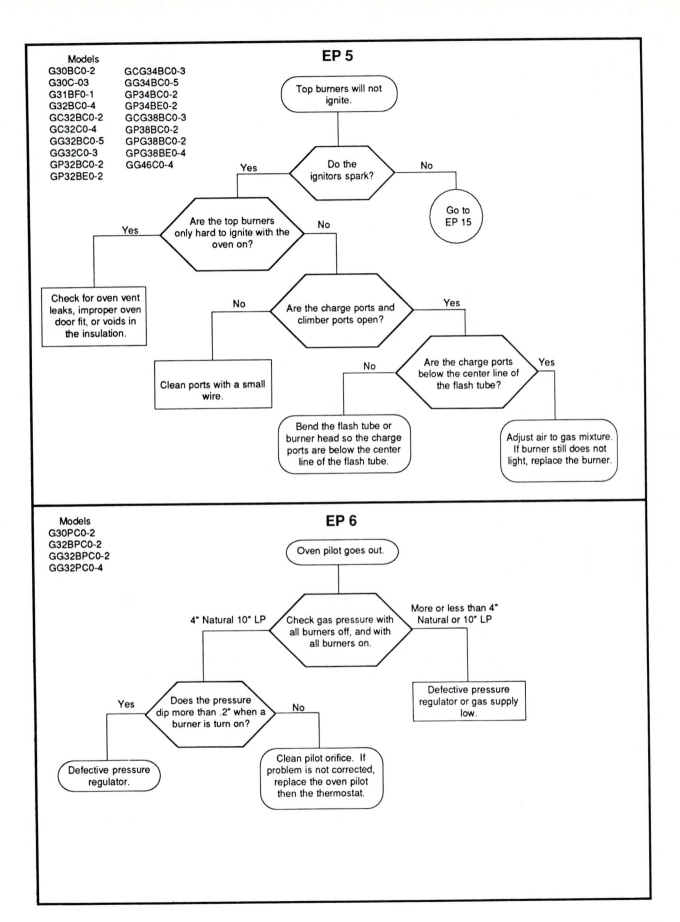

EP 5

Models
G30BC0-2
G30C-03
G31BF0-1
G32BC0-4
GC32BC0-2
GC32C0-4
GG32BC0-5
GG32C0-3
GP32BC0-2
GP32BE0-2

GCG34BC0-3
GG34BC0-5
GP34BC0-2
GP34BE0-2
GCG38BC0-3
GP38BC0-2
GPG38BC0-2
GPG38BE0-4
GG46C0-4

Top burners will not ignite.

Do the ignitors spark?

No → Go to EP 15

Yes → Are the top burners only hard to ignite with the oven on?

Yes → Check for oven vent leaks, improper oven door fit, or voids in the insulation.

No → Are the charge ports and climber ports open?

No → Clean ports with a small wire.

Yes → Are the charge ports below the center line of the flash tube?

No → Bend the flash tube or burner head so the charge ports are below the center line of the flash tube.

Yes → Adjust air to gas mixture. If burner still does not light, replace the burner.

EP 6

Models
G30PC0-2
G32BPC0-2
GG32BPC0-2
GG32PC0-4

Oven pilot goes out.

Check gas pressure with all burners off, and with all burners on.

4" Natural 10" LP → Does the pressure dip more than .2" when a burner is turn on?

Yes → Defective pressure regulator.

No → Clean pilot orifice. If problem is not corrected, replace the oven pilot then the thermostat.

More or less than 4" Natural or 10" LP → Defective pressure regulator or gas supply low.

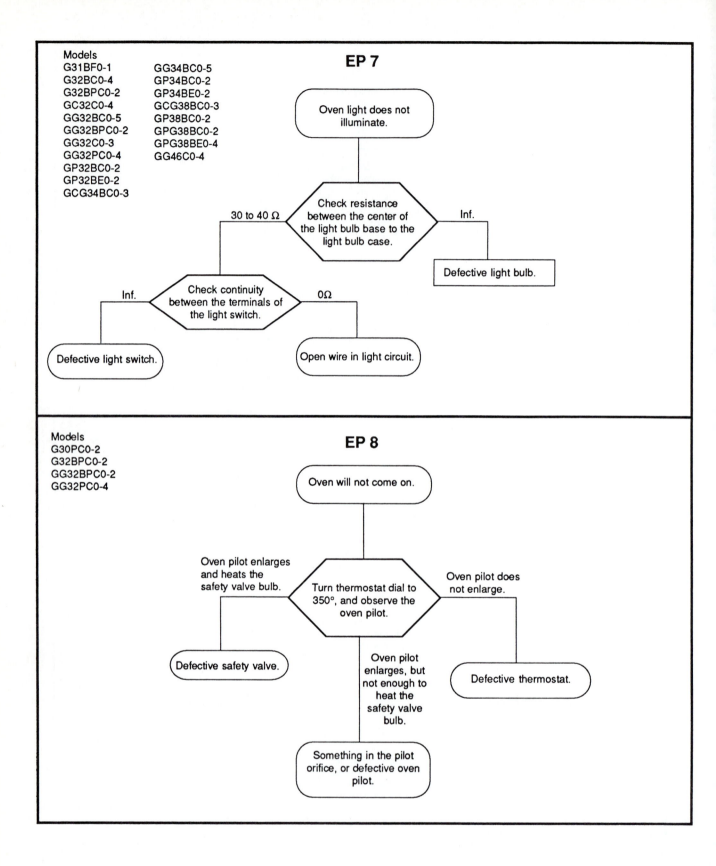

Models
G31BF0-1
G32BC0-4
G32BPC0-2
GC32C0-4
GG32BC0-5
GG32BPC0-2
GG32C0-3
GG32PC0-4
GP32BC0-2
GP32BE0-2
GCG34BC0-3

GG34BC0-5
GP34BC0-2
GP34BE0-2
GCG38BC0-3
GP38BC0-2
GPG38BC0-2
GPG38BE0-4
GG46C0-4

EP 7

Oven light does not illuminate.

Check resistance between the center of the light bulb base to the light bulb case.

30 to 40 Ω

Inf.

Defective light bulb.

Check continuity between the terminals of the light switch.

Inf.

0Ω

Defective light switch.

Open wire in light circuit.

Models
G30PC0-2
G32BPC0-2
GG32BPC0-2
GG32PC0-4

EP 8

Oven will not come on.

Turn thermostat dial to 350°, and observe the oven pilot.

Oven pilot enlarges and heats the safety valve bulb.

Oven pilot does not enlarge.

Defective safety valve.

Oven pilot enlarges, but not enough to heat the safety valve bulb.

Defective thermostat.

Something in the pilot orifice, or defective oven pilot.

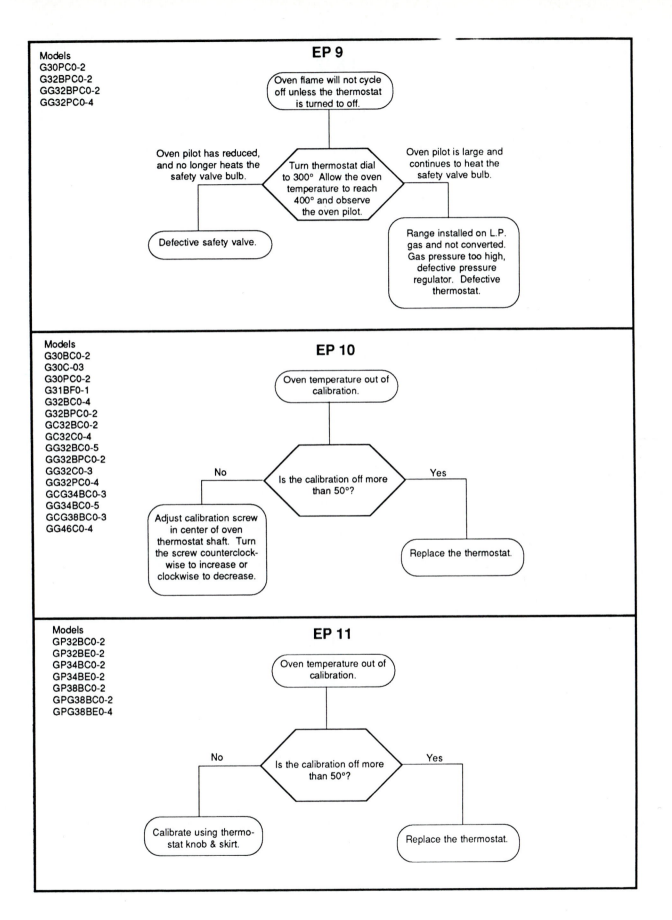

EP 9

Models
G30PC0-2
G32BPC0-2
GG32BPC0-2
GG32PC0-4

Oven flame will not cycle off unless the thermostat is turned to off.

Turn thermostat dial to 300° Allow the oven temperature to reach 400° and observe the oven pilot.

Oven pilot has reduced, and no longer heats the safety valve bulb.

Defective safety valve.

Oven pilot is large and continues to heat the safety valve bulb.

Range installed on L.P. gas and not converted. Gas pressure too high, defective pressure regulator. Defective thermostat.

EP 10

Models
G30BC0-2
G30C-03
G30PC0-2
G31BF0-1
G32BC0-4
G32BPC0-2
GC32BC0-2
GC32C0-4
GG32BC0-5
GG32BPC0-2
GG32C0-3
GG32PC0-4
GCG34BC0-3
GG34BC0-5
GCG38BC0-3
GG46C0-4

Oven temperature out of calibration.

Is the calibration off more than 50°?

No

Adjust calibration screw in center of oven thermostat shaft. Turn the screw counterclockwise to increase or clockwise to decrease.

Yes

Replace the thermostat.

EP 11

Models
GP32BC0-2
GP32BE0-2
GP34BC0-2
GP34BE0-2
GP38BC0-2
GPG38BC0-2
GPG38BE0-4

Oven temperature out of calibration.

Is the calibration off more than 50°?

No

Calibrate using thermostat knob & skirt.

Yes

Replace the thermostat.

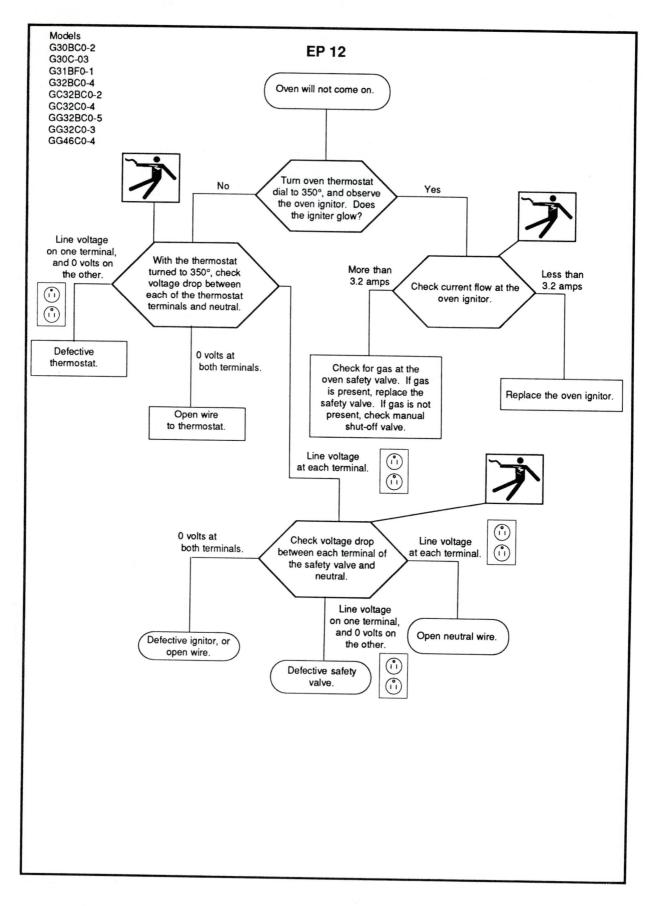

Models
G30BC0-2
G30C-03
G31BF0-1
G32BC0-4
GC32BC0-2
GC32C0-4
GG32BC0-5
GG32C0-3
GG46C0-4

EP 12

Oven will not come on.

Turn oven thermostat dial to 350°, and observe the oven ignitor. Does the igniter glow?

No

Yes

Line voltage on one terminal, and 0 volts on the other.

With the thermostat turned to 350°, check voltage drop between each of the thermostat terminals and neutral.

Defective thermostat.

0 volts at both terminals.

Open wire to thermostat.

More than 3.2 amps

Check current flow at the oven ignitor.

Less than 3.2 amps

Check for gas at the oven safety valve. If gas is present, replace the safety valve. If gas is not present, check manual shut-off valve.

Replace the oven ignitor.

Line voltage at each terminal.

0 volts at both terminals.

Check voltage drop between each terminal of the safety valve and neutral.

Line voltage at each terminal.

Line voltage on one terminal, and 0 volts on the other.

Defective ignitor, or open wire.

Defective safety valve.

Open neutral wire.

528

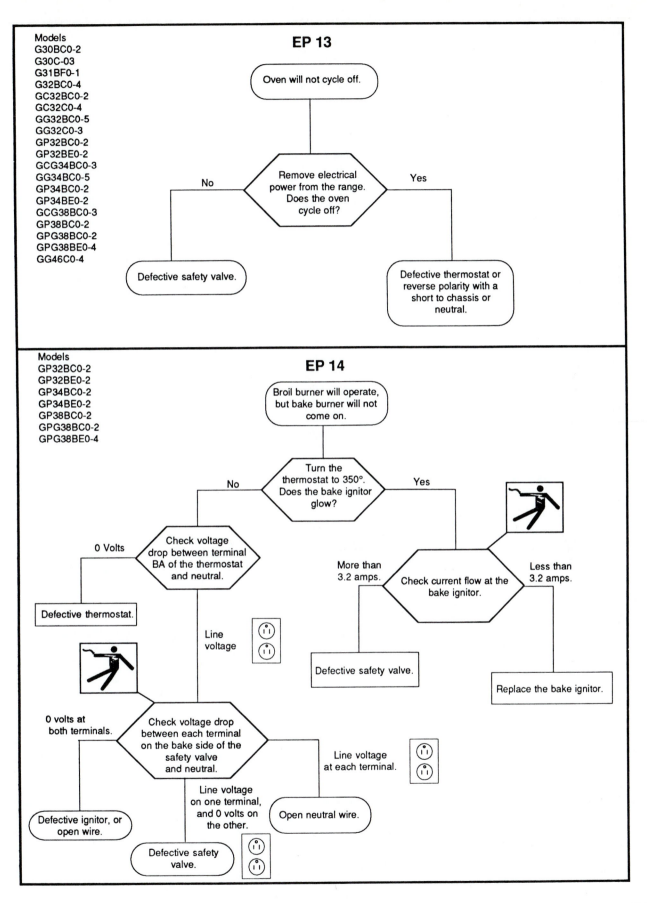

Models
G30BC0-2
G30C-03
G31BF0-1
G32BC0-4
GC32BC0-2
GC32C0-4
GG32BC0-5
GG32C0-3
GP32BC0-2
GP32BE0-2
GCG34BC0-3
GG34BC0-5
GP34BC0-2
GP34BE0-2
GCG38BC0-3
GP38BC0-2
GPG38BC0-2
GPG38BE0-4
GG46C0-4

EP 13

Oven will not cycle off.

Remove electrical power from the range. Does the oven cycle off?

No

Yes

Defective safety valve.

Defective thermostat or reverse polarity with a short to chassis or neutral.

Models
GP32BC0-2
GP32BE0-2
GP34BC0-2
GP34BE0-2
GP38BC0-2
GPG38BC0-2
GPG38BE0-4

EP 14

Broil burner will operate, but bake burner will not come on.

Turn the thermostat to 350°. Does the bake ignitor glow?

No

Yes

Check voltage drop between terminal BA of the thermostat and neutral.

0 Volts

Defective thermostat.

Line voltage

Check current flow at the bake ignitor.

More than 3.2 amps.

Less than 3.2 amps.

Defective safety valve.

Replace the bake ignitor.

Check voltage drop between each terminal on the bake side of the safety valve and neutral.

0 volts at both terminals.

Defective ignitor, or open wire.

Line voltage on one terminal, and 0 volts on the other.

Line voltage at each terminal.

Open neutral wire.

Defective safety valve.

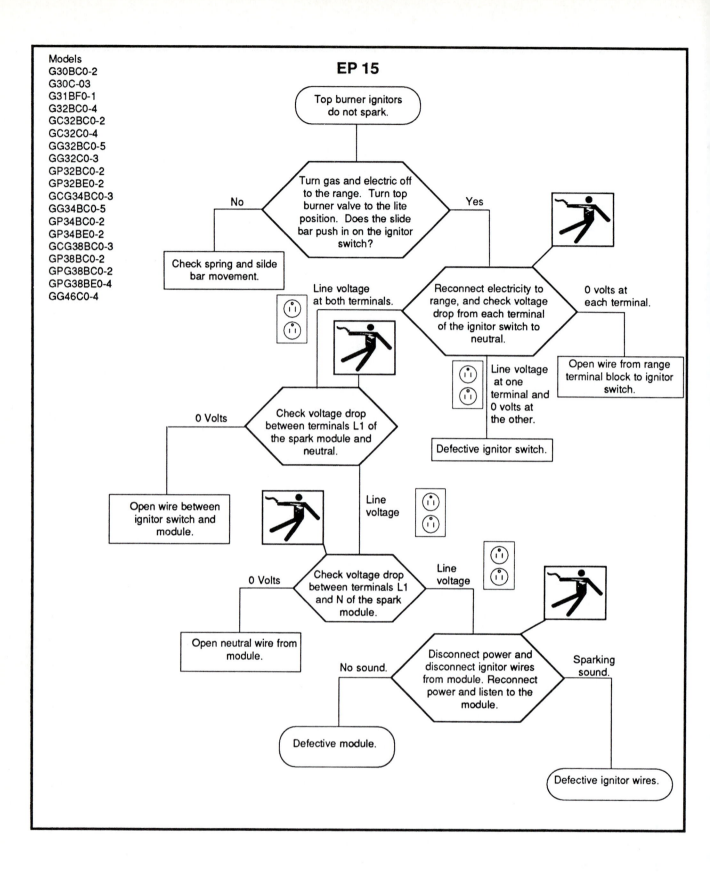

Models
G30BC0-2
G30C-03
G31BF0-1
G32BC0-4
GC32BC0-2
GC32C0-4
GG32BC0-5
GG32C0-3
GP32BC0-2
GP32BE0-2
GCG34BC0-3
GG34BC0-5
GP34BC0-2
GP34BE0-2
GCG38BC0-3
GP38BC0-2
GPG38BC0-2
GPG38BE0-4
GG46C0-4

EP 15

Top burner ignitors do not spark.

Turn gas and electric off to the range. Turn top burner valve to the lite position. Does the slide bar push in on the ignitor switch?

No — Check spring and silde bar movement.

Yes

Reconnect electricity to range, and check voltage drop from each terminal of the ignitor switch to neutral.

Line voltage at both terminals.

0 volts at each terminal. — Open wire from range terminal block to ignitor switch.

Line voltage at one terminal and 0 volts at the other. — Defective ignitor switch.

Check voltage drop between terminals L1 of the spark module and neutral.

0 Volts — Open wire between ignitor switch and module.

Line voltage

Check voltage drop between terminals L1 and N of the spark module.

0 Volts — Open neutral wire from module.

Line voltage

Disconnect power and disconnect ignitor wires from module. Reconnect power and listen to the module.

No sound. — Defective module.

Sparking sound. — Defective ignitor wires.

530

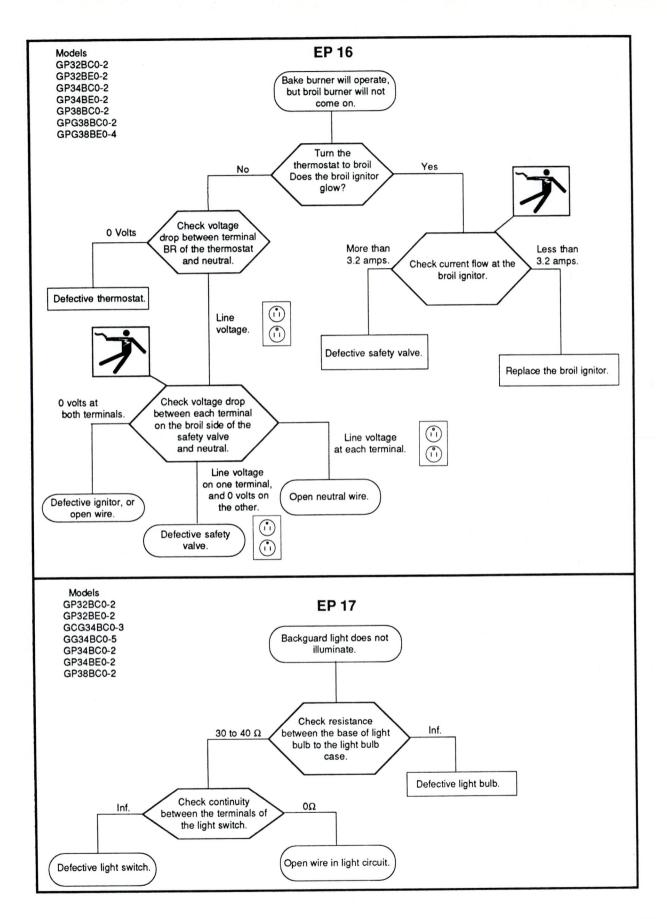

EP 16

Models
GP32BC0-2
GP32BE0-2
GP34BC0-2
GP34BE0-2
GP38BC0-2
GPG38BC0-2
GPG38BE0-4

Bake burner will operate, but broil burner will not come on.

Turn the thermostat to broil Does the broil ignitor glow?

No — Check voltage drop between terminal BR of the thermostat and neutral.

Yes — Check current flow at the broil ignitor.

0 Volts — Defective thermostat.

Line voltage.

More than 3.2 amps. — Defective safety valve.

Less than 3.2 amps. — Replace the broil ignitor.

Check voltage drop between each terminal on the broil side of the safety valve and neutral.

0 volts at both terminals. — Defective ignitor, or open wire.

Line voltage on one terminal, and 0 volts on the other. — Defective safety valve.

Line voltage at each terminal. — Open neutral wire.

EP 17

Models
GP32BC0-2
GP32BE0-2
GCG34BC0-3
GG34BC0-5
GP34BC0-2
GP34BE0-2
GP38BC0-2

Backguard light does not illuminate.

Check resistance between the base of light bulb to the light bulb case.

30 to 40 Ω — Check continuity between the terminals of the light switch.

Inf. — Defective light bulb.

Inf. — Defective light switch.

0Ω — Open wire in light circuit.

531

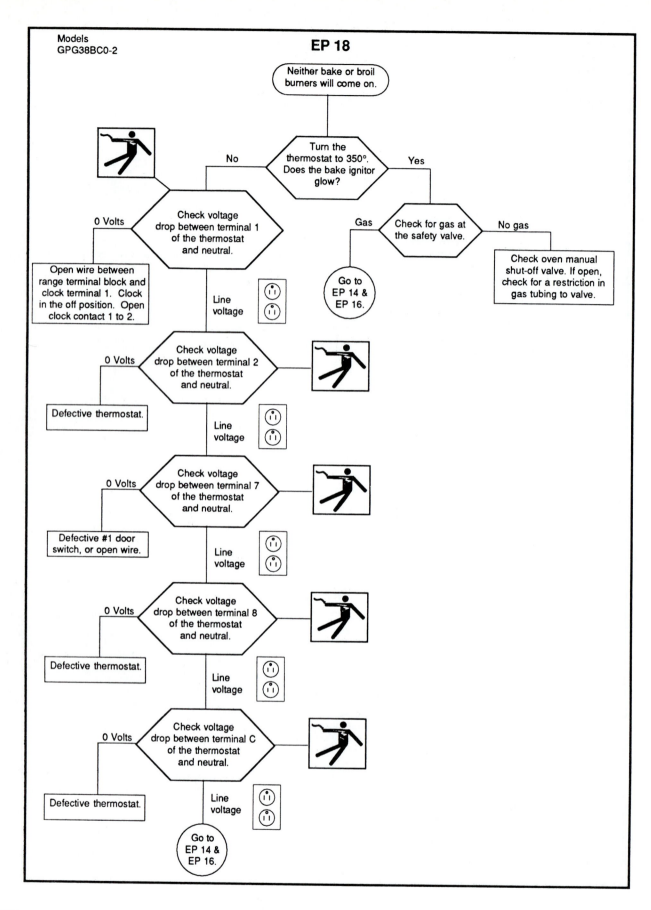

Models
GPG38BC0-2

EP 18

Neither bake or broil burners will come on.

Turn the thermostat to 350°. Does the bake ignitor glow?

No

Yes

Check voltage drop between terminal 1 of the thermostat and neutral.

0 Volts

Open wire between range terminal block and clock terminal 1. Clock in the off position. Open clock contact 1 to 2.

Line voltage

Check for gas at the safety valve.

Gas

No gas

Go to EP 14 & EP 16.

Check oven manual shut-off valve. If open, check for a restriction in gas tubing to valve.

Check voltage drop between terminal 2 of the thermostat and neutral.

0 Volts

Defective thermostat.

Line voltage

Check voltage drop between terminal 7 of the thermostat and neutral.

0 Volts

Defective #1 door switch, or open wire.

Line voltage

Check voltage drop between terminal 8 of the thermostat and neutral.

0 Volts

Defective thermostat.

Line voltage

Check voltage drop between terminal C of the thermostat and neutral.

0 Volts

Defective thermostat.

Line voltage

Go to EP 14 & EP 16.

532

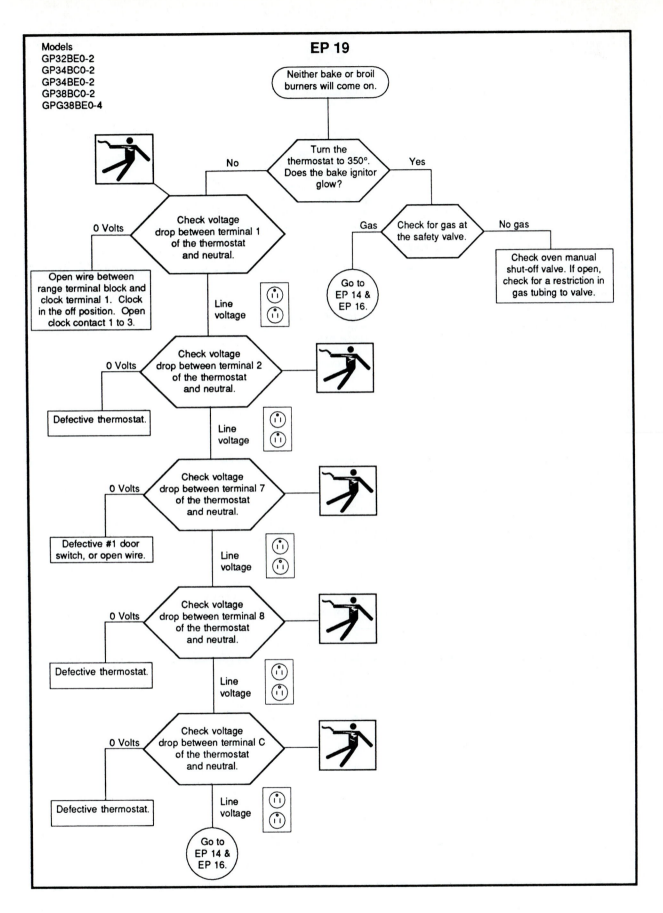

Models
GP32BE0-2
GP34BC0-2
GP34BE0-2
GP38BC0-2
GPG38BE0-4

EP 19

Neither bake or broil burners will come on.

Turn the thermostat to 350°. Does the bake ignitor glow?

No

Yes

Check voltage drop between terminal 1 of the thermostat and neutral.

0 Volts

Open wire between range terminal block and clock terminal 1. Clock in the off position. Open clock contact 1 to 3.

Line voltage

Check for gas at the safety valve.

Gas

No gas

Go to EP 14 & EP 16.

Check oven manual shut-off valve. If open, check for a restriction in gas tubing to valve.

Check voltage drop between terminal 2 of the thermostat and neutral.

0 Volts

Defective thermostat.

Line voltage

Check voltage drop between terminal 7 of the thermostat and neutral.

0 Volts

Defective #1 door switch, or open wire.

Line voltage

Check voltage drop between terminal 8 of the thermostat and neutral.

0 Volts

Defective thermostat.

Line voltage

Check voltage drop between terminal C of the thermostat and neutral.

0 Volts

Defective thermostat.

Line voltage

Go to EP 14 & EP 16.

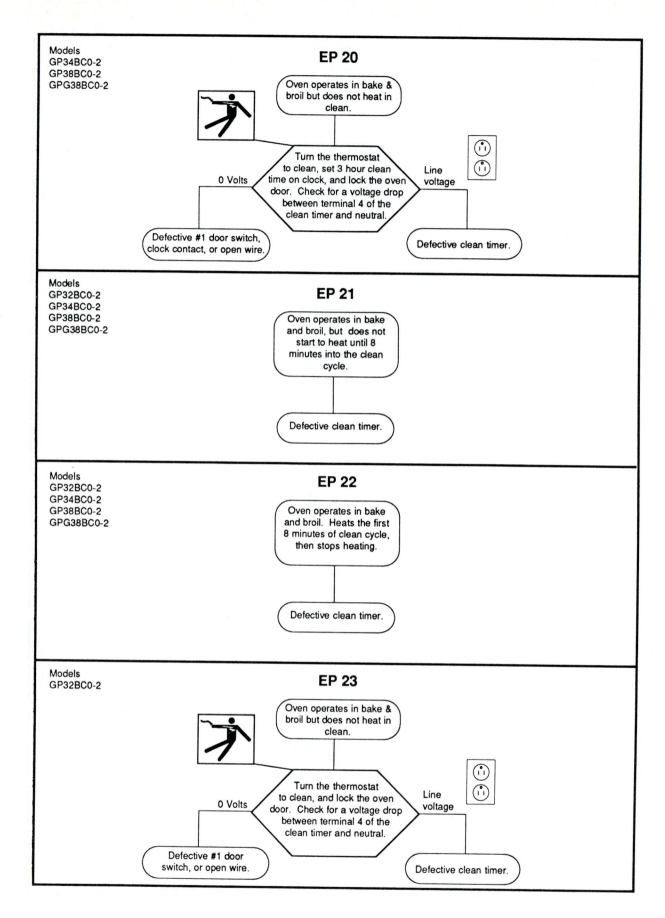

Models
GP34BC0-2
GP38BC0-2
GPG38BC0-2

EP 20

Oven operates in bake & broil but does not heat in clean.

Turn the thermostat to clean, set 3 hour clean time on clock, and lock the oven door. Check for a voltage drop between terminal 4 of the clean timer and neutral.

0 Volts

Line voltage

Defective #1 door switch, clock contact, or open wire.

Defective clean timer.

Models
GP32BC0-2
GP34BC0-2
GP38BC0-2
GPG38BC0-2

EP 21

Oven operates in bake and broil, but does not start to heat until 8 minutes into the clean cycle.

Defective clean timer.

Models
GP32BC0-2
GP34BC0-2
GP38BC0-2
GPG38BC0-2

EP 22

Oven operates in bake and broil. Heats the first 8 minutes of clean cycle, then stops heating.

Defective clean timer.

Models
GP32BC0-2

EP 23

Oven operates in bake & broil but does not heat in clean.

Turn the thermostat to clean, and lock the oven door. Check for a voltage drop between terminal 4 of the clean timer and neutral.

0 Volts

Line voltage

Defective #1 door switch, or open wire.

Defective clean timer.

534

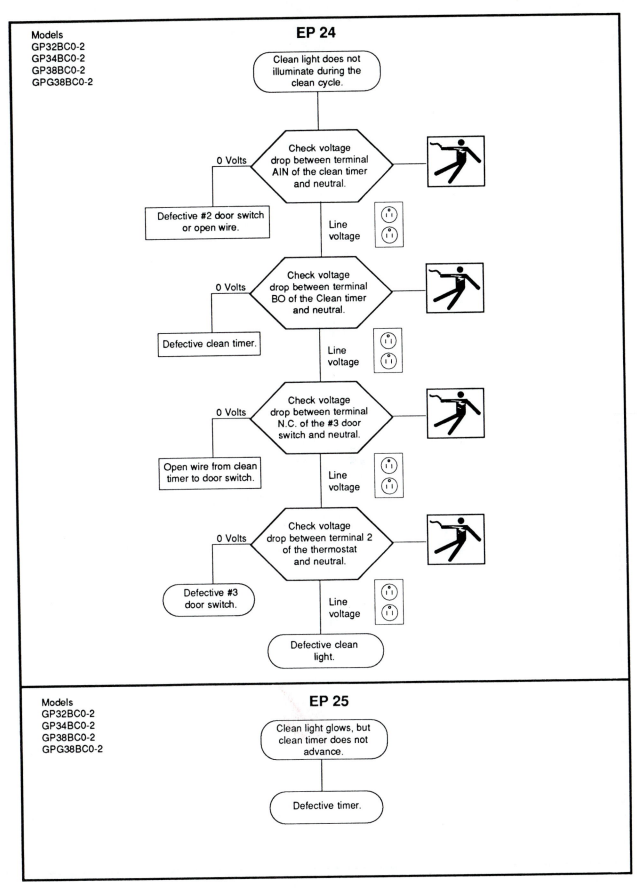

EP 24

Models
GP32BC0-2
GP34BC0-2
GP38BC0-2
GPG38BC0-2

Clean light does not illuminate during the clean cycle.

Check voltage drop between terminal AIN of the clean timer and neutral.

0 Volts → Defective #2 door switch or open wire.

Line voltage

Check voltage drop between terminal BO of the Clean timer and neutral.

0 Volts → Defective clean timer.

Line voltage

Check voltage drop between terminal N.C. of the #3 door switch and neutral.

0 Volts → Open wire from clean timer to door switch.

Line voltage

Check voltage drop between terminal 2 of the thermostat and neutral.

0 Volts → Defective #3 door switch.

Line voltage

Defective clean light.

EP 25

Models
GP32BC0-2
GP34BC0-2
GP38BC0-2
GPG38BC0-2

Clean light glows, but clean timer does not advance.

Defective timer.

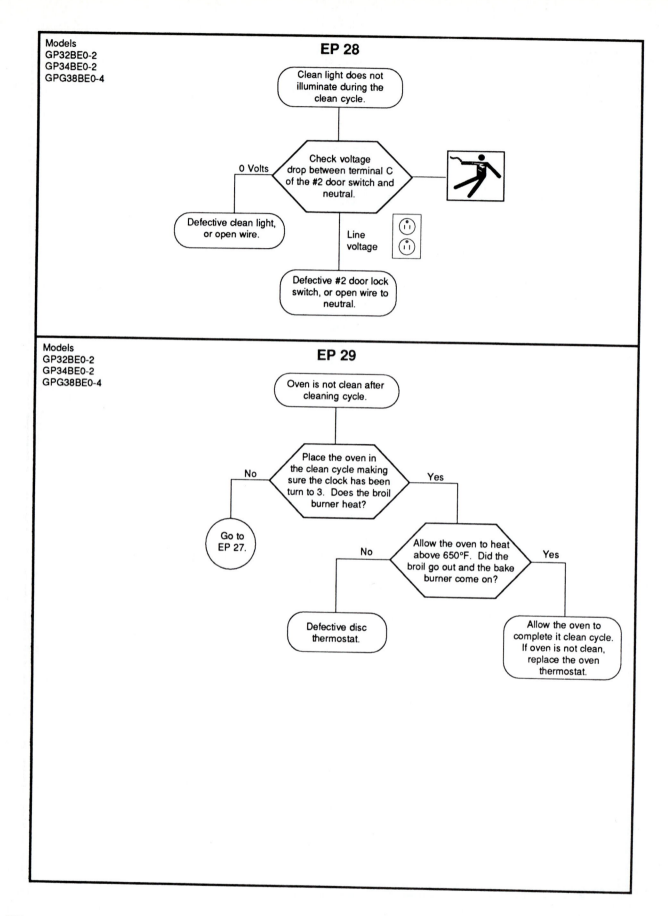

Models
GP32BE0-2
GP34BE0-2
GPG38BE0-4

EP 28

Clean light does not illuminate during the clean cycle.

Check voltage drop between terminal C of the #2 door switch and neutral.

0 Volts

Defective clean light, or open wire.

Line voltage

Defective #2 door lock switch, or open wire to neutral.

Models
GP32BE0-2
GP34BE0-2
GPG38BE0-4

EP 29

Oven is not clean after cleaning cycle.

Place the oven in the clean cycle making sure the clock has been turn to 3. Does the broil burner heat?

No

Go to EP 27.

Yes

Allow the oven to heat above 650°F. Did the broil go out and the bake burner come on?

No

Defective disc thermostat.

Yes

Allow the oven to complete it clean cycle. If oven is not clean, replace the oven thermostat.

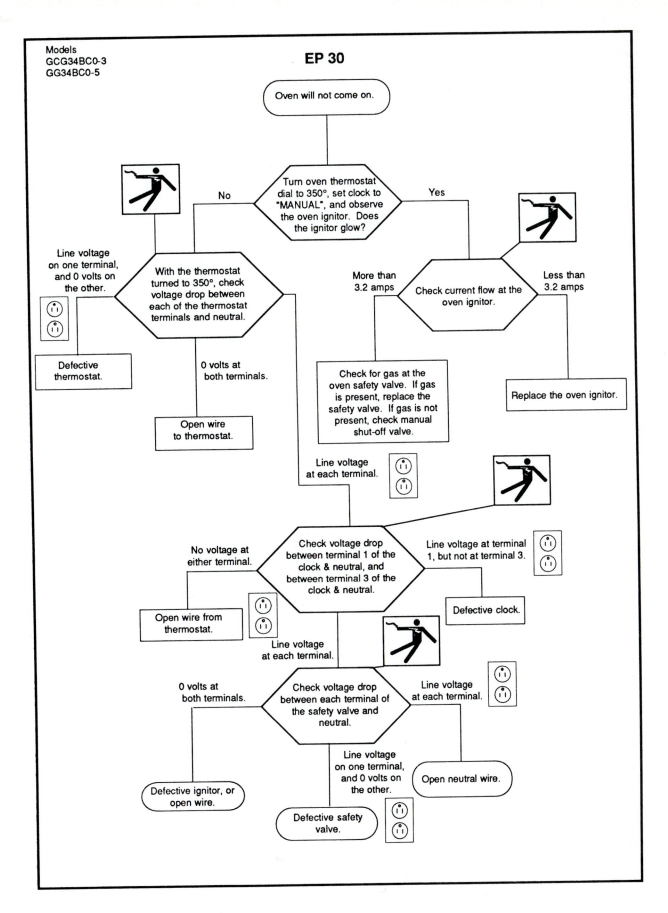

Models
GCG34BC0-3
GG34BC0-5

EP 30

Oven will not come on.

Turn oven thermostat dial to 350°, set clock to "MANUAL", and observe the oven ignitor. Does the ignitor glow?

No

Yes

Line voltage on one terminal, and 0 volts on the other.

With the thermostat turned to 350°, check voltage drop between each of the thermostat terminals and neutral.

More than 3.2 amps

Check current flow at the oven ignitor.

Less than 3.2 amps

Defective thermostat.

0 volts at both terminals.

Check for gas at the oven safety valve. If gas is present, replace the safety valve. If gas is not present, check manual shut-off valve.

Replace the oven ignitor.

Open wire to thermostat.

Line voltage at each terminal.

Check voltage drop between terminal 1 of the clock & neutral, and between terminal 3 of the clock & neutral.

No voltage at either terminal.

Line voltage at terminal 1, but not at terminal 3.

Open wire from thermostat.

Defective clock.

Line voltage at each terminal.

0 volts at both terminals.

Check voltage drop between each terminal of the safety valve and neutral.

Line voltage at each terminal.

Defective ignitor, or open wire.

Line voltage on one terminal, and 0 volts on the other.

Open neutral wire.

Defective safety valve.

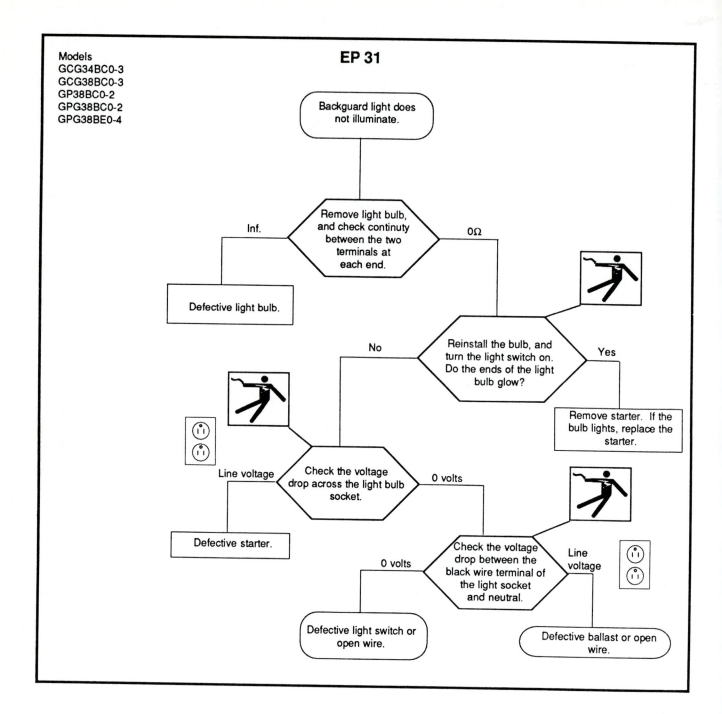

EP 31

Models
GCG34BC0-3
GCG38BC0-3
GP38BC0-2
GPG38BC0-2
GPG38BE0-4

Backguard light does not illuminate.

Remove light bulb, and check continuty between the two terminals at each end.

Inf. — Defective light bulb.

0Ω — Reinstall the bulb, and turn the light switch on. Do the ends of the light bulb glow?

Yes — Remove starter. If the bulb lights, replace the starter.

No — Check the voltage drop across the light bulb socket.

Line voltage — Defective starter.

0 volts — Check the voltage drop between the black wire terminal of the light socket and neutral.

0 volts — Defective light switch or open wire.

Line voltage — Defective ballast or open wire.

538

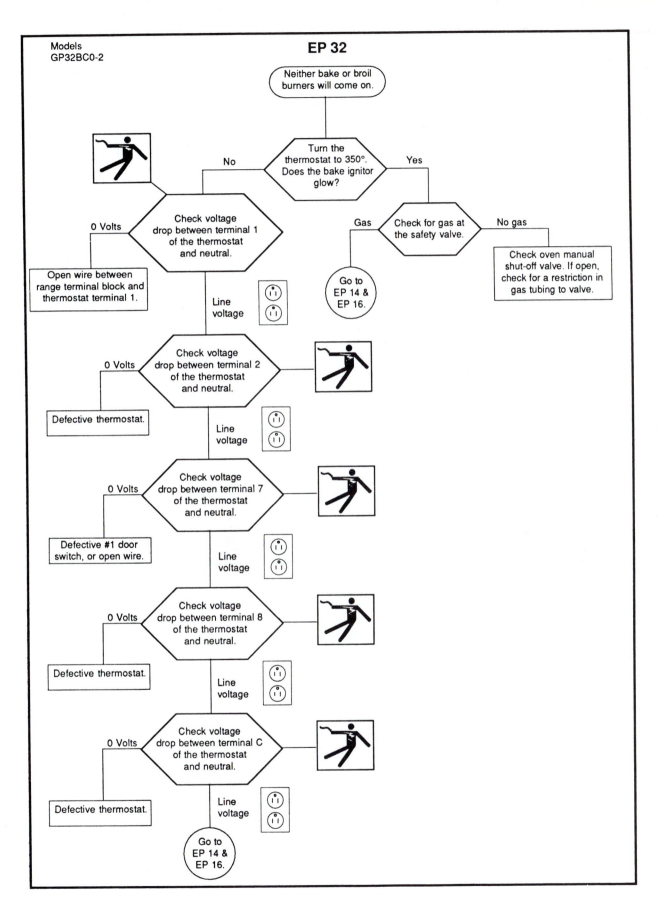

Models
GP32BC0-2

EP 32

Neither bake or broil burners will come on.

Turn the thermostat to 350°. Does the bake ignitor glow?

No

Yes

Check voltage drop between terminal 1 of the thermostat and neutral.

0 Volts

Open wire between range terminal block and thermostat terminal 1.

Line voltage

Check for gas at the safety valve.

Gas

No gas

Go to EP 14 & EP 16.

Check oven manual shut-off valve. If open, check for a restriction in gas tubing to valve.

Check voltage drop between terminal 2 of the thermostat and neutral.

0 Volts

Defective thermostat.

Line voltage

Check voltage drop between terminal 7 of the thermostat and neutral.

0 Volts

Defective #1 door switch, or open wire.

Line voltage

Check voltage drop between terminal 8 of the thermostat and neutral.

0 Volts

Defective thermostat.

Line voltage

Check voltage drop between terminal C of the thermostat and neutral.

0 Volts

Defective thermostat.

Line voltage

Go to EP 14 & EP 16.

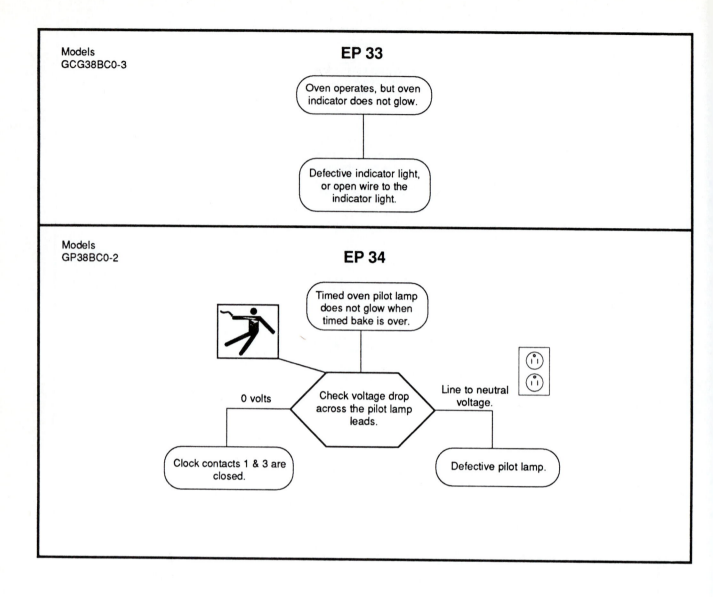

Models
GCG38BC0-3

EP 33

Oven operates, but oven indicator does not glow.

Defective indicator light, or open wire to the indicator light.

Models
GP38BC0-2

EP 34

Timed oven pilot lamp does not glow when timed bake is over.

Check voltage drop across the pilot lamp leads.

0 volts

Line to neutral voltage.

Clock contacts 1 & 3 are closed.

Defective pilot lamp.

Index